EAI/Springer Innovations in Communication and Computing

Series Editor

Imrich Chlamtac, European Alliance for Innovation, Ghent, Belgium

The impact of information technologies is creating a new world yet not fully understood. The extent and speed of economic, life style and social changes already perceived in everyday life is hard to estimate without understanding the technological driving forces behind it. This series presents contributed volumes featuring the latest research and development in the various information engineering technologies that play a key role in this process. The range of topics, focusing primarily on communications and computing engineering include, but are not limited to, wireless networks; mobile communication; design and learning; gaming; interaction; e-health and pervasive healthcare; energy management; smart grids; internet of things; cognitive radio networks; computation; cloud computing; ubiquitous connectivity, and in mode general smart living, smart cities, Internet of Things and more. The series publishes a combination of expanded papers selected from hosted and sponsored European Alliance for Innovation (EAI) conferences that present cutting edge, global research as well as provide new perspectives on traditional related engineering fields. This content, complemented with open calls for contribution of book titles and individual chapters, together maintain Springer's and EAI's high standards of academic excellence. The audience for the books consists of researchers, industry professionals, advanced level students as well as practitioners in related fields of activity include information and communication specialists, security experts, economists, urban planners, doctors, and in general representatives in all those walks of life affected ad contributing to the information revolution.

Indexing: This series is indexed in Scopus, Ei Compendex, and zbMATH.

About EAI - EAI is a grassroots member organization initiated through cooperation between businesses, public, private and government organizations to address the global challenges of Europe's future competitiveness and link the European Research community with its counterparts around the globe. EAI reaches out to hundreds of thousands of individual subscribers on all continents and collaborates with an institutional member base including Fortune 500 companies, government organizations, and educational institutions, provide a free research and innovation platform. Through its open free membership model EAI promotes a new research and innovation culture based on collaboration, connectivity and recognition of excellence by community.

Shishir Kumar Shandilya • Agni Datta •
Yash Kartik • Atulya Nagar

Digital Resilience: Navigating Disruption and Safeguarding Data Privacy

Springer

EAI
RESEARCH MEETS INNOVATION

Shishir Kumar Shandilya
School of Data Science and Forecasting
Devi Ahilya Vishwavidyalaya (DAVV)
Indore, Madhya Pradesh, India

Yash Kartik
SECURE - Center of Excellence in Cyber Security
VIT Bhopal University
Bhopal, Madhya Pradesh, India

Agni Datta
SECURE — Centre of Excellence in Cyber Security
VIT Bhopal University
Bhopal, Madhya Pradesh, India

Atulya Nagar
School of Mathematics, Computer Science and Engineering
Liverpool Hope University
Liverpool, UK

ISSN 2522-8595 ISSN 2522-8609 (electronic)
EAI/Springer Innovations in Communication and Computing
ISBN 978-3-031-53289-4 ISBN 978-3-031-53290-0 (eBook)
https://doi.org/10.1007/978-3-031-53290-0

This Springer imprint is published by the registered company Springer Nature Switzerland AG
The registered company address is: Gewerbestrasse 11, 6330 Cham, Switzerland

It doesn't matter how beautiful your theory is, it doesn't matter how smart you are. If it doesn't agree with the experiment, it's wrong.

Richard Phillips Feynman

To my lifelines, Smita, Samarth, and Nityaa

Shishir

To my precious mummum and daddy who are forever young in mind and soul, my dearest and irreplaceable sister, and to the one who has departed

Agni

To my dearest parents and beloved brother

Yash

To Jyoti, and lovely daughters, Kopal and Priyel.

Atulya

Preface

Audere est faucere.

Scope

The goal of the book is to provide readers with a detailed exploration of the concepts regulating digital resilience as well as their practical applications in reducing technological disruptions. The principal objective of this book, which covers a various issues related to digital infrastructure, cybersecurity, and approaches for encouraging flexibility and resilience, is to provide readers with the knowledge and abilities required for understanding and implementing the subject matter of digital resilience in life. This contribution is expected to provide significant value to a large target audience, including a diverse demography of individuals, including those enrolled in undergraduate and postgraduate programmes. It is not only designed for students seeking academic degrees in computer science, engineering, cybersecurity, information technology, and related fields. It is also relevant to professionals working in information technology (IT), cybersecurity, system administration, and technology management.

The ideas given in the book provide novel views and practical solutions for enhancing the resilience of digital systems and infrastructure inside businesses. The detailed examination of digital resilience concepts presented here has the potential to greatly improve the understanding and comprehension of students and researchers working in the fields of cybersecurity, digital infrastructure, and technology management. This fundamental knowledge, in turn, serves as a platform for informing and directing future research activities in these particular fields. Individuals entrusted with developing technology policies and regulations, in particular, stand to gain practical insights into the need for digital resilience in ensuring the security and stability of critical infrastructure.

Because of its thorough coverage, this material is highly recommended for use in beginning cybersecurity courses and other information security (infosec) instructional programmes. It offers a detailed examination of key ideas and strategies relevant to the disciplines of cybersecurity and information security, while acting as an essential primer for students beginning a cybersecurity study, providing a solid and basic reference to complete their educational requirements.

Motivation

This book examines how closely connected both domains are as a complete resource for people with a strong interest in cybersecurity, as well as students and professionals seeking assistance on the issue of digital resilience. Originally conceived as a study on digital resilience, it became apparent that any book addressing this subject would explain the field of cybersecurity due to the link between the two. Due to the COVID-19 pandemic, there has been a collective adaptation leading to the adoption of several new behavioural patterns. However, with the wide diversity of threats and the availability of potential bad actors, a flexible policy is required to ensure that our safety while using the Internet is not compromised. The concept of digital resilience is therefore broken down and examined in great detail, providing a tool that can assist individuals, communities, and even entire societies in coping with the numerous dangers present in our lives when we are online.

The challenges posed by cyber threats are numerous, including hackers purposefully breaking into our computer systems, covert data theft schemes, and unplanned havoc in our technological infrastructure by natural disasters. Broadly speaking, the dangers faced in the digital world present us with a unique and diverse set of challenges that cannot be dealt with by a singular set of plans but rather require a more comprehensive approach. Hence in this regard, the concept of digital resilience is similar to a protective barrier, equipping us with the resources necessary to bounce back fast from failures and the tenacity to persevere in the face of challenges. It provides us with the same kind of determination in the electronic domain that we possess in the physical sphere.

Maintaining an awareness of digital resilience is emerging as a critical method to protect our digital life against the rising frequency of cyberattacks, data breaches, and associated risks. As a result, the primary emphasis of this book is on investigating the numerous aspects that contribute to digital resilience. Understanding its subtleties gives us the tools we need to increase our electronic involvement and digital lives in the face of growing threats. Recent technology breakthroughs have enabled us to see first-hand the enormous potential of digital resilience. This barrier acts as a protective addition having an impact that extends beyond its immediate users to include enterprises and the broader social environment. The consequence of digital resilience in our everyday lives cannot be emphasized, and this book is an essential resource for specialist audiences wanting a thorough grasp of the problem.

The intelligent computers, the learning machines, and a one-of-a-kind digital system known as blockchain are the ones who are in charge of this digital universe. These smart innovations can detect and avoid problems in real time, which is a huge advantage. This is the equivalent, in the world of digital technology, of having a team of superheroes at your command. These technical pioneers are paving the way for a new future in which we can head off problems before they become more serious. Regardless, digital resilience is a lot more than just a technical definition. It is constructed in the shape of a wall that encircles and safeguards all of society. By adopting what is called a 'digitally resilient' mindset, individuals, corporations, and even society as a whole can strengthen their defences against the numerous online threats that are currently present. We enhance our capacity to mitigate the adverse consequences of cyberattacks, such as financial instability and disruptions to the social order, by possessing such resilience.

However, the scope of digital resilience extends well beyond simple repair; it also develops confidence in the stability of the regenerative capabilities of the digital environment. This sense of safety acts very similarly to a safety net for us while we are navigating the virtual world. The effects of digital resilience are far-reaching, penetrating not only the technological but, crucially, the social and economic underpinnings of our society as well. As the power of the digital economy continues to develop, it is important to keep in mind two required factors: innovation and growth. In this context, digital resilience functions as a catalysing force, making it possible for these components to develop to their maximum potential. In a similar vein, the complex social framework, which is driven by the interaction of a great number of distinct pieces, necessitates the presence of this power.

Digital resilience serves as a cornerstone, somewhat in the same way that a solid foundation does, and it is essential to the symbiotic relationship between the dynamics of society. The capability to recover quickly from technological setbacks is necessary for ensuring that everything works together without any difficulties. Its influence is felt not only in the technology but also in the larger social picture, where it acts as an essential variable in promoting both collaborative advancement and cohesive functionality. Its influence is felt not only in technology but also in the larger social picture. Within the pages of this academic initiative, we have all come together to emphasize how important it is to establish digital resilience.

The increasing level of sophistication in technology and the complexities of the Internet world make digital resilience an important problem. Our investigation takes a comprehensive and intellectual approach, exploring the potential of novel concepts. As we explain the many strands that contribute to the significance of digital resilience, we will conduct comprehensive research across different sectors to shed light on the numerous ways in which it influences a variety of elements. Ultimately, we aim to gain a better understanding of why digital resilience is critical in today's society and encourage its widespread adoption. We expect this book to act as a guide to the components of digital resiliency, highlighting its prevalence in the world of technology and its relevance.

Prerequisites

To fully engage with and comprehend the content of this book, readers must possess a set of essential prerequisites. These prerequisites contain a fundamental grasp of technological principles, including computer systems, networks, and software. Acquiring this level of proficiency is considered necessary for readers to understand the concepts and principles elaborated in this book. To understand the chapter related to quantum computing, readers must possess a requisite level of mathematical proficiency. While not mandatory, proficiency in fundamental mathematical principles, including algebra and statistics, is welcome in this context. The utilization of mathematical formalism is evident in very few specific sections of the book, necessitating readers to be adequately equipped to understand and interact with these mathematical components. However, it is worth noting that most readers who have taken introductory discrete mathematics or mathematics for computer science classes should be able to follow along without any issues.

While not obligatory, possessing a certain level of familiarity with cybersecurity ideas and practices might provide a significant benefit. A foundation of previous knowledge will enhance the grasp of specific chapters that explore subjects related to cybersecurity. It is expected that readers will maintain critical thinking and problem-solving abilities. The book promotes active involvement by using case studies, and reflection. The acquisition of these skills will be important for the proficient application of the principles explained to readers who possess a genuine curiosity about delving into the cross-disciplinary aspects of digital resilience. This concept spans a range of disciplines, including economics, technology, psychology, and policy design. Developing a proclivity for understanding the links and collaborative contributions of numerous fields to digital resilience will boost the reader's engagement with the book.

Pedagogy

The present book is structured into eight chapters, each of which is devoted to a specific piece of the digital world, with the aim of providing an all-round and tailored instructional approach for diverse audiences, while ensuring clarity and coherence throughout the entire book. It is worth noting that the chapters contained in this book are arranged in a deliberate and logical order, such that they encompass the constantly evolving field of computer science and technology. Each of these chapters serves as a reminder of the global significance of digital resilience and cybersecurity. What is particularly noteworthy about the structure of the book is that it has been designed to accommodate a range of teaching techniques, catering to readers of varying levels of expertise, thereby aligning with the book's primary purpose of serving as a guide to the recent developments in computer science. Besides, each chapter of the book offers an in-depth breakdown of the concepts and phenomena that shape the digital world, while simultaneously emphasizing the complex relationship between digital resilience and cybersecurity, and highlighting their combined significance. This emphasis on their intersection enhances readers' understanding of how these distinct dimensions interact and reinforce one another.

The pedagogical structure of this text has been carefully crafted to facilitate a broad understanding and practical application of digital resilience principles. To achieve this, the book starts by laying a solid foundation in the fundamental concepts relating to digital resilience, introducing the reader to the essential concepts of digital infrastructure, cybersecurity, and adaptability, and, where necessary, presenting mathematical models and formal definitions to clarify these ideas. The pedagogical structure of the book follows a logical progression from basic to advanced topics, with each chapter building on the previous one's content, ensuring a continuous learning journey. To illustrate the applicability of digital resilience principles, the book includes case studies from the real world, which are analysed, enabling readers to apply their theoretical understanding to practical scenarios.

The book's interdisciplinary perspective is another distinguishing feature of its approach to education. To explain the essence of digital resilience, the book draws on a range of disciplines, including computer science, engineering, applied economics, and applied psychology. The implementation of digital resilience strategies in practice is a central focus of this book, and to facilitate hands-on learning, the book guides readers through the process of applying these principles and methodologies to real-world situations. The structure of the book is designed to accommodate varying levels of curiosity and expertise, as readers can select which chapters to read first after the introductory chapter, maximizing their initial time investment, while more complex portions can be deferred until later, allowing for a thorough but idiomatic understanding.

Indore, India — Shishir Kumar Shandilya
Bhopal, India — Agni Datta
Bhopal, India — Yash Kartik
Liverpool, UK — Atulya Nagar

Acknowledgement

We humbly believe that saying ‘thank you’ is a better way to show our gratitude for someone’s assistance than just acknowledging it, so that is what we will try to do. Our expertise in cybersecurity and digital resilience has been honed through the teachings and mentorship of Dr Shishir Kumar Shandilya, our co-author and mentor whom we were fortunate enough to meet during our graduate studies. He has guided and instructed us throughout the entire process of writing this book, providing invaluable support and insights. We are grateful for his constant and kind guidance, and his input has undoubtedly been a significant contributor to the book’s success.

Our journey began on a freezing winter night in our dormitory room, where we engaged in a collaborative brainstorming session with our co-authors. It was an interactive session that eventually led to the creation of a textbook and several other initiatives. We would like to take this opportunity to express our deepest gratitude to the numerous individuals who contributed significantly to the production of this book. Their valuable input and ideas have helped us to bring this project to fruition. We are mindful of the importance of not causing any kind of offence and, therefore, wish to explicitly acknowledge every individual who contributed to the project’s success. We have made a conscious effort to avoid overlooking anyone and want to express our appreciation to all those who have helped us accomplish our objectives.

We would like to express our deep gratitude towards the individuals who played a key role in the publication of our work. Eliska Valková deserves a special mention for her persistent and invaluable assistance in the process of editing and production, as well as for keeping us on track with the deadlines. Professor Imrich Chlamtac deserves major recognition for his significant contributions and innovative ideas that have not only enriched the project but also several other fields of computer science. We are truly grateful to him for his immense support. Lastly, we would like to extend our heartfelt appreciation to all the people who were essential in the successful creation of this textbook. Their contributions have been key in making this project a success.

It is important to acknowledge the help and leadership that Kadhambari S. Viswanathan, Assistant Vice President of VIT Bhopal, has consistently provided.

We would also like to express our deep appreciation and gratitude to the Chancellor, Vice President, Trustees, and Vice Chancellor of VIT Bhopal University. Further, we would like to thank our research collaborators, Ajit Kumar, Gaurav Choudhary, David (Bong Jun) Choi, Saket Upadhyay, and Chirag Ganguli, for their significant contributions to this project and other research initiatives we have collaborated on.

It is with utmost gratitude and appreciation that we express our deepest thanks to our parents, loved ones, friends, and colleagues whose encouragement and invaluable contributions have been vital in our personal and professional growth, and without whom we would be nothing more than a mere shadow of ourselves. We are particularly grateful to the dedicated research team at SECURE – Centre of Excellence (CoE) in Cyber Security, VIT Bhopal University, which we are a part of, and of which we feel privileged and honoured to be members, for their tireless efforts and commitment to excellence in the field of cybersecurity.

The students enrolled in the cybersecurity and digital forensics programme at VIT Bhopal University are deserving of special recognition for their vibrant and intellectually stimulating opinions, as well as their genuine interest in the subject matter being discussed. We extend our sincere gratitude to everyone who contributed to the identification and resolution of errata and typographical errors, including but not limited to the valuable intellectual assistance of Devangana Sujay A. and Sidharth Panda, two of our close friends and colleagues on the SECURE-CoE team, whose support and mentorship have been an integral part of our journey.

Their dedication and commitment to excellence in the field of cybersecurity have been an inspiration to us, and we are proud to be a part of this esteemed institution. We are grateful for the opportunities that we have been afforded at VIT Bhopal University, and we look forward to continuing our journey of discovery and innovation in the field of cybersecurity, thanks to the support and mentorship of our colleagues and mentors on the SECURE-CoE team. We remain determined to the quest for excellence in the field of cybersecurity and to make a beneficial contribution to the community through our research.

As we reflect on the process of creating this book, we are filled with immense gratitude towards the creators of LaTeX, the honourable Donald E. Knuth and Leslie Lamport, whose clever creation has been an inspiration to us since our undergraduate days, and has aided us in typesetting the contents of this book. We cannot overlook the integral part that Comprehensive TeX Archive Network (CTAN's) vast collection of macros and packages played in facilitating the creation of mathematical equations and other fancy compositions all while maintaining consistent formatting throughout the book.

Indore, India — Shishir Kumar Shandilya
Bhopal, India — Agni Datta
Bhopal, India — Yash Kartik
Liverpool, UK — Atulya Nagar

Contents

Part I Understanding Digital Resilience

1 What Is Digital Resilience? 3
- 1 Introduction 3
- 2 Context 5
- 3 What Is Cybersecurity? 6
- 4 History 8
- 5 Definition 9
  - 5.1 Technical Measures 11
  - 5.2 Operational Measures 12
  - 5.3 Cybersecurity Awareness 13
  - 5.4 Effective Incident Response 14
- 6 Digital Resilience and Cyber Resilience 15
- 7 Significance 19
- 8 Benefits 21
- 9 Challenges Encountered in Adoption 24
  - 9.1 Challenges in the Industries and the Business Sectors 25
  - 9.2 Challenges in the Education Sector and for Minors 26
  - 9.3 Challenges for Critical Infrastructures 27
  - 9.4 Challenges in the Socioeconomic Sphere 28
  - 9.5 Challenges on the Legal Front 29
  - 9.6 Challenges in the Technical Sphere 30
- 10 Characteristics of a Cyber Resilient Organization 31
  - 10.1 Seven Strategies to Improve Digital Resilience 31
- 11 Embracing a Digitally Resilient Future 40
- 12 Concluding Remarks 41

2 Achieving Digital Resilience with Cybersecurity 43
- 1 Introduction 43
- 2 Nature of Cyber Threats 44
  - 2.1 Motives and Tactics of Attackers 44
  - 2.2 Evolving Nature of Threats and Ongoing Vigilance 50

3 The Impact of Cyber Threats on Digital Resilience 51
3.1 Cybersecurity in Maintaining Trust and Resilience 52
4 Types of Cyber Threats 53
4.1 Malware 54
4.2 Social Engineering 56
4.3 Supply Chain Attacks 57
4.4 Man-in-the-Middle 58
4.5 Denial-of-Service 59
4.6 Injection Attacks 60
4.7 Zero-Day Vulnerabilities 61
5 Threat Intelligence 63
5.1 Defining Threat Intelligence 64
5.2 Attributes of Threat Intelligence 65
5.3 Lifecycle of Threat Intelligence 66
5.4 Types of Threat Intelligence 69
5.5 Threat Intelligence with Machine Learning 70
6 Strategies for Building Digital Resilience 71
6.1 Designing Systems with Resilience 72
6.2 Enforcing Multi-layered Security Mechanisms 74
6.3 Developing a Culture of Security Awareness and Resilience .. 75
7 Trustworthy System 76
7.1 Fundamental Design Principles 77
7.2 Benefits of Trustworthy Computing Systems 79
7.3 Procedure 80
7.4 Importance 81
7.5 Concept of Zero-Trust Systems 84
7.6 Advantages of Zero-Trust Systems 85
7.7 Drawbacks of Zero-Trust Systems 86
8 Balancing Security and Resilience 87
8.1 Security and Usability in Digital Systems 88
8.2 Architecting Resilient Systems 89
9 Safeguarding Digital Assets 90
9.1 Value of Digital Assets and Impact of Loss or Compromise... 90
10 Role of Encryption in Safeguarding Digital Assets 91
10.1 Homomorphic Encryption 91
10.2 Zero-Knowledge Proof 94
10.3 Multi-party Computation 96
11 Enhancing the Robustness of Digital Assets 98
11.1 Robust Access Controls and Authentication Protocols 98
11.2 Safeguarding Data Through Backup Strategy 99
12 Incident Response 101
12.1 Digital Forensics and Incident Response (DFIR) 103
12.2 Advantages of Incident Response 106
12.3 Disadvantages of Incident Response 109

13 Digital Resilience Through Effective IRP ... 110
13.1 Concrete Illustrations ... 110
13.2 Steps Towards Designing Scalable IRPs ... 111
14 Significance of Swift Response and Recovery ... 112
15 Role of Regular Drills and Exercises ... 113
16 DevSecOps ... 114
16.1 DevSecOps and Digital Resilience ... 116
16.2 Advantages ... 118
16.3 Disadvantages ... 119
16.4 Security Challenges ... 119
16.5 Research Avenues ... 120
17 Concluding Remarks ... 122

Part II Building Digital Resilience

3 Navigating the Regulatory Landscape ... 127
1 Introduction ... 127
2 Regulations and Impact on Shaping Technology ... 129
3 Importance of Legal Frameworks ... 131
4 Understanding Regulatory Authorities ... 133
4.1 Government Agencies ... 133
4.2 Industry-Specific Bodies ... 135
5 Adaptations for Legal Frameworks ... 137
6 Cybersecurity and Privacy Laws, USA ... 139
6.1 FCRA, 1970 ... 140
6.2 FERPA, 1974 ... 141
6.3 ECPA, 1986 ... 143
6.4 VPPA, 1988 ... 144
6.5 HIPAA, 1996 ... 146
6.6 COPPA, 1998 ... 147
6.7 GLBA, 1999 ... 148
6.8 CCPA, 2018 ... 150
6.9 ADPPA, 2022 ... 152
7 Cybersecurity and Privacy Laws, UK ... 156
7.1 Computer Misuse Act, 1990 ... 156
7.2 Privacy and Electronic Communications Regulations, 2003 ... 157
7.3 Investigatory Powers Act, 2016 ... 158
7.4 Payment Services Regulations, 2017 ... 159
7.5 Network and Information Systems Regulations, 2018 ... 160
7.6 Data Protection Act, 2018 ... 161
7.7 UK-GDPR, 2021 ... 162
8 Cybersecurity and Privacy Laws, Australia ... 163
8.1 The Privacy Act, 1988 ... 164
8.2 Telecommunications Act, 1997 ... 166
8.3 Electronic Transactions Act, 1999 ... 167

8.4 Spam Act, 2003 168
8.5 DNCR Act, 2006 169
8.6 HI Act, 2010 170
8.7 MHR Act, 2012 171
8.8 TOLA Act, 2018 173
8.9 NDB Scheme, 2018 174
9 Cybersecurity and Privacy Laws, EU 176
9.1 GDPR, 2018 176
9.2 DSA, 2022 179
9.3 DMA, 2022 180
9.4 DPF, 2023 181
9.5 AI Act, 2023 182
9.6 ePR, Expected 2023 185
10 Cybersecurity and Privacy Laws, India 186
10.1 The Information Technology Act, 2000 188
10.2 Know Your Customer, 2002 189
10.3 Information Technology (Amendment) Act, 2008 190
10.4 Information Technology Rules, 2011 192
10.5 Indian SPDI Rules for Reasonable Security Practices, 2011 193
10.6 National Cyber Security Policy, 2013 195
10.7 National Cyber Security Strategy, 2020 196
10.8 IT Rules, 2021 197
10.9 The Digital Personal Data Protection Act, 2023 198
11 Cybersecurity Regulations 200
11.1 NIST 201
11.2 ISO 202
11.3 NIST Cybersecurity Framework 203
11.4 ISO Standards 206
11.5 UKCIS Digital Resilience Framework 212
12 Digital Resilience and Regulatory Compliance 215
12.1 Intersection 216
12.2 Impact of Non-compliance 217
12.3 Benefits of Regulatory Compliance 218
13 Compliance Challenges 221
13.1 Complexity of Regulatory Landscape 223
13.2 Cross-Border Compliance Issues 225
13.3 Rapidly Changing Regulations 227
13.4 Compliance Costs 230
14 Regulatory Landscape 232
14.1 Building a Regulatory Compliance Team 232
14.2 Regulatory Risk Assessment 233
14.3 Compliance Audits and Assessments 233
14.4 Compliance Monitoring and Reporting 234
15 Conforming AI to Regulatory Frameworks 235
15.1 Safe and Effective Systems 237

15.2 Algorithmic Discrimination and Bias ... 237
15.3 Data Privacy ... 238
15.4 Notice and Explainability ... 238
15.5 Human Alternatives, Consideration, and Fallback ... 239
16 Concluding Remarks ... 239

4 Nurturing Resilience in Minors ... 241
1 Introduction ... 241
2 Digital Environment and Cybercrimes ... 243
2.1 Cyberbullying Against Minors ... 243
2.2 Offensive Name-Calling ... 244
2.3 Spreading of False Rumour ... 244
2.4 Receiving Explicit Images Without Solicitation ... 245
2.5 Cyberstalking ... 246
2.6 Physical Threats ... 246
2.7 Unauthorized Dissemination of Explicit Images ... 247
2.8 Exposure to Inappropriate Content ... 248
3 Digital Environment and Minors ... 248
4 Digital Environment and Supervision ... 250
5 Digital Environment and Guidelines ... 252
6 Digital Environment and Government's Role ... 254
7 Stance of the World's Governments ... 255
7.1 Australia ... 257
7.2 European Union (EU) ... 257
7.3 India ... 258
7.4 Japan ... 259
7.5 The United Kingdom (UK) ... 260
7.6 The United States of America (USA) ... 262
8 Stance of Multinational Corporations ... 263
8.1 Microsoft ... 263
8.2 Apple ... 264
8.3 Facebook ... 265
8.4 Google ... 265
9 Stance of Non-governmental Organizations ... 266
9.1 ICMEC ... 266
9.2 ASACP ... 267
9.3 WePROTECT Global Alliance ... 268
10 Limitations of Global Digital Resilience Frameworks ... 269
10.1 One-Size-Fits-All Approach ... 269
10.2 Lack of Empirical Evidence ... 270
10.3 Rapidly Changing Technology ... 270
10.4 Limited Parental Involvement ... 270
10.5 Insufficient Resources ... 271
11 Guidelines for Guardians ... 271
11.1 Tailoring Social Media Use ... 272

11.2 Adult Monitoring and Coaching 272
11.3 Minimizing Exposure to Harmful Content 273
11.4 Combatting Cyberbullying and Hate 273
11.5 Monitoring Screen Time 274
11.6 Encouraging Social Media Literacy 274
12 Concluding Remarks 277

5 A Study in Attack and Breaches 279
1 Introduction 279
2 Implications and Repercussions 280
3 Ukraine Powergrid Attack, 2015 282
3.1 Synopsis 282
3.2 Ramifications 284
4 WannaCry Attack, 2017 286
4.1 Synopsis 286
4.2 Ramifications 288
5 NotPetya Attack, 2017 289
5.1 Synopsis 289
5.2 Ramifications 293
6 Marriott Breach, 2018 294
6.1 Synopsis 295
6.2 Ramifications 298
7 RockYou 2021 Breach, 2021 298
7.1 Synopsis 299
7.2 Ramifications 300
8 Colonial Pipeline Attack, 2021 301
8.1 Synopsis 301
8.2 Ramifications 302
9 AIIMS Ransomware Attack, 2022 303
9.1 Synopsis 303
9.2 Ramifications 305
10 Implications of Digital Resilience 306
11 Concluding Remarks 308

Part III Multidimensional Digital Resilience

6 Role of Artificial Intelligence and Machine Learning 313
1 Introduction 313
2 History 314
3 Defining AI and ML 316
3.1 Artificial Intelligence (AI) 316
3.2 Machine Learning (ML) and Deep Learning (DL) 317
4 Applications 321
4.1 Agriculture 321
4.2 Autonomous Systems 324
4.3 Education 326

4.4 Cybersecurity ... 329
4.5 Energy ... 332
4.6 Entertainment ... 334
4.7 Finance ... 338
4.8 Healthcare ... 341
4.9 Manufacturing ... 344
4.10 Weather Forecasting ... 345
5 Advancing Cybersecurity Through AI ... 348
6 Advantages and Benefits ... 350
6.1 AI and Availability ... 350
6.2 AI and Scalability ... 352
6.3 AI and Accuracy ... 354
6.4 AI and Big Data ... 356
6.5 AI and Automation ... 358
6.6 AI and Hazard Mitigation ... 360
7 Challenges and Considerations ... 361
7.1 AI and Data ... 362
7.2 AI and System Integration ... 363
7.3 AI and Professional Competence ... 365
7.4 AI and Creativity ... 367
7.5 AI and Resources ... 370
7.6 AI and Ethics ... 371
8 Explainable AI ... 374
8.1 Application ... 377
8.2 Advantages ... 380
8.3 Disadvantages ... 381
9 Responsible AI ... 383
9.1 Principles ... 385
9.2 Application ... 387
9.3 Advantages ... 389
9.4 Disadvantages ... 392
10 Digital Resilience with AI ... 393
11 Concluding Remarks ... 396

7 Thriving in the Quantum Era ... 401
1 Introduction ... 401
1.1 Quantum Superposition ... 404
1.2 Quantum Entanglement ... 405
1.3 No-Cloning Theorem ... 407
2 Quantum Computing ... 410
2.1 Development of Quantum Computing ... 411
2.2 Classical versus Quantum Computing ... 413
2.3 Challenges in Quantum Computing ... 414
3 Qubits and Quantum Gates ... 415
3.1 Single-Qubit Gates ... 417

3.2 Multi-Qubit Gates ... 419
3.3 Quantum Algorithms ... 419
4 Quantum Cryptography ... 421
4.1 Lattice Cryptography ... 425
4.2 Code-Based Cryptography ... 428
4.3 Hash-Based Cryptography ... 430
4.4 Isogeny-Based Cryptography ... 432
4.5 Unbalanced Oil and Vinegar Schemes ... 433
5 Quantum Key Distribution (QKD) ... 434
6 Quantum Applications ... 437
6.1 Quantum Machine Learning (QML) ... 438
6.2 Computational Chemistry ... 439
6.3 Financial Portfolio and Asset Optimization ... 440
6.4 Industrial-Scale Logistics and Scheduling ... 442
7 Challenges and Future Prospects ... 443
7.1 Error Correction and Scaling Quantum Systems ... 443
7.2 Ethical and Societal Implications ... 445
7.3 Challenges in Contemporary Cryptography ... 446
8 Dangers to Conventional Cryptography ... 447
8.1 Factorization Problem ... 448
8.2 Discrete Logarithm Problem ... 450
8.3 Shor's Algorithm ... 452
9 Concluding Remarks ... 456

8 Advancing Security and Resilience ... 459
1 Introduction ... 459
2 Paradigms of Computing ... 461
2.1 Centralized Computing ... 461
2.2 Decentralized Computing ... 463
2.3 Distributed Computing ... 463
2.4 Blockchain ... 466
3 Cloud Computing ... 467
3.1 Cloud Security ... 469
3.2 Advantages ... 471
3.3 Security Challenges ... 473
3.4 Research Avenues ... 475
4 Hardware Security ... 476
4.1 Significance ... 476
4.2 Security Challenges ... 478
4.3 Future ... 480
5 Internet of Things (IoT) ... 481
5.1 Advantages ... 482
5.2 Drawbacks ... 483
5.3 Security Challenges ... 484
5.4 Research Avenues ... 485

6 Cyber-Physical Systems (CPS) 486
6.1 Significance 487
6.2 Challenges 491
6.3 Research Avenues 493
7 Autonomous Systems 494
7.1 Significance 495
7.2 Challenges 496
7.3 Research Avenues 498
8 Adaptive Defence 499
8.1 Significance 501
8.2 Security Challenges 503
8.3 Research Avenues 504
9 Cyber-Security in an Open World: New Space 507
9.1 Significance 507
9.2 Security Challenges 508
9.3 Research Avenues 510
10 Computational Psychology and Cyberpsychology 512
10.1 Significance 514
10.2 Ethical Considerations 516
10.3 Psychological Factors 516
10.4 Designing Resilient Systems 516
10.5 Ethical Considerations 517
10.6 Research Approaches 518
11 Consideration of Human Factors 519
11.1 Human-Centric Approaches 520
11.2 Recognizing Human Error and Behaviour 521
11.3 Training and Education 521
11.4 Interdisciplinary Collaboration 521
12 Metrics for Resilience 522
12.1 Defining and Measuring Resilience 522
12.2 Quantitative Metrics 523
12.3 Qualitative Metrics 523
12.4 Challenges in Resilience Metrics 524
12.5 Role of Metrics in Resilience Strategies 525
13 Concluding Remarks 526

Glossary 531

References 537

Index 553

About the Authors

Shishir Kumar Shandilya, DSc, PhD, is an Associate Professor in the School of Data Science and Forecasting at Devi Ahilya University in India. He is also a Visiting Professor at Liverpool Hope University in the United Kingdom. He is a Cambridge University-certified professional teacher and trainer, a TEDx speaker, an Association for Computing Machinery (ACM) Distinguished Speaker, and a senior member of Institute of Electrical and Electronics Engineers (IEEE). Shandilya is a National Association of Software and Service Companies (NASSCOM)-certified master trainer for security analysts in Security Operations Center (SOC) (SSC/Q0909: National Vocational Education Qualification Framework (NVEQF) Level 7). He has received the International Development Association (IDA) Teaching Excellence Award for distinctive use of technology in teaching by the Indian Didactics Association in Bangalore (2016), and the Young Scientist Award for two consecutive years, 2005 and 2006, by the MP Science Congress and MP Council of Science and Technology. He is a highly regarded author with published research works with reputable academic publishers such as Springer, IGI-USA, River Denmark, and Prentice Hall of India. Shandilya also has international and national patents and copyrights granted for adaptive cyber defence methods. His research interest includes adaptive defence, advanced digital investigation and forensic methods, explainable artificial intelligence, privacy-preserving computing, nature-inspired cryptography (NIC), and nature-inspired cyber security (NICS).

Agni Datta is a cryptologist working with SECURE – Centre of Excellence in Cyber Security and a researcher who specializes in theoretical computer science at VIT Bhopal University in India. He is a member of several prestigious professional associations, including ACM, IEEE, American Mathematical Society (AMS), and Society for Industrial and Applied Mathematics (SIAM). His research interests include various cryptographic primitives like zero-knowledge proofs, multiparty computation, and lattice cryptography. He is particularly focused on security primitives for computing, as well as computational complexity and probabilistic proofs.

Yash Kartik is a security researcher working at SECURE – Centre of Excellence in Cyber Security at VIT Bhopal University in India. He is affiliated with esteemed organizations, such as SIAM, ACM, IEEE, and AMS. His research focuses on security methods such as cybersecurity for space systems, anonymous networking, and cyber resilience. He also has a keen interest in cryptographic primitives like zero-knowledge proofs, secure multi-party computation, and homomorphic encryption.

Atulya Nagar Foundation Professor of Computer and Mathematical Sciences at Liverpool Hope University, and Head of the Department of Computer Science, is a distinguished mathematician with expertise spanning computational science, bioinformatics, operations research, and systems engineering. Awarded a Commonwealth Fellowship, he earned his doctorate in applied non-linear mathematics from the University of York in 1996. Nagar is internationally recognized for his work in theoretical computer science, applied mathematical analysis, operations research, and industrial systems engineering. Leading the Intrusion Detection System (IDS) research group, he focuses on strategic and applied research, particularly in advancing applications of engineering, computational systems, and biological systems. The group's notable contributions include innovative DNA sequence analysis using sophisticated computational techniques. Nagar has published extensively in reputable outlets like the IEE and IEEE, and he serves on editorial boards for prestigious journals. His leadership extends to chairing and participating in international conferences. In addition to his academic roles, Nagar supervises PhD projects, serves as an external examiner, and holds visiting and adjunct professorships. His teaching expertise encompasses Applied Analysis, Systems Engineering, and Computational Biology. Before joining Liverpool Hope University, he made significant contributions as a Senior Research Scientist at Brunel University, focusing on Engineering and Physical Sciences Research Council (EPSRC)-sponsored research projects in mathematical sciences and systems engineering.

List of Figures

Fig. 1.1 Cybersecurity mindmap 8
Fig. 1.2 Digital resilience 12
Fig. 1.3 Critical infrastructure 27
Fig. 1.4 Seven-step strategy 32
Fig. 2.1 Cyber kill chain 47
Fig. 2.2 A basic firewall architecture 49
Fig. 2.3 An example of the basic types of cyber threats 54
Fig. 2.4 Categorization of malware 55
Fig. 2.5 Social engineering attack 56
Fig. 2.6 Anatomy of supply chain attacks 57
Fig. 2.7 Normal connection versus man-in-the-middle 58
Fig. 2.8 Usage of botnet in DoS 59
Fig. 2.9 Categorization of injection attacks 61
Fig. 2.10 Anatomy of zero-day vulnerability exploitation 62
Fig. 2.11 Cyber threat intelligence cycle 66
Fig. 2.12 Homomorphic encryption in cloud 92
Fig. 2.13 Zero-knowledge proof 94
Fig. 2.14 NIST incident response cycle 102
Fig. 2.15 NIST digital forensics framework 104
Fig. 2.16 DevSecOps versus DevOps 116
Fig. 6.1 AI and its domains 318
Fig. 6.2 Interrelation of XAI 375
Fig. 6.3 White box versus black box 376
Fig. 8.1 Distributive computing 464
Fig. 8.2 Blockchain 467
Fig. 8.3 Cloud computing paradigm 468
Fig. 8.4 Hardware security components 476
Fig. 8.5 IoT system 482
Fig. 8.6 Cyber-Physical Systems (CPS) 486
Fig. 8.7 Interaction of CPS 486
Fig. 8.8 Intersection of cyberpsychology with computational intelligence 514

List of Tables

Table 1.1 Cybersecurity measures for cyber resiliency [Ros+21] 18
Table 1.2 Digital resilience vs cyber resilience 19
Table 2.1 Incident response classifications 102
Table 2.2 Comparison of digital forensics and incident response 107
Table 2.3 Comparison of DevOps and DevSecOps 115
Table 4.1 Risks, guidelines, and effects of screen time 275
Table 6.1 Comparison of machine learning and deep learning 320
Table 6.2 Comparison of XAI and RAI 384
Table 7.1 Post-quantum cryptographic methods 424
Table 8.1 Comparison of centralized, decentralized, and distributed computing 462
Table 8.2 Comparison of CPS and IoT 488
Table 8.3 Computational psychology and cyberpsychology 513

Part I
Understanding Digital Resilience

Chapter 1
What Is Digital Resilience?

1 Introduction

The digital world can be compared to an expansive and interconnected ecosystem, similar to the complex balance observed among various species and organisms residing in a coral reef. The concept of digital resilience refers to the capacity of the digital ecosystem to endure and overcome unforeseen challenges. This involves the adoption of policies and safeguards designed to improve the authenticity, confidentiality, and uninterrupted operation of digital operations, similar to conservation efforts that safeguard the welfare of a coral reef. By promoting the development of digital resilience, both organizations and individuals may successfully protect the stability and flexibility of their digital systems and technologies in response to the constantly changing digital space. This phenomenon might be compared to the resilience exhibited by coral reefs, which possess the capacity to endure and recuperate from diverse disturbances.

This example serves as a significant representation of the present state of cybersecurity, emphasizing the need for individuals, enterprises, and other societal entities to possess the capacity to effectively surmount the wide range of problems associated with cyber threats in the present-day scenario. To fulfil this requirement, it is important to prioritize the development of our collective *digital resilience*. This involves the establishment and implementation of robust protocols for cybersecurity, the formulation and implementation of efficient strategies to mitigate incidents, and the development of a phase of education and consciousness surrounding cybersecurity concerns. By operating together, we can enhance our ability to face and adjust to the challenges presented by the digital age. This includes safeguarding our interests as well as the broader welfare of our communities from the risks associated with cyber threats.

The development of digital resilience presents numerous significant advantages that incorporate various dimensions of personal, organizational, and societal

S. K. Shandilya et al., *Digital Resilience: Navigating Disruption and Safeguarding Data Privacy*, EAI/Springer Innovations in Communication and Computing,
https://doi.org/10.1007/978-3-031-53290-0_1

domains. The seamless continuation of company operations in the face of cyber threats is an integral component of digital resilience. Economic operations can continue uninterrupted as a result of this, which helps businesses cope with digital interruptions. The development of digital resilience serves to enhance the preservation of one's reputation, which is a very valuable resource for both individuals and companies.

Similarly, the impacts of cyberattacks on critical infrastructure, a recurring issue in today's networked world, might be lessened by the construction of digital resilience. The potential negative consequences for critical social infrastructures such as power grids, transportation systems, and communication networks might be reduced by enhancing the capability to confront and recover from cyberattacks, therefore strengthening societal stability and security. The concept of digital resilience is essential in protecting the integrity and legitimacy of democratic processes. The growing use of digital platforms for elections and civic participation has raised concerns about their vulnerability to cyber manipulation, presenting a significant danger to the integrity of democratic processes. The goal of improving digital resilience is also to create a strong defensive system against malicious influences, maintaining the integrity of democratic processes.

Attaining digital resilience is contingent upon collaborative undertakings and efforts of individuals, organizations, and governmental entities. The successful execution of a strategy requires the allocation of resources towards cybersecurity tools and technology, the development of a cybersecurity-conscious culture through extensive awareness and instructional initiatives, and the facilitation of information exchange and collaborative initiatives.

Governments play a foundational role in facilitating and advancing the promotion and development of digital resilience. Government entities have the authority to adopt and enforce legislation and regulations that mark the parameters of cybersecurity practices and standards, thereby promoting a cohesive and uniform approach across various industries. It is important to recognize the role that governments play in distributing money for the progress of research and development in state-of-the-art cybersecurity measures. The implementation of this proactive approach aims to strengthen the range of strategies utilized in addressing cyber threats.

Also, the distribution of aid and resources to organizations and communities, especially in the early stages of their efforts to develop digital resilience, holds special significance. Governments are critical and integral in promoting the establishment and implementation of robust cybersecurity plans by offering support and, therefore, enabling a conducive environment for the development and execution of those initiatives. Accordingly, this facilitates firms' efficient management of and recovery from cyber disturbances.

Marking today's widespread technology integration and the increase in sophisticated cyberattacks, it is crucial to highlight the development of digital resilience. To meet this importance, stakeholders must develop and implement a strategy that comprises the development and deployment of robust cybersecurity initiatives, the establishment of effective strategies for managing cyber incidents, and the promotion of a cultural environment that prioritizes cybersecurity awareness and education.

Collaborative efforts are also necessary for establishing digital resilience. The previously indicated requirement involves the integration of resources and specialist skills, as well as the facilitation of awareness campaigns and the promotion of information exchange. Governments are most impactful in creating the cybersecurity ecosystem by participating in policy and regulatory formation, promoting innovation, and assisting individuals and organizations wanting to increase their digital resilience. Prioritization of joint efforts aimed at developing a more secure digital space is of essence.

2 Context

The growing dependence of individuals, corporations, and governments on digital infrastructure and data-driven operations has made them more vulnerable to cyberattacks. The task at hand is to critically analyse the security challenges encountered in today's digital environment, demonstrating how interconnected these challenges are, the ever-changing essence of the threats, and the implications for individuals, businesses, and society as a whole. The interconnectedness with systems and devices is a major security problem in today's society. The notion of the Internet of Things (IoT), in which a rising number of machines are networked to the Internet, exemplifies the problem at issue. A broad variety of technology, including smart household appliances and industrial equipment, usually lack proper security protections, making it susceptible to unscrupulous individuals.

The connectivity of many vulnerable devices creates a large attack surface that may be controlled to launch coordinated attacks, causing major security concerns. The increasing reliance on cloud computing services and remote work arrangements worsens the problem of connectivity. Because of their potential to centralize data storage and processing, cloud services have become a target for hackers. The introduction of remote work within an organization introduces many devices and networks into its infrastructure, which may lack the same degree of security as typical office settings. The complicated relationship between these components creates complex security challenges as the job of protecting the perimeter gets more challenging.

The threats' changing nature poses a substantial security challenge. Cybercriminals are always coming up with new methods and approaches to get around current security procedures. Cyberattackers take advantage of flaws in software, hardware, and human behaviour. The all-presence of ransomware attacks demonstrates criminal actors' constant adaptation and inventiveness. Ransomware is a kind of malicious software that employs encryption methods to make an organization's data unavailable, forcing the victim to pay a ransom to recover access. This conduct causes major disruptions and financial consequences for the impacted organization. As cybersecurity defences improve, threat actors react by inventing more complex and evasive techniques. State-sponsored cyberattacks have increased, posing substantial security difficulties.

Nation-states engage in cyber espionage, cyber warfare, and cybercrime for a variety of reasons, including economic gain, political leverage, and information gathering. The consequences of these attacks may be far-reaching, affecting not just economic stability but also national security and geopolitical stability. The attribution of culpability for these attacks is typically complex, further complicating the security environment.

Individuals are vulnerable to various security risks today. Phishing attacks, which include the false imitation of legitimate companies to trick individuals into disclosing sensitive information, are common. This condition jeopardizes the revelation of personal information, financial assets, and privacy. The massive data breaches of personal information performed by notable businesses highlight individuals' vulnerability to the dangers of identity theft and fraudulent activity.

Organizations face security issues in a variety of areas. Aside from the financial consequences of data breaches and hacking, the potential for major reputational loss should not be disregarded. Organizations are expected by consumers and stakeholders to ensure the safeguarding of their data and privacy. The occurrence of a security incident that leads to a breach of trust can have substantial repercussions on a brand's reputation and result in client attrition.

Given the importance of digital infrastructure in providing essential services and systems, security concerns have a broad influence on society. Cyberattacks against critical infrastructure, such as power grids, water treatment facilities, and healthcare systems, have the potential to interrupt daily operations, jeopardize public safety, and undermine national security. The ongoing challenge of protecting critical infrastructure from cyberattacks involves collaboration among government agencies, industry partners, and cybersecurity specialists.

The modern digital world provides a complex set of interrelated security challenges. Cybersecurity is important because of the interconnection of systems and devices, the dynamic threat environment, and the wide-ranging effects on individuals, organizations, and society. These issues need a plan that includes technological solutions, legislative frameworks, and cybersecurity education and awareness. To tackle the ever-present and ever-changing dangers to the digital environment, protection must be strengthened as technology progresses.

3 What Is Cybersecurity?

Cybersecurity comprises a range of components that are essential for ensuring an individual's online security. The term 'cybersecurity' refers to a complex integration of technologies, protocols, and processes that have been carefully designed to safeguard networks, devices, applications, and data from malicious attacks, destruction, or unauthorized access. The term 'cybersecurity' is occasionally used interchangeably with 'information technology security' in certain contexts.

Definition 1 (Cybersecurity) Cybersecurity is the act of averting harm to, safeguarding, and restoring computers, electronic communication services and systems, wire communication, and electronic systems, including the enclosed information, ensuring its continued availability, unaltered condition, verification, confidentiality, and prevention of denial of participation.

Implementing cybersecurity measures must be necessary for protecting digital information systems (DIS) against adverse cyberattacks. These measures operate as a powerful defensive mechanism, preventing potential attackers from exploiting system weaknesses. Physical security practices, on the other hand, rely on the use of conventional locks and security systems, with law enforcement agencies in charge of safeguarding individuals' physical well-being.

The significance of cybersecurity stems from the acquisition, alterations, and preservation of enormous volumes of data at exceptional rates by political, military, economic, financial, and medical institutions utilizing computers and similar technologies. A considerable proportion of the contested data consists of confidential or other forms of information, including intellectual property, financial records, personal data, and other similar categories of data. The compromise of sensitive information could potentially result in severe consequences.

Organizations transport sensitive data across networks and devices as part of their operations. In this context, cybersecurity emerges as a unique discipline aimed at protecting both information and the technology used to process and store it. Because of the increasing frequency and complexity of cyberattacks, organizations, particularly those in charge of protecting sensitive information related to national security, health, or finance, must implement security measures to safeguard their confidential business data and personnel records. Significantly, major intelligence officers in the country said as early as the beginning of this decade that cyberattacks and digital espionage pose a greater threat to national security than even terrorism.

The ever-changing nature of security challenges is a fundamental difficulty in cybersecurity. Historically, both institutions and governments have emphasized perimeter security, with an emphasis on protecting their most critical system components and mitigating recognized threats. This technique has been ineffectual because of threats' rapid expansion and evolution, which has outpaced the organization's capability to keep up. As a result, consulting firms recommend using proactive and adaptive cybersecurity solutions. Similarly, the National Institute of Standards and Technology (NIST) has developed guidelines and an assessment methodology to promote the adoption of real-time evaluations and continuous monitoring. This approach denotes a shift away from the previous perimeter-oriented notion and towards a data-centric security paradigm.

To further enhance understanding, individuals may refer to the representation of the cybersecurity mindmap, as shown in Fig. 1.1. This graphical representation

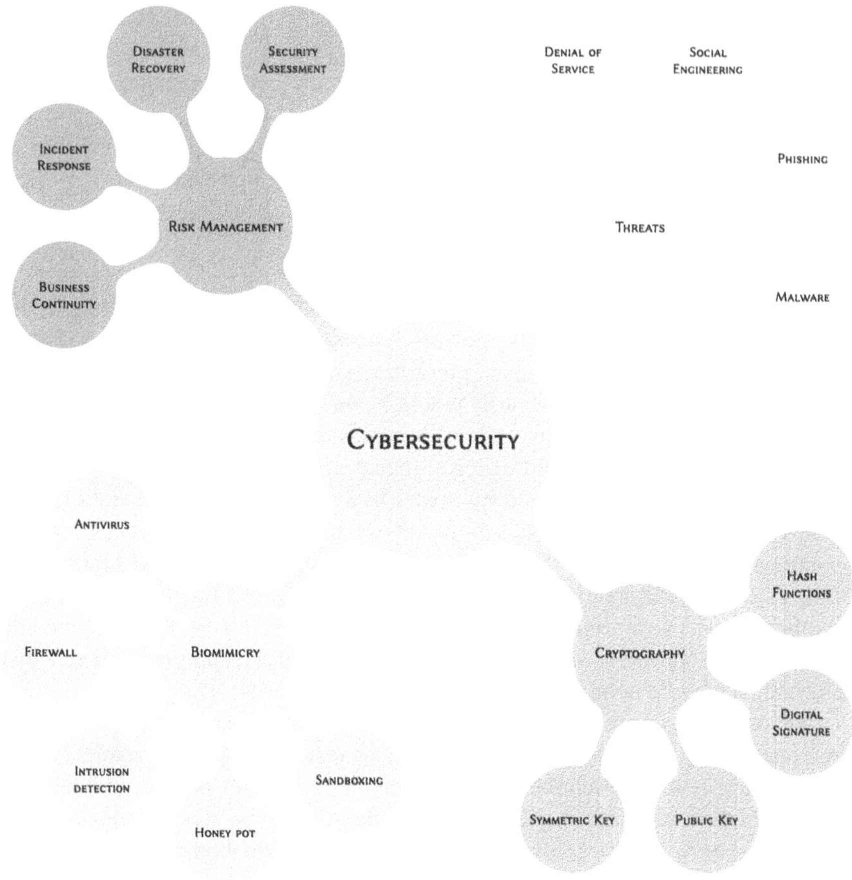

Fig. 1.1 Cybersecurity mindmap

facilitates an understanding of the wide-ranging and diverse attributes associated with cybersecurity, demonstrating its inherently multidisciplinary character.

4 History

The notion of digital resilience has received a noteworthy spotlight in the technology industry for a prolonged period. The necessity for firms to be sufficiently prepared for unanticipated disruptions in their digital infrastructure has become increasingly apparent. The COVID-19 pandemic has had a substantial effect on the adoption and execution of digital resilience solutions, resulting in a fundamental revolution. It exacerbated an unprecedented worldwide economic and societal disruption, revealing the weaknesses of countless businesses and organizations that were ill-equipped to handle this crisis due to a lack of necessary resources.

The post-pandemic years experienced significant development and expansion in digital resilience. In the past, digital resilience was mostly associated with cybersecurity and online competence. However, it has now evolved to comprise a range of strategies that organizations use to effectively recover from or handle unexpected disasters. During the COVID-19 pandemic, several organizations resorted to utilizing digital technologies to maintain certain levels of functioning within the crisis. The scope of digital resiliency involved the adoption of digital platforms by small food establishments, the explosive development of Internet commerce, and the incorporation of new manufacturing technology.

The swift and widespread shift towards remote work, digital education, telemedicine, and other online pursuits has highlighted the importance of digital resilience. Organizations that had previously made investments in digital transformation and had implemented effective strategies to enhance their digital resilience were better equipped to handle the difficulties arising from the pandemic and maintain their operational capabilities.

Accordingly, there has been an increasing trend in the implementation of strategies that promote digital resilience in diverse industries such as healthcare, finance, education, and manufacturing. Organizations are currently increasing their resource allocations towards cybersecurity, cloud computing, data analysis, and other technological advancements that enable them to effectively address emergent situations and maintain seamless corporate operations.

The concept of digital resilience is undergoing continuous evolution, expanding its scope beyond the conventional focus on cybersecurity and disaster recovery. The concept has evolved to embrace a wider range of skills and abilities, including agility, flexibility, creativity, and sustainability. To maintain a competitive edge in the digital age, organizations must exhibit the capability to effectively utilize emerging technology and quickly adapt to evolving situations. The COVID-19 pandemic has significantly transformed enterprises' perceptions and strategies around digital resilience. The incorporation of tactics that enhance digital resilience will continue to be an important aspect of organizations' operations and growth strategies as they navigate unprecedented difficulties and uncertainties.

5 Definition

Let us now try a more thorough and theoretical explanation of digital resilience. In this section, we shall undertake a more thorough study of its complexity with more precision and meticulousness. The idea of digital resilience is very important today, and it is necessary to accurately describe it and appreciate its relevance. Individuals may successfully confront and negotiate threats connected with digital technology, eventually leading to personal development and progress via hard situations.

In today's world of pervasive technological integration, the idea of digital resilience serves an important purpose in supporting both individuals and organizations in efficiently reacting to and recovering from cyber-attacks and disruptions. A full knowledge of digital resilience is required before developing measures to promote and improve its effectiveness. As a result, before delving into a complete exploration of its implications, a clear and thorough definition of digital resilience is required.

Definition 2 (Digital Resilience) The organization's or individual's ability to maintain the confidentiality, integrity, and availability of crucial information and systems, achieved through a cohesive integration of technical, operational, and organizational actions while withstanding persistent threats, by implementing robust cybersecurity controls, effective incident response strategies, and a culture promoting cybersecurity awareness and education, all while recognizing the inevitability of security incidents and necessitating rapid and efficient counteractions, is defined as 'digital resilience'.

The efficient collaboration of governments, organizations, and individuals is required for the effective implementation of digital resilience. This includes allocating resources to the creation and deployment of cybersecurity procedures and technology, promoting a culture that values and educates on cybersecurity, and facilitating information exchange and collaborative efforts. Considering the key function of technology and the increasing frequency and complexity of cyber threats, the modern digital world necessitates joint efforts.

Digital resilience is a critical quality that individuals ought to develop by actively engaging in appropriate online challenges and opportunities, as opposed to evading them. In light of the growing prevalence of technology, individuals must develop the capacity to identify and proficiently handle the dangers that exist within digital spaces. This requires the development of self-control and the acquisition of the capacity to respond judiciously to adverse circumstances. Secure and supervised online experiences are crucial in facilitating the adjustment process by equipping people with the necessary knowledge, skills, and self-assurance to effectively navigate negative interactions on the Internet. Also, it is equally critical to offer the required guidance and support to aid in the restoration and re-engagement of individuals with digital opportunities.

Digital resilience is a complex aspect of an individual's nature that develops through the interaction of personal experiences and introspective reflection. The state under consideration does not possess a static or immutable quality. On the flip side, the way it appears can differ significantly among individuals as a result of contextual variables, personal backgrounds, chronological progression, and additional pertinent aspects. There is a shared responsibility among various stakeholders, including but not limited to families, educators, policymakers, frontline care providers, and industries, to promote an environment that enhances resilience rather than undermines it. It is crucial to acknowledge that online interactions involve complex processes that are impacted by a variety of internal and external factors that affect users. These components include vulnerabilities such as social isolation, the pursuit of personal identity, and obstacles to mental and cognitive health.

The creation of environments that enhance resilience requires the precise balance of numerous components. Communities and organizations can enhance digital resilience through the promotion of secure online experiences and the support of individuals in developing self-regulation skills and judicious reactions to challenging circumstances. This involves offering assistance and direction to individuals who have experienced negative stressors on the Internet, as well as creating opportunities for self-reflection and education.

The significance of digital resilience is heightened in the present times when individuals engage in social interactions, exploration, innovation, and labour within digital spaces. The development of digital resilience empowers individuals to effectively identify and handle potential risks, extract valuable lessons from challenging situations, recover from setbacks, and maintain their overall well-being. This promotes a cultural environment distinguished by personal responsibility and autonomy, wherein participants possess the expertise and competencies essential for safeguarding themselves in the ever-changing and complex digital world with confidence and adaptability.

Digital resilience can be achieved through an approach that includes technical, operational, and organizational measures, as well as increased cybersecurity awareness and an efficient incident response framework; as shown in Fig. 1.2, we will go into detail for each topic in the following subsections. Technical measures include the use of strong security protocols and technology to strengthen digital infrastructure. The implementation of procedural protections and threat management frameworks and optimizing operational procedures to minimize risks are examples of operational measures. Resilience must be integrated into organizational culture, governance structures, and strategic planning as part of organizational measures. Cybersecurity awareness is instilling a cautious mentality in employees so that they can recognize and neutralize any threats. Finally, robust incident response procedures are critical for resolving and limiting the effects of security problems as soon as possible, hence improving overall digital resilience.

5.1 Technical Measures

Technical measures comprise a range of strategies, tactics, and technologies that have been deliberately designed to strengthen the protection of digital assets against unauthorized access, modification, pilferage, or degradation. In the context of digital resilience, the implementation of these procedures is critical for preserving the availability, confidentiality, and integrity of data. Technical measures comprise hardware and software components that serve as the fundamental building blocks

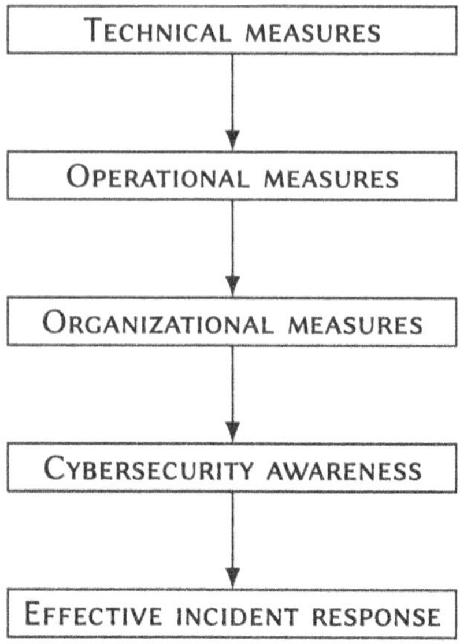

Fig. 1.2 Digital resilience

of security measures. Technical safeguards consist of access controls, firewalls, antivirus software, and encryption methods, among others.

To protect against cyberattacks, the primary responsibility of constructing multiple layers of defence—technical measures—is to increase the difficulty for potential attackers to penetrate a system. These methods greatly improve their efficacy by facilitating the immediate detection and resolution of attacks, thereby providing observant system administrators with alerts and notifications.

Technical measures are of greatest significance not only for their ability to mitigate cyberattacks but also for ensuring the protection of data's three fundamental qualities—accessibility, integrity, and confidentiality. Preserving the integrity of sensitive information by preventing unauthorized access or disclosure is the foundation of the principle of confidentiality. The fundamental essence of integrity lies in guaranteeing the accuracy, exhaustiveness, and uniformity of information. Conversely, availability refers to the continuous accessibility of data at the required moment, ensuring its punctual provision.

5.2 Operational Measures

Operational measures must be put in place to ensure digital resilience. The practice requires developing and implementing effective rules and standards to ensure responsible governance of digital assets inside organizations. These procedures include several tasks such as performing system backups, developing disaster recovery plans, and applying software updates.

The main purpose of operational measures is to accelerate and improve the process of restoring organizational systems after digital interruptions or intrusions. To reach this goal, businesses must develop plans for effectively managing their digital resources. This usually involves using carefully built methods to safeguard critical data and applications via systematic backup operations. Besides, well-defined procedures are required to efficiently address the consequences of cyberattacks and other sorts of digital disruptions.

The use of patch management emphasizes the need for operational measures even more. Routine software updates and security improvements are crucial to preserving system security and increasing resistance to new attacks. This approach's successful execution requires careful planning and coordination, as well as excellent communication with programme providers and stakeholders. Collaborative projects encourage the study and successful implementation of improvements.

It is impossible to emphasize the value of adopting effective operational strategies for attaining digital resilience. Businesses may mitigate the effects of cyber disasters and improve the efficiency of their recovery operations by using strategic measures. Implementing strong operational processes enables firms to improve their capacity to mitigate cyber risks, boosting their overall resilience in an age typified by increased digitalization.

5.3 Cybersecurity Awareness

Cybersecurity competence and awareness are extremely significant for digital resilience. The major goal of the concept is to improve individuals' and organizations' capacity to detect many types of threats and dangers that exist in cyberspace, as well as their ability to adopt suitable processes to mitigate these risks. The idea of cybersecurity awareness includes a vast variety of knowledge, skills, and attitudes that, when combined, serve as a key defence against the ever-changing cyber threats.

The basic idea of cybersecurity awareness revolves around the critical factor of detecting and comprehending the wide range of cyber risks that exist in the digital space. Malicious activities such as phishing attempts, malware infections, social engineering strategies, and other sorts of cyberattacks are among the threats. It is critical to provide individuals and organizations with the required abilities to actively detect and assess these potential dangers since this allows them to adopt preventative actions before any serious damage is done.

One important part of cybersecurity knowledge is using strong passwords and the observance of fundamental security rules. These practices, although seemingly mundane, are really important. The consistent creation and use of strong passwords, as well as the practice of keeping software up to date, using credible antivirus software, and exercising caution when encountering emails or hyperlinks of questionable origin form the foundation of robust digital defence mechanisms. These modest actions operate as barriers against current cyberattacks, considerably reducing the probability of successful intrusions.

Understanding cybersecurity goes beyond the scope of fundamental technology knowledge. It involves creating a set of attitudes and behaviours that highlight security as a basic feature of all parts of the digital ecosystem. This includes exercising care when sharing personal information on online platforms, selecting safe means for communicating sensitive data on purpose, and being aware of the potentially far-reaching consequences of cyberattacks. Implementing a complete cybersecurity awareness plan creates an attentive and strategic attitude, effectively encouraging a culture of awareness of digital security.

The need to boost cybersecurity awareness to jointly improve digital resilience becomes obvious in an exhaustive way. Individuals and corporations may develop considerable defences against vulnerabilities by creating a universal culture that encourages cybersecurity awareness. The proactive implementation of cybersecurity awareness serves as a preventive strategy, considerably lowering the probability of successful cyber intrusions. When events occur, the long-term consequences of these issues become apparent. The establishment of a deeply ingrained cybersecurity culture promotes the swift identification and effective management of cyber incidents, resulting in a significant reduction of their consequences and the mitigation of possible breaches' secondary impacts.

The creation of cybersecurity awareness is not a restricted or localized activity within the greater framework of social growth, but rather a shared commitment with global implications. As the digital world becomes more linked, the reach of cyberattacks expands across borders, necessitating a coordinated and collaborative approach to resilience development. Individuals and organizations who gain a deep grasp of cybersecurity establish a robust system of protection for the digital space. The global digital infrastructure consequently becomes more secure and resilient.

5.4 Effective Incident Response

Within the complicated context of digital resilience, the deployment of a strong incident response plan is critical. The digital world is rife with cyber phenomena such as cyberattacks, data breaches, and other disturbances. These instances have the potential to do significant damage to an enterprise's digital assets, interrupt its operational rhythm, and destroy its brand.

To combat the growing danger of cyber incidents, businesses must develop an incident response strategy that is competent and timely in execution. The core of this approach should be a defined set of procedures that are specifically targeted to the goals of incident reporting, incident analysis, and incident repair. It is critical to swiftly alert key stakeholders within the business, such as information technology workers or the designated incident response team, during the early phase of event reporting. The rapid and efficient sharing of information is essential for launching the response process' following phases.

As the event analysis evolves, the emphasis shifts to conducting a study of the incident's location and its subsequent repercussions, as well as determining the underlying cause of the incident. The thorough assessment of these factors contributes to an enhanced understanding of the many components of the incident and subsequent prioritization of the cause that precipitated its occurrence.

After conducting a thorough analysis and assessment of the complex elements involved in the incident, the organization's attention transitions to the stage of incident remediation. This situation demonstrates the deliberate coordination of efforts to restrict the situation to feasible limits and mitigate the escalation of its adverse repercussions. These precautions include a wide range of strategies that include mitigating any vulnerabilities that the incident may have exposed and eliminating any malicious software that may have infiltrated computer systems. Likewise, an integral aspect of incident remediation that guarantees the resumption of digital operations is the methodical retrieval of data from archives that have been closely monitored.

The successful execution of an incident response plan is conditional upon its capacity to lessen the ramifications of cyber incidents and assist the organization's restoration to a state of operational stability. The absence of a well-structured incident response strategy could potentially have major consequences. The potential for exacerbating damages resulting from a cyber event and prolonging operational disruption exists when there is a delayed and fragmented response to such events.

Besides, the absence of a thorough incident response strategy exposes an organization to the risk of repeating similar situations. They are vulnerable to the recurrence of such incidents because they have failed to implement the required preventative measures to avert their repeat. Given the current situation, the need to develop incident response plans emerges within the framework of efforts to improve digital resilience.

6 Digital Resilience and Cyber Resilience

The comparison between cyber resilience and digital resilience highlights the differences that exist between these two enduring concepts. In contrast to the traditional understanding of cyber resilience, which merely requires the ability to detect weaknesses in information technology (IT), effectively counter intrusions, and protect IT assets, digital resilience transcends these boundaries. Digital resilience, in its all-including sense, involves the utilization of modern technologies to effectively manage a wide range of situations while preserving the ability to function. It also incorporates change management and risk mitigation in operational contexts, IT durability, strategic planning for competitive advantage, business continuity management, and recovery from calamitous events, in addition to cyber threats.

Digital resilience refers to an organization's capacity to successfully operate digital technology to manage risks connected with hardware failures, natural disasters, and cyber vulnerabilities. A solid cybersecurity strategy, fast incident response mechanisms, the implementation of business continuity planning (BCP), and the protection of digital assets against potential threats and attack vectors are the core components of this system. Organizations are now faced with the critical need to develop digital resilience as a strategy to effectively manage disruptions, maintain operational capacity in the face of disturbances, and adapt to changing conditions. By using this strategy, an organization not only enhances its capacity to endure disruptions but also strategically positions itself to capitalize on beneficial conditions while protecting its customers, reputation, and financial status. The definition of cyber resilience is as follows:

Definition 3 (Cyber Resilience) An organization's or individual's competence in averting, identifying, and reacting to cyber threats and attacks, all the while securing critical information and upholding a culture of cybersecurity consciousness among its employees and oneself while implementing technical measures and incident response plans to mitigate the damage caused by cyberattacks.

The integration of cyber resilience strategies is crucial to digital resilience, which represents a significant shift away from the previous practice of conflating the two concepts. The effective implementation of cyber resilience strategies is critical for maintaining an organization's adaptability in providing timely and efficient responses to ever-changing threats. These strategies include analytical monitoring; security coordination that utilizes defence-in-depth strategies; deception implementation to trick adversaries; dynamic network positioning to expedite recovery from common mode failures; continuous awareness achieved through moving representation; non-persistence strategies to reduce risks; privileged access restrictions determined by the user, system, and environmental factors; and realignment of privileges.

Amidst the current business environment, the integration of cyber and digital resilience becomes a critical strategic need. Organizations that adopt both types of resilience—which are unique but interconnected—are more adept at handling the unpredictable threats to the digital space. By using this strategy, any potential disruptions are effectively mitigated, and consumer confidence, which is important for the long-term success of a business, is maintained. Digital resilience, as a proactive framework, contains the skilful administration and surmounting of digital obstacles via secure network architectures, resilient data backup and recovery systems, and an organizational workforce that embraces digital literacy. The proactive nature of this approach allows for the detection and reduction of potential issues before they become substantial disruptions, thereby granting a competitive edge and protecting the reputation of the brand and the confidence of customers.

Conversely, cyber resilience is an approach that prioritizes safeguarding digital assets against cyber threats, focusing specifically on the availability, confidentiality, and integrity of vital data. To defend against this modality, many technical safeguards are required, including intrusion prevention and detection systems, network firewalls, and security incident management systems. Notwithstanding the strength of these technological fortifications, the human element continues to be required. The significance of cybersecurity awareness training is emphasized in cyber resilience to equip personnel with the ability to identify threats and take preventative measures; thus, this contributes to the general cyber resilience of an organization.

The associated characteristics of cyber resilience can be observed in the well-structured design of a strategy for responding to incidents. Included in this plan are procedures for identifying and containing the cyberattack, minimizing its impact, and recovering critical systems and data in the case of an attack. Organizations strengthen themselves against the ever-changing threats by adopting cyber resilience. This method not only develops a strong sense of trust among stakeholders and customers but also enhances the resilience of the organization's digital infrastructure, making it resistant to unpredictable disruptions.

Organizations should stress the need to provide training and educational initiatives aimed at enhancing the digital and cyber resilience of their workforce. An essential element of this initiative is the provision of regular and organized cybersecurity awareness training to employees, leading to an understanding of potential dangers and the ability to identify hidden threats. It is necessary to build a consistent process for acquiring knowledge and enhancing skills, a goal that may be achieved by judicious investments in continuous professional development initiatives. This initiative will facilitate the development of a dynamic and knowledgeable labour force. The systematic implementation of risk assessments and audits is of utmost importance in effectively carrying out organizational strategies. It is recommended that these exercises be included in the organizational framework as routine practices and executed with caution to identify latent vulnerabilities and enhance the defensive stance of the company. Table 1.1 showcases some measures that can be taken to achieve cyber resiliency.

These assessments should provide a review of an organization's whole digital and cyber environment, which includes various components such as hardware, software, network infrastructure, and interactions with third-party suppliers. Organizations may build a solid foundation of digital and cyber resilience by taking a proactive strategy. This helps them to face the complex issues of the digital age successfully. These disasters can result in the formation of stronger and more recovered businesses. A comprehensive compilation of prominent cyber resilience solutions, which are frequently discussed within the field of systems security engineering, is presented in the table below [Ros+21].

Let's examine the differences between digital and cyber resilience. Gaining an understanding and establishing connections between various concepts is crucial to effectively engage in the implementation and meaningful discussions surrounding mechanisms aimed at addressing and safeguarding systems and organizations.

Table 1.1 Cybersecurity measures for cyber resiliency [Ros+21]

Cybersecurity measure	Objective
Adaptive response	To enhance the ability to effectively and appropriately address emergent events promptly
Analytic monitoring	To promptly and effectively monitor and identify harmful activities and situations, providing actionable insights in a timely way
Coordinated protection	To deploy a defence-in-depth approach that requires enemies to overcome several barriers
Deception	To intentionally mislead, confuse, conceal important assets from, or covertly reveal contaminated assets to opponents
Diversity	To use heterogeneity to mitigate the occurrence of common mode failures, particularly those that exploit shared vulnerabilities
Dynamic positioning (DP)	To improve the capacity to quickly recover from non-adversarial catastrophes, such as natural disasters, by dispersing and diversifying the network distribution
Dynamic representation	To ensure the continuous maintenance of an up-to-date depiction of the network, enhancing the comprehension of interdependencies between cyber and non-cyber resources and revealing any discernible patterns or trends in enemy behaviour.
Non-persistence	To generate and retain resources for a specific purpose or a limited duration, aiming to minimize the risk of corruption, alteration, or compromise of these resources
Privilege restriction	To impose limitations on privileges by considering the characteristics of users and system components, together with environmental conditions
Realignment	The aim is to reduce the interdependencies between mission-critical and noncritical services, hence decreasing the probability of noncritical service failures affecting mission-critical services
Redundancy	To ensure the availability of many instances of essential resources that are safeguarded against potential failures or disruptions
Segmentation	To establish clear definitions and distinctions among system components, mostly based on their level of criticality and trustworthiness
Ensuring integrity via substantiation	Determination of the presence of corruption in essential system components
Unpredictability	To introduce random and unexpected changes, heightening an opponent's uncertainty about encountered system safeguards and posing a greater challenge for the adversary in determining the proper course of action

Table 1.2 presented below offers an analysis of the respective subjects, highlighting both their points of divergence and their notable similarities. It is crucial to comprehend that both elements are typically used in conjunction to enhance clarity and establish a legitimate, safeguarded setting.

Table 1.2 Digital resilience vs cyber resilience

Digital resilience	Cyber resilience
The concept refers to the organization's overarching capacity to adjust and rebound from digital disturbances, regardless of the root cause	The concept refers to an organization's capacity to fend off, identify, and take action against cyber threats and attacks
Requires a deep understanding of an organization's entire digital environment, including hardware, software, and network infrastructure	Requires a deep understanding of the specific cyber threats facing an organization, as well as the technical measures required to defend against them
Involves implementing data backup and recovery systems to ensure that critical information can be restored in the event of a disruption	Involves the deployment of technical controls, such as access controls, intrusion detection and prevention systems, firewalls, and other similar measures, to prevent or mitigate cyberattacks
Requires a culture of digital literacy among employees to ensure that they understand the importance of digital resilience and can contribute to the organization's efforts	Requires a culture of cybersecurity awareness among employees to ensure that they understand the risks and can help prevent attacks
Helps organizations maintain their competitive advantage by enabling them to quickly recover from disruptions and maintain customer trust, thereby enabling business continuity	Helps organizations maintain the confidentiality, integrity, and availability of their critical information
Considers both the technical and human aspects of an organization's digital environment	Considers both internal and external threats, including threats from third-party vendors and supply chain partners

7 Significance

In today's business world, the importance of digital services for the operations of any organization is critical. In any case, as our reliance on these digital systems grows, there is an increased risk of interruptions that might impede our operations. These interruptions have the potential to undermine the efficacy of service delivery, data management, and the security of critical information. As a result, we must concentrate our efforts on improving digital resilience.

The capacity of an organization to foresee, plan for, and successfully deal with changes and difficulties in the digital space can be referred to as digital resilience. The idea includes different aspects of risk management, such as the potential threats associated with cyberattacks, technological failures, and unanticipated occurrences. Organizations that prioritize digital resilience not only efficiently handle the negative repercussions of operational interruptions but also boost their brand and build confidence among their consumers.

To put the concept of digital resilience into action, businesses must create a strategy that successfully mitigates a broad variety of digital risks. A systematic and detailed technique for identifying possible risks, a constant review of weaknesses or susceptibilities, and the creation of backup plans to successfully handle unanticipated obstacles should all be part of the suggested strategy. Organizations must build a state of readiness to successfully manage possible dangers while keeping a high degree of awareness of new challenges and the accompanying methods to solve them.

Everyone in the company is responsible for understanding and actively participating in the enhancement of digital resilience. This necessitates raising awareness among individuals about probable scenarios in the event of an accident, encouraging the open discussion of identified difficulties, and encouraging a common commitment to preserving the organization's digital resources. An organization's capacity to confront and overcome difficulties in the digital space has a substantial influence on its prospects. Organizations that prioritize digital resilience may ensure smooth operations, protect client information confidentiality, and establish their supremacy in the digital world.

The growing reliance on digital systems has resulted in an increase in sensitivity to disruptions. Various disruptions, such as technological faults and cyberattacks, have the potential to affect a company's operating performance. As a consequence, such activities may damage the trust of both consumers and stakeholders, having a detrimental influence on the organization's position and possibly having a major impact on its financial success. As a result, developing and strengthening digital resilience is a strategic need rather than a discretionary option.

In its most basic form, digital resilience includes the essential idea of preparedness, flexibility, and responsiveness when faced with digital adversity. This includes the capacity to anticipate potential disruptions, devise thorough mitigation strategies, and demonstrate agility in handling the changing digital world. The vast scope of digital resilience includes aspects such as threat management, technological failure mitigation, and dealing with unanticipated occurrences. A strategy for risk management is required to ensure that the firm is sufficiently prepared for a wide range of potential interruptions.

To develop digital resilience, organizations should implement a risk management framework that serves as a navigational tool, leading them through the complex terrain of digital uncertainties. This architecture is made up of several distinct types of components that work together to protect against possible interruptions. One of the most important parts in this context is a thorough risk assessment technique that carefully discovers flaws and possible threats to the company's digital operations. Routine vulnerability testing is necessary because it helps companies proactively discover and fix areas of weakness, reducing the chance of these vulnerabilities being exploited as possible entry points for disruptive occurrences.

If outages occur despite precautionary efforts, an incident response plan becomes an essential component of the digital resilience structure. This plan provides a detailed framework defining the tactics and processes that will be used by the organization to successfully react to and restore operations in the event of an interruption. This plan details the several roles, responsibilities, and processes that must be followed to mitigate the consequences of the disruption and resume normal operations. The goal of this responsiveness is to reduce times of inactivity, limit reputational harm, and hasten the recovery process.

Significantly, the idea of digital resilience implies a culture shift inside the organization. Professionals at all levels, from top executives to frontline individuals, must shift their perspectives and behaviours. Creating a culture of digital resilience requires equipping employees with the information and tools they need to successfully detect and report possible risks. This is followed by a proactive mindset that encourages continuous learning, adaptability, and a shared feeling of responsibility for safeguarding the organization's digital resources.

The increasing importance of digital services in current organizational practices highlights the critical need to implement digital resilience as a basic requirement. Potential disruptions pose huge risks that might undermine growth and stability as businesses navigate the complex environment of digital operations. Organizations may establish strong defences against disruptions by promoting a culture of digital resilience and implementing risk management techniques.

The proactive approach has far-reaching implications that extend beyond just reducing critical disruptions. The convergence of these variables results in the construction of a stronger brand reputation, the development of long-term customer trust, and the preservation of consistent financial success, all of which reinforce the organization's position in the digital world. Today, the idea of digital resilience is essential in directing businesses towards stability and success. It encourages a resilient attitude capable of efficiently overcoming barriers and capitalizing on the benefits provided by the digital world.

8 Benefits

The concept of digital resilience provides enormous advantages to both individuals and businesses and hence merits careful consideration. Individuals may secure their personal data and valuable digital assets by creating digital resilience, successfully limiting potential cyberattacks. Because of the omnipresent nature of technology in modern society, the ability to anticipate, plan for, and react to disruptions in the digital space has become critical.

The importance of digital resilience in businesses cannot be emphasized, as it is essential in maintaining operations, securing private information, and promoting customer trust. Organizations with excellent digital resilience are well-prepared to manage any disruptions that may arise, thereby limiting the negative impact on their operations, reputation, and customer loyalty. As a result, it can be deduced that having such expertise adds to the overall performance and competitiveness in the digital marketplace.

One significant benefit of prioritizing digital resilience is an increase in brand reputation and a clear display of commitment to safeguarding customer data. Firms may proactively develop a unique and beneficial position within the competitive environment of the marketplace by emphasizing the creation of digital resilience. This method is particularly successful in acquiring customers who value both digital security and privacy. To attain this goal, it is thought necessary to create a risk management system that includes the full digital infrastructure. This method allows for the rapid identification of hidden dangers and the subsequent deployment of essential mitigation measures.

Digital resilience gives businesses the required flexibility to change and react to emerging threats and fast-evolving technologies. Their adaptability gives them a distinct edge in maintaining development and driving innovation. Organizations cannot only preserve a market competitive edge but also assure long-term sustainability.

As stated, the following points highlight many advantages that result from the application of digital resilience solutions in today's technological world.

(1) **Enhanced Security and Protection:** The deployment of digital resilience enhances security measures, resulting in increased protection against a range of cyber threats, including viruses, malware, and unauthorized access. Enhanced defences in digital spaces contribute to a more secure infrastructure benefiting both individuals and enterprises by shielding them from potential digital threats.

(2) **Continuous Functionality:** The creation of digital resilience techniques is critical to maintaining system operating capabilities even in the face of disruptions like power outages or hardware failures. This skilful maintenance of functioning enables a continuous flow of processes, which contributes to organizational stability.

(3) **Adaptive Response to Challenges:** The concept of digital resilience helps systems become more adaptable and agile in responding to unexpected challenges, changes in technology trends, or new threats. This adaptability helps to protect against external changes and ensures sustained system performance even when the external environment is unstable.

(4) **Reduced Downtime:** The careful implementation of digital resilience strategies results in the reduction of potential periods of inactivity that might occur from technological failures or cyberattacks. The use of this cautious strategy subsequently reduces the negative impact on productivity and revenue, consequently enhancing overall operational effectiveness.

(5) **Data Integrity and Recovery:** The core principles of digital resilience revolve around prioritizing the preservation of data integrity while simultaneously implementing effective means for data recovery. This principle is distinguished by its significance, as it prevents data loss and corruption, thereby enabling the restoration of data to its accurate state.

(6) **Long-Term Sustainability:** Organizations that prioritize the protection of privacy in the implementation of digital resilience enhance their prospects for long-term sustainability. This investment is realized through establishing a technologically relevant, efficient, and functioning infrastructure that is capable of adapting to technological advancements without becoming obsolete.

(7) **Risk Mitigation and Compliance:** Digital resilience, as a strategic tool, serves as a vanguard against potential risks through its facility to identify vulnerabilities and enact redressal measures. Also, its important role in

following industry rules and protecting data makes it even more useful for an organization.

(8) **Improved Customer Trust:** Enterprises that demonstrate a consistent commitment to digital resilience build a higher degree of confidence among their customers. This trust is promoted by the guarantee of security and continuity in digital service offerings, which builds client loyalty and confidence.

(9) **Cost Efficiency:** The adoption of digital resilience as a strategic approach can result in cost efficiencies over time. The effective path of this trajectory can be attributed to the amplification of the consequences arising from disruptions, hence eliminating the need for emergency reactions and subsequent recovery initiatives that are frequently accompanied by substantial costs.

(10) **Innovation Facilitation:** The promotion of digital resilience contributes to the advancement of innovation by providing a solid foundation of stability that supports the experimentation and exploration of emerging technologies. Organizations are therefore granted the freedom to develop innovative solutions without compromising their fundamental digital infrastructure.

The significance of addressing digital resilience extends beyond immediate benefits, extending into risk management. Organizations proficient in digital resilience acknowledge a connection between a risk management framework and the reinforcement of their digital infrastructure. The implementation of this proactive strategy allows for the detection and improvement of possible weaknesses, thereby guaranteeing that the organization operates in a digital setting with caution and effectiveness.

Likewise, the concept of digital resilience includes the aspect of adaptation. The dynamic progression of technologies and the simultaneous appearance of new threats require a setting that promotes prompt and efficient adaptation. The concept of organizational agility within a digital space refers to the ability to engage in innovation, thereby facilitating a lasting competitive edge and long-term viability.

In a time when human life and digital platforms are merging, the choice to focus on the development of digital resilience is considered a reasonable option. A comfortable coexistence with the digital paradigm is accomplished if individuals increase their digital defences and organizations accept resilience as strategically important.

The development of digital resilience offers several advantages that benefit both individuals and enterprises. This primary initiative not only assigns individuals the responsibility of protecting their personal information as well as their digital assets, but it also enables enterprises to secure the continuity of their operations, safeguard sensitive data, and promote trust. Individuals and organizations may efficiently traverse the digital space with a better degree of knowledge and certainty if digital resilience is carefully integrated into the strategy framework and a complete risk management framework is implemented.

9 Challenges Encountered in Adoption

Beyond a doubt, the broad use of digital technology has resulted in several advantages in the current age. However, it is critical to understand that these advantages are outweighed by several difficult issues that have the potential to undermine the underlying concept of digital resilience. This section will go over the key issues that stand in the way of achieving digital resilience. The enormous and impending existence of cybersecurity risks, which constitute a considerable danger within digital technology, is one of the key problems that demand special attention. The serious consequences of cyberattacks, such as their propensity to inflict financial damage and destroy reputations, highlight the significance of developing effective digital resilience solutions.

Simultaneously, the fast pace of technical innovation, while promoting the growth of the digital world, creates unexpected weaknesses that malicious cyber actors may exploit. The aforementioned phenomenon needs a detailed plan for creating digital resilience that goes beyond just using technology developments. The existence of major disparities in digital accessibility sometimes referred to as the digital divide creates a crucial crossroad. This unsettling issue can exacerbate pre-existing imbalances in social, economic, and political sectors, demanding a difficult drive to achieve fair and strong digital resilience.

The rising prevalence of disinformation and false news is an equally substantial impediment, undermining the basic foundations of confidence and legitimacy. Inaccurate information has far-reaching consequences, affecting both the individual and social components. The little but large susceptibility of human mistakes emerges as a critical impediment to achieving digital resilience. Instances of inadequate password strength and failure to keep software up to date serve as harsh reminders of the flaws in digital security systems.

It is clear that a complete strategy is required, one that combines proactive techniques to increase network, device, and data security holistically. Critical to the success of staying safe in a digital space is establishing and maintaining a culture that emphasizes lifelong learning. Such a culture helps create a resilient mindset. To guarantee that the advantages of digital resilience are accessible to all individuals and not just a select few, addressing digital inequality is an essential element of this journey.

Similarly, a greater emphasis on digital literacy emerges as a critical component in increasing digital resilience. Individuals are enabled to distinguish between credible sources and disinformation by gaining a sophisticated awareness of digital technologies and their implications, hence lessening the negative impacts of lying. Along with promoting literacy, there is a focus on building critical thinking abilities, which allow individuals to interact with digital material in a deliberate, analytical, and careful way.

The significance of collaborative efforts among individuals, organizations, and politicians is enormous and greatly anticipated. Digital resilience can only be strengthened via collaboration since each member brings a unique perspective and function to the table. The interaction of numerous aspects promotes the establishment of a strong and resilient human civilization in face of the ever-growing list of digital threats and disruptions.

9.1 Challenges in the Industries and the Business Sectors

It is difficult to overrate the significance of digital resilience in the context of businesses and industries. This refers to the critical role it plays in defending companies from numerous threats while also assuring continuous operations. However, organizations and sectors face a slew of complex obstacles that impede the seamless implementation of digital resilience.

Lack of knowledge is a significant barrier that companies must overcome. The lack of awareness of this area, along with the specific characteristics of digital resilience, is a significant impediment to its effective implementation. An additional crucial aspect to think about relates to the complexities of the digital world. The formulation and implementation of effective strategies to ensure digital resilience become challenging due to their complexity, necessitating the use of advanced solutions that align with the subtleties of modern technology and infrastructure.

The monetary factor also emerges as an essential component impeding development towards digital resilience. Due to the high expenses of buying hardware, software, and the requisite training and resources, some firms may find it difficult to prioritize digital resilience. These costs are often expensive and represent roadblocks in the goal of digital resilience. Cultural resistance worsens the situation. Organizational cultures, which are often deeply rooted, may also be resistant to change, hindering the process of adopting the paradigm shift necessary for digital resilience.

Aside from these issues, the continually changing list of threats is a major worry. To keep their digital resilience plans effective, companies must constantly monitor the ever-changing nature of cyber threats. Addressing these complex issues demands a diverse strategy that involves education, training, resource allocation, and an unrelenting commitment to continual growth.

A key pillar is to give workers a complete understanding of digital resilience, including its complexities and implementation approaches. Education is crucial for raising awareness of digital resilience and assisting stakeholders in understanding and dealing with its extreme complexities. At the same time, resource allocation is critical, the acquisition of necessary hardware, software, and training materials is a strategic investment in the creation of a highly secure infrastructure.

The radical cultural change necessary for digital resilience is one essential issue that must be highlighted. Implementing new work practices and operational modalities necessitates a change in employee attitudes, which may be greeted with resistance. This emphasizes the need to develop new management strategies that not only allow the introduction of new approaches but also inspire a shift in stakeholders' thinking.

9.2 Challenges in the Education Sector and for Minors

The work being done to integrate digital resilience in the educational arena is critical, especially in terms of protecting minors and creating a secure digital environment for their growth. Having stated that it is crucial to recognize many obstacles to implementing digital resilience in education, especially for minors.

One of the most pressing problems is the widespread lack of digital literacy among the younger population. A big obstacle to effectively incorporating digital resilience into education is the lack of awareness about it and the complexity of the concept. This argument is worsened by a scarcity of technical resources. The inaccessibility of technological resources may hinder the equitable distribution of digital resilience technologies and knowledge, emphasizing the need to correct this mismatch.

The current issue highlights the need to select content that is acceptable for diverse age groups, a process that requires careful curation and attentive monitoring. The digital world, with its vast amount of information, is not immune to content that does not correspond to the developmental requirements and sensitivities of individuals under the age of majority. This emphasizes the tough challenge of striking a delicate balance between giving access to information and ensuring that the material strengthens the emotional and cognitive development of minors.

Aside from that, the distressing phenomenon of cyberbullying is a major obstacle to minors' digital resilience. The sensitivity of minors to the negative consequences of cyberbullying highlights the need to develop methods to reduce these risks and establish a supportive online environment. Together, privacy issues are crucial, particularly given the increasing use of online learning platforms and digital technology. The safety of children's data is essential, requiring the implementation of strict data protection procedures.

A plan is necessary to solve these complex issues. Education, awareness, legislation, and technical solutions all come together to form a vital component. Education is essential not only for giving minors a thorough understanding of digital resilience but also for assisting them in distinguishing between safe and unsafe digital practices. Increasing awareness among educators, parents, and other stakeholders is required since their advice is important in effectively traversing the digital space.

This strategy is supplemented and supported by strategic policies. Adoption of legislation regulating suitable material for different age groups, combatting cyberbullying, and stressing personal data protection may promote the development of digital resilience among youngsters. In symbiotic partnership with these strategies, technological solutions play an essential role. Incorporating technology that facilitates safe online interactions, monitors content, and preserves data security may all help to strengthen digital resilience.

9.3 Challenges for Critical Infrastructures

Particularly challenging is the implementation of digital resilience in critical infrastructures due to the characteristics of such systems. Critical infrastructure, comprising essential systems such as healthcare facilities and electrical utilities, encounters unique challenges compared to other sectors. An enumeration of critical infrastructure can be found in Fig. 1.3. The problems stem from the complexities of critical infrastructures, such as their interdependence, human factors, regulatory frameworks, and resource constraints.

The biggest obstacle is the obvious complexity that serves as the backbone of critical infrastructure. These systems provide particular problems in digital resilience due to their complexity. The existence of these dependencies complicates the work at hand. Because these infrastructures are interrelated, a cyberattack on one component might cause a chain reaction that affects other interdependent systems. Because of the complexities of this scenario, a smart solution that extends beyond individual considerations is required.

The existence of human influence is a complicated factor that requires careful and detailed consideration. The challenge of appropriately alerting human operators about important cyber threats and providing them with the best cybersecurity practices warrants special consideration. The integration of human cognitive processes with the digital setting needs a strategy that promotes awareness, offers training, and facilitates the development of a resilient mentality capable of successfully handling the complexities connected with cyber risks.

Critical Infrastructure

- Agriculture and Food
- Chemical and Hazardous Materials Industry
- Defense Industrial Base
- Government Facilities
- Nuclear Reactors, Materials, and Waste
- Communications
- Financial
- Critical Manufacturing
- Emergency Services
- Information Technology
- Transportation
- Commercial Facilities
- Dams
- Energy
- Healthcare and Public Health
- Water and Water Treatment Systems

Fig. 1.3 Critical infrastructure

At the same time, regulatory systems provide a difficult position. Achieving a balance between adherence to legal frameworks and the required flexibility to steer ever-changing digital settings is a difficult task. These infrastructures, noted for their high degree of sensitivity, need to strictly follow the norms. The urgency of digital change necessitates a responsive strategy to new threats, necessitating a delicate balancing of two competing needs.

The domain in issue is distinguished by the existence of resource constraints, which cause major difficulties. The process of updating or replacing obsolete systems, which is usually required to improve digital resilience, may be hampered by financial and technical restrictions. The lack of resources devoted to investing in digital resilience measures is a visible impediment, emphasizing the need for careful allocation and deliberate prioritization.

To handle these complex issues, a collaborative and coordinated approach is required. Collaboration is encouraged because the inclusion of diverse expertise and opinions may improve understanding of the complexities and help in the creation of complete solutions. Central to this plan is the distribution of funds for the improvement and expansion of existing systems. The process of replacing or updating obsolete technologies is critical in establishing a strong digital infrastructure capable of handling new cybersecurity issues.

The human component needs training and development. The aim of equipping human operators with the requisite expertise and abilities to recognize and mitigate cyber threats is a continuing initiative. To stay ahead of the ever-changing threats, it is critical to acquire information and skills via initial training as well as ongoing education. It is critical to strike a realistic balance between obedience to laws and flexibility. Aligning regulatory compliance with adaptability is vital in allowing critical infrastructures to efficiently handle the numerous difficulties posed in the digital age.

9.4 Challenges in the Socioeconomic Sphere

Despite many benefits of implementing digital resilience, there are complicated socioeconomic barriers that must be thoroughly deconstructed. The major emphasis of study in this scenario is on the financial consequences of developing digital resilience, which may pose a significant barrier for some groups of society. At the same time, there is a significant and visible discrepancy in Internet access and expertise, revealing inequalities among socioeconomic groups.

Financial constraints make it difficult to deploy digital resilience solutions effectively. The financial expenses of digital resilience may be a barrier for individuals in certain socioeconomic groups, limiting their capacity to benefit. This highlights the need to research solutions to reduce financial barriers and promote more inclusive participation.

The modern paradigm incorporates the concept of the digital divide, which emphasizes the unequal distribution of digital resources and knowledge across socioeconomic strata. Achieving digital resilience requires a thorough awareness of a variety of topics, including cybersecurity, privacy protection, and the development of critical thinking skills. However, it is critical to acknowledge that there is a demographic subgroup that lacks the financial wherewithal to pursue education and training, resulting in an unequal circle of possibilities.

The complexities of regulatory structures become an impediment. Handling complex regulatory concerns that differ across countries is often required for digital resilience. The introduction of such a mismatch has the potential to cause misunderstanding, hence impeding the smooth implementation of digital resilience solutions. Cultural resistance may worsen the problem, since ingrained traditions and practices may obstruct the acceptance of seemingly revolutionary digital paradigms.

For the successful execution of a complete plan, a combination of education, training, resource allocation, policy initiatives, and a commitment to continuous improvement is required. Education is critical to the equal distribution of digital literacy and the skills that go with it.

Training is an important part of this process because it provides individuals with the knowledge and skills they need to traverse the digital space with ease. The proper deployment of resources is critical in creating an environment in which individuals' ability to connect and engage is not hampered by financial restrictions. Policy actions play a vital part in building regulatory framework harmonization and the development of a single path to digital resilience.

9.5 Challenges on the Legal Front

The need to connect all the above-mentioned efforts with the parameters of present law and regulations is closely related to the integration of digital resilience in the legal sector. However, the effort faces several challenges, including a constantly changing legal environment, the complexities of transferring data across borders, the need to address issues of liability and accountability, privacy concerns, and a lack of standardized practices. These elements, taken together, add to the field's complexity.

The ever-changing nature of the legal environment is one of the most critical obstacles. The rapid growth of digital resilience rules and regulations creates a scenario in which methods may accidentally lag behind newly developed legal standards. The many complexities of cross-border data flows compound the problem. The existence of several legal systems and regulatory frameworks makes it difficult to deploy uniform digital resilience measures. As a result, it is critical to take a smart strategy that tackles these complications effectively.

In this case, accountability and responsibility are critical considerations. The dynamic interaction between digital resilience measures and accountability distribution may get complicated at times, demanding careful adaptation to ensure that accountability is distributed equally. The focus on privacy worsens the issues. Personal data protection is crucial in the context of digital resilience, emphasizing the need for privacy preservation methods that go beyond basic regulatory compliance.

The lack of established procedures adds to the difficulties. The absence of defined digital resilience standards or norms produces a complicated ecosystem with no clear principles. As a consequence, a strategy that goes beyond specific indicators is required. To address these issues, a coordinated plan including legal, regulatory, and technological factors is required.

The importance of cooperation among various stakeholders is at the heart of this approach. To connect digital resilience plans with ever-changing legal constraints, legal, regulatory, and technology specialists must coordinate. The importance of investing in training and creating awareness cannot be emphasized, as having a well-informed and competent group of individuals is vital for ensuring that digital resilience not only follows the law but is also responsive to growing legal changes.

The importance of a commitment to constant development and adaptability cannot be emphasized. It is critical to take a proactive approach that constantly reassesses and adapts digital resilience methods to keep up with changing legal dynamics. Concurrently, the careful implementation of optimum approaches and new legal frameworks needs attention and constant adaptation.

9.6 Challenges in the Technical Sphere

Significant technical measures for defending against cyberattacks must be established to embrace digital resilience. Still, it's riddled with challenges that highlight the difficulties innate in technological characteristics. These challenges are examined in terms of their complexity, compatibility, scalability, interoperability, and the ever-changing nature of emergent threats. One of the most pressing challenges is the apparent complexity of technical measures. Because of the complexity of cyber threats, developing and implementing effective technology initiatives requires a high degree of skill. A synchronous concern is the question of compatibility. The faultless synchronization of several technological metrics might often need major time, resource, and expertise inputs. The technological environment is also influenced by scalability and interoperability. Resources must be continually allocated to information technology infrastructure, systems, and personnel to support the ever-expanding variety of digital resilience measures. This necessitates a continual commitment to development and modification, in which the processes of scalability and interoperability are handled to suit changing demands.

The quick and changing appearance of threats is a major issue. The digital space is distinguished by the ongoing growth of threats and vulnerabilities. This problem necessitates digital resilience strategies that may change with the evolving threats and vulnerabilities while still being effective. To predict and confront the ever-changing list of risks requires ongoing attention and a proactive approach. It is critical to have broad and firm commitment while negotiating these technical challenges. Both businesses and individuals must have a constant monitoring and improvement approach. The continuous enhancement of technological measures allows for the handling of many obstacles. The key to tackling these difficulties is a consistent commitment to development and improvement.

10 Characteristics of a Cyber Resilient Organization

There is a considerable discrepancy across enterprises in their capacity to recognize and address threats and dangers arising from disruptions in digital services, as well as in their ability to restore operations thereafter. The observed disparity possesses the capacity to exercise a substantial influence on the satisfaction of consumers as well as the overall worth and reputation of corporations. In light of the escalating levels of innovation and competition within the market, it is important to acknowledge the potential consequences of digital transformation on the consumer experience. To uphold a competitive edge in the digital space, companies must strategize and execute a 'digital resilience' initiative.

We must recognize that digital resilience comprises more than just the ability to sustain strong operations and security in the face of business disruptions. Instead, it involves the effective management of all threats that arise from interruptions in the digital environment. Executives can effectively mitigate the consequences of expensive service disruptions by prioritizing defence, detection, repair, recovery, diagnosis, and refinement strategies. Hence, the inclusion of digital resilience as a fundamental element of both strategic planning and operational activities is important. Organizations must develop a holistic strategy towards digital resilience, which includes multiple aspects including technological and operational proficiencies, governance, risk management, and regulatory compliance. Prioritizing digital resilience at the highest level of corporate governance is crucial to ensuring its integration into the organizational culture.

10.1 Seven Strategies to Improve Digital Resilience

The combination of the seven strategies has been experimentally proven to be effective across several industries and may be modified to meet the precise needs of a business. The introduction and use of these strategies are expected to reduce potential dangers, increase organizational capabilities, and build overall resilience. By combining these strategies into the operational framework appropriately, a company may improve its capacity to overcome unanticipated problems, reduce possible risks, and build a path to long-term success.

It is critical to stress that the adoption of these strategies may need careful planning, coordinated efforts, and sensible resource allocation. However, the multiple benefits of building an organization, such as enhanced resilience, make the investment in effort incredibly profitable. Organizations may achieve long-term

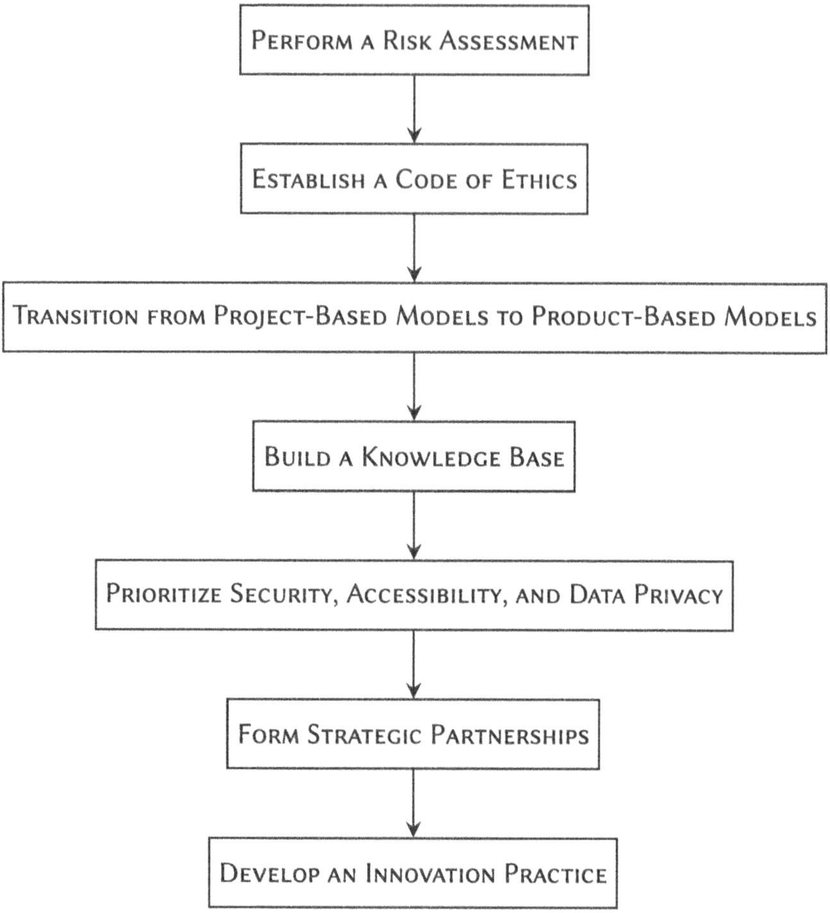

Fig. 1.4 Seven-step strategy

success by predicting possible risks and overcoming unanticipated difficulties by applying strategic approaches.

For enhanced comprehension, refer to Fig. 1.4 which illustrates the seven-step process.

10.1.1 Perform

The organization must do a risk assessment during the initial stage. The fundamental goal of this systematic procedure is to discover, analyse, and prioritize any risks and vulnerabilities that can impair the organization's operations and harm its valued assets. A variety of challenges, such as cyber threats that specifically target the organization's digital infrastructure, natural disasters that can disrupt physical operations, and unexpected disruptions that have the potential to undermine operational effectiveness, are among the potential threats.

The assessment should not be characterized by a feeling of haste; rather, it demands a thorough and methodical analysis of a broad variety of possible threats. This involves a detailed examination of the possible consequences associated with each identified risk, especially their impact on the company's overall operational performance. As a result, it is critical to put these found threats through a thorough rating approach that assesses the probable consequences. The company acquires the ability to precisely evaluate the degree of each risk by using this process. As a consequence, smart resource allocation becomes feasible, allowing the execution of mitigation options or efforts to avoid the problem. The fundamental purpose of these efforts is to guarantee that the organization's operating capability remains constant as well as to maintain its irreplaceable assets.

The significance of the first step of a risk assessment cannot be overstated. It serves as the foundation upon which the operational resilience structure is created. The method serves as a proactive tool, helping the company to anticipate potential disruptions and design steps to reduce or prevent their negative consequences. The company faces a wide range of potential threats, ranging from the digital domain to the physical environment, all of which create a large variety of challenges. The current danger of cyber threats manifests itself in efforts to compromise the integrity of an organization's digital infrastructure. Natural catastrophes, which have the power to disrupt physical operations and incapacitate essential resources, match the severity of these threats. The terrain is rife with unplanned interruptions that can impede the smooth working of operations, emphasizing the need to undertake a risk assessment.

The evaluation procedure's methodical approach emphasizes its nature. Every possible risk is thoroughly analysed using a methodology that assesses its influence on the all-around performance of the company. This analytical approach helps the organization to make well-informed resource allocation decisions. The organization focuses on the most severe dangers by prioritizing risks based on their potential consequences, allowing for the strategic deployment of resources to effectively manage and mitigate these vulnerabilities.

10.1.2 Establish

The construction of an effective code of ethics occupies a position of utmost significance to encourage a culture of ethical behaviour and responsible activities within the organizational framework. This fundamental effort requires the development of an easily understandable collection of principles and guidelines outlining the anticipated ethical norms for all relevant individuals and groups involved. These stakeholders include not just individuals and partners but also other related entities associated with the organization. These ethical code functions as a navigational tool, directing the course of ethical decision-making and conduct across a broad range of situations.

On the other hand, the mere articulation of these ethical standards in writing format is insufficient to fully accomplish the undertaking. The third crucial stage is the distribution of this ethical code to all individuals associated with the organizational ecosystem. This extends beyond the simple act of distributing something; it covers the practice of effectively communicating in a way that emphasizes ethical obligations and their importance. Also, it is necessary that the organization assume the obligation of regularly evaluating and modifying the ethical code. The iterative assessment procedure is essential to guarantee the ongoing congruence of the system with the dynamic ethical environment and its effectiveness in directing ethical conduct.

The purposeful adoption of a robust ethical code represents a strategic effort to solidify an organization's values and ideals through the establishment of explicit rules. These rules function as a proclamation of the criteria that establish appropriate conduct and decision-making inside the organizational domain. The establishment of these expectations by the ethical code serves to create a shared framework that promotes coherence and standardization in behaviour.

That said, the concrete impact of the ethical code can only be fully realized when it is efficiently distributed and internalized by all individuals associated with the organization. The act of disseminating information goes beyond simply distributing physical documents. It requires the implementation of a communication strategy that guarantees a thorough understanding of the ethical principles involved. By employing efficient communication strategies, the organization emphasizes the importance of ethical conduct, strengthening the belief that ethical considerations are not secondary but rather fundamental to the organization's basic values.

Despite this, it is consequential to acknowledge that an ethical code, regardless of its comprehensiveness, is susceptible to changes in the ethical principles of the time. Ethical paradigms exhibit a dynamic nature as they transform in reaction to alterations in society, cultural dynamics, and the emergence of new difficulties. Therefore, the implementation of regular evaluation and modification is necessary. To ensure the ongoing relevance and resonance of the code, the organization must participate in introspective practices. The evaluation procedure serves as a protective measure against the possible outdatedness of the ethical code, guaranteeing its ongoing conformity with the dynamic ethical principles.

The ever-changing nature of ethical considerations requires organizations to demonstrate a firm dedication to caution and adaptability. Through consistent participation in ongoing evaluations and revisions, the company showcases its commitment to maintaining elevated ethical norms. This demonstrates a dedication to advancement, guaranteeing the continued relevance and efficacy of ethical principles capable of guiding behaviour and choices in present-day circumstances.

10.1.3 Transition

It is advised that the business make a strategic shift from old-fashioned project-centric models to more adaptive and flexible product-centric frameworks to improve operational efficiency, decrease wasteful practices, and speed up the development of creative initiatives. The process of transformative evolution involves a departure from the traditional linear concentration on individual projects and instead adopts a more adaptable and iterative approach that centres around products. Within the framework of this unique organizational structure, various departments within the company are instructed to concentrate their efforts on conceptualizing products and services that directly confer substantial value to clients.

The implementation of this novel operational methodology enables the company to harness enhanced agility and flexibility. Consequently, the organization becomes sufficiently equipped to rapidly respond to the dynamic market demands and the shifting preferences of customers. The conventional methodology for project management frequently involves a sequence of discrete projects, each characterized by its own lifecycle, resource allocation, and managerial supervision. Despite the widespread use of this specific organizational architecture, it possesses the capacity to give rise to inefficiencies, resource duplication, and misalignment with overarching strategic objectives. In contrast, the implementation of an agile, product-centric framework facilitates a holistic approach.

Within the framework of the product-centred paradigm, there is a shift in focus from singular initiatives to cohesive products that offer enduring value. The establishment of this alignment facilitates collaboration between different departments and encourages the ongoing enhancement of existing products as well as the development of new ones. The ongoing advancement of this product possesses inherent adaptability, enabling the firm to promptly modify its strategy, following fluctuating market situations or evolving consumer desires. In a fast-paced corporate climate characterized by frequent and abrupt changes, the concept of improved flexibility is critical. Adopting a product-centric strategy helps the organization to increase its agility, allowing for fast changes such as product refinement or the accelerated launch of new ones. The organization's capacity to swiftly adapt allows it to capitalize on new opportunities and effectively solve obstacles with a competitive advantage.

Again, this transformation promotes a symbiotic relationship between the company and its customers. The goods developed under this framework are automatically personalized to fit the consumers' particular demands and preferences. The greater alignment indicated above increases customer satisfaction and develops long-term loyalty, contributing to the organization's continuous success. The transition from project-centred models to product-centric paradigms is more than just a change in operations; it is a movement in the organization's thinking and approach. The statement signifies a change away from rigid and compartmentalized efforts and towards a unified and adaptable approach that is in line with the changing nature and needs of today's business environment.

10.1.4 Build

Businesses must create and maintain a consistent repository of information to improve the effectiveness of informed decision-making and to allow continuing learning. Implementing this reservoir is critical for effectively accommodating and overseeing critical information, skills, and ideal approaches. This initiative intends to allow unrestricted information transmission and accessibility throughout the organizational spectrum. The reservoir's content, which contains critical knowledge and expertise, should be updated and maintained regularly. The reservoir's continuous maintenance is critical for guaranteeing its continued relevance, accuracy, and effectiveness in enabling the organization's various projects.

Adopting a centralized information repository offers several advantages. First and foremost, it promotes an atmosphere in which decision-making is considerably aided by the provision of trustworthy information. This circumstance arises as a consequence of the availability of a large collection of validated information and established methods. The availability of relevant information considerably improves the decision-makers' ability to develop strategic plans based on complete understanding and useful viewpoints.

Similarly, the idea of continuous education is indistinguishably related to the production and upkeep of this store of information. The repository is a technique for keeping knowledge and information that has been acquired and developed over a certain period. It essentially comprises the aggregate cognitive ability and experiential knowledge of the organization. The company supports a culture of knowledge-seeking by actively building and extending this repository inside its operational structure. Team members are free to apply a variety of tried-and-true tactics, fresh ideas, and previous events, resulting in an environment that encourages ongoing education and progress.

In any case, the creation of this archive is just the first step; its long-term significance is related to its continuing upkeep. The content housed inside the repository requires regular examination, updates, and verifications to maintain its reputation as a reputable source of knowledge. Adopting a proactive maintenance strategy ensures that the information contained therein is constantly aligned with the dynamic features of the organization and its external surroundings. The repository's effectiveness is dependent on its relevance and accuracy, both of which are strengthened by regular maintenance.

10.1.5 Prioritize

The operational objectives of the company, as well as the basic principles of security, accessibility, and data privacy, must be granted the most essential significance. To secure personal and sensitive information from possible breaches or unauthorized access, precautions must be implemented. Simultaneously, it is critical to develop mechanisms that provide unrestricted access to all systems and processes, promoting a culture of inclusion that is crucial for developing an environment of cohesiveness and greater efficiency.

Within the organizational setting, strictly following the applicable regulations and norms governing data protection is essential. By strictly following these regulations, the corporation accepts a huge responsibility that includes not only protecting its brand but also keeping the sanctity of data integrity. This dedication is shown by a proactive and unrelenting focus on the confidentiality and protection of sensitive and personal information. This phenomenon, in turn, strengthens the organization's credibility and trustworthiness, building a reputation acknowledged by both stakeholders and supporters.

Security concerns must be incorporated thoroughly across all aspects of the operation in the context of modern organizational dynamics. The deployment of proactive steps to strengthen information security is distinguished by a complete strategy. Among the precautions described above are advanced encryption mechanisms, strict access control procedures, and vigilant monitoring systems. When these processes are coupled, they provide a strong barrier against possible breaches, ensuring the security and integrity of sensitive data.

However, the pursuit of the organization's principles cannot be performed in a lonely way. It is also critical to create an inclusive environment in which the democratization of access to systems and procedures is prioritized. By removing unnecessary barriers to participation, the organization may promote a sense of inclusion and active engagement among its employees. The idea of inclusion extends beyond physical accessibility to include digital interfaces, ensuring that all personnel within the organization can successfully explore and engage with technology platforms without difficulty.

The organization's careful commitment to data privacy legislation and standards is important to its ethical and operational duties. The digital age has raised unprecedented concerns about the security of individuals' personal information. Recognizing the importance of this problem, regulatory bodies have developed frameworks that define the permitted parameters for data processing. The organization's compliance to these norms not only fulfils its legal obligations but also reflects its commitment to ethical conduct.

It is vital to maintain the security and privacy of personal information in light of increased awareness and scrutiny of data practices. The organization's reputation is related to its ability to assure data integrity. By building a culture based on data privacy protection, the organization embarks on a road marked by transparency, accountability, and the establishment of long-term trust with its stakeholders.

10.1.6 Form

To continue to promote collaborative efforts, encourage innovative thinking, and facilitate growth, it is important for the organization to actively develop strategic partnerships with key stakeholders, suppliers, customers, and even competitors in the sector. The establishment of these coalitions is firmly grounded in a foundation of common aspirations and beliefs. The interconnections between these ties can drive the organization into unexplored areas, developing technological domains, and sources of specialized expertise. Besides, these linkages offer opportunities for pioneering collaborative initiatives and innovative breakthroughs.

The necessity of establishing strategic alliances arises from the understanding that no entity operates in isolation. In contrast, the current state of the corporate environment is characterized by a complex network of connections and mutually beneficial associations. By promoting collaborative relationships with key stakeholders, including suppliers, consumers, and even competitors, the organization enters a domain where mutual advantages and shared goals serve as guiding principles.

The fundamental basis of these relationships is rooted in the notion of mutually held beliefs and objectives. The establishment of congruence between the objectives and principles of different organizations serves as the basis for a collaborative effort characterized by unity and synergy. These partnerships transcend simple contractual arrangements and develop into strategic connections that facilitate a dynamic interchange of ideas, resources, and skills. The collective momentum generated has the potential to drive the organization towards hitherto uncharted markets, areas, and new opportunities that were previously not considered.

By establishing collaborative alliances with suppliers, the organization can develop supply chains that possess both resilience and ingenuity. The collaboration and exchange of ideas between an organization and its suppliers can result in the creation of innovative goods, improved production methods, and a more efficient use of resources. Collaborative efforts of this nature enhance the competitive advantage of the organization and its ability to effectively adapt to changing market conditions.

Beyond that, the partnership with clients extends beyond the mere exchange of goods or services. By actively participating in collaborative engagement, the business can gain significant and crucial knowledge regarding client preferences, developing trends, and hidden needs. The increased comprehension of customer needs and preferences is the foundation for improving and optimizing products and services, ultimately leading to heightened levels of customer satisfaction and loyalty. The company transcends its role as a mere provider and assumes the role of a collaborative partner, actively contributing to the success of its clients.

Interestingly enough, competitors can form partnerships within this paradigm. Collaborations with competitors have the potential to boost the establishment of universally accepted standards within an industry, facilitate the exchange of technological innovations, and enable the consolidation of collective resources to address mutual obstacles. This practice not only develops a collective sentiment of cohesion within the sector but also possesses the capacity to unleash novel opportunities for innovation and advancement that surpass the limitations of individual organizations.

As these collaborations progress, they serve as channels for the advancement of innovative ideas and practices. The exchange and integration of ideas, experiences, and expertise contribute to the emergence of novel views and innovative solutions. Consequently, this phenomenon revitalizes the organization's ability to adjust, develop, and introduce innovative products or services.

10.1.7 Develop

To address rising impediments and capitalize on emerging opportunities, the company must create a unified innovation framework. This framework must be deeply established in the organizational culture and operational procedures. This approach's primary goal is the systematic identification, prioritization, and implementation of unique ideas, technologies, and solutions.

This paradigm is built on a methodology that includes many critical steps. The first stage of the process comprises identifying and analysing new trends and technologies with the potential to fundamentally change the industry's existing condition and structure. This is a thorough assessment of the external environment to uncover changes that may influence the organization's direction.

Following that, the chosen ideas are submitted to a detailed assessment process that includes examining their potential impact and applicability within the context of the firm. This evaluation examines the ideas' alignment with the company's strategic goals as well as their capacity to deal with existing challenges or capitalize on forthcoming opportunities.

Prioritization is an essential component of innovation practice. It is critical to acknowledge that not all ideas have equal potential, needing a careful review procedure to determine which ideas merit extra resources and efforts. Ideas that have the greatest potential to provide substantial value to the company are chosen for further development.

The last step in the innovation process is to put the selected ideas into action. To ensure the transfer of conceptual ideas into actual and quantifiable achievements, this phase requires thorough strategic planning, appropriate resource allocation, and attentive monitoring. The implementation process goes beyond technical elements and necessitates the collaboration of teams from different disciplines, effective project management, and the ability to deal with unexpected hurdles.

Incorporating this novel practice into the organizational structure provides multiple substantial benefits. It primarily supports a culture marked by continuous growth and adaptability. By embedding innovation into the core principles of the organization, there is a greater proclivity to actively address and adapt to changes. The capacity to adapt fast and effectively is critical for navigating the ever-changing business environment successfully.

More significantly, the use of this systematic approach to innovation efficiently steers the company away from haphazard or arbitrary decision-making methods. A study of the feasibility and possible implications of suggestions, on the other hand, informs choices. This method decreases the likelihood of creating businesses that lack strategic alignment or are only focused on temporary trends.

In the long term, developing an innovation practice facilitates an environment in which creativity is valued and the pursuit of novel ideas is encouraged. This paradigm shift promotes a culture in which workers at all levels of the hierarchy are encouraged to actively engage by sharing their ideas, promoting a feeling of ownership and enhanced participation. The organization provides an atmosphere favourable to idea generation and exploration, where strategic risks are accepted in the quest for substantial improvements.

11 Embracing a Digitally Resilient Future

The global community has been faced with a catastrophe since the onset of the pandemic in 2020, prompting extensive deliberations on digital transformation and the development of the digital economy. Throughout that particular year, the difficulties arising from the pandemic led to the emergence of an unfamiliar paradigm known as the *new normal*. The prevalence of online platforms and the use of automated and intelligent technology were the main factors driving this shift. The swift changes observed have served as a compelling incentive for organizations to develop digital resilience and enhance their competitive edge. There was a growing realization that organizations that embraced proactive planning and avoided reactive approaches were able to generate innovative value and improve their competitive advantage.

Businesses have been forced to reallocate their investments to strengthen organizational resilience and prepare for digital transformation, given the present state of things. Governments and industry leaders must provide clear guidelines and promises to improve digital skills. The relevance of digital goals in both socioeconomic recovery and basic economic revitalized programmes was obvious.

High-speed Internet in facilitating trustworthy telecommuting and education should be stressed, emphasizing the need for significant investments in this important area. The issues caused by the pandemic have highlighted the need to commit to digital transformation and the progress of the digital economy. Following the outbreak, some industries saw the need to concentrate investments in digital technology to increase their long-term resilience and competitiveness.

During the COVID-19 pandemic, approximately one hundred million students worldwide were permitted to engage in online education in elementary and secondary educational institutions. This achievement was made possible by the development of a scalable cloud computing architecture that allowed for the efficient storage, analysis, and distribution of data and services.

More importantly, modern technologies such as the Internet of Things (IoT) and big data have greatly assisted in the adoption of automation technology in business operations and services at a reduced cost. This has had a significant impact on enhancing supply chain resilience and decision-making processes. The contemporary workplace has undergone a significant shift towards touchless operations, owing mostly to the incorporation of automation technologies in industrial facilities and the deployment of autonomous vehicle logistics services in leisure settings. Because of the rapid advancement of the digital revolution, information technology today serves a dual purpose in lowering carbon emissions.

Some may argue that information technology may effectively reduce carbon emissions, construct high-performance networks, and encourage energy saving while being environmentally conscious and sustainable. From another perspective, this innovative technology has the potential to increase energy conservation and decrease emissions in a range of industries, supporting a symbiotic and mutually beneficial interaction between scientific progress and the natural environment.

Given the fast growth of technology, it is necessary to define this project within the context of digital resilience, emphasizing a strategic approach led by business principles and backed by information technology (IT). A hierarchical approach driven by senior executives in top-tier roles within the organization is required to achieve these goals. The leader's articulation of a thorough strategy is essential in establishing a unified and logically clear knowledge of the initiative, hence improving general acceptance and execution. Leadership practices should show a progressive and inventive attitude, as well as a dedication to adaptation.

12 Concluding Remarks

It becomes clearer as time goes on that the reinvigorated outlook on the future has been fuelled by lessons garnered from the COVID-19 pandemic and subsequent transformational occurrences. The intersection of digital resilience and the expansion of the digital economy offers an overload of possibilities and problems that need careful consideration. The effect of the pandemic on companies has highlighted the concept that preemptive preparation provides more benefits than reactive responses. This insight emphasizes the necessity of building a business climate that proactively incorporates technology solutions to assure process continuity and anticipates future interruptions. The continuing move to digital transformation is not a passing trend, but rather a significant project requiring thorough and strategic planning.

Meanwhile, the link between digital transformation and economic recovery has commenced a new age in which the digital economy has become a crucial engine for social progress. The combination of business and government usage of digital technology offers a tangible possibility to boost economic regeneration. According to the concept underpinning this statement, promoting digital ecosystems may raise productivity, provide new market possibilities, and boost global competitiveness.

The increased popularity of online learning in educational settings during the global pandemic has sparked a debate on the usefulness and inclusivity of digital learning platforms. While these platforms have increased the accessibility of education to individuals around the globe, worries about equal technological access and the possibility of a digital gap remain. Governments and educational institutions must identify and solve these inequities if all segments of society are to benefit equally from the advantages of digital education.

As a result, the integration of automation technologies and the introduction of touchless operations across a variety of industries represents a significant paradigm change in production techniques and service delivery. Although technology developments have been shown to boost efficiency and resilience, they have also raised ethical issues and prompted labour-related queries. Achieving a healthy cohabitation of the advantages of automation and human employment, as well as the growth of a competent workforce, is a complex problem that requires the participation of labour groups, governments, and corporations.

The obligation to reduce carbon emissions via the use of digital technologies reveals a potentially fruitful method in the quest for environmental sustainability. However, it is vital to approach this prospect with caution and to perform an examination of the environmental consequences of digital technology. Attaining a condition of balance that takes into account both the good elements of digitalization and their possible negative environmental implications requires careful thought and continual improvement.

Recognizing that digital transformation has cultural and strategic elements in addition to technical ones has gained traction in talks about corporate culture and leadership. When challenged with change, executives must exemplify transformational leadership skills to motivate and urge their subordinates to embrace and apply digital capabilities. Creating a business culture that encourages innovation, constant learning, and adaptation is critical for effectively tackling the complex challenges that a digitally resilient environment presents.

Chapter 2
Achieving Digital Resilience with Cybersecurity

1 Introduction

Digital resilience was introduced in the preceding chapter; in this chapter, we shall examine how digital resilience can be attained. The digital world has become deeply integrated with the physical world to the extent that virtually every aspect of our routine behaviours is now linked to its digital components. Cyberspace has emerged as a foundation, facilitating instantaneous worldwide communication, enabling financial transactions via online banking, and catering to segments of business and entertainment via e-commerce platforms and digital entertainment hubs.

However, although the expanding integration of cyberspace has facilitated communication and enhanced convenience, it has also augmented the susceptibilities that threaten its secure functioning. Unwittingly, our reliance on technology has paved the way for an increase in the prevalence of cybersecurity threats that threaten the digital space. The potential repercussions of these cyberattacks are severe, including the deliberate compromise of critical infrastructure, which supports societal operations, and the theft of personal information, which exposes individuals to a variety of cyber-fraudulent activities. The roots of cybersecurity concerns can be traced back to the nascent phase of computer technology, when safeguarding physical assets, such as the robust protections that accompanied mainframe systems at the time, was of primary importance. However, with the progression of technology and its increased accessibility, there was a transition in focus towards protecting intangible assets, including vital database servers and sophisticated software systems.

The Internet being public in the 1990s initiated a paradigm shift concerning security concerns in digital spaces. In this context, malicious software, phishing attempts, and denial-of-service attacks have surfaced as formidable foes capable of infecting and crippling digital systems. Due to malicious intent, the sophistication and complexity of these cybersecurity threats have increased. For instance, malware

S. K. Shandilya et al., *Digital Resilience: Navigating Disruption and Safeguarding Data Privacy*, EAI/Springer Innovations in Communication and Computing,
https://doi.org/10.1007/978-3-031-53290-0_2

has developed complex code structures, whereas phishing operations methodically exploit human vulnerabilities. The expansion of the Internet of Things (IoT), cyber-physical systems (CPS), and autonomous technologies has given rise to novel susceptibilities that present substantial risks. The importance of robust cybersecurity measures is widely acknowledged.

To address the issue of cybercrime, governments are implementing proactive measures, including strengthening domestic cybersecurity frameworks, promoting coordination among diverse authorities, and engaging in international collaboration. Simultaneously, enterprises are allocating resources towards technological advancements aimed at strengthening their digital security and oversight. To increase their security, individuals are implementing proactive measures such as employing two-factor authentication, utilizing robust passwords, and exercising caution when traversing the digital world. Cyber threats continue to evolve, resulting in a fundamentally unresolved issue.

Every entity is endowed with cybersecurity obligations. Governments are entrusted with the development of policies that create a secure digital environment. Businesses are expected to safeguard their digital assets through the implementation of state-of-the-art technologies and optimal methodologies. Individuals are provided with opportunities to acquire knowledge of cybersecurity to avert concealed threats. The integration of cybersecurity education into academic curricula is necessary, given that it provides students with the necessary abilities and understanding required to stay secure in the digital spaces. The potential for augmenting awareness and resistance to cyber threats could be actualized through the integration of artificial intelligence and blockchain technology. Regardless of the degree of accomplishment, it is critical to remember that cybersecurity is an ongoing journey that requires consistent adjustment and focus to effectively address emergent threats resulting from technological progress.

2 Nature of Cyber Threats

This section will centre its attention on the dynamic characteristics of cyber threats and the related techniques developed for their mitigation. Our objective is to establish a foundation for the future by examining the motives and tactics of attackers, as well as the effects, countermeasures, and broader impact on society.

2.1 Motives and Tactics of Attackers

During the continuous digitalization process, there are persistent and ever-changing cybersecurity risks that pose a continual danger to both individual organizations and major companies. As a result, it is critical to build a strong state of digital resilience that not only recognizes the continued presence of these digital vulnerabilities but also seeks to thoroughly grasp the attackers' motivations, approaches, and attack techniques.

Understanding the motivations that drive these malicious actors, as well as the strategies they use to emerge, is vital for creating a resilient digital posture. Simply stated, understanding the psychology and approach of an attacker allows us to figure out what he is after; this is crucial as it solves half of our problem. Once one is aware of what is in danger, one can work to defend it. So, say, a greater understanding of the unknown surrounding cyber threats may be reached by exploring the minute complexities of the reasons that inspire these offenders. A detailed examination of the strategies used by malicious actors gives an oversupply of information about how vulnerabilities are exploited.

There is a strong focus on the practical advantages of investigating these elements. Understanding the fundamental motives and techniques that characterize malicious actors allows for a better understanding of the asymmetrical environment that defines cyber conflicts. Decisions about the development of essential defensive measures can be based on this information. Finally, having a keen awareness of the motives and strategies used by attackers helps both people and organizations to anticipate possible attack paths and design tailored ways for defending against, reducing, and recovering from such threats.

2.1.1 Motives of Attackers

In this section, let us understand why individuals engage in cyberattacks, which is very important for building a strong digital defence system. This understanding helps those in charge of protecting digital systems predict the types of attacks that might happen and then put measures in place to stop or reduce their effects. Figuring out the different reasons behind cyberattacks gives valuable information about the way these attacks happen. We are well aware the prominent among these reasons are making money, having political or personal beliefs, and having personal interests.

- **Money**: Firstly, the desire to make money is a powerful motivation and usually the most common motivation for cybercriminals to do their harmful activities. These bad actors target weak points in digital systems to get sensitive information or money illegally. They do this using many different tricks, including things like ransomware attacks and dishonest strategies. When they do these activities, cybercriminals use advanced methods to get into private data stores or financial systems without permission. Knowing about their financial goals is important for those defending digital systems to get ready for these money-focused attacks.
- **Political or Ideological Motivations**: Another strong reason for cyberattacks is politics or beliefs. In this situation, individuals or groups who are called 'hacktivists' or are supported by governments have a preference for getting into digital networks. They do this because they want to disturb important systems or steal private information to promote their political or personal beliefs. These actions have a big impact, as they can cause problems in various areas, such

as government, business, and communication. Knowing about these political and personal reasons helps defenders make strong plans to protect against these kinds of attacks.

- **Personal Gain**: Again, cyberattacks that come from personal motivations are also very significant to understand. Specifically, individuals who have special access to digital systems might misuse their positions for personal gain or revenge. This kind of bad behaviour depends on secretly taking advantage of trust. Individuals with proper access to digital resources secretly misuse their rights. The possible bad outcomes are many, from getting private information to deliberately damaging digital systems. Fully understanding this tendency for personal gain helps those protecting the systems make rules that stop these inside threats effectively.

Seeing the many reasons behind cyberattacks, those who are responsible for digital systems are well-prepared to create plans that reduce or stop these bad actions. For example, organizations can follow strict rules about who can access systems, keep an eye on things, and teach individuals to stop threats from inside. These actions make digital systems more secure by preventing unauthorized access from within. Also, to deal with outside threats, using strong tools to stop malware and keep information safe is a good strategy. These actions make the security system strong by stopping outside attacks, which makes digital systems more protected against many possible attacks.

Effective comprehension of the motives fuelling cyberattacks is a linchpin in the effort to construct a resilient digital defence. This observant understanding enables defenders to anticipate the diverse modalities that attackers may use and correspondingly implement measures to obstruct or abate their impact. The triad of motivations including financial gain, political or ideological aspirations, and personal interests reveals various incentives that drive individuals to embark upon cyberattacks. Armed with this awareness, defenders can deftly carve their defensive strategies, thereby augmenting the resilience of digital infrastructure against the various threats that may manifest.

2.1.2 Tactics of Attackers

The tactics of attackers contain distinct approaches utilized to compromise digital systems, data, and resources. These approaches are strategically orchestrated sequences of actions that form what is referred to as the 'kill chain'. The objective of attackers is to breach security measures, gain unauthorized access, and potentially exploit the compromised system for their advantage. This discourse describes the varied types of attacker approaches, outlines the components of the kill chain, describes the gains associated with each approach, highlights the advantages attributed to each style, and acknowledges the attendant drawbacks.

The concept of the 'cyber kill chain' [YR15] represents a structured framework that outlines the distinct stages or steps that malicious actors or cyberattackers traverse while directing a successful cyberattack. This model, often used in cybersecurity analysis, provides an understanding of the various phases an attacker must navigate to breach a target system, compromise its integrity, and achieve their malicious objectives. By breaking down the attack process into discrete steps, the cyber kill chain aids cybersecurity professionals in identifying and detecting potential threats effectively. Figure 2.1 showcases a pictorial representation of the cyber kill chain model.

The cyber kill chain consists of several sequential phases, each representing a critical point where countermeasures can be implemented to halt or mitigate the progress of an attack. These phases are as follows:

(1) **Reconnaissance**: In this initial phase, attackers gather intelligence about the target system and potential vulnerabilities. Methods like network scanning can help find possible entry points, or gathering information from publically accessible sources like social media can also be part of the picture.

Fig. 2.1 Cyber kill chain

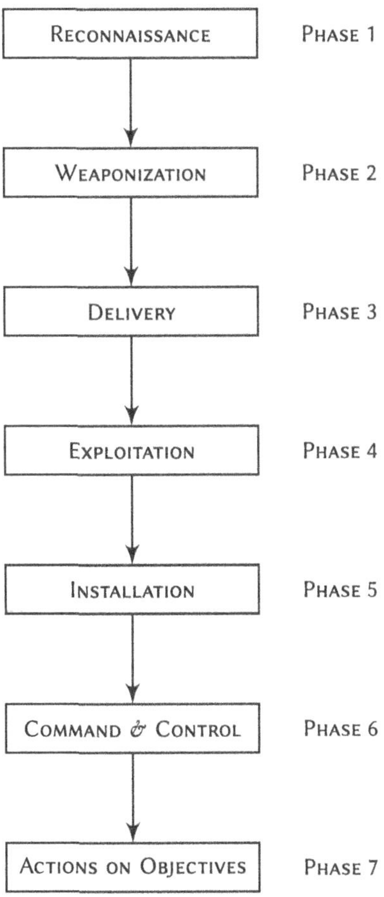

(2) **Weaponization**: In this phase, the attackers craft or obtain malicious tools, such as malware, exploit kits, or phishing emails, which they plan to use in the attack. These tools are designed to exploit specific vulnerabilities discovered during the reconnaissance phase.
(3) **Delivery**: During this stage, the attackers deliver the malicious payload to the target system. This could occur through methods like sending phishing emails, infecting websites, or distributing malicious attachments.
(4) **Exploitation**: In this phase, the attackers execute the payload on the target's system, taking advantage of the identified vulnerabilities. Once executed, the malicious code gains a foothold within the system, granting the attacker initial access.
(5) **Installation**: After gaining access, the attackers proceed to install their tools or establish a persistent presence within the compromised system. This could involve creating backdoors, establishing remote access, or installing additional malware.
(6) **Command and Control (C2)**: At this point, a communication channel is established by the attackers between the breached system and their infrastructure. This channel, known as the command and control server, enables the attacker to remotely control the compromised system, exfiltrate data, or issue further commands.
(7) **Actions on Objectives**: In the final phase, the attackers aim for their intended goals. Theft of data, financial fraud, espionage, disruption of systems, or any other form of invasive action that aligns with their goals could be involved.

The cyber kill chain model serves as a guide for security professionals to implement proactive measures at various stages to prevent or disrupt an attack. By understanding each phase, defenders can develop strategies and deploy defences to detect and block threats. For example, robust perimeter defences, such as firewalls and intrusion detection systems, can help prevent the delivery and exploitation stages. Security awareness training can aid in minimizing the effectiveness of social engineering tactics during the reconnaissance and weaponization phases.

The tactics of attackers include various approaches that can be broadly categorized into four main styles: social engineering, malware attacks, network-based attacks, and physical attacks. Each style corresponds to a distinct modus operandi used by attackers to infiltrate systems and compromise their integrity.

- The social engineering approach primarily exploits human psychology and social interactions. Attackers use psychological manipulation to trick victims into giving up sensitive information or completing actions that allow them to get unauthorized access. This approach preys on trust, fear, or curiosity, thereby tricking users into divulging sensitive information such as passwords or personal data.
- Malware attacks involve the dissemination of malicious software to exploit vulnerabilities in software or hardware. Malware can take diverse forms, such as viruses, worms, Trojans, or ransomware. Once the malware is executed, it can

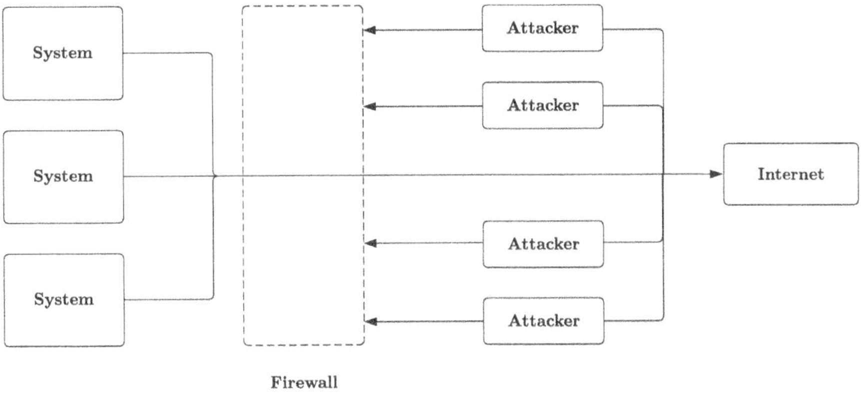

Fig. 2.2 A basic firewall architecture

disrupt system operations, steal sensitive data, or enable remote control of the compromised system.

- Network-based attacks target vulnerabilities in network infrastructure. These attackers exploit weaknesses in routers, firewalls (Fig. 2.2), or servers to gain unauthorized access to a system. This approach often includes activities like sniffing, eavesdropping, or man-in-the-middle attacks, where attackers intercept communication between two parties and can potentially extract sensitive data.
- Physical attacks involve direct manipulation of hardware components or physical entry to breach security. Attackers may physically steal devices, manipulate hardware configurations, or insert malicious devices into the target system. This approach is often used to gain access to systems that are considered secure from remote attacks.

The returns an attacker gains from utilizing these approaches are varied. Gaining unauthorized access allows attackers to obtain sensitive data, potentially leading to identity theft or financial fraud. Besides, attackers can exploit compromised systems to launch further attacks, create botnets, or engage in cryptocurrency mining. Attackers may also seek to disrupt critical services or infrastructure, causing significant economic and operational damage.

Several advantages are associated with each style of attack. Social engineering leverages human psychology, often exploiting the most vulnerable linkage in the security chain: an individual human mind. This approach requires minimal technical expertise and can be highly effective against unsuspecting users. Malware attacks enable automation, allowing attackers to target a large number of systems simultaneously. Network-based attacks exploit vulnerabilities within digital architecture and can be conducted remotely, providing anonymity to attackers. Physical attacks, though requiring proximity, can be highly targeted and difficult to detect, especially in environments with lax physical security.

However, each approach carries its inherent drawbacks. Social engineering attacks heavily depend on the manipulation of human behaviour and are contingent on the attacker's ability to create convincing scenarios. Malware attacks can trigger countermeasures such as antivirus software, potentially limiting their effectiveness. Network-based attacks may encounter firewalls and intrusion detection systems, impeding unauthorized access. Physical attacks require proximity to the target, which can be a significant constraint.

2.2 Evolving Nature of Threats and Ongoing Vigilance

Both enterprises and individuals encounter an ever-changing set of cyber threats. The emergence of interconnected devices, cyber-physical systems (CPS), the Internet of Things (IoT), and autonomous systems and the growing integration of critical infrastructure into digital platforms have resulted in a pressing need for sustained alertness in response to these attacks. The concept of digital resilience stands as the linchpin for upholding the capacity to endure and rebound from cyberattacks. A fundamental aspect of constructing this resilience involves the recognition of the evolving characteristics inherent in these threats.

Digital resilience, denoting the capability of an organization to uphold its fundamental operations and rapidly recuperate from cyberattacks, includes various aspects. These include guaranteeing the secrecy, consistency, and accessibility of data, in addition to upholding the availability of crucial systems and services. For genuine resilience, entities must exhibit the capability to promptly differentiate and counteract cyberattacks while also deploying measures to stall future attacks.

Modern businesses are extremely concerned with the dynamic nature of cyber threats. Cybercriminals consistently develop novel methodologies and tactics to penetrate established security protocols. Companies must therefore be prepared to react to these fluctuations to preserve their resilience. It is key to maintain constant surveillance and stay informed about emerging cyber threats and trends, both current and future.

The establishment of a threat intelligence programme emerges as a key tactic for recognizing the dynamic nature of threats. This involves amassing and dissecting information about the most recent threats and vulnerabilities. This information is derived from both internal sources, like security logs, and external sources, such as industry consortia and governmental bodies. By remaining apprised of the most recent threats and trends in cybersecurity, entities can better prepare themselves for potential attacks and proactively adopt protective measures.

The introduction of an enduring security awareness training regimen for personnel is another important tactic. Employees are frequent targets of cybercriminals, particularly through techniques like phishing and other forms of social engineering. Thus, the provision of information to personnel regarding the latest tactics and optimal practices for evading such attacks contributes significantly to diminishing the probability of a successful breach.

Entities should contemplate integrating a perpetual security monitoring scheme that includes the routine scanning of systems and networks to identify vulnerabilities and other security-related concerns. This proactive approach assists in the identification of potential threats before malicious actors can capitalize on them. Consequently, it furnishes valuable insights into the security stance of the organization.

The construction of digital resilience necessitates a strategy that includes individuals, processes, and technology. By acknowledging the mutable attributes of threats and sustaining ongoing attention, entities can heighten their defences against cyberattacks and preserve their ability to promptly recuperate from breaches. Digital resilience is not a solitary effort but rather a sustained commitment towards establishing a secure and resilient digital world.

3 The Impact of Cyber Threats on Digital Resilience

As the world becomes increasingly interconnected through digital networks, the idea of digital resilience has arisen as a crucial area of study and concern. This refers to the ability of systems, organizations, and individuals to endure, adapt, and rebound or recover from disruptive events in the virtual domain, ensuring the continuity of essential functions and services. One of the foremost challenges to digital resilience in this context is posed by cyber threats, which include a spectrum of malicious activities orchestrated through digital means to compromise the confidentiality, integrity, and availability of digital systems and data.

The increasing frequency and complexity of cyber threats have underlined the critical nature of understanding their diverse effects for the resilience of digital infrastructure. Investigating the elaborate linkages between cyber threats and digital resilience is therefore crucial to contemporary academic comprehension. This section examines the nature of cyber threats and the extensive effects they have on the concept of digital resilience.

The rationale behind probing into this subject matter is twofold: first, to enhance the understanding of the complex mechanisms through which cyber threats target and undermine digital systems and, second, to reveal the strategies and measures that can be devised and implemented to strengthen digital resilience in the face of such threats. To this end, an exploration of a diverse set of cyber threats, containing a range of attacks such as malware, phishing, and distributed denial of service (DDoS) attacks, is essential.

Against this backdrop, the objective of this section is to contribute by providing an analysis of the complex connection between cyber risks and digital resilience. This part aims to explore the tactics and countermeasures that can enhance digital resilience against cyber threats, utilizing an examination of empirical data, theoretical frameworks, and practical case studies. The primary objective of this section is to offer a thorough understanding of the various factors involved, enabling individuals and organizations to effectively address the difficulties presented by cyber threats. By doing so, it aims to enhance the ability of modern systems and societies to withstand and recover from cyber-attacks.

3.1 Cybersecurity in Maintaining Trust and Resilience

The expansion and saturation of cyber threats have become more extensive and complex, making the concept of digital resilience of central value for organizations. Safeguarding digital systems against such threats is important for maintaining operational continuity and reliability. Cybersecurity plays an important role in upholding digital resilience by including measures designed to safeguard computer networks, systems, and data from unauthorized access, cyberattacks, and potential harm. The objective of cybersecurity is to guarantee the confidentiality, integrity, and availability of digital assets, as a result sustaining trust and the seamless operation of digital systems.

A fundamental aspect of cybersecurity's contribution to enhancing trust and resilience within digital frameworks is its role in securing sensitive data. Data breaches that expose sensitive information can profoundly impact an organization's digital resilience. Financial data, personally identifiable information, and intellectual property may be compromised, eroding the organization's credibility. By implementing robust cybersecurity measures such as data backups, encryption, and access controls, organizations can effectively protect their sensitive data, a critical step in promoting trust among stakeholders.

Safeguarding crucial assets is a significant responsibility within cybersecurity. Ensuring the security of critical infrastructure, including sectors like transportation, healthcare, and energy networks, becomes vital as these sectors are susceptible to cyberattacks with potential impacts on both economic advancement and public safety. For instance, the cyberattack on the Colonial Pipeline [EF23] led to its shutdown, resulting in gasoline shortages and price hikes across the eastern United States. Organizations can enhance the security of their critical assets and reinforce digital resilience by implementing strict cybersecurity conventions, comprising intrusion detection systems, firewalls, and routine vulnerability assessments.

Trust among stakeholders, including customers, partners, and investors, rests on the foundation of cybersecurity's influence. Ensuring data protection and demonstrating a genuine commitment to cybersecurity through robust measures are essential for acquiring this trust. Organizations can exhibit their dedication to cybersecurity and establish trust among stakeholders by adopting practices like regular security evaluations and audits.

Consequently, the importance of cybersecurity in developing both trust and resilience in digital systems becomes apparent. The consequences of cyber threats on an organization's digital resilience are severe, highlighting the importance of investing in cybersecurity to safeguard sensitive data and critical infrastructure while sustaining trust among stakeholders. The significant cost of managing data breaches and the extended timeframes needed to identify and contain such breaches emphasize the importance of cybersecurity in maintaining digital resilience. Robust cybersecurity measures empower organizations to effectively counteract cyber threats and expedite recovery following breaches.

Enforcing resilient security measures serves to deter unauthorized access attempts and prevent data breaches, establishing trust among stakeholders such as customers, investors, and partners. This nurtures a secure environment for digital interactions and transactions. Cybersecurity enhances resilience by identifying vulnerabilities and implementing strategies to mitigate their impact, thereby reducing the consequences of potential cyber incidents and ensuring uninterrupted essential system functioning.

However, it is crucial to acknowledge that implementing complex security protocols and measures necessitates financial investment and resource allocation. The financial burden associated with this effort can pose challenges, particularly for smaller enterprises. The ever-evolving essence of cyber threats also requires continuous adaptation of security measures, highlighting the need for continual vigilance and resource allocation to maintain their effectiveness against evolving threats.

Differing viewpoints exist regarding the centrality of cybersecurity in the digital age. Advocates argue that it forms the foundation of trust in digital interactions and communications. They stress its indispensable integration into organizational practices, essential for improving customer confidence, investor trust, and consistent service provision. Conversely, some contend that an excessive focus on cybersecurity might lead to excessive regulation, stifle innovation, and hinder the seamless flow of digital exchanges. These proponents advocate a balanced approach that harmonizes security procedures with the importance of convenience and innovation.

4 Types of Cyber Threats

To preserve digital resilience in the face of an ever-expanding and increasingly sophisticated list of cyber threats, organizations must prioritize the development and implementation of defence mechanisms against an extensive variety of cyber threats. Significant repercussions may result from cyberattacks, including financial loss, reputational harm, data intrusions, and legal liability. Thus, organizations must possess an extensive comprehension of the prevailing categories of cyber threats and formulate tactics to reduce their associated risks. This section will examine several commonly encountered categories of cyber threats, the resulting consequences for organizations, and corresponding approaches to efficiently decrease the risks they pose. For an outline of the various categories of cyber threats, refer to the following reference: Fig. 2.3.

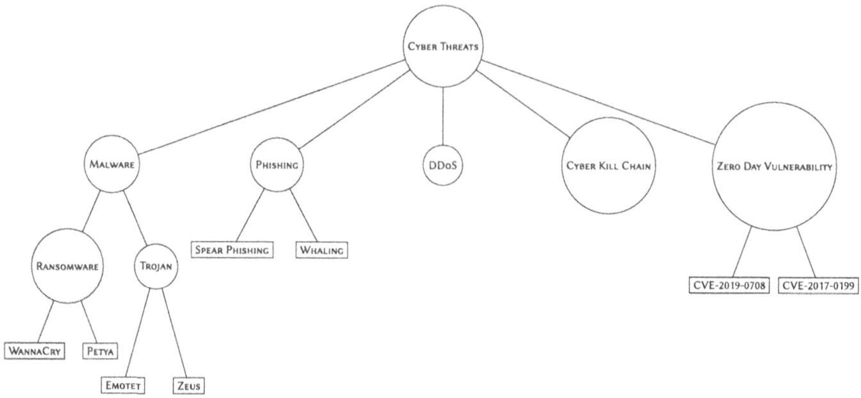

Fig. 2.3 An example of the basic types of cyber threats

4.1 Malware

Malware, a contraction of the term *malicious software,* contains various damaging digital elements such as viruses, trojans, worms, ransomware, and spyware, constituting the dominant form of cyberattack. Infiltrating systems, usually through hyperlinks in unreliable websites or emails or through unsolicited software downloads, malware establishes a presence within the target system. Subsequently, it proceeds to acquire sensitive information, manipulate and obstruct access to network constituents, and, in certain cases, effectuate data obliteration or system shutdown. Refer to Fig. 2.4 for an insight into some malware schematics. Several principal categories of malware attacks are discernible:

(1) **Viruses**: A piece of code finds its way into a program. The malicious code begins to carry out its predetermined tasks as soon as the application is launched.
(2) **Worms**: Malware infiltrates an operating system by taking advantage of vulnerabilities in the software and hidden pathways. As soon as it's part of the network, it can launch DDoS attacks and other hostile operations.
(3) **Trojans**: To evade detection, malicious code or software masquerades as a harmless application, game, or email attachment. By downloading these trojans, unsuspecting users grant the malware access to their devices.
(4) **Ransomware**: Using encryption to deny a user or entity the ability to access their systems or data. The offender commonly requests a monetary payment as a condition for providing a decryption key that enables the restoration of access. Yet, sending the ransom does not ensure the full recovery of unrestricted access or optimal performance.
(5) **Cryptojacking**: Attackers install software on a target's device, secretly utilizing their computing power to engage in cryptocurrency mining, without the victim's knowledge. The affected systems may encounter reduced efficiency, and these cryptojacking toolkits have the potential to undermine system stability.

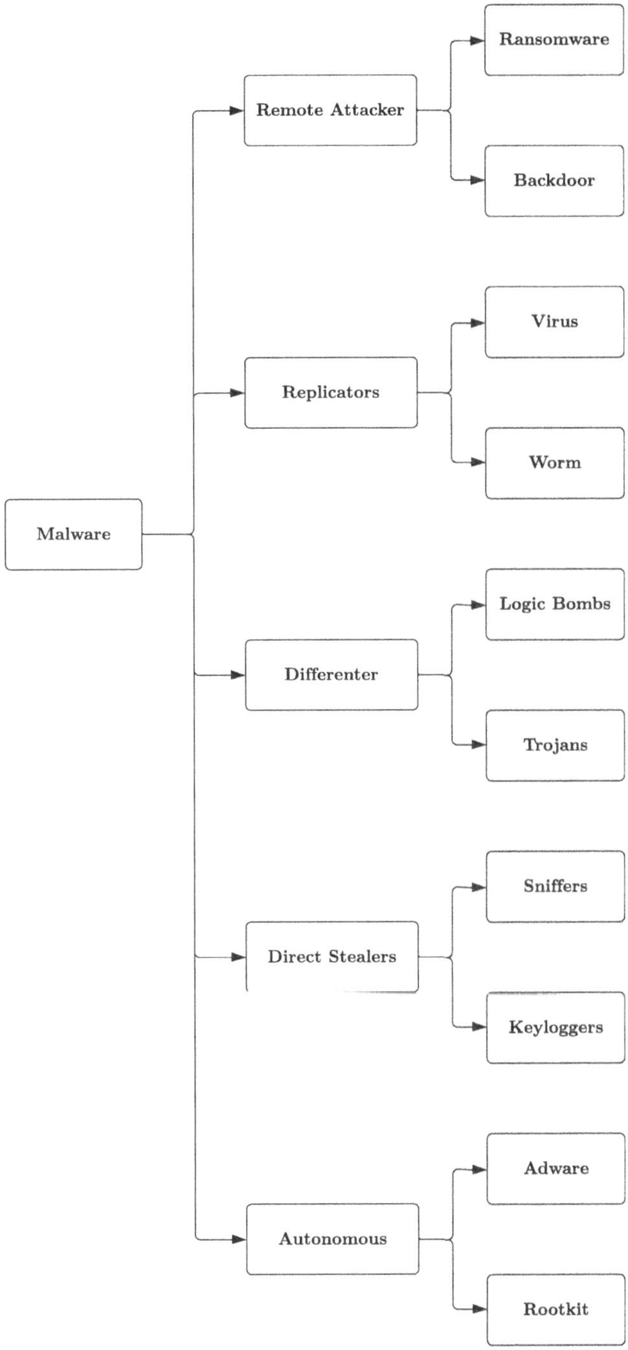

Fig. 2.4 Categorization of malware

(6) **Spyware**: An unauthorized software gains illegitimate access to an unwary user's information, including sensitive data such as passwords and payment details. Spyware can affect multiple platforms, including mobile devices, desktop browsers, and programs.
(7) **Adware**: To enable targeted advertising, adware tracks a user's online activities to identify patterns of behaviour and interests. Adware is similar to spyware, except it doesn't install any software on the user's device and could not be dangerous. However, privacy might be jeopardized if it is implemented without user agreement.
(8) **Fileless malware**: The operating system does not have any software installed in this version. Malicious functionality is instead enabled via manipulating inherent files like WMI and PowerShell. Because it masquerades as valid files, this kind of assault is characterized by its stealthiness and evades traditional detection systems, such as antivirus software.
(9) **Rootkits**: By infiltrating a computer's software, kernel, operating system, or hypervisor, rootkits grant remote administrative access. As soon as the operating system is started in a compromised setting, the criminal gains control of the system and can spread other malware.

4.2 Social Engineering

Social engineering denotes the act of deceiving users into furnishing a route for malware entry. Deceived into divulging critical information or unknowingly installing malware on their device, the victim falls prey to the impostor's disguise as a genuine actor. Refer to Fig. 2.5 for an insight into some social engineering schematics. Several key variants of social engineering attacks are evident:

(1) **Baiting**: In this scheme, the perpetrator entices a user into a snare of social engineering, often through the allure of enticing incentives such as complimentary gift cards. The victim, ensnared in the ruse, surrenders sensitive data like credentials to the malicious actor.
(2) **Pretexting**: Analogous to baiting, this approach coerces the target to relinquish information under fictitious pretexts. The tactic frequently involves masquerading as an authoritative figure, for instance, a police officer or an IRS, whose stature compels the target's compliance.
(3) **Phishing**: In this approach, the perpetrator sends out emails that appear to have come from a reliable source. Phishing frequently involves dispatching deceitful

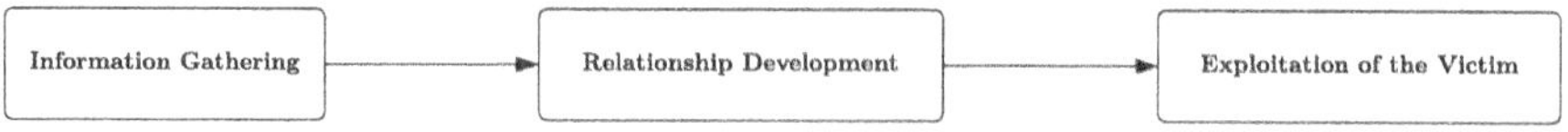

Fig. 2.5 Social engineering attack

emails to a wide spectrum of users, but it may also adopt a targeted approach. For instance, 'spear phishing' personalizes the message or email content to tailor it to a specific individual, whereas 'whaling' elevates the technique by concentrating on individuals of high value, like CEOs.

(4) **Vishing (voice phishing)**: Imposters often use phone scams to trick their targets into giving up private information or granting them access to their systems. Primarily focusing on senior individuals, vishing can, however, be wielded against any category.
(5) **Smishing (SMS phishing)**: Employing text messages as the medium of deceit, the attacker efforts to beguile the target.
(6) **Piggybacking**: Someone else can get physical access sanctioned by an authorized user by riding on the back, piggybacking, of the authorized user's credentials. An example of this would be an employee granting access to a person pretending to be a new hire who has lost their ID.
(7) **Tailgating**: A non-authorized user trails an authorized individual into a designated area, oftentimes by opportunistically infiltrating a secured portal immediately after the authorized user's activation. This scheme mirrors piggybacking, save for the fact that the individual being trailed is oblivious to their exploitation as an instrument by the other individual.

4.3 Supply Chain Attacks

Software suppliers and developers are now the targets of a new kind of threat known as supply chain attacks. Subtle goals include infiltrating legitimate apps and spreading malware via build processes, software update systems, or source code. The bad guys look for security holes in programming practices, server architecture, and network protocols. By taking advantage of these flaws, attackers can tamper with source code, hide harmful content, and get access to build and update protocols. Refer to Fig. 2.6 for insights into the anatomy of supply chain attacks.

Supply chain attacks have increased in seriousness, owing to the subversion of applications carried out by attackers with the authorization of trusted suppliers. There is also a software supply chain assault where the software vendor does not

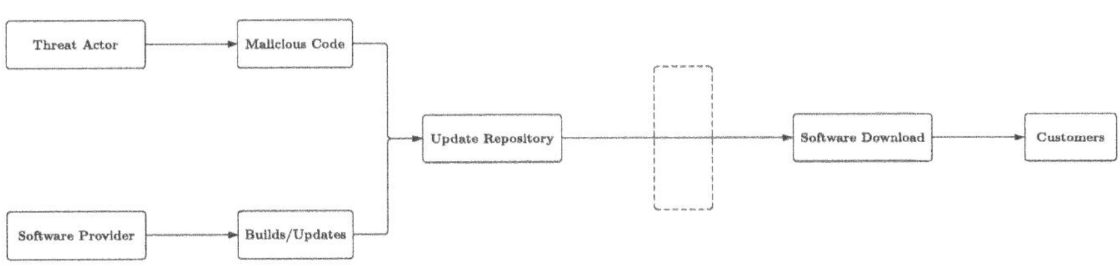

Fig. 2.6 Anatomy of supply chain attacks

know that their updates or programs have been infected with harmful viruses. The compromised programme and the malicious code share the same rights and trust and run hand in hand. The following are examples of supply chain attacks:

(1) Construction device or development pipeline compromise
(2) Development accounts or code signing method compromise
(3) Malicious code distribution through firmware or hardware component automated updates
(4) Pre-inclusion of physical devices with malicious code

4.4 *Man-in-the-Middle*

A man-in-the-middle (MitM) attack refers to the act of intercepting communication between two terminals, namely, a user and an application. Figure 2.7 shows a connection with a man-in-the-middle. In this particular scenario, the attacker gains the capacity to covertly observe the topic at hand, secretively acquiring confidential information while also assuming the identity of one of the individuals engaged in the interaction. MitM operations include the following:

(1) **Wi-Fi eavesdropping**: The perpetrator orchestrates a simulated Wi-Fi connection, feigning legitimacy as an entity like a commercial establishment, thus luring users into connection. An attacker can steal sensitive information, such as login credentials and credit card details, by monitoring the activities of users on this malicious Wi-Fi channel.
(2) **Email hijacking**: The perpetrator masquerades as a genuine organization, for instance, a financial organization, through the falsification of the email address. Subsequently, the attacker exploits this deception to manipulate users into relinquishing sensitive data or facilitating monetary transfers to their advantage. Users, under the impression of following the directives of the authentic organization, are unknowingly complying with the attacker's schemes.

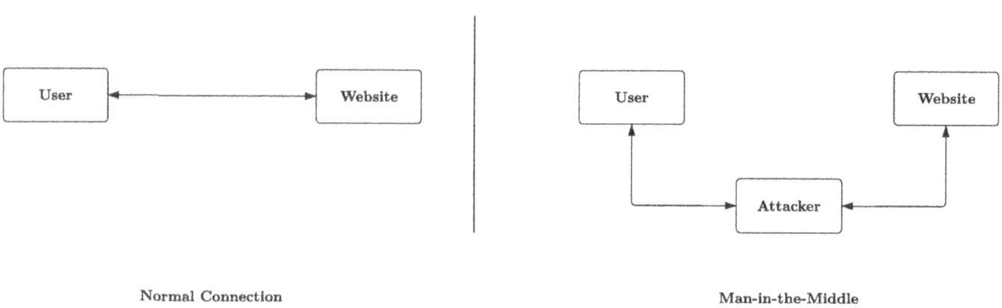

Fig. 2.7 Normal connection versus man-in-the-middle

(3) **DNS spoofing**: By simulating a domain name server (DNS), the attacker reroutes users to a malicious website impersonating a lawful counterpart. In such scenarios, the perpetrator might redirect traffic from the genuine website or purloin user credentials.
(4) **IP spoofing**: An Internet protocol (IP) address functions as the conduit connecting users to a specific website. Through IP spoofing, the attacker counterfeits an IP address to adopt the guise of a website, thereby misleading users into believing they are interacting with the authentic site.
(5) **HTTPS spoofing**: Although HTTPS is often thought of as the safer version of HTTP, it can also be used to trick browsers into thinking a malicious website is secure when in fact it isn't. The malicious website's properties are concealed when the attacker includes 'HTTPS' in the URL.

4.5 Denial-of-Service

A denial-of-service (DoS) attack occurs when a victim system is overwhelmed by a significant increase in data traffic, resulting in the system's inability to operate normally. When a variety of devices are used to carry out an attack, it assumes the characteristics of a distributed denial-of-service (DDoS) attack. Figure 2.8 showcases the schematics of a DOS attack using botnets. The compilation of denial-of-service (DoS) attack techniques contains a collection of methods and strategies used to disrupt or disable targeted computer systems or networks.

(1) **HTTP flood DDoS**: In this modality, the perpetrator sends out what appear to be legitimate HTTP requests to overwhelm a web server or application. Notably, this approach does not necessitate copious bandwidth or malformed packets.

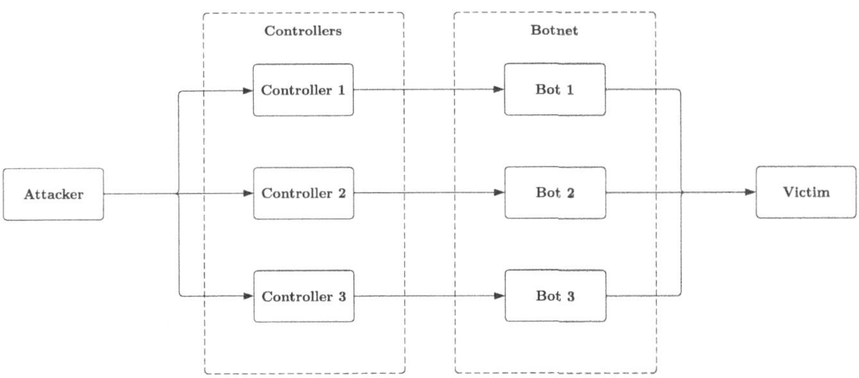

Fig. 2.8 Usage of botnet in DoS

Instead, its scheme revolves around compelling the target system to allocate maximal resources for every request.

(2) **SYN flood DDoS**: The initiation of a transmission control protocol (TCP) connection sequence mandates the transmission of an SYN request to which the host must reciprocate with an SYN-ACK, acknowledging the solicitation. Subsequently, the requester responds with an ACK. Through the execution of SYN requests without acknowledging the host's SYN-ACKs, the attackers conceal server resources by utilizing this sequence.

(3) **UDP flood DDoS**: Attackers force a remote host to look for programs matching the affected ports by flooding it with user datagram protocol (UDP) packets sent towards random ports. In response, the host emits 'destination unreachable' packets, effectively consuming host resources.

(4) **ICMP flood**: To exhaust the target's incoming and outgoing bandwidth, this tactic utilizes an inflow of ICMP echo request packets. Despite servers' best efforts, system performance degrades because they are unable to keep up with the influx of requests, even when they try to respond to every request through an ICMP echo reply packet.

(5) **NTP amplification**: By leveraging network time protocol (NTP) servers, which are publicly available, the attackers can orchestrate voluminous UDP traffic directed towards a designated server. This practice qualifies as an amplification attack due to the query-to-response ratio, oscillating between 1:20 and 1:200. Consequently, this vulnerability permits malicious actors to harness open NTP servers in executing large-scale, high-bandwidth DDoS attacks.

4.6 Injection Attacks

Injection attacks can insert malicious code into a web application's code structure by taking advantage of many vulnerabilities. If carried out effectively, these attacks can compromise sensitive data, launch a denial-of-service (DoS) attack, or undermine the entire system. The subsequent sections outline many prevalent injection attack vectors, some of which are also listed out in Fig. 2.9.

(1) **SQL injection**: The attacker inserts a SQL query into a web form or comment box, for example, to steal information from the endpoint. An exploitable application will send the perpetrator's data to the database, where it will execute any SQL commands put into the query. Web applications are susceptible to SQL injection attacks because they depend on databases built on structured query language (SQL) principles. NoSQL attacks are when threat actors try to attack databases that don't use the usual way of organizing data, making them less safe.

(2) **Code injection**: This ploy facilitates the attacker's insertion of code into a vulnerable application. Subsequently, the web server executes the malicious code under the pretence of it being an integral component of the application.

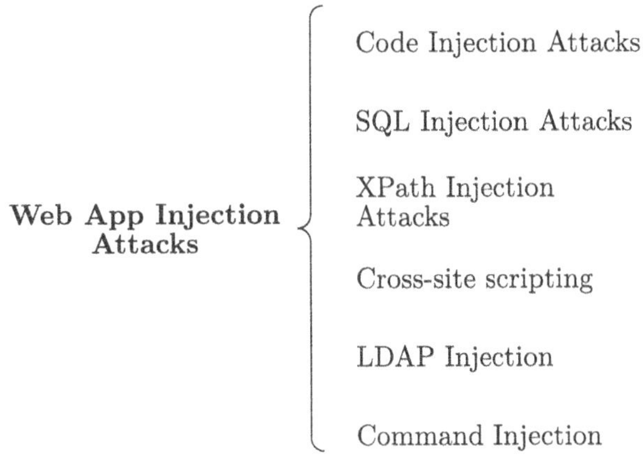

Fig. 2.9 Categorization of injection attacks

(3) **OS command injection**: Capitalizing on a vulnerability about command injection, the attacker inputs commands intended for execution by the operating system. This makes it possible for the attack to take over the system or steal data from the operating system.
(4) **LDAP injection**: By inputting specific characters, the attacker manipulates lightweight directory access protocol (LDAP) queries. Unsanitized LDAP requests leave a system vulnerable and open to attacks. These attacks assume a heightened severity due to the potential for LDAP servers to contain user accounts and organizational credentials.
(5) **XML external entities (XXE) injection**: This attack exploits specially constructed XML documents, diverging from other vectors in that it capitalizes on innate vulnerabilities within legacy XML parsers, in place of user inputs that have not been validated. Manipulated XML documents permit path traversal, remote code execution, and the enactment of server-side request forgery (SSRF).
(6) **Cross-site scripting (XSS)**: The perpetrator enters a string of characters that contains harmful JavaScript. By injecting this code into the target's browser, an attacker can hijack their session, send visitors to a malicious website, or redirect their traffic to another URL. An application is susceptible to XSS if it neglects to sanitize user inputs, neglecting the removal of JavaScript code.

4.7 Zero-Day Vulnerabilities

A zero-day vulnerability, also known as a 0-day, refers to a security flaw within computer software or hardware that remains unknown to the parties responsible for its mitigation, such as the software vendor. In the absence of awareness, cyberattackers can exploit this vulnerability to detrimentally impact programs, data, additional computers, or an entire network. This type of exploit is termed a zero-day exploit or zero-day attack.

Definition 4 A zero-day attack, also referred to as a zero-day vulnerability exploit, refers to a cybersecurity incident where malicious actors exploit a software vulnerability or weakness that is unknown to the software vendor and, consequently, has no official patch or fix available.

The origin of the term 'zero-day' traces back to the interval since the release of a new software piece to the public; hence, 'zero-day software' was procured by infiltrating a developer's computer prior to the official release. Over time, this expression extended its meaning to include the vulnerabilities enabling such hacking efforts, along with the duration that vendors possess to rectify these vulnerabilities.

Upon detection of the vulnerability, vendors typically respond by creating patches or suggesting workarounds to ease the issue. The timeliness of the vendor's knowledge regarding the vulnerability governs the probability of the existence of a remedy or mitigation. For recently identified vulnerabilities where no fix or mitigation has yet been formulated, the probability of successful exploitation remains high.

As vendors develop and distribute fixes, the efficacy of the exploit wanes as more users apply the solution over time. With zero-day exploits, however, unless the vulnerability is inadvertently remedied through an unrelated update, the probability of users applying a vendor-supplied patch remains null, thereby perpetuating the exploit's availability. Refer to Fig. 2.10 for a clearer view of the anatomy of zero-day vulnerability exploitation.

Zero-day attacks involve a significant threat, given their capacity to bypass existing security measures and directly impact vulnerable systems. A zero-day vulnerability denotes a security gap within a system or device that has been disclosed but remains unaddressed by a patch. Consequently, an attack that capitalizes on a zero-day vulnerability is termed a zero-day exploit. Such vulnerabilities pose a heightened risk to users due to their discovery before security experts and software developers can address them through patches. This scenario creates a window of opportunity for cybercriminals to exploit these vulnerabilities for their gains, as well as a period during which vulnerable systems remain susceptible until a suitable patch is deployed by the vendor.

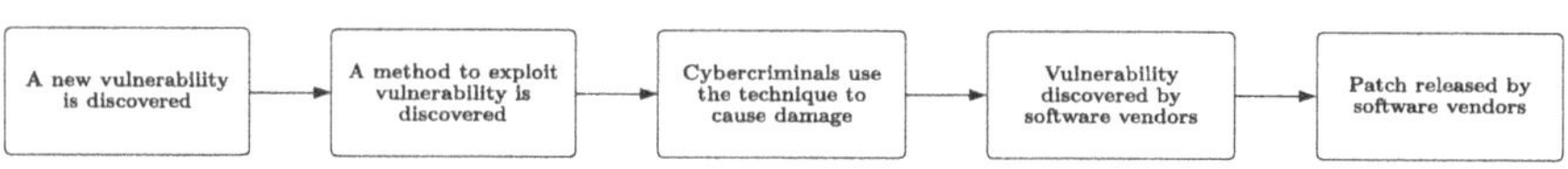

Fig. 2.10 Anatomy of zero-day vulnerability exploitation

Although zero-day vulnerabilities are frequently associated with targeted attacks, many campaigns still rely on older vulnerabilities. These vulnerabilities pose a heightened level of risk due to their novelty. In such cases, cyberattackers can use these undiscovered weaknesses to infiltrate systems, exfiltrate sensitive data, compromise system integrity, and gain unauthorized access to critical systems and information. The inherent stealth associated with zero-day vulnerabilities allows attackers to exploit these weaknesses covertly before defensive measures can be taken.

To mitigate the risks presented by zero-day vulnerabilities, organizations can adopt various strategic measures. Implementing a robust patch management process is a fundamental approach. Regular updates involving security patches are crucial, as they address known vulnerabilities and rectify exploitable weaknesses. Consistent application of these updates ensures minimal exposure to potential zero-day attacks.

The utilization of intrusion detection systems (IDS) proves valuable. IDS actively monitors network activities and system behaviour, promptly flagging any irregular patterns that may indicate ongoing attacks or unauthorized access attempts. This expedient alert system enables administrators to respond swiftly to potential zero-day exploits, reducing potential damage. Conducting regular vulnerability assessments remains a critical strategy. By continually evaluating systems and applications, organizations can proactively identify latent zero-day vulnerabilities and initiate preemptive measures to mitigate the associated risks.

5 Threat Intelligence

Threat intelligence also referred to as 'threat intel' or 'cyber threat intelligence' (CTI) includes an exhaustive reservoir of information that complexly dissects cybersecurity threats that specifically target an organization. This key dataset empowers security teams to proactively embrace a vigilant stance, thereby equipping them with the acumen required to preemptively detect cyberattacks before their materialization.

Also, the potency of threat intelligence extends to developing an organization's skill in identifying and promptly countering ongoing attacks. The ensuing elucidation carefully expounds upon the various dimensions of threat intelligence, spanning its genesis, defining attributes, and pronounced relevance. The advent of automation and heightened interconnectivity has orchestrated transformative shifts across global economic and sociocultural paradigms. Nonetheless, this digital revolution has concurrently ushered in a new dimension of risk, encapsulated in the form of cyberattacks.

Threat intelligence, often interchangeably aligned with open source intelligence (OSINT), embodies a corpus of knowledge designed to empower proactive measures or reduce the impact of such attacks. Rooted in data, threat intelligence serves to provide context – unveiling the identities of the aggressors, unravelling their motivations and capabilities, and elucidating the specific indicators of compromise entrenched within an organization's systems. This contextual insight acts as the foundation for informed decisions concerning security strategies. The various applications of TI render it indispensable for cross-functional teams within any organization. While its immediate value lies in preventing attacks, TI serves as a critical component in triage, risk assessment, vulnerability management, and strategic decision-making.

The function of threat intelligence is essential in the present day's cybersecurity scene. Organizations can proactively identify cyber threats, accelerate detection processes, and lessen the financial repercussions of data breaches when equipped with insights. The complex lifecycle of threat intelligence integrates the processes of analysis, distribution, and iterative feedback loops to generate practical insights that are customized to meet the unique needs of various organizations. The classification of threat intelligence into tactical, operational, and strategic tiers further empowers stakeholders to protect their digital infrastructures from an ever-changing set of cyber threats.

5.1 Defining Threat Intelligence

Presently, cybersecurity confronts a series of formidable obstacles. These include the escalating tenacity and cunning of malicious actors, a constant surge of data laden with excessive information, and erroneous alerts originating from disjointed security systems. The scope of vulnerable points that organizations must understand and safeguard has expanded substantially. These threats emerge from various angles, necessitating organizations to understand business risks arising from cyberattacks, operational disruptions, damage to reputation, physical security breaches, and more.

Certain entities effort to integrate threat data feeds into their networks; however, they grapple with effectively managing the surplus data. Consequently, this places an additional burden on analysts who lack the tools to determine which elements warrant prioritization and which can be disregarded. A feasible solution to these predicaments can be found in the implementation of a threat intelligence framework.

The most effective of such solutions utilizes a fusion of ML to automate the collection and processing of data. They seamlessly integrate with existing systems, assimilate unstructured data from diverse sources, and subsequently establish connections by supplying contextual insights concerning indicators of compromise (IoCs) as well as the methodologies used by threat actors, including their tactics, techniques, and procedures (TTPs). This form of threat intelligence is actionable and is characterized by its timeliness, contextual relevance, and accessibility to decision-makers.

Central to the production of threat intelligence are security analysts. Their efforts include the aggregation of raw threat data and security-related information from a wide set of sources. These disparate data points later undergo strict correlation and analysis, facilitating the identification of trends, patterns, and interrelationships. Ultimately, these revelations coalesce to yield an understanding of existing or potential threats. Significantly, the resultant threat intelligence is carefully tailored to the unique context of the organization. This approach circumvents generic insights and instead directs attention towards vulnerabilities within the organization's attack surface, the corresponding exploitative tactics, and the assets left exposed to potential risks.

5.2 Attributes of Threat Intelligence

The widespread misconception that threat intelligence is solely utilized by analysts fails to take into account its practical applications in diverse security functions across organizations of varying scales. A notable disadvantage arises when threat intelligence is compartmentalized as an independent entity within the larger security team rather than being acknowledged as an implicit enhancement to all aspects of security. This is because numerous individuals who have the most to gain from threat intelligence are unable to obtain it on time. This discrepancy restricts the capacity of security operations teams to efficiently handle the abundance of alerts that they are bombarded with. By autonomously prioritizing and filtering signals and other possible threats, threat intelligence integration with existing security solutions circumvents this difficulty.

Vulnerability management teams can notably enhance their efficacy by accurately assessing and ranking the most critical vulnerabilities. This is achieved through the incorporation of external insights and contextual information supplied by threat intelligence. Understanding the recent threat types is made easier with threat intelligence, which enriches higher-level security procedures like risk analysis and fraud prevention. Included in this is an abundance of supplementary data collected from various online sources, along with crucial insights on threat actors and their strategies, tactics, techniques, and procedures.

Exploring the use cases below provides a comprehension of how each security role stands to gain from the application of threat intelligence. Notably, threat intelligence exhibits distinctive attributes that accentuate its essential status within cybersecurity. Its hallmark feature is its orientation towards the organization's unique vulnerabilities as opposed to standardized parameters such as generic catalogues of prevalent malware strains. This tailored focus highlights its relevance and pertinence. The intelligence itself boasts granularity and context, including not only the specific threats targeted at the organization but also the identities of potential threat actors instigating these attacks.

This contextual depth extends further to include the spectrum of tactics, techniques, and procedures (TTPs) used by these threat actors. It also encapsulates indicators of compromise (IoCs) that could potentially herald a particular cyberattack. Ultimately, the essence of threat intelligence resides in its actionable nature, furnishing information that empowers information security teams to confront vulnerabilities, prioritize and counteract emerging threats, and even evaluate the efficacy of existing or novel cybersecurity tools.

5.3 Lifecycle of Threat Intelligence

How is threat intelligence then created? Threat intelligence is the outcome of a comprehensive six-stage process involving data acquisition, processing, and analysis; it should not be confused with raw data. New inquiries and knowledge deficits are discovered throughout the intelligence development process, which in turn results in the establishment of fresh collection requirements. Thus, this procedure is cyclical. Iteratively improving, an effective intelligence programme develops its capabilities over time.

To optimize the worth of the threat intelligence you provide, it is necessary to first identify your use cases and establish your objectives before proceeding with any other actions. The lifecycle of threat intelligence is an iterative and ongoing process integral to the sustenance of robust cybersecurity. Spanning six distinct stages, this process underpins the production, dissemination, and continuous enhancement of threat intelligence. One can get a clear view of the cyber threat intelligence cycle in Fig. 2.11.

5.3.1 Planning Phase

Effective threat intelligence begins with a collaborative effort involving security analysts and key stakeholders within the organization. This cooperation facilitates the identification of intelligence requirements, which includes inquiries that decision-makers consider critical for addressing. These inquiries define the scope of necessary insights for cybersecurity. The initial step towards generating actionable threat intelligence is to formulate pertinent questions. The most effective questions for implementing actionable threat intelligence are those that focus on singular

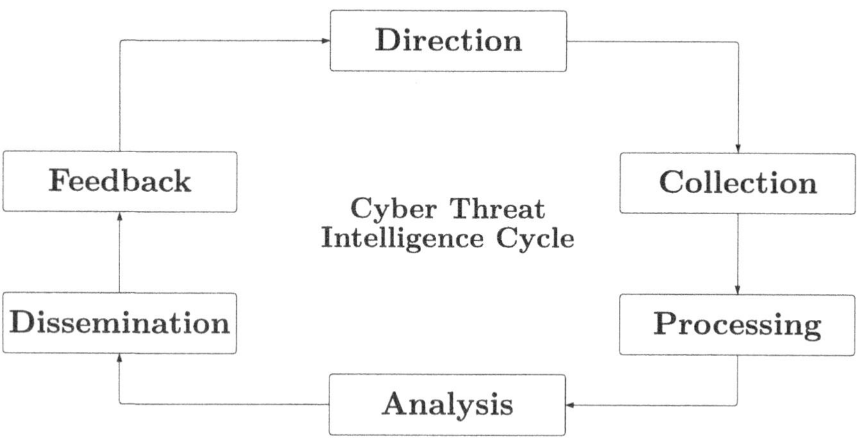

Fig. 2.11 Cyber threat intelligence cycle

facts, events, or activities. It is generally advisable to avoid formulating broad and open-ended questions. When prioritizing intelligence objectives, factors such as alignment with the organization's core values, the potential impact on decisions, and the urgency of decisions are important. A necessary consideration at this stage is understanding the intended recipients of the final product. Whether the intelligence is destined for technically proficient analyst teams in need of swift reports on emerging exploits or executives seeking trends for strategic security investment decisions, recipient profiling guides this stage.

5.3.2 Data Collection Phase

During this phase, security teams gather raw threat data that has the potential to address the identified intelligence requirements. This data is sourced from diverse outlets, including threat intelligence feeds, information-sharing communities, and internal security logs. After identification, the next step involves assembling raw data that fulfils the criteria established in the planning phase. Data must be gathered from a variety of sources to formulate an approach. These sources may include technical repositories, the dark web, and internal resources like historical incident response records and network event logs. IoCs, which consist of domain names, malicious IP addresses, and file hashes, are common components of threat data. However, personally identifiable information (PII), vulnerability data, textual content from social media or news sources, and raw code extracted from paste sites may also be included.

5.3.3 Processing Phase

In this stage, security analysts engage in aggregating, standardizing, and correlating the collected raw data. These preparatory measures enhance the data's readiness for subsequent analysis. After the completion of raw data accumulation, organization and refinement steps come into play. These include tagging data with metadata, eliminating redundant information, and filtering out false positives and negatives. Daily, organizations amass hundreds of thousands of indicators and millions of log events, which frequently result in substantial volumes of data. Given the inefficiency of human analysis at this scale, automation becomes important for data collection and processing to yield meaningful insights.

Security information and event management (SIEM) systems offer an ideal starting point, enabling streamlined data structuring through correlation rules tailored for specific use cases. However, these kinds of devices can only hold certain kinds of data. Stronger methods are needed to deal with the collection of different unstructured data from many sources both inside and outside.

5.3.4 Analysis Phase

This important phase entails the transformation of raw threat data into actionable threat intelligence. Security analysts undertake validation of trends, patterns, and insights that meet the intelligence requirements outlined by stakeholders. This process culminates in the formulation of recommendations aimed at addressing identified vulnerabilities and potential threats. After that, making sense of the processed information comes next. The main goal of the analysis is to find possible security issues and let the right teams know about them in a way that fits with the information needs that were set during the planning phase. Threat intelligence can show up in several different ways, depending on the original goals and target groups. Formats vary from simple lists of threats to studies that have been carefully looked over.

5.3.5 Dissemination Phase

Insights and recommendations arising from the analysis are shared with relevant stakeholders. These insights serve as a foundation for informed decision-making, potentially resulting in the adoption of security measures or adjustments to existing protocols. The final product is subsequently distributed to its intended recipients. To render threat intelligence actionable, timely dissemination to the appropriate parties is essential. Effective tracking ensures continuity between successive intelligence cycles and the retention of acquired insights. Leveraging ticketing systems that integrate with other security platforms facilitates tracking throughout the intelligence cycle. This includes the submission, documentation, review, and fulfilment of intelligence requests by multiple teams, thus centralizing the process.

5.3.6 Feedback Phase

Stakeholders and analysts engage in reflective evaluation post-analysis to assess the effectiveness of the recent threat intelligence cycle. Insights retained from this introspection influence subsequent cycles and address any existing gaps in intelligence. This last stage marks the closing of the intelligence cycle and is very similar to the planning stage and the direction stage that came before it. When the person who made the original request gets the final intelligence product, they look it over and decide if the questions they asked have been answered properly. The evaluation sets the goals and methods for the next intelligence cycle, which emphasizes how important it is to keep records and keep things going.

5.4 Types of Threat Intelligence

The variability of the final output is apparent in the intelligence lifecycle, which is contingent upon the initial intelligence requirements, information sources, and intended recipients. The categorization of threat intelligence according to these factors is advantageous. Threat intelligence is frequently divided into the following three subcategories: The strategic classification begins with more general trends that are appropriate for a non-technical audience. Following that, the tactical classification serves to inform a more technically proficient audience by outlining the strategies, methods, and protocols employed by threat actors. Lastly, operational classification provides complex technical understandings of particular campaigns and attacks; this is also known as technical threat intelligence. The lifecycle of threat intelligence generates distinct forms of intelligence contingent upon the participation of stakeholders, specified requirements, and overall objectives. These distinct categories, including three primary types, effectively address the operational necessities of diverse factors within an organization.

5.4.1 Tactical Threat Intelligence

Tactical threat intelligence serves as a fitting support for the security operations centre (SOC) by facilitating real-time identification and response to ongoing cyberattacks. This form of intelligence concentrates on prevalent indicators of compromise (IoCs), such as IP addresses related to command and control servers, established malware attacks associated with file hashes, and email subject lines exhibiting characteristics of phishing efforts. Beyond mere incident reaction, tactical threat intelligence proves valuable to teams involved in threat exploration, enabling the unmasking of hidden attackers. This category of intelligence describes the tactics, techniques, and procedures (TTPs) used by threat actors. It aims to equip defenders with a specific understanding of potential modes of attack against their organization and the optimal strategies for countering or mitigating such threats. Typically comprising technical context, it finds utility among personnel directly engaged in organizational defence, including system architects, administrators, and security personnel.

5.4.2 Operational Threat Intelligence

Operational threat intelligence, often known as 'technical threat intelligence', is designed to anticipate and detect attacks. This aspect of intelligence expounds upon the tactics, techniques, and procedures (TTPs) characteristic of well-established threat actors. This involves their chosen avenues of attack, exploited vulnerabilities, and targeted assets. Decision-makers, including chief information security officers (CISOs) and chief investigating officers (CIOs), leverage operational threat intelligence to enhance the security of their organizations and preempt potential attacks. Operational intelligence comprises information regarding cyber campaigns, incidents, or events; it offers incident response teams specialized insights that aid in the comprehension of the nature, timing, and intent of particular attacks. Security leaders, security operations centre (SOC) managers, cyber threat intelligence teams, threat hunters, and incident responders are stakeholders and consumers of this form of intelligence.

5.4.3 Strategic Threat Intelligence

Strategic threat intelligence offers a wide perspective, furnishing elevated insights into the global threat outlook and an organization's positioning within it. This category of intelligence, extending its benefits beyond IT giants, enables executives and CEOs to grasp imminent cyber threats. Strategic threat intelligence aligns broader organizational risk management strategies including geopolitical dynamics, industry-specific cyber threat trends, and the vulnerabilities of strategic assets with the prevailing cyber threats. The purpose of strategic threat intelligence is to inform the decisions of directors and other key decision-makers by providing a comprehensive view of the threat to an organization. Despite the generally less technical nature of the material, it is communicated via reports or briefings. Effective threat intelligence should shed light on crucial domains, including significant geopolitical occurrences and trends, risks associated with particular courses of action, and overall patterns in the tactics and targets of threat actors. The C-suite (CISO, CIO, CSO, CTO), senior vice presidents, board members, and intelligence executives in the physical and cyber domains are stakeholders and recipients of strategic intelligence.

5.5 Threat Intelligence with Machine Learning

Data processing, the scale at which threat intelligence (TI) data is handled, necessitates an automation approach. This entails amalgamating data from various sources including the open web, dark web, and technical outlets, to construct the most robust depiction attainable. To this end, Recorded Future leverages ML methodologies across four key domains to refine data collection and aggregation: data categorization, multilingual text analysis, risk assessment, and predictive modelling.

5.5.1 Categorizing Data into Entities and Events

The notion of ontology concerns itself with the classification and division of concepts. Attributes, identities, and interrelationships are utilized by data scientists to designate groups of entities within ontologies, which facilitate the classification of entities into hierarchical structures. Notably, entities such as 'Delhi', 'London', and 'Berlin' are distinct yet collectively fall under the superset category of 'city'. While entities categorize distinct concepts, events organize them temporally. Recorded Future's events transcend language barriers.

5.5.2 Multilingual Text Structuring

Through natural language processing (NLP), events and entities extend beyond mere keywords, converting unstructured text from diverse languages into a structured database. By leveraging machine learning, this procedure accurately differentiates and categorizes advertising from primary content; categorizes content into genres such as code, prose, or data logs; and disambiguates homonymous terms (e.g. 'Apple' or 'Blackberry' refers to the company rather than 'Apple' or 'Blackberry' the fruit) through the examination of contextual cues. This enables the parsing of text across seven languages from millions of documents daily – a task that would otherwise need an impractical human workforce.

5.5.3 Event and Entity Classification

To further rank entities and events according to their significance, ML and statistical techniques are applied to designate risk scores to potentially malicious elements. These scores result from two systems: one governed by human-derived rules and intuition, and another driven by ML using validated data. Classifiers, such as risk scores, provide both a verdict (e.g. 'this event is critical') and a contextual justification (e.g. 'this IP address is malicious according to multiple sources').

5.5.4 Predictive Modelling

ML is also leveraged to generate predictive models that anticipate future occurrences with greater precision than human analysts. Drawing from extensive, categorized data repositories, this approach capitalizes on the 'law of large numbers', resulting in increasingly accurate predictive models as data sources expand.

6 Strategies for Building Digital Resilience

Cybersecurity strategies specifically focus on their important role in the establishment of digital resilience. In the age characterized by ubiquitous digital interconnectedness and interdependence, the vulnerability of information systems to many malicious cyber threats necessitates the formulation and implementation of robust strategies aimed at safeguarding digital infrastructures. This section attempts to describe the foundational principles and pragmatic methodologies that underlie effective cybersecurity strategies, thereby contributing to the enhancement of organizations' abilities to withstand and recover from cyber incidents. Through a combination of theoretical analysis and empirical insights, we shall provide an explanation of the dimensions surrounding cybersecurity strategies and their role in building digital resilience.

6.1 Designing Systems with Resilience

The design and implementation of resilient systems are critical factors in the strategic efforts of organizations and their operations. The indispensability of a resilient system surfaces when confronted with unforeseen adversities such as cyberattacks, disruptions in the supply chain, and other unanticipated challenges. Designing for resilience includes a diverse approach, including various dimensions aimed at strengthening systems against disruptions and preserving operational continuity.

6.1.1 Discerning Potential Risks and Threats

At the core of designing resilient systems lies the process of recognizing potential risks and threats that could undermine their functionality. This requires an analysis of the system's operations, taking into account a multitude of factors that possess the capacity to impede its smooth functioning. These factors include a broad spectrum, ranging from cyberattacks that could potentially compromise data security and system availability to supply chain disturbances capable of disrupting the flow of essential components and human errors that might lead to system failures. Effectively discerning potential risks and threats necessitates a risk assessment process. This involves an exhaustive review of all system constituents, including hardware, software, networks, and data. Also, a deep analysis of the organization's operational processes and their interdependence on the system's functionality is important. Following the identification of potential risks, the subsequent step involves an in-depth evaluation of testing outcomes, thereby pinpointing areas for enhancement. These insights serve as foundational groundwork for the formulation of strategies aimed at mitigating risks and reducing their impact.

6.1.2 Devising a Disruption Response Plan

Once potential risks have been identified, the subsequent strategic effort refers to the formulation of a plan to effectively respond to disruptions. This plan should include a collection of strategies geared towards minimizing the repercussions of disruptions and expeditiously reinstating operational normalcy. Strategies could include the deployment of backup systems, redundancy mechanisms, disaster recovery protocols, and other measures calibrated to ensure the perpetuity of operations. An essential component of this approach is guaranteeing that every stakeholder is thoroughly informed about the developed strategy and comprehends their obligations and duties in the case of a disturbance. In order to formulate a robust response strategy to disruptions, it is recommended that organizations perform a business impact analysis. The recognition of vital organizational processes and systems, evaluation of their dependencies and vulnerabilities, and development of strategies to accelerate their recovery in the event of a disruption comprise this procedure. Establishing unambiguous and concise communication strategies that guarantee all stakeholders are kept well-informed and consistently updated during the response phase is of equal significance.

6.1.3 Efficacious System Testing

Regular testing serves as a standard for upholding the resilience of systems. This involves the testing of backup systems and recovery strategies to ascertain their effectiveness and efficiency. Testing protocols should span a set of conditions, including simulated disruptions and worst-case scenarios, to comprehensively assess the system's response. This process is important in revealing potential weaknesses or gaps within the system, thereby facilitating preemptive adjustments before the actualization of real-world disruptions. To optimize the effectiveness of testing, organizations should define clear testing objectives and protocols. Testing efforts should unfold within a controlled environment and include all relevant stakeholders. Rigid documentation and analysis of testing results ensure fertile ground for the identification of areas necessitating improvement.

6.1.4 Continuous System Monitoring

The deployment of monitoring tools stands as an indispensable mechanism for the continuous evaluation of system performance, permitting the identification of new problems before they evolve into substantial predicaments. These tools serve to detect anomalies, including irregular traffic patterns, anomalous activities, or performance deterioration, which could serve as early indicators of impending disruptions. A regular and unswerving monitoring regimen augments the potential for organizations to discern trends and patterns that contribute to the refinement of system resilience over time. To optimize the efficacy of monitoring, the establishment of explicit monitoring objectives and procedures is critical. Monitoring undertakings should be continuous and inclusive of all relevant stakeholders. An analysis of monitoring results should be undertaken to uncover avenues for system enhancement.

6.1.5 Developing a Culture of Resilience

Developing a culture of resilience within the organization holds importance in ensuring that all personnel are aware of potential risks and collaborate to address them. This effort includes the promotion of open communication, transparency, and collaboration across all stakeholders. All members of the organization must grasp the significance of resilience and possess a clear understanding of their roles and responsibilities in building it. This objective can be actualized through consistent communication, recurring training initiatives, and recognition and reward mechanisms directed towards individuals and teams that experience resilience in their professional efforts.

6.1.6 Continual Adaptation and Enhancement

Lastly, the effort of perpetually evaluating and augmenting the resilience of the system is generally significant. This involves the revision and refinement of plans and procedures, necessitated by evolving risks or changing circumstances. This may further include the integration of novel technologies or strategies conceived to develop the system's resilience quotient. Continuous enhancement mandates an ongoing course of action characterized by regular reviews, evaluations of performance, and the solicitation of feedback from stakeholders.

6.2 Enforcing Multi-layered Security Mechanisms

To counteract the escalating complexity of cyber threats, enterprises must adopt a security approach that utilizes a range of defence layers. This strategy, known as a multi-layered defence strategy, involves the utilization of multiple security measures to establish a series of barriers against cyberattacks. Each distinct layer is deliberately tailored to address various categories of threats and collaboratively construct a robust and efficacious defence mechanism. The conventional multi-layered defence strategy is composed of five distinct layers, which are as follows:

- **Perimeter Layer**: The initial layer of defence includes the enforcement of stringent access controls to detect unauthorized entry into the system. These controls may include elements such as password policies, multi-factor authentication, and access management based on user roles and permissions. To exemplify, Google's G Suite utilizes two-factor authentication to provide an additional tier of security, shielding user accounts against phishing and other forms of attack.
- **Network Layer**: The second defence layer revolves around the deployment of security tools at the network level, which may include firewalls, intrusion detection systems, and intrusion prevention systems. These sophisticated software solutions are adept at preemptively identifying and deterring malicious activities

like denial-of-service attacks and malware outbreaks. These attacks are aimed at infiltrating the network and disrupting the operational integrity of the system. For instance, the SonicWall Next-Generation Firewall presents diverse protection against advanced threats such as ransomware, viruses, and spyware.
- **Host Layer**: The third defence layer involves the implementation of endpoint security mechanisms such as endpoint detection and response (EDR) tools, antivirus software, and mobile device management (MDM) software. These tools are adept at identifying and stopping cyberattacks directed at endpoints, including devices like laptops, mobile devices, and servers.
- **Application Layer**: The fourth defence scheme refers to the integration of application security measures, which include regular software updates, secure coding practices, and vulnerability assessments. These practices are effective at identifying and averting cyberattacks aimed at exploiting vulnerabilities within applications and software. For instance, Microsoft Office 365 adopts secure coding methodologies to counter frequently occurring weaknesses like cross-site scripting (XSS) and SQL injection.
- **Data Layer**: The ultimate defence tier incorporates incident response and business continuity measures, such as disaster recovery planning, backup and recovery protocols, and incident response strategies. The deployment of these measures furnishes remedies against the detrimental outcomes of cyberattacks and safeguards the continuous functionality of enterprises during security breaches. By way of illustration, IBM's Resiliency Services[1] furnish an all-including suite of business continuity and disaster recovery services. This suite equips enterprises to prepare for, manage, and restore their operations following an attack or any form of disruption.

The execution of a multi-layered defence strategy is important for organizations to mitigate the diverse cyber threats. By employing an assortment of security measures that confront various threat categories, organizations can develop a more robust and secure digital infrastructure.

6.3 Developing a Culture of Security Awareness and Resilience

Establishing a security culture that prioritizes alertness and robustness is vital for safeguarding both individuals and enterprises from possible threats. Creating a secure space to prevent and mitigate cyberattacks, educating individuals and organizations about the significance of cybersecurity, and conducting training are all components of building a culture of heightened security awareness and resilience. The pedagogical aspect is key in increasing awareness regarding potential risks and the nature of cybersecurity. Organizations and individuals must have a complete

[1] www.ibm.com/consulting/business-continuity.

understanding of prevalent cyberattack categories such as phishing, malware, and ransomware, in addition to the potential repercussions of these dangers. Ongoing cybersecurity education is critical to ensure that individuals remain informed about the most recent threats and preventative measures.

The provision of training is a key component in developing a culture of security awareness and resilience. Password security, two-factor authentication, and data encryption should all be included in cybersecurity training. Employees must be skilled in identifying and dealing with possible threats. To guarantee that personnel are well-versed in reacting successfully to possible threats, consistent training and simulations should be implemented. It is essential to recognize that cybersecurity relates to the entire organization and not solely to the IT departments. Cybersecurity training should be provided to all individuals, regardless of their position in the organization.

Using a safe workplace is also important for developing a culture of security awareness and resilience. This involves putting in place security mechanisms like firewalls and antivirus software, as well as keeping software and systems up to date with frequent updates. As discussed before, companies should develop a contingency plan to address possible risks, including protocols for incident response and protecting duplicates of key data. The development of a secure place makes it far more difficult for bad actors to get unauthorized access to an organization's data and information, lowering the likelihood of success in cyberattacks.

It is worth noting that developing a culture of security awareness and resilience necessitates a paradigm change in cybersecurity thinking. Individuals must be aware of their online actions and understand their role in improving the organization's cybersecurity posture. This requires refraining from using weak passwords, refraining from clicking on questionable links or files, and refraining from disseminating sensitive data online. Individual proactivity and watchfulness about self-preservation and the protection of the organization against possible dangers are enhanced by recalibrating the attitude towards cybersecurity.

7 Trustworthy System

The concept of trustworthy systems assumes vital consequences within digital resilience. These systems are formulated and implemented to incorporate traits of security, dependability, and the ability to withstand the complex challenges posed by cyber threats. An example of a trustworthy system instils confidence in its users by executing its designated tasks as envisaged, safeguarding sensitive information, and producing outcomes that are both dependable and precise. Similarly, these systems operate in a manner that is open and responsible, allowing users to understand their inner workings and holding those responsible for their operation accountable.

Definition 5 (Trustworthy System) A trustworthy system is characterized by its capacity to reliably fulfil its intended functions devoid of errors, malfunctions, or prejudiced behaviour; such systems ensure security and operational transparency throughout their functioning.

The relevance of reliable systems has increased due to the growing number and complexity of cyber threats. Cyberattacks have far-reaching consequences, including financial repercussions, damage to an organization's reputation, and loss of consumer confidence. In the current times, reliable systems serve as a foundation for security and consistency, supporting the operations of numerous industries in a progressively digital society.

7.1 Fundamental Design Principles

The establishment of systems characterized by their trustworthiness assumes a position of significance in the strengthening of digital resilience. This significance is derived from the foundational role these systems play in a robust framework for the provision of secure and unwaveringly dependable digital services and operational functionalities. The fundamental crux of a system's dependability is inherently tied to the precision and rigidity with which its design follows the most elevated standards, completely surrounding domains such as security, privacy, reliability, transparency, and user-friendliness.

7.1.1 Security Aspect

Security constitutes a key aspect of a dependable system, acting as a shield against diverse cyber risks like malware, ransomware, and phishing attacks. The system should integrate multiple layers of security measures to detect unauthorized access, ensuring the confidentiality, integrity, and availability of data. Illustrative of secure systems are messaging platforms such as WhatsApp, Signal, and Telegram, which implement end-to-end encryption.

7.1.2 Privacy Element

Privacy forms another milestone of trustworthy systems. Honouring user privacy and safeguarding personal data from unauthorized access and exposure are important. A robust data protection framework should govern the system, in compliance with privacy laws like GDPR and CCPA. Notable examples of privacy-oriented systems include DuckDuckGo, a search engine devoid of user search tracking, and ProtonMail, an encrypted email service that abstains from data collection.

7.1.3 Reliability Factor

Dependability also hinges on reliability. A dependable system must exhibit consistency and operational correctness across all circumstances. Redundancy and failover mechanisms should underpin the system, ensuring uninterrupted service even amidst system failures or catastrophic events. Illustratively, Amazon Web Services offers cloud computing services characterized by high availability and fault tolerance, and Tesla's self-driving technology relies on redundant sensors and computational systems to ensure safe and reliable functioning.

7.1.4 Transparency Aspect

Transparency and accountability are equally fundamental for trustworthy systems. The system must articulate transparent policies and protocols governing user data and security incidents. Users should possess access to their data and the capability to regulate privacy settings. One specific instance of transparent systems includes Google's Transparency Report,[2] which provides information about government demands for user data and content removal. These systems enable users to have control over their privacy settings and the ability to download their data.

7.1.5 User-Friendliness Criterion

Trustworthy systems must exhibit user-friendly characteristics. They should feature easy-to-use interfaces that facilitate user interactions with minimal effort. Besides, the system should refrain from complex technical jargon, favouring clear and succinct instructions to reduce cognitive burdens on users. Examples of user-centric systems include Apple's iOS, marked by its straightforward and consistent interface across devices, and Dropbox, which seamlessly synchronizes and shares files across various devices and platforms.

For the achievement of digital resilience, organizations must accord primacy to the design and deployment of trustworthy systems. This effort includes embedding security, privacy, and resilience into the system's foundational design, subjecting it to security assessments and testing, and enforcing robust access controls and authentication mechanisms. An effective incident response and recovery strategy are equally indispensable to curtail the impact of cyberattacks. In guarding against established attacks, trustworthy systems must also be architected to mitigate

[2] https://transparencyreport.google.com/.

emerging threats, such as zero-day vulnerabilities and advanced persistent threats. A sustained focus on the design and implementation of trustworthy systems remains important to upholding digital resilience.

7.2 *Benefits of Trustworthy Computing Systems*

Dependable computing systems are important due to their vital role within modern society. The widespread integration of technology across multiple aspects of our lives, including communication, financial transactions, and daily routines, emphasizes the indispensability of these systems. Foremost among these factors is their capacity to assure the security and privacy of user data, as well as their capability to counteract vulnerabilities in the digital infrastructure.

Trustworthy computing systems play a central role in safeguarding sensitive information and preserving personal privacy. This is especially relevant when confidential data is routinely exchanged and stored in digital formats. The dependability of these systems substantially contributes to the prevention of unauthorized access, breaches of data, and instances of identity theft. Consequently, their presence acts as a foundational defence against potential cyber threats that possess the ability to cause substantial harm to both individuals and organizations.

The stress placed on trustworthy computing systems is fundamentally grounded in their aptitude to uphold the integrity and continuity of diverse operations. The reliability of these systems ensures the availability and functionality of crucial services, even in the face of unforeseen disruptions or malicious attacks. This resilience is important for mitigating the potential consequences of cyber incidents and for sustaining the seamless operation of critical functions such as online transactions, communication platforms, and data storage.

- **Preservation of Confidential Data**: A preeminent benefit attributed to trustworthy systems is the security of sensitive data. These systems are engineered to ensure the confidentiality and integrity of data. Their operational mechanisms are potent in diminishing the probability of data breaches, which can incur financial losses and reputational harm and diminish consumer trust.
- **Mitigation of Cyberattacks**: Another advantage inherent to trustworthy systems is their proficiency in anticipating cyberattacks. The increasing complexity of cyberattacks necessitates the existence of systems that possess the capability to detect and repel such threats. Trustworthy systems incorporate layered defensive mechanisms, including firewalls, intrusion detection systems, and encryption, to forestall unauthorized access and provide safeguards against malware, ransomware, and phishing attacks.
- **Enhanced Faith in Digital Infrastructures**: When users are confident in the protection of their data and the respect for their privacy, their trust in digital systems grows. This trust is essential for the expansion and adoption of digital innovations like electronic commerce, online banking, and cloud computing.

- **Compliance with Regulatory Frameworks**: Trustworthy systems facilitate compliance to regulatory frameworks, exemplified by GDPR and CCPA. These regulations mandate the safeguarding of users' personal information and privacy. Non-compliance with such stipulations may lead to grave legal consequences and financial penalties. Dependable systems integrate robust data protection protocols and procedures to ensure compliance with these regulatory mandates.
- **Uninterrupted Operations in Diverse Sectors**: Trustworthy systems contribute to the seamless operation of assorted sectors, including domains like healthcare, finance, and governance. These sectors heavily rely on secure and reliable information exchange, which trustworthy systems facilitate, ensuring uninterrupted processes.

The various benefits inherent in trustworthy systems include the protection of sensitive data, the deterrence of cyberattacks, the building of trust in digital infrastructures, compliance to regulatory standards, and the unhindered functionality of various sectors. Enterprises that place importance on the establishment and integration of trustworthy systems adeptly shield their data, clientele, and standing from potential harm.

7.3 Procedure

The establishment of a dependable system necessitates a course of action including design, development, and testing. This approach is essential to ensure both the system's reliability and its alignment with users' expectations. The essential aspects of this effort involve the integration of robust security measures, the upholding of system integrity, and the assurance of transparency within operational processes. Crafting systems with trustworthiness as a core attribute demands a methodical and cooperative effort, drawing upon the collective input of all involved stakeholders.

- **Definition of Requirements**: The foundational step in developing a trustworthy system involves the precise articulation of its requisites. This task involves describing the system's objectives, identifying its end-users, characterizing the nature of the data it will manage, and specifying the importance of security and privacy.
- **Architectural Design**: After defining the requirements, the system is subjected to the design phase. This phase involves the conception of a high-level architecture that outlines the constituent elements of the system, their interrelations, and the protocols governing their interactions. The architectural design also requires the inclusion of robust security and privacy mechanisms, key in safeguarding the system from potential threats.
- **System Implementation**: After finalizing the design, the system undergoes the implementation process. This stage involves coding of the system components, followed by their seamless integration into the overall framework. Strict compliance to the best security practices, inclusive of secure coding techniques,

encryption protocols, and strict access controls, is important throughout the implementation effort.

- **Validation and Verification**: Before transitioning to deployment, the system undergoes validation and verification procedures. These measures are in place to ascertain the system's alignment with stipulated requirements, as well as its robustness and security. Testing procedures including functional, security, and performance evaluations are important during this phase.
- **Deployment**: Following successful validation and verification, the system proceeds to the deployment stage. This requires the installation of the system within the production environment, coupled with meticulous configuration to align with end-users' operational needs. The deployment must follow the best security and privacy practices, including the strengthening of the system's defences and the organization of effective access controls.
- **Continual Maintenance and Vigilance**: Post-deployment, the system necessitates ongoing maintenance and vigilant monitoring. This ongoing effort includes consistent surveillance to detect and address potential security and performance issues. Timely rectification of identified issues, coupled with proactive updates to mitigate evolving threats and vulnerabilities, is major. This maintenance initiative is rooted in best practices about both security and privacy, inclusive of the timely implementation of software updates and patches.

The procedural framework stands as a guidepost for the creation of trustworthy systems within the domains of mathematics and computer science. Its emphasis on a systematic and collaborative approach is unequivocally meritorious, as it highlights the significance of involving all stakeholders in the goal of achieving trustworthiness. The structured approach outlined above, characterized by dedication to security, privacy, and robustness, resoundingly aligns with the foundational principles of cybersecurity and business risk management. Its assertive assertion of methodical rigour and cooperative engagement serves as a model for the creation of trustworthy systems.

7.4 Importance

With a society filled with cyber threats and privacy violations, the concept of trustworthiness functions as a fundamental principle that directs us towards technological solutions that maintain their robustness despite encountering obstacles. From securing sensitive data from hostile breaches to assuring regulatory compliance, the features of trustworthy systems resound across sectors, improving not just the security of our digital interactions but also the amount of faith we place in them. We strive to convey the multiple reasons why trustworthiness is the key basis upon which our technological growth must be built by a detailed investigation of practical instances and multifarious characteristics.

The rising digitalization of several industries has provided considerable ease, but it has also exposed us to countless threats. As we move towards a data-driven world, the critical need for developing systems that are both safe and trustworthy becomes abundantly evident. The following discussion will provide a thorough study of the basic issues that highlight the importance of trustworthiness in systems.

7.4.1 Safeguarding of Sensitive Data

Trustworthy systems are designed to provide impenetrable protection for sensitive data against unauthorized access, exploitation, or misuse. These systems are resilient due to a complex set of strong security mechanisms, such as encryption, firewalls, and intrusion detection systems. This robust security architecture serves as an impenetrable prevention to cyber hostile actors attempting to breach private data. This problem is especially important in firms entrusted with the custody of personal or financial data about their customers, such as financial institutions, healthcare providers, and e-commerce businesses.

To illustrate, financial organizations and healthcare establishments are burdened with the supervision of clients' personal and fiscal information. An intrusion into their digital infrastructure can give rise to severe consequences, including financial losses, identity theft, and medical fraud. A major example of this vulnerability surfaced in 2017, with the Equifax breach [Zou+18, Gre17] that resulted in the exposure of sensitive data of approximately 143 million individuals. This compromised information included their social security numbers, dates of birth, and residential addresses, resulting in a settlement of $700 million.

7.4.2 Mitigation of Cyberattacks

Systems grounded in trustworthiness are devised to handle cyberattacks, including but not restricted to malware propagation, ransomware schemes, and phishing stratagems. These systems are strengthened by a diverse set of defensive layers, including antivirus software, network surveillance utilities, and periodic security updates. This combination of defensive mechanisms collectively serves as a defence against the designs of malicious cyber actors. Given the escalating incidence of cyber intrusions, it becomes important for organizations to strengthen their digital infrastructure with systems that possess the tenacity to withstand such threats. For instance, in the year 2020, Microsoft coordinated the detection and prevention of the Democratic People's Republic of Korea cyber group that sought to exploit a zero-day vulnerability to infiltrate an entity operating within the cryptocurrency domain. Also, Apple's pioneering Face ID technology demonstrates the ability of utilizing advanced biometric authentication to repel unauthorized access to its devices.

7.4.3 Incubation of User Confidence

Systems characterized by trustworthiness are essential in infusing a sense of user assurance in the sanctity of their data and privacy. This proposition acquires significance in business enterprises reliant on digital substrates for the execution of their operations. The breach of this trust causes ramifications extending to damage to reputation and decrease in trust. The presence of trustworthiness within these systems provides users with a sense of security, thereby promoting heightened adoption and engagement levels. To describe, the utilization of end-to-end encryption within messaging platforms like WhatsApp and Signal develops secure communication channels. This, in turn, augments user trust and thereby catalyzes the increased adoption of these platforms.

7.4.4 Conformance with Regulatory Edicts

Systems characterized by trustworthiness are important for compliance to statutory mandates such as the General Data Protection Regulation (GDPR) [Com16] and the California Consumer Privacy Act (CCPA) [Sta18]. These regulations enjoin organizations to guarantee the safeguarding of personal data and privacy. Non-compliance to these mandates attracts penal measures spanning financial penalties and reputational impairment. The implementation of trustworthiness within systems efficiently ensures the organization's compliance with these regulatory importants while simultaneously saving them against penal consequences.

For instance, the GDPR mandates explicit consent acquisition from individuals before the procurement, storage, or processing of their data. Trustworthy systems, symbolized by examples like ProtonMail – a secure email service[3] based on encryption – seamlessly resonates with the GDPR's rule, and regulations, furnishing users with augmented layers of privacy while assuring compliance.

7.4.5 Facilitation of Seamless Sectoral Operations

Integral to the smooth operation of numerous industries – including finance, transportation, healthcare, and operations – are systems that are distinguished by their reliability. For example, in the domain of healthcare, trustworthiness promotes the protection and privacy of patient information. Within the domain of finance, these systems prevent occurrences of fraudulent activity and guarantee the confidentiality of financial transactions. Trustworthiness is an important factor in enhancing the safety and reliability of self-driving automobiles and similar autonomous vehicles within the transportation sector.

[3] proton.me.

For instance, Amazon Web Services (AWS)[4] offers cloud computing services that display high availability and fault tolerance, thereby furnishing enterprises with uninterrupted operations. Parallel to this, Apple's iOS interface presents a coherent and consistent navigational user experience (UX) across its many devices, resulting in an intuitive and seamless user experience.

7.5 Concept of Zero-Trust Systems

Zero-trust systems represent a paradigm shift in the field of cybersecurity. As opposed to traditional trustworthy systems that rely on the establishment of trust within a secure perimeter, zero-trust systems are based on the fundamental principle that no device or user should be trusted without first being validated.

This perspective presents an alternative to the conventional idea of dependable systems, which centres on the formation of trust within predetermined boundaries. In contrast, zero-trust systems function on the fundamental principle that trust cannot be automatically bestowed upon any user or device; thus, they necessitate the utilization of verification protocols. This introductory section provides an in-depth analysis of zero-trust systems, highlighting their core principles, operational mechanisms, and significance in ensuring digital resilience. A comprehension of zero-trust systems becomes crucial as we try to learn about present-day information security and to withstand the overflowing surge of cyberattacks and interruptions.

Zero-trust systems are defined as a security framework grounded in the postulation that every user or device demands verification before being accorded trust. This approach highlights the perpetual requirement for ongoing authentication and monitoring of all users and devices. Within the framework of a zero-trust system, all inbound network traffic undergoes scrutiny and authentication before gaining admittance to the network. Users are subject to authentication and authorization procedures before being endowed with access to sensitive data or systems. This methodology contributes an additional layer of security against both internal insider threats and external attacks.

Notable illustrations of zero-trust systems constitute Google's BeyondCorp,[5] which leverages identity and context-driven access controls to ensure secure access to applications and data, as well as Microsoft's Zero Trust Security model,[6] which utilizes a fusion of identity, devices, applications, network parameters, and data to substantiate and safeguard all endpoints.

In digital resilience, both trustworthy systems and zero-trust systems wield essential roles in safeguarding against cyber threats and the assurance of the availability, integrity, and confidentiality of digital systems and data. By embracing

[4] aws.amazon.com.

[5] cloud.google.com/beyondcorp.

[6] www.microsoft.com/en-in/security/business/zero-trustp.

these systems, entities can enhance their capacity to endure and recuperate from cyberattacks and other disruptive incidents.

Trustworthy systems constitute an indispensable aspect of digital resilience, serving as the foundation for secure, dependable, and accountable digital systems. The substantial financial consequences of cybercrime and the potential repercussions of cyberattacks emphasize the importance of primacy in the design and implementation of trustworthy systems. By adopting a risk-centric approach, conducting exhaustive security evaluations, and integrating best practices into system design and deployment, organizations can attain digital resilience even amidst evolving cyber threats.

7.6 Advantages of Zero-Trust Systems

The adoption of zero-trust systems in digital resilience heralds a series of distinct advantages and effects that collectively enhance the security posture and operational robustness of modern organizations. By deviating from the conventional perimeter-based trust model, zero-trust systems offer a more proactive and adaptable approach to countering ever-evolving cyber threats. This section outlines the compelling advantages and transformative effects that zero-trust systems bestow on digital environments.

- **Mitigation of Insider Threats**: One of the foremost advantages of zero-trust systems lies in their capacity to mitigate insider threats. By refraining from the assumption of implicit trust for users and devices within the network, these systems impose a continuous cycle of authentication and monitoring. This thorough scrutiny significantly reduces the potential for malicious actions by privileged insiders, thereby enhancing data integrity and confidentiality.
- **Resilience Against External Attacks**: Zero-trust systems strengthen an organization's resilience against external attacks by enforcing strict access controls and authentications. By necessitating the verification of each user and device, these systems establish a deterrent against unauthorized access attempts. Consequently, potential breaches are detected at the onset, diminishing the probability of successful cyber intrusions.
- **Adaptability in Dynamic Environments**: In contrast to traditional security models that struggle to accommodate the dynamism of modern IT environments, zero-trust systems exhibit inherent adaptability. As the digital infrastructure evolves with the integration of cloud services, remote work, and IoT devices, zero-trust principles remain firm. This adaptability ensures that security measures remain effective even in the face of organizational growth and technological shifts.
- **Enhanced Data Privacy and Compliance**: Zero-trust systems contribute to enhanced data privacy and regulatory compliance. By mandating continuous authentication and access validation, these systems assist organizations in adher-

ing to strict data protection regulations. Also, fine-grained access controls enable organizations to align their security practices with the principles of least privilege, restricting access to sensitive information solely to authorized entities.

- **Improved Incident Response and Detection**: The continuous authentication and monitoring of zero-trust systems to develop incident response and threat detection capabilities is one of the principal tasks. Suspicious activities are promptly identified through real-time monitoring, enabling swift countermeasures to be deployed. This proactive approach minimizes the dwell time of threats within the network and limits potential damages.
- **Shift Towards a Risk-Based Approach**: Embracing zero-trust principles facilitates a paradigm shift towards a risk-based approach to security. Organizations transition from a reactionary stance to a proactive stance, concentrating resources where vulnerabilities and threats are most likely to manifest. This strategic allocation optimizes resource utilization and enhances the overall efficacy of cybersecurity efforts.

7.7 Drawbacks of Zero-Trust Systems

While zero-trust systems have gained prominence as a great approach to cybersecurity and digital resilience, it is essential to acknowledge that their implementation is not without its share of drawbacks and consequential effects. This section discusses some of these drawbacks and their broader impacts on organizational operations and security strategies.

- **Complex Implementation Challenges**: A notable drawback of zero-trust systems lies in the complexity associated with their implementation. The shift from a perimeter-based trust model to a continuous authentication framework demands a substantial restructuring of existing IT infrastructure. This transition requires complex integration processes, and potential compatibility issues, and necessitates employee training to ensure seamless adoption.
- **Operational Overhead and Performance Impact**: The authentication and monitoring protocols inherent to zero-trust systems can impose an operational overhead on network resources. Real-time verification processes may lead to latency concerns, impacting user experience and system performance. Striking a balance between strict security measures and maintaining optimal system operations can be a delicate undertaking.
- **User Experience and Accessibility Challenges**: While the heightened security measures of zero-trust systems are designed to safeguard sensitive data, they can inadvertently introduce challenges to user experience and accessibility. The requirement for continuous authentication, particularly in remote or mobile scenarios, may lead to frustration and hinder productivity. Striking a balance between security and user convenience is a perpetual concern.

- **Resource-Intensive Nature**: The resource-intensive nature of zero-trust systems can be a drawback, particularly for organizations with limited budgets or technological capacities. Implementing and maintaining continuous monitoring tools, authentication mechanisms, and access controls necessitate substantial financial investments and ongoing operational commitments.
- **Potential for False Positives**: The continuous monitoring inherent to zero-trust systems may result in the generation of false positives – instances where legitimate user activities are flagged as suspicious. These false alarms can strain incident response teams, diverting attention from genuine threats and potentially creating a sense of complacency over time.
- **Integration with Legacy Systems**: For organizations relying heavily on legacy systems, the integration of zero-trust principles can be complex. Retrofitting older technology to align with modern security paradigms may pose compatibility challenges, potentially leaving critical components of the infrastructure vulnerable to exploitation.
- **Organizational Resistance and Cultural Shifts**: The adoption of zero-trust systems often requires a cultural shift within organizations. Convincing stakeholders of the necessity to question long-held assumptions about trust and embrace continuous authentication can cause resistance. Overcoming this resistance demands effective change management strategies and clear communication.

8 Balancing Security and Resilience

One can contemplate the present-day situation, wherein technology and digital interfaces are integrated with all aspects of life. We engage in a variety of digital activities, including social networking and online banking. Nonetheless, due to the widespread dependence on digital technologies, we must protect our digital assets from malicious individuals who aim to cause damage in the digital society. One significant obstacle in guaranteeing the protection of our digital assets involves finding an acceptable compromise between implementing strong security protocols and ensuring an intuitive user experience. One of the primary reasons why a strong level of security has significance is to prevent cybercrimes. On the contrary, an excessive emphasis on complexity for security may result in systems that are difficult to use, which would be an unintended consequence.

Throughout history, specific security protocols have been overly focused on security rather than user experience, resulting in interfaces that are complicated for users to use. However, modern sensibilities acknowledge the equal importance of security and user-friendliness, which has led to a fundamental change in the way we tackle digital security. For example, consider mobile finance applications. During previous versions, users were required to complete complicated authentication procedures that involved a multitude of codes and passwords. Despite the robust financial security provided, the functionality of these applications resembled a conundrum. As a result, numerous people pursued more straightforward alternatives.

However, forward-thinking entities within the financial sector implemented cutting-edge functionalities into their applications, including biometric scanning and facial recognition. Through this integration, account security is not only enhanced but also the user experience is optimized. As a result, the present-day setting has been enhanced with digital artefacts that seamlessly integrate accessibility and security. The continuous struggle between security and user-friendliness becomes increasingly critical as we develop digital systems that are prepared to withstand cyber threats. Although security functions as a barrier against digital misconduct, an excessive amount of complexity causes complications. Given the broadening range of activities that occur on the Internet, the pursuit of desirable stability has become more and more important.

8.1 Security and Usability in Digital Systems

Our culture is now heavily immersed in the world of the Internet, which covers various activities such as social networking and online financial transactions. Because of the rising prominence of digital systems in our daily lives, increased security standards must be implemented to counteract the immediate threats presented by cyberattacks. The delicate balance that must be maintained between security and usability is a fundamental difficulty that emerges during the construction of secure digital systems. Although robust security measures are essential for identifying cyber breaches, too complex security frameworks may generate usability concerns that impair the system's operating effectiveness.

Throughout history, a large number of security systems have been designed with an overemphasis on security, usually at the expense of usability. This strategy often resulted in complex systems that were difficult for users to use, resulting in user discontent and low user approval. However, there has been a growing acknowledgement of the critical role of usability in digital systems in recent years, prompting a paradigm change in the way security is treated. The growth of mobile financial apps is an appropriate exemplification of this transformation. Historically, a lot of mobile banking apps require users to go through a lengthy verification process that includes entering many security codes and passwords. While these security features effectively secured user information, they also made the program difficult to operate. As a result, a significant number of users chose more intuitive replacements, resulting in the discontinuance of those programs.

Financial institutions have improved their mobile banking apps to overcome the problems that result from the trade-off between security and usability. These businesses have effectively utilized biometric authentication technologies, such as face and fingerprint recognition, to offer secure account access while providing an amazing customer experience. This technique has resulted in the development of digital systems that efficiently blend user-centric design with thorough security measures. When building cyber-resistant digital systems, the combination of security and usability needs key considerations. Although security measures are necessary to guard against intrusions, excessive complexity in security systems may result in usability concerns that jeopardize the system's performance. As a result, establishing an optimum balance between usability and security is a crucial effort in the creation of digital systems that successfully convey both security and convenience of use, both of which are essential concerns today.

8.2 *Architecting Resilient Systems*

In response, the importance arises for the formulation of systems that possess the dexterity to swiftly rebound from such incidents while sustaining their availability and usability. This concept is encapsulated within the framework of digital resilience. Digital resilience signifies the capacity of digital systems to endure and recuperate from cyberattacks and security breaches. It includes the strategic design of systems capable of promptly detecting and responding to security incidents, expediently restoring services, and forestalling the recurrence of analogous incidents. A major motivation underpinning the construction of resilient systems is the mitigation of the repercussions of security incidents on business operations. To construct a resilient system, the following steps ought to be undertaken:

- Identification of conceivable security threats and vulnerabilities
- Implementation of multi-tiered security measures to detect and identify security incidents
- Formulation of incident response protocols and procedures
- Regular execution of security assessments and vulnerability appraisals
- Provision of training to employees concerning optimal cybersecurity practices and the importance of reporting security incidents

Microsoft is a compelling instance of an organization that has embraced a resilient system. The company has established a Security Development Lifecycle (SDL)[7] regimen that includes threat modelling, security testing, and incident response strategies. Microsoft has developed an extensive network of security professionals and collaborators to ensure the safety of their systems against emerging threats.

The blueprinting of systems capable of swift recovery from security incidents while concurrently preserving availability and usability emerges as an indispensable aspect of digital resilience. Establishing a reliable system that can quickly recover from security breaches is essential for protecting sensitive information, preventing cyber threats, and building confidence in digital infrastructures. Consequently, organizations have an obligation to guarantee the resilience and robustness of the systems they construct, which will enable them to face the challenges of the modern society with efficacy.

[7] microsoft.com/en-us/securityengineering/sdl.

9 Safeguarding Digital Assets

The term 'digital assets' refers to valuable electronic resources owned by individuals, organizations, or entities. These include various assets, spanning from sensitive personal data and financial information to proprietary software codes and creative content. Ensuring the security and integrity of these digital assets is important due to the potential consequences of their compromise, including unauthorized access, data breaches, financial losses, and infringement upon intellectual property rights.

9.1 Value of Digital Assets and Impact of Loss or Compromise

Digital assets have evolved into a key component for both organizations and individuals. These assets include sensitive data, intellectual property, financial records, personal information, and other vital assets that are indispensable for businesses, governments, and individuals to operate and remain competitive within the global economy. The dependency on digital assets has created an urgent necessity to recognize their value and the potential effects of their loss or compromise.

The loss or compromise of digital assets can result in grave consequences for organizations and individuals alike. Consequences such as financial losses, damage to reputation, legal accountability, and even personal harm can arise as a result of cyberattacks, data breaches, and other digital attacks. The repercussions of such incidents can be severe, leading to significant disruptions in business operations and causing a decline in revenue, productivity, and customer confidence. The loss or theft of personal information can lead to identity theft, financial fraud, and other detrimental outcomes for individuals.

To mitigate these risks, it is important to implement robust security measures and uphold digital resilience. Organizations and individuals must formulate strict security policies, regularly conduct security audits, and offer security training to personnel to safeguard digital assets against cyber threats. These measures must be ingrained in the organization's culture and be an integral part of its overall business strategy.

One of the fundamental security measures for protecting digital assets is encryption, which will be thoroughly expounded upon in the subsequent section. Encryption is a widely used security measure for safeguarding sensitive data, including financial information, personal data, and other critical assets. Encryption serves to shield digital assets from unauthorized access, theft, and interception by malicious actors. Similarly, it preserves the privacy and confidentiality of sensitive information, establishing trust between entities that handle such data. Encryption is necessary in ensuring compliance to regulatory mandates concerning data protection and privacy. Non-compliance with these regulations can result in substantial fines, legal liabilities, and impairment of reputation.

10 Role of Encryption in Safeguarding Digital Assets

Cryptographic encryption stands as a foundational technique used to develop data security, safeguard digital assets, and develop trust among individuals and entities. By employing this method, plaintext data is transformed into ciphertext, an unintelligible format that requires a confidential key to decipher. The significance of encryption in shielding sensitive data from unauthorized access, theft, and interception by malicious actors cannot be overstated. The encryption process ensures that even if a cybercriminal manages to obtain the ciphertext, they are unable to interpret the information without the corresponding secret key. This confers considerable potency to encryption as a security measure, extending protection to financial data, personal particulars, and other critical assets.

Beyond its role in safeguarding digital assets, encryption also contributes to upholding the privacy and confidentiality of sensitive information. This assumes particular prominence given the escalating prevalence of data breaches and cyber intrusions. Through data encryption, both individuals and organizations can ensure that solely authorized entities possess the capability to access and peruse the information.

The implementation of encryption is essential not only for digital asset protection but also for compliance to regulations governing data privacy and security. Notably, regulations like GDPR and HIPPA [Com16, Con96] mandate the adoption of robust encryption protocols to safeguard sensitive data against unauthorized access and ensure its secure retention. Neglecting these regulatory stipulations can bring about severe consequences, including legal actions and substantial financial penalties.

To institute robust encryption measures, various factors warrant consideration by organizations and individuals alike. These factors include the establishment of secure communication channels like SSL/TLS, the utilization of potent encryption algorithms such as AES or RSA, and the application of encryption technologies like disk encryption and file-level encryption to secure data storage. Ensuring the encryption of data transmission, particularly when transmitting sensitive information across public networks, is significant.

10.1 Homomorphic Encryption

Homomorphic encryption emerges as an advanced cryptographic technique offering the ability to conduct secure computations on encrypted data without necessitating decryption. This innovative strategy addresses a significant challenge in data security: the ability to perform computations while preserving the confidentiality of the underlying information. Figure 2.12 showcases the schematics of holomorphic encryption for better comprehension.

The central concept of homomorphic encryption is its capability to perform mathematical operations on ciphertext, producing results consistent with those from

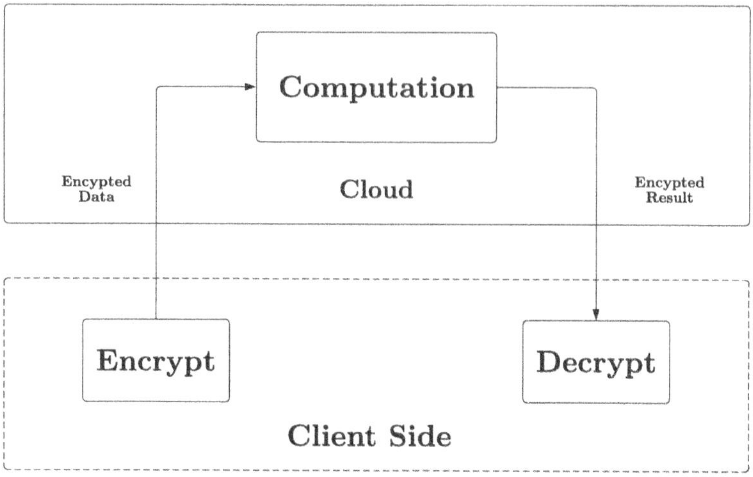

Fig. 2.12 Homomorphic encryption in cloud

plaintext operations. This paradigm shift disrupts the conventional requirement for decryption for computation purposes, thereby minimizing data exposure during computation and its associated vulnerabilities. Homomorphic encryption enables mathematical operations on ciphertext, yielding results similar to analogous plaintext operations. This transformative approach overturns the traditional mandate of decrypting data before computation, effectively mitigating vulnerabilities tied to plaintext exposure during processing.

Homomorphic encryption facilitates the delegation of computations to a third party while preserving the confidentiality of sensitive information. Formally, let E denote the encryption function, D the decryption function, and M the message space. For a plaintext message m in M, the ciphertext c is obtained as follows: $c = E(m)$. The decryption of c yields the original plaintext: $m = D(c)$. The homomorphic property is expressed as $E(m_1 \oplus m_2) = E(m_1) \oplus E(m_2)$, where $\oplus$ denotes a bitwise addition modulo 2.

In a homomorphic encryption scheme, there exist three fundamental operations: addition, multiplication, and a negation operation for additive homomorphism. Let $E(m_1)$ and $E(m_2)$ represent the encrypted forms of messages m_1 and m_2, respectively. The homomorphic property for addition is expressed as $E(m_1 + m_2) = E(m_1) \oplus E(m_2)$, and for multiplication as $E(m_1 \cdot m_2) = E(m_1) \otimes E(m_2)$, where $\otimes$ denotes an operation dependent on the specific homomorphic encryption scheme.

The security of homomorphic encryption is often grounded in the presumed hardness of mathematical problems, such as the integer factorization problem or the decisional learning with errors (DLWE) assumption. The security parameter is typically denoted by λ, and the scheme is considered secure if an adversary with polynomial computational resources cannot distinguish between encrypted messages or learn any information about the plaintext.

Homomorphic encryption operates through various schemes tailored to specific computation categories. Partial homomorphic encryption allows addition or multiplication operations on encrypted data. In contrast, fully homomorphic encryption (FHE) extends this to cover both addition and multiplication. FHE stands out as a significant advancement, enabling complex calculations on encrypted data, and includes tasks such as sorting, searching, and advanced ML algorithms.

The significance of homomorphic encryption becomes evident in scenarios where sensitive data requires computational processing while preserving its confidentiality. Industries like healthcare, finance, and data analysis stand to benefit significantly from this breakthrough. For instance, in healthcare, researchers can collaboratively analyse encrypted patient data, upholding the confidentiality of individual health records. Similarly, financial organizations can perform encrypted computations involving transactional data, enhancing security while deriving valuable insights.

However, it's important to acknowledge that applying homomorphic encryption introduces computational overhead due to complex mathematical operations. The inherent complexity of these operations can impact processing times, making the technique less efficient compared to traditional non-encrypted approaches. Despite this limitation, ongoing research and technological advancements aim to optimize and expedite homomorphic encryption techniques. The ultimate goal is to strike a balance between robust data security and computational efficiency. FHE extends this capability to include both addition and multiplication. FHE represents a groundbreaking advancement, as it permits complex calculations to be conducted on encrypted data, spanning functions as diverse as sorting, searching, and advanced ML algorithms.

The practical significance of homomorphic encryption becomes particularly pronounced in scenarios where computations on sensitive data are important while upholding strict privacy mandates. Industries such as healthcare, finance, and data analysis stand to reap substantial benefits from this technology. For instance, medical researchers can collaboratively analyse encrypted patient data, thereby facilitating data-driven insights without compromising individual medical records' confidentiality. In the financial sector, encrypted computations can be applied to transactional data, developing security while enabling valuable analytical insights.

It is important to acknowledge that homomorphic encryption introduces computational overhead due to its complex mathematical operations. This computational complexity can impact processing times, rendering the technique less efficient when compared to traditional non-encrypted computations. Despite this trade-off, ongoing research efforts and technological advancements are focused on optimizing and accelerating homomorphic encryption techniques, aiming to strike a balance between robust security and computational efficiency.

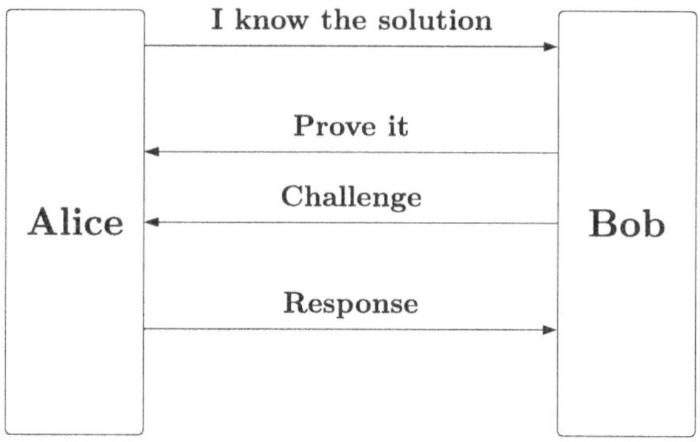

Fig. 2.13 Zero-knowledge proof

10.2 Zero-Knowledge Proof

Zero-knowledge proof (ZKP) is a collection of cryptographic techniques based on interactive proofs that facilitate the confirmation of information without revealing the actual data. The inception of zero-knowledge proof dates back to 1985, when Goldwasser-Micali-Rackoff [Gol+89] introduced it as a classical cryptographic concept. Historically, due to its considerable computational complexity, zero-knowledge proof had been deemed impractical. However, recent strides have enhanced its efficiency, unlocking novel paradigms in the development of trustless and distributed systems. This evolution positions zero-knowledge proof as a critical technology for the future trajectory of blockchain systems. The most basic working of the zero-knowledge proof has been showcased in Fig. 2.13 for better understanding.

Within a zero-knowledge proof protocol, a 'verifier' is presented with evidence by a 'prover' that a computational statement holds, all while abstaining from revealing the actual data. This process is executed through a sequence of cryptographic algorithms that permit the verification of computations without disclosing the underlying data. Zero-knowledge proof presents various potential applications across diverse domains, including cryptography, cybersecurity, and privacy. An eminent benefit of zero-knowledge proof is its capacity to ensure the accuracy of data and computations while preserving the confidentiality of information. This aspect is particularly key in contexts where sensitive information exchange transpires among untrusting parties.

The black-box definition of ZKPs highlights the information-theoretic aspect of the proof, considering the scenario where the verifier is modelled as a probabilistic polynomial-time algorithm with access to an oracle. Let P denote the prover, V the verifier, and x the statement in question. In a black-box ZKP, the verifier interacts with the prover by sending queries to an oracle that simulates the computation of the prover. The oracle provides responses based solely on the information available to the prover. For a statement x to be proven by ZKP, there exist three essential properties:

- **Completeness**: If the statement x is true, an honest prover P can convince the verifier V with high probability.
- **Soundness**: If the statement x is false, no efficient adversary, even one with access to the oracle, can convince the verifier V with more than a negligible probability.
- **Zero Knowledge**: The verifier V learns nothing about the statement x other than its validity. Formally, for any efficient adversary A attempting to learn additional information about x, there exists a simulator S that can simulate the interaction between the prover P and the verifier V in a way that is indistinguishable from real interaction.

In the black-box model, the emphasis is on the limited information available to the verifier, treating the prover's computation as a black box. The security of ZKPs often relies on the hardness of certain mathematical problems, such as discrete logarithms or factorization.

To illustrate, consider the application of zero-knowledge proof in password authentication systems. In this scenario, the prover can convince the verifier of their knowledge of a password without actually disclosing the password itself. Broadly, zero-knowledge proof embodies a key cryptographic instrument that sanctions the verification of computations without divulging sensitive data. Its applications span from password authentication mechanisms to safeguarding the privacy of data analytics. The advancements and breakthroughs within zero-knowledge-proof technology are poised to exert an influence on the establishment and enhancement of secure and reliable systems.

The foundational premise of a 'zero-knowledge proof' rests upon the principle of validating knowledge without unveiling it. To merit classification as an authentic zero-knowledge proof, the prover must fulfil three critical criteria. Firstly, the prover must establish completeness, convincing the verifier of their possession of the claimed knowledge. Secondly, the proof must comply to soundness, ensuring the verifier's inability to deceive him with false information. Lastly, the proof must follow the principle of zero knowledge, ensuring that no additional information beyond the ultimate result is disclosed to the verifier. Zero-knowledge proofs empower the substantiation of specific data truths without transmitting them to a third party. As an illustration, users can validate their eligibility for a service without divulging their exact age or demonstrate sufficient income without disclosing the precise monetary amount. This distinctive attribute renders zero-knowledge proofs a potent tool for identity authentication systems, where personal data is not required to validate an individual's identity.

Zero-knowledge proofing can be defined into two categories: interactive and non-interactive. Interactive zero-knowledge proofs necessitate the prover to execute a series of actions to persuade the verifier of a certain assertion. However, this class of proofing exhibits limitations in terms of transferability, as the entire process must be repeated to substantiate the same assertion to a different verifier. Non-interactive zero-knowledge proofs, on the other hand, enable the prover to provide evidence that is accessible to all. In this approach, the prover is tasked with solving a challenge chosen at random by the verifier. By employing a hash function to determine the challenge, an interactive proof can be converted to a non-interactive one, as demonstrated by cryptographers Fiat and Shamir. This eliminates the need for iterative communication between the prover and verifier, allowing the proof to be contained in a single message from the prover to the verifier.

Zero-knowledge proof also finds utility in online gaming and gambling, certifying the fairness of game outcomes and precluding manipulation by any involved party. Zero-knowledge proof is pertinent in identity verification systems, where individuals can substantiate their identity without disclosing personal information. Privacy-preserving cryptocurrencies are leveraging zero-knowledge proof for their development, ensuring transactional privacy. Through zero-knowledge proofs, users can validate the adequacy of funds for transactions without revealing their account balances. Also, zero-knowledge proofs are advancing blockchain scalability through concepts like zkRollups and zkEVMs, enabling the processing of numerous transactions without necessitating universal validation by all network nodes.

The consequences of zero-knowledge proofs extend across a spectrum of domains, including finance, healthcare, and identity verification. In the financial domain, zero-knowledge proofs enable confidential transactions without disclosing sender, receiver, or transaction amount details. In healthcare, they facilitate secure data sharing among stakeholders while upholding patient privacy. Identity verification benefits from zero-knowledge proofs by permitting secure authentication without involving the exchange of sensitive personal data.

Zero-knowledge proof emerges as a potent technology with the capacity to revolutionize trust establishment and privacy assurance mechanisms. Its adeptness in ascertaining data and computational validity without compromising confidentiality renders it invaluable across diverse applications, from identity validation to online gaming and beyond. Given the evolving trajectory in this domain, zero-knowledge proofs are poised to play a vital role in the future trajectory of distributed systems and blockchain technology.

10.3 Multi-party Computation

Multi-party computation (MPC) [Yao86, BW88] is a cryptographic protocol designed to facilitate the secure computation of distributed data while preserving its confidentiality. Its beginning is rooted in the importance of secure computation in scenarios where sensitive data is involved. Consider a scenario wherein a group of five individuals within a corporate context aims to collectively compute their average salary while safeguarding the privacy of their salary figures. Traditional methodologies might lead to the intervention of a trusted third party to facilitate the calculation; however, concerns about trust and potential data exposure can undermine the efficacy of such an approach. Multi-party computation offers a solution by fragmenting the computation process itself as opposed to exposing the raw data necessary for computation.

Within multi-party computation, multiple parties collaboratively engage in computation to derive a result while shielding their respective inputs. Each participant oversees a specific component of the computation procedure, yet no single entity possesses access to the entire set of inputs. The final output becomes accessible to all participants, but the individual inputs remain undisclosed, thereby ensuring the data's confidentiality. Multi-party computation has notably gained traction in digital assets, where it can supplant the reliance on individual private keys for transaction signing. This involves the distribution of the signing process across multiple machines, each holding a portion of private data representing a fractional key segment. This method guarantees that no single entity holds full access to the private key, thereby curtailing the vulnerability to compromise.

MPC involves defining the protocol in terms of parties, input spaces, a function to be computed, and a secure computation mechanism. Let $P_1, P_2, \ldots, P_n$ represent n parties participating in the MPC protocol. Each party P_i holds a private input x_i from an input space X_i. The objective is to jointly compute a function $f : X_1 \times X_2 \times \ldots \times X_n \rightarrow Y$ over the private inputs, where Y is the output space. Hence, the MPC protocol is defined by a set of communication and computation steps:

(1) **Input Sharing**: Each party P_i privately splits their input x_i into shares or parts, denoted as $x_{i_1}, x_{i_2}, \ldots, x_{i_k}$, where k is a security parameter. These shares are distributed among the parties in such a way that no single party can reconstruct the input on its own.
(2) **Communication Rounds**: The parties engage in a series of communication rounds to exchange information. This communication is typically done securely to prevent any individual party from learning more than what is revealed by the protocol.
(3) **Secure Computation**: The parties perform secure computations using the shared inputs to compute intermediate results. This involves using cryptographic techniques such as secret sharing, oblivious transfer, or homomorphic encryption to ensure that no party gains information about the inputs of others.
(4) **Output Reconstruction**: Finally, the parties collaboratively reconstruct the output y without revealing any additional information about their private inputs. The reconstructed output is then jointly held or revealed based on the application.

The security of MPC is often defined in terms of the adversary's computational power and its ability to gain information about the private inputs. The goal is to ensure that the protocol is secure against malicious adversaries attempting to deviate from the prescribed steps or gain unauthorized information. A distinctive advantage of multi-party computation is its obviation of the need to repose trust in third parties for safeguarding data integrity and security. By retaining data within internal firewalls while upholding privacy, multi-party computation enables the exhaustive use of analytical capabilities. Also, encrypted data used in multi-party computation remains resilient against quantum attacks, bestowing an additional layer of security. However, the technique does require computational overhead and elevated communication costs due to factors like random number generation and the requisite relationship among participating entities. Despite these challenges, multi-party computation has successfully materialized for data scientists, offering a fusion of data privacy and precision while aligning with GDPR and sovereign data privacy mandates.

The intensifying digitization of technology has precipitated the rapid dissemination of sensitive information. Multi-party computation is emerging as a safeguarding measure, functioning similarly to a digital non-disclosure agreement to govern information exposure. For instance, in genetic testing, multi-party computation allows individuals to scrutinize their genetic profiles without divulging sensitive data to governmental entities or insurers. In sealed-bid auctions, multi-party computation guarantees the complete confidentiality of each submitted bid. Similarly, sensitive research empowers secure data collection and analysis without the need to share sensitive information with external parties.

Consequently, secure multi-party computation has transitioned from a theoretical proposition to practical implementation, endowing data privacy and accuracy in collaboration with compliance to GDPR and sovereign data privacy requisites. The escalating deployment of multi-party computation across various sectors serves as a bastion for protecting sensitive data and modulating the extent of information disclosure. While computational overhead and elevated communication expenses persist, multi-party computation provides a secure avenue for data analysis that does not compromise privacy. With the progression of technology, the prevalence of multi-party computation is poised to flourish within industries necessitating secure computation, thereby guaranteeing the confidentiality of sensitive data.

11 Enhancing the Robustness of Digital Assets

The need to guarantee the consistent longevity of digital assets has acquired crucial relevance. The complex relationship between the widespread use of technology and the increasing range of risks highlights the need to strengthen digital assets to mitigate possible vulnerabilities. Through a thorough examination of the complex aspects of resilience and its interconnectedness with digital assets, this discussion aims to develop a full comprehension of the essential processes that contribute to enhancing the resilience of these vital components in our modern society.

11.1 Robust Access Controls and Authentication Protocols

In the modern society, the increase in cybersecurity threats has rendered the implementation of robust access controls and authentication protocols indispensable. These measures are fundamental in strengthening the security and robustness of digital systems and in shielding sensitive information from cyberattacks.

Access controls serve as the mechanisms that reduce and regulate user entry to digital resources, including information systems, networks, and applications. Through this process, they ensure that access to these resources is restricted to authorized users only, thus effectively reducing the potential for unauthorized intrusion and data breaches. Authentication protocols, conversely, are used to validate the identity of users and to substantiate their claimed identities.

The deployment of potent access controls and authentication protocols enables organizations to diminish the threat of unauthorized intrusion into their digital systems and confidential information. Consequently, the utilization of robust authentication measures such as dual-factor authentication or biometric authentication can markedly mitigate the vulnerability to cyberattacks.

An example that demonstrates the effectiveness of robust access controls and authentication protocols is discerned in Google's Advanced Protection Program.[8] This initiative imparts supplementary security attributes to shield high-risk users, including journalists, activists, and political figures. The program incorporates hardware security keys, necessitating physical access to the user's device, alongside sophisticated anti-phishing safeguards and automated security updates.

Another illustrative case involves the integration of biometric authentication, which relies on distinct physiological or behavioural attributes to verify a user's identity. This technology is increasingly being embraced by mobile devices and financial applications, imparting an additional layer of security that surpasses traditional password-based authentication.

The implementation of strong access controls and authentication protocols is vital in upholding digital robustness and countering cyber threats. Given the sustained reliance on digital systems by organizations and the increasing scale of sensitive data at stake, it is important to ensure that access is granted solely to authorized users and that their identities are unequivocally corroborated. Through the implementation of these measures, organizations can control the threat of data breaches and improve their overall cybersecurity stance.

11.2 Safeguarding Data Through Backup Strategy

The need to implement data backup procedures cannot be overstated. It is an essential component of digital resilience, with the ability to mitigate the consequences of data loss caused by cyber breaches, hardware problems, or unforeseen circumstances. The primary objective of this discussion is to look into the relevance of backing up critical information and to highlight the several channels through which it contributes to the reinforcement of digital resilience.

[8] landing.google.com/advancedprotection.

11.2.1 The Pervasive Ramifications of Data Loss

The repercussions of data loss are universally consequential, impacting both individuals and entities at large. The forfeiture of essential data can precipitate financial detriment, legal entanglements, and even besmirchment of reputation. Hence, individuals and organizations alike must undertake proactive measures against data loss by embracing a systematic approach to backing up critical data. Data loss can manifest through diverse conduits, such as hardware breakdowns, software deterioration, human fallibility, cyberattacks, natural catastrophes, or theft. Irrespective of the aetiology, the repercussions of data loss are severe and far-reaching.

11.2.2 Implementation of Robust Data Backup Measures

Deploying robust critical data backup measures is one of the most productive strategies to preempt data loss and expedite recuperation in the event of catastrophic circumstances. The practice of data backup involves the creation of duplicate copies of important data, stored in a distinct location from the primary repository. This protocol ensures that, in the aftermath of data loss, the duplicated data can be reclaimed for data restoration.

Numerous methodologies for executing critical data backups exist, including the utilization of external hard drives, cloud-based storage services, and network-attached storage (NAS) devices. Each approach possesses its various merits and demerits, thereby warranting careful selection based on the specific requisites of individuals or organizations.

For instance, an individual might opt to utilize an external hard drive to preserve backups of their personal computer's data. By interfacing the external hard drive with their computer and employing specialized software, they can create backups of their files. Subsequently, the external hard drive can be securely stored in a designated location, such as a safe or a bank vault, to preclude damage or theft.

Conversely, an organization might opt for a cloud-based storage service to safeguard its critical data. Prominent services like Dropbox, Google Drive, and OneDrive present a seamless and cost-effective solution for data storage and accessibility across diverse geographical locations. These platforms also furnish advanced security features, including encryption and multi-factor authentication, to prepare against potential cyber threats.

11.2.3 Backing Up Critical Data

The act of backing up critical data serves as a foundational pillar of digital resilience, bringing about quick recovery from data loss incidents and minimizing operational downtime. To illustrate, consider the scenario of a medical organization that houses critical patient data within its server infrastructure. In the event of server failure without adequate data backup, the outcome would lead to data loss, protracted downtime, and an imminent threat to patient well-being.

11.2.4 Safeguarding Essential Data via Backup

Organizations possess the capability to adopt a variety of optimal methodologies, including the subsequent key aspects:

(1) **Regular Backup Procedures**: Entities should perform scheduled backups of vital data to ensure the availability of the most recent version of the data.
(2) **Placement in External Repositories**: Backup files should be stored in a separate repository, ideally located offsite, as a precautionary measure against potential losses caused by events such as fires, floods, or theft.
(3) **Implementation of Cryptography**: Data backups must undergo encryption protocols to serve as a security measure against unauthorized access.
(4) **Assessment and Authentication**: Organizations are encouraged to periodically evaluate and verify their backups to confirm their suitability for recovery purposes and alignment with current datasets.
(5) **Integration of Automation**: Incorporating automated backup solutions offers a practical approach for organizations, enabling them to save time while maintaining a consistent and regular backup routine.

The initiative of safeguarding critical data through backup strategies is vital for digital resilience. This practice entrusts organizations to swiftly recover from data breaches, thereby reducing instances of operational downtime. This essential undertaking forms the foundation for ensuring uninterrupted business operations, instilling confidence among clients, and fulfilling legal and regulatory obligations. By adhering to the optimal practices concerning the protection of critical data, organizations are well-prepared to confront challenges with readiness and adaptability.

12 Incident Response

The concept of 'incident response' is significant, ready to detect and mitigate the impact of security breaches and data compromises. This section discusses the complexity of the incident response, classifications, and strategic execution.

Incident response refers to the methodical procedure of identifying, confining, and reducing the ramifications of a security incident or data breach. It involves a coordinated effort involving diverse stakeholders, including IT and security teams, law and regulations professionals, and external vendors. Their actions are aimed at quickly addressing the incident and mitigating its consequences for the organization. Refer to Fig. 2.14 for the NIST incident response cycle to understand how a basic IRC should function.

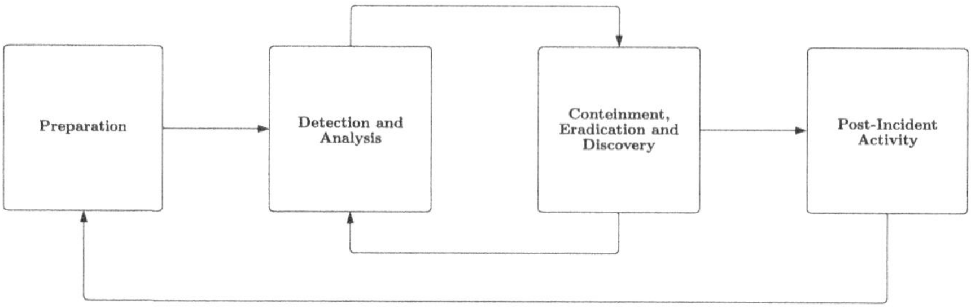

Fig. 2.14 NIST incident response cycle

Table 2.1 Incident response classifications

Incident response classification	Definition	Example
Level 1: minor incident	An incident yielding low impact on the organization and amenable to swift and facile resolution	A minor phishing attack affecting a solitary employee was promptly contained
Level 2: moderate incident	An incident causing a moderate impact on the organization, necessitating a commensurate level of response	A malware infection impacting multiple systems, requiring extensive cleansing and recovery efforts
Level 3: major incident	An incident precipitating noteworthy impact on the organization, demanding a heightened level of response	A ransomware attack encrypts crucial data, disrupting business operations, necessitating a coordinated effort to mitigate the attack and reinstate data
Level 4: catastrophic incident	An incident resulting in severe repercussions for the organization, mandating immediate and all-including response	A cyberattack culminating in the loss of sensitive data or critical systems, resulting in substantial financial and reputational detriment

Incident response is an essential component of digital resilience, granting organizations the ability to mitigate the consequences of cyber threats. It involves the implementation of a methodical approach to identify, mitigate, and recover from security incidents, such as violations of confidentiality and cyberattacks. The importance of incident response has become increasingly apparent in light of the growing frequency and complexity of cyberattacks. When incident response is addressed in organizational scenarios, it is commonly categorized using a threat matrix; Table 2.1 illustrates the following.

A well-thought-out incident response plan should include the following stages:

- **Preparation:** This stage involves establishing the parameters of the incident response strategy, determining the significance of critical resources and information, and assigning duties and obligations to members of the incident response team. Protocols for communicating with stakeholders, such as partners, customers, and regulatory bodies, should also be incorporated into the plan.

- **Detection and Analysis:** Herein, systems are vigilantly monitored for indicators of security incidents, such as anomalous network activities or unauthorized access efforts. Incident response team members should be adept at promptly identifying potential threats and evaluating their repercussions for the organization.
- **Containment and Eradication:** Post-identification of a security incident, the incident response team is tasked with constraining the impairment and eliminating the threat. This may include isolating affected systems, purging malicious code, and restoring data from backups.
- **Recovery and Post-incident Analysis:** Following containment and eradication, the organization pivots to restoring normal operations and conducting post-incident analysis. This appraisal should uncover the underlying cause of the incident, gauge the efficacy of the incident response plan, and ascertain avenues for enhancement.

An illustrative instance of effective incident response is the countermeasure to the 2017 WannaCry ransomware attack. This cyber onslaught had far-reaching consequences, touching over 200,000 computers across 150 nations. The aftermath wrought extensive disruptions to critical systems and business entities. The response efforts of organizations like the UK's National Health Service (NHS) [Col17] proved key in reducing the malware's propagation and mitigating its fallout.

Incident response stands as an indispensable aspect of digital resilience, serving to reduce the impact of cyber threats on organizations. By following a methodical incident response approach, organizations can adeptly identify, confine, and recuperate from security incidents, thereby shielding vital assets and upholding the trust of stakeholders.

12.1 Digital Forensics and Incident Response (DFIR)

The primary priority in cybersecurity is mostly implementing preventive measures to deter any security breaches. The significance of these preventive measures is undeniable, yet currently the cyberattacks have a rising occurrence rate and increased complexity. As a result, the occurrence of a cybersecurity event within an organization is now a matter of inevitability rather than uncertainty.

Given the current circumstances, it is important to redirect our focus towards understanding the essential measures required for the management and recuperation of these occurrences. This section provides an examination of digital forensics and incident response (DFIR) in cybersecurity. It explores the fundamental aspects of DFIR, including its definition, procedural components, significance in cybersecurity, necessary skills, and indispensable technologies.

Digital forensics and incident response is a field of study and practice that includes the collection, preservation, analysis, and presentation of digital evidence to investigate and respond to cyber incidents. DFIR involves the application of scientific and investigative techniques to identify, extract, and interpret digital artefacts and data from various sources. DFIR represents a highly specialized domain that is closely engaged in the identification, mitigation, and investigation of cybersecurity problems. DFIR, short for digital forensics and incident response, comprises two essential components: digital forensics and incident response.

This particular component relates to the methodical gathering, safeguarding, and thorough examination of forensic evidence obtained from digital sources. The basic aim of digital forensics is to uncover significant evidence, similar to the approach taken in traditional forensic science but specifically inside the digital domain. In distinction, incident response refers to the quick implementation of containment, cessation, and preventive measures both during and following a cyberattack. The scope of incident response typically includes tactical measures, but digital forensics and incident response (DFIR) specifically emphasizes the application of digital forensics procedures within this framework.

Digital forensics and incident response effectively combines traditional incident response tasks, including response planning, architectural documentation, and playbook development, with the methodology utilized in digital forensics. The investigative aspect is emphasized, positioning it as the central component of the response approach. Through the integration of digital forensics and incident response (DFIR) practices, organizations can effectively and efficiently recover from cyberattacks while simultaneously strengthening their ability to withstand and mitigate potential future threats. For a fundamental understanding of the NIST Digital Forensics framework, consult the following reference: Fig. 2.15.

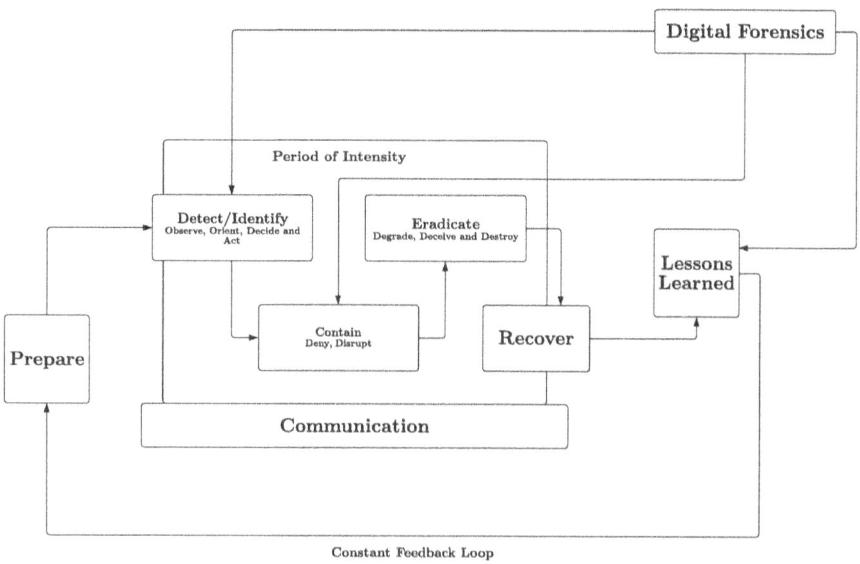

Fig. 2.15 NIST digital forensics framework

Digital forensics and incident response provides the necessary evidentiary foundation for commencing legal actions against individuals involved in cybercrime or validating assertions within cyber insurance. Given the significant financial costs and resulting harm that are frequently associated with a cyberattack, it is important to learn about appropriate response strategies and effectively traverse the legal avenues available for redress.

Digital forensics is a subfield of forensic science that involves the examination and analysis of digital data and devices in order to collect evidence. The concept of 'investigation' refers to the act of revealing or making something known. Digital forensics, a fundamental component of digital forensics and incident response (DFIR), is a specialized field within the wider subject of forensic science that is specifically designed to address the complexities associated with digital technology.

Forensic analysts who specialize in this field are responsible for the retrieval, study, and investigation of digital data found on a variety of digital devices. The primary objective of digital forensics is to gather and protect persuasive evidence, which serves as a crucial resource in prospective legal proceedings against hackers involved in an attack. Digital forensics activities aim to achieve four main objectives inside organizations:

(1) **Confirmation of Cyberattacks**: The verification of cyberattacks is a primary goal within digital forensics. This is the process of finding and documenting the digital footprints left behind by threat actors.
(2) **Assessment of Impact**: The evaluation of impact extends beyond mere verification in digital forensics, aiming to assess the overall consequences of a cyber incident. Understanding the magnitude of the harm is crucial for making well-informed decisions in the aftermath of an attack.
(3) **Determining the Cause**: Uncovering the underlying factor responsible for a cyberattack is an essential investigative effort. The process involves the investigation of the breach's origins, the identification of vulnerabilities, and the comprehension of the strategies utilized by the attackers.
(4) **Evidence Collection**: The compilation of indisputable data that can verify the incidence of a cyberattack is often regarded as one of the most crucial functions of digital forensics. This evidence assumes great significance in judicial processes.

Emphasizing the significance of quick action within digital forensics is crucial. The importance of time becomes particularly significant in a continuing cyberattack. The quick implementation of intervention measures has the potential to effectively impede the advancement of an attack and minimize its overall repercussions on the organization being targeted. In digital technology, various devices, networks, and systems produce an uninterrupted flow of data, a significant portion of which holds necessary value for investigative efforts, even during periods of device inactivity. Nevertheless, as the passage of time progresses, the likelihood of data deletion, overwriting, or alteration increases considerably. Numerous forensic artefacts rely on the initial state of a computer or network following an incident. Hence, it is important for forensic investigators to quickly undertake measures to safeguard this crucial information before its degradation. Before we conclude here is a comprehensive list of differences (Table 2.2) between incident response versus digital forensics.

Digital forensics and incident response (DFIR) has become an essential and fundamental component within cybersecurity. This technique is characterized by its complex nature, as it incorporates both digital forensics and incident response. It is strategically utilized to effectively tackle and reduce the escalating risk posed by cyberattacks. Digital forensics, as an integral component of digital forensics and incident response (DFIR), assumes a crucial function in the verification, evaluation, examination, and acquisition of data about cyber incidents. The proactive and time-sensitive characteristics of this approach guarantee that organizations are adequately prepared to effectively address cyberattacks, thereby reducing possible harm and enabling legal action if required. The significance of digital forensics and incident response (DFIR) in cybersecurity remains steady as the digital society undergoes continuous transformation. DFIR provides a structure that enhances the ability to withstand and recuperate from cyberattacks.

12.2 Advantages of Incident Response

Embracing a robust incident response framework bestows several undeniable advantages upon organizations. These advantages, underpinned by analysis and empirical evidence, highlight the importance of adopting a proactive stance in the face of mounting cyber threats.

(1) **Mitigation of Financial Losses**: Incident response lays the groundwork for swift containment and resolution of security breaches. By reducing the duration of an incident and minimizing its impact, organizations can substantially reduce financial losses arising from downtime, data theft, and operational disruptions.
(2) **Preservation of Reputation**: Instant communication and amplified social media reach, maintaining a pristine reputation, is integral. A carefully executed incident response strategy, by demonstrating swift and effective action in the aftermath of an incident, helps safeguard an organization's reputation and preserves the trust of stakeholders.
(3) **Regulatory Compliance**: The regulations governing data security is becoming increasingly strict. An established incident response plan, designed with legal and regulatory considerations in mind, ensures that organizations remain in compliance to relevant regulations, thus preempting potential legal liabilities.
(4) **Enhanced Preparedness**: An incident response plan, continually updated and stress-tested, nurtures a culture of preparedness, enabling organizations to be prepared for the ever-changing threats.

Table 2.2 Comparison of digital forensics and incident response

Digital forensics	Incident response
Definition	Definition
Digital forensics involves the systematic investigation and analysis of digital evidence to uncover facts suitable for legal proceedings	Incident response is a set of procedures aimed at managing and mitigating the impact of a security incident, focusing on minimizing damage and restoring normal operations
Objective	Objective
The primary objective of digital forensics is to collect, preserve, analyse, and present digital evidence in a forensically sound manner for legal purposes	Incident response aims to detect, respond to, and recover from security incidents promptly to minimize the impact on the organization
Timing	Timing
Digital forensics is typically a post-incident activity, conducted after an event has occurred, to gather evidence for legal proceedings	Incident response is a real-time or near-real-time activity, focusing on immediate detection and response to ongoing security incidents
Scope	Scope
Digital forensics has a broader scope, covering the comprehensive analysis of digital systems and data to reconstruct events and establish facts	Incident response has a narrower focus, concentrating on the specific steps needed to contain and eradicate a security incident
Legal admissibility	Legal admissibility
Digital forensics outputs are often intended to be legally admissible, requiring strict compliance to forensic methodologies and chain of custody procedures	Incident response focuses more on immediate containment and eradication, with legal considerations being secondary
Skill set	Skill set
Digital forensics requires specialized skills in evidence collection, preservation, and analysis, often necessitating a deep understanding of computer systems and file structures	Incident response demands a skill set emphasizing rapid decision-making, communication, and coordination to address and mitigate active security threats
Involvement of law enforcement	Involvement of law enforcement
Digital forensics investigations often involve collaboration with law enforcement agencies, as the gathered evidence may be used in legal proceedings	Incident response may or may not involve law enforcement, depending on the nature and severity of the incident
Timeline	Timeline
Digital forensics investigations tend to have a longer timeline, as the analysis of evidence requires time for accuracy and completeness	Incident response demands swift action within a compressed timeframe to contain and mitigate the impact of the incident effectively
Data collection	Data collection
Digital forensics involves the systematic collection of digital evidence, ensuring its integrity and preservation for legal purposes	Incident response focuses on collecting real-time data about the ongoing incident to make informed decisions and take immediate action

(continued)

Table 2.2 (continued)

Digital forensics	Incident response
Proactive vs reactive	Proactive vs reactive
Digital forensics is often reactive, triggered by the occurrence of an incident or suspicion of malicious activity	Incident response can be both proactive, with measures in place to detect incidents before they occur, and reactive, responding to active incidents
Prevention vs response	Prevention vs response
Digital forensics is primarily concerned with responding to incidents after they occur, rather than preventing them	Incident response places a strong emphasis on both preventing and responding to security incidents, with proactive measures to minimize vulnerabilities
Documentation	Documentation
Digital forensics necessitates detailed documentation of the entire investigative process, ensuring transparency and reproducibility of findings in a court of law	Incident response requires documentation for internal analysis, improvement, and potential legal or regulatory reporting
Nature of findings	Nature of findings
Digital forensics aims to provide a comprehensive understanding of past events, detailing what happened and how	Incident response focuses on immediate issues, such as the identification of active threats and the steps needed to neutralize them
Post-incident analysis	Post-incident analysis
Digital forensics often involves a thorough post-incident analysis to understand the root causes and contributing factors	Incident response may include post-incident analysis, but the primary focus is on containment, eradication, and recovery during the incident
Training emphasis	Training emphasis
Digital forensics training emphasizes deep technical knowledge of forensic tools, data recovery, and legal procedures	Incident response training emphasizes rapid decision-making, communication, and collaboration to address dynamic and evolving security incidents

(5) **Strategic Learning and Improvement**: Post-incident analysis serves as a crucible for growth. Through careful examination of incident causes, responses, and outcomes, organizations can extract invaluable insights to refine their strategies, enhancing their overall cybersecurity posture.

(6) **Stakeholder Confidence**: A well-planned incident response framework radiates a sense of professionalism and diligence. This not only establishes confidence within the organization but also reassures clients, partners, and investors that the organization is proactive in safeguarding their interests.

(7) **Resource Optimization**: Rather than resorting to ad-hoc responses during a crisis, incident response streamlines the deployment of resources. This optimization extends to personnel, technology, and financial assets, ultimately minimizing wastage and enabling a more efficient resolution process.

(8) **Strategic Advantage:**. Organizations that can effectively detect and recover from cyber incidents position themselves as trusted partners, attracting collaborators and customers alike.

12.3 *Disadvantages of Incident Response*

While incident response holds undeniable advantages for organizations, it is crucial to acknowledge that this strategic approach is not devoid of challenges and potential drawbacks. By confronting these disadvantages head-on, organizations can better handle the complex execution of incident response more effectively.

(1) **Resource Intensiveness**: Crafting and maintaining a robust incident response plan demands substantial resources, including financial investments, personnel training, and technology infrastructure. For organizations with limited budgets, allocating these resources might pose a strain on other critical initiatives.
(2) **Complexity and Expertise**: The complications of incident response necessitate an understanding of diverse cybersecurity domains. Building an internal team with the requisite expertise can be daunting, requiring careful recruitment and continuous training efforts.
(3) **False Positives and Negatives**: The discernment between actual security incidents and benign anomalies can be complex. An overzealous incident response approach might lead to resource wastage on false positives while understating genuine threats could result in severe consequences.
(4) **Operational Disruption**: Implementing incident response protocols during a live security breach can inadvertently disrupt normal operations. Balancing the urgency of addressing the breach with the need to maintain business continuity requires deft manoeuvring.
(5) **Evolving Threats**: The rapid evolution of cyber threats demands a dynamic incident response strategy. Sticking to a static plan might render it obsolete, necessitating continuous updates and adaptability, which can strain resources.
(6) **Legal and Regulatory Challenges**: Incident response involves navigating a labyrinth of legal and regulatory considerations. Mishandling data during response efforts could lead to legal liabilities or regulatory penalties, emphasizing the need for compliance expertise.
(7) **Delayed Responses**: Overcomplicating the incident response process could lead to delays in addressing security incidents, affording attackers more time to cause damage. Striking the right balance between thorough analysis and swift action is a perpetual challenge.
(8) **Employee Morale**: Consistently focusing on the negative aspects of cybersecurity, including incidents and potential breaches, could dampen employee morale. An overtly cautious culture might hamper creativity and innovation within the organization.

(9) **Cost-Effectiveness**: For smaller organizations, the expenses associated with incident response measures might outweigh the potential risks they face. Balancing the cost of implementation against the probability and impact of incidents is a critical consideration.

(10) **Unforeseen Consequences**: Unforeseen consequences might arise during incident response efforts, such as accidental data loss or accidental disruption of critical systems. These unintended outcomes can exacerbate the damage caused by the original incident.

13 Digital Resilience Through Effective IRP

With the rapidly expanding influence of technology, the incidence and complexity of cyberattacks are consistently on the ascent. As a direct consequence, it is now important for organizations to establish a robust and well-structured incident response plan (IRP) to effectively counter cyber threats. The IRP includes a collection of protocols that an organization adopts to handle cyber incidents, such as malware infiltrations, denial-of-service (DoS) attacks, and data breaches.

However, adopting a uniform approach to incident response is insufficient, given that distinct organizations possess varying risk profiles, threat environments, and resource constraints. Consequently, the design of IRPs must prioritize scalability, enabling adaptation to a range of scenarios and requirements. A scalable IRP demonstrates the capacity to manage incidents of diverse types, scales, and complexities, all the while upholding the quality of the response. In digital resilience, the existence of a scalable IRP emerges as an important component in strengthening an organization against cyber threats. By promptly and efficiently responding to incidents, organizations can reduce the impact of attacks, diminish downtime, and sustain the trust and assurance of their clientele, collaborators, and stakeholders.

13.1 Concrete Illustrations

The year 2019 witnessed Capital One grappling with a substantial data breach [Net+20] that compromised personal information belonging to over a hundred million customers and applicants. Regardless, the organization's carefully prepared incident response plan proved its efficacy by swiftly identifying and containing the breach. Capital One's investments in cloud-based technologies expedited analysis and remediation, thereby mitigating the impact on its reputation and customer base. Marriott's incident response plan [Bre18] in 2018 was marked by a lack of scalability, resulting in a considerable time lag before the breach was detected and controlled. The subsequent delay incurred noteworthy consequences, including financial losses, legal actions, and reputational harm. Nevertheless, Marriott gleaned insights from this incident and subsequently introduced a more scalable incident response plan, featuring real-time monitoring and automated alerts to forestall similar occurrences.

The SolarWinds supply chain attack [Nas21] of 2020 highlighted the importance of scalable incident response plans. This attack affected a multitude of entities, including government agencies and prominent private companies like Microsoft and FireEye. Its orchestrated nature, involving the injection of malicious code into SolarWinds' software updates, emphasized the indispensability of responses from diverse organizations, thereby accentuating the significance of scalable incident response plans. The Colonial Pipeline ransomware attack [EF23] in 2021, leading to the shutdown of a major US fuel pipeline, further emphasized the critical role of scalable incident response plans. Organized by a criminal group, this attack necessitated a retort involving various government agencies and private enterprises. The incident substantiated the urgency of preparedness and the importance of possessing a scalable plan capable of mitigating the repercussions of such incidents.

13.2 Steps Towards Designing Scalable IRPs

For organizations trying to fashion scalable IRPs, the following steps hold relevance:

(1) **Defining Scope and Objectives**: The foremost undertaking is the identification of systems, assets, and data necessitating protection, coupled with the determination of requisite response levels for each type of incident. Organizations should articulate the aims and priorities of the IRP, including damage minimization, evidence preservation, service restoration, and stakeholder communication.

(2) **Developing a Response Framework**: Subsequently, a response framework should be constructed, assigning roles, responsibilities, and procedures for distinct phases of the incident lifecycle, including detection, analysis, containment, eradication, and recovery. The framework should address communication, escalation, decision-making, and documentation, among other considerations.

(3) **Conducting Risk Assessments and Testing**: Conducting periodic risk assessments is important to identify and prioritize potential threats and vulnerabilities. The IRP should be subjected to tabletop exercises, simulations, and drills to validate and enhance its effectiveness and efficiency, revealing gaps and shortcomings that warrant attention.

(4) **Ensuring Resource Availability**: Ensuring that the IRP can scale according to requirements mandates having adequate resources in place, including personnel, tools, and infrastructure. Establishing partnerships with external stakeholders, such as incident response service providers, law enforcement agencies, and industry associations, augments preparedness.

(5) **Monitoring and Review**: Continual monitoring and evaluation of the IRP's performance, coupled with adaptations based on feedback and insights garnered from prior incidents, is key. This ongoing enhancement facilitates improved

incident response capabilities and the maintenance of a state of preparedness vis-à-vis evolving threats.

The creation of scalable incident response plans is a key aspect in the construction of digital resilience. By equipping organizations to respond adeptly and swiftly to cyberattacks of varying scales and natures, the efficacy of such attacks can be mitigated. Employing the outlined key steps, along with lessons learnt from real-life scenarios, enables organizations to craft effective incident response plans, implementing resilience against the ever-evolving panorama of cyber threats.

14 Significance of Swift Response and Recovery

Rapid reaction and persistent recovery are two critical factors that contrast sharply yet are mutually exclusive. While one needs immediate intervention to calm the turbulence caused by a breach, the other requires deliberate repair to bring back normality. The heart of digital resilience originates from this requirement for urgency and alertness. On the one hand, immediate identification and control of cyber intrusions, also known as rapid reaction, acts as an indicator of warning against the black clouds of digital malicious intent. This preventive strategy proves essential in the ever-changing arena of cyber threats, where the speed of attack significantly surpasses defence capabilities. Rapid identification and response, comparable to a digital guardian on high alert, may avert catastrophic interruptions and preserve the confidence required for digital transactions.

Contrary to popular belief, the complicated process of persistent healing appears as an intervention that improves digital system damages. The precise synchronization of repair, reconstruction, and restoration converts a damaged system back into a stronghold of operations. Sustained recovery is a form of resilience because it seeks to reconstruct the underlying foundation, thereby rendering it more robust to future shocks rather than only repairing what is obvious damage. Quick reaction and successful rehabilitation are critical components of digital resilience, especially in the face of cyber violations and other disruptive incidents. When a system has been compromised or left unavailable, having the capacity to quickly detect and address the problem becomes essential for maintaining reliance on the affected digital frameworks.

Safeguarding reliance on digital systems holds significance within the modern technological age, wherein cyberattacks have the potential to inflict substantial harm upon entities and individuals. The expeditious identification and recovery from disturbances are foundational aspects of digital resilience, facilitating organizations in promptly pinpointing and addressing security incidents while minimizing their repercussions. To introduce and sustain digital resilience, organizations should incorporate the subsequent actions within their response and recovery framework:

- **Identification and Preparatory Measures for Incidents**: Organizations should enact a scheme for addressing cyber incidents. This includes the identification of potential threats, the establishment of incident response units, the assigning of roles and responsibilities, and the periodic execution of training and simulations.

- **Detection and Thorough Analysis**: Organizations should institute mechanisms for the real-time detection and analysis of cyber incidents. This includes vigilant monitoring of systems to detect anomalous activities, the application of forensic scrutiny to unveil the nature and extent of the incident, and the assessment of the potential effects on the entity.
- **Confinement, Elimination, and Rehabilitation**: Following the identification and analysis of an incident, the organization must embark on measures to reduce and obliterate the threat. This process includes isolating compromised systems, expunging malicious code, and reinstating data from archived backups. Subsequently, after neutralizing the threat, the entity can initiate the process of recovery and normalization of operations.
- **Effective Communication and Harmonization**: During the occurrence of a cyber incident, adept communication and harmonization play important roles. Organizations should devise a clear communication strategy to ensure stakeholders, including employees, clients, suppliers, and regulatory entities, are informed of developments throughout the incident.
- **Continuous Enhancement**: Ultimately, organizations should continually assess and better their response and recovery blueprint, informed by insights from previous incidents. This involves conducting post-incident analyses to identify areas necessitating refinement, revising policies and protocols, and routinely executing training and simulations to sustain the efficacy of the strategy.

Organizations need to implement an established incident response strategy to ensure rapid reaction and recovery. The strategy should specify each team member's roles and responsibilities, as well as lay out clear protocols for dealing with cyber occurrences. Again, it should contain a communication plan stating how stakeholders, customers, and regulatory organizations will be notified about the occurrence. Swift reaction and skilled recovery are core elements of digital resilience, and they are critical for maintaining confidence in digital systems. Given the increasing frequency of cyberattacks, businesses must have a well-structured incident response plan that promotes quick identification, mitigation, and relief of cyber event aftermath. This preventative approach may mitigate the financial and reputational consequences of breaches, ensuring the allegiance of customers and stakeholders.

15 Role of Regular Drills and Exercises

Safeguarding digital resilience requires preparedness to tackle cyber incidents. A potent approach to achieving this state is through recurrent drills and exercises designed to assess incident response readiness. The fundamental purpose of these exercises lies in identifying latent vulnerabilities within systems and response strategies, thus enhancing the efficacy of an organization's incident response team. This includes examining the team's capacity to identify, confine, and mitigate cyber incidents, as well as liaising with stakeholders and synchronizing with external resources, such as law enforcement agencies or incident response contractors.

Beyond shoring up response capabilities, these drills and exercises play a critical role in developing awareness among employees about cyber threats and the importance of adhering to security protocols. This proactive approach contributes to the prevention of incidents in the first place. An example of the significance of routine drills is evident in the annual 'Cyber Storm [CIS22]' exercise organized by the US Department of Homeland Security. This collaborative exercise engages federal, state, and local government bodies, alongside private sector organizations, to evaluate their capacity to respond to synchronized cyberattacks. The exercise not only identifies gaps in response strategies but also streamlines coordination across diverse agencies and organizations.

16 DevSecOps

As we reach the end of this chapter, let's discuss the paradigm of DevSecOps. It is not only widely used in the software development industry but also the preferred way of creating fast and secure applications while being able to deploy them with the same efficiency as the DevOps approach for software development. DevSecOps is an approach that combines the concepts of development, security, and operations to create seamless integration. Table 2.3 provides a comparison of DevOps and DevSecOps. The pursuit to increase security in software development is a large and difficult task. To create a resilient security posture, it is required to have a thorough understanding of the various sorts of potential vulnerabilities and threats inside the defined region. The inherent abstractness of security efficacy, which is measured largely by the absence of detrimental occurrences rather than visible accomplishments, may result in a diminished impression of its value. DevSecOps, an approach that tries to integrate security, engineering, and operations to enable a complete security view across the DevOps lifecycle, may provide one feasible response to this problem (Fig. 2.16).

The fundamental principle of DevSecOps revolves around the objective of enabling engineers to create web applications that integrate an extensive level of security. This objective is attained through a set of practical and realizable strategies, which provides a substantial opportunity to gain an understanding of the foundational principles of DevSecOps. It is important to undertake an exhaustive inquiry to comprehensively scrutinize these practices.

A fundamental requirement that emerges during developmental processes is the integration of security as an innate component that is complexly intertwined with the essence of progress. It is essential to incorporate security standards from the earliest stages of the development lifecycle and maintain consistency throughout all phases to guarantee their successful implementation. Equal weight should be

Table 2.3 Comparison of DevOps and DevSecOps

DevOps	DevSecOps
1. Focus	1. Focus
DevOps primarily focuses on collaboration and communication between development and operations teams to streamline the software delivery process	DevSecOps extends the focus to include security, integrating it into the entire development lifecycle, ensuring a secure-by-design approach
2. Security integration	2. Security integration
DevOps emphasizes automation and efficiency but may not inherently address security concerns, leading to potential vulnerabilities	DevSecOps integrates security practices from the beginning, ensuring security is an integral part of the development process
3. Lifecycle involvement	3. Lifecycle involvement
DevOps spans the entire software development lifecycle but may leave security considerations to later stages	DevSecOps involves security throughout the entire lifecycle, from planning and coding to testing and deployment
4. Collaboration	4. Collaboration
DevOps encourages collaboration between development and operations teams for faster and more efficient delivery	DevSecOps extends collaboration to include security teams, fostering a collaborative culture among all stakeholders
5. Automation	5. Automation
DevOps heavily relies on automation for continuous integration, delivery, and deployment	DevSecOps extends automation to include security testing and regulatory compliance checks, ensuring security measures are automated
6. Speed and agility	6. Speed and agility
DevOps aims for rapid and agile development, focusing on quick releases and continuous improvement	DevSecOps maintains speed and agility but adds security checks, ensuring that speed does not compromise security
7. Risk management	7. Risk management
DevOps may not explicitly address security risks in the early stages, leading to potential vulnerabilities	DevSecOps actively manages and mitigates security risks throughout the development process, reducing the likelihood of security breaches
8. Culture of responsibility	8. Culture of responsibility
DevOps instils a culture of shared responsibility between development and operations teams	DevSecOps extends the culture of responsibility to include security, making everyone accountable for security measures
9. Continuous monitoring	9. Continuous monitoring
DevOps focuses on continuous monitoring for performance but may not prioritize continuous security monitoring	DevSecOps incorporates continuous security monitoring, ensuring real-time visibility into security threats and vulnerabilities
10. Compliance and governance	10. Compliance and governance
DevOps may address regulatory compliance as an afterthought, potentially leading to non-compliance issues	DevSecOps integrates regulatory compliance and governance from the start, ensuring that security measures align with regulatory requirements

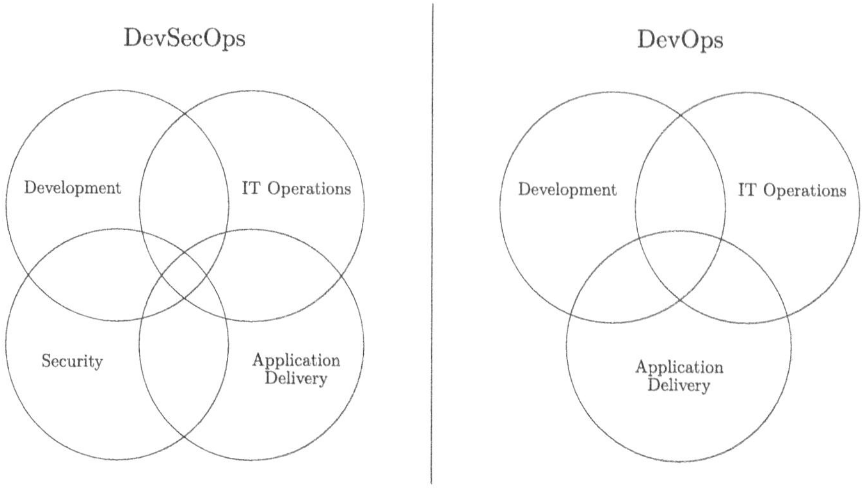

Fig. 2.16 DevSecOps versus DevOps

given to the importance of security requirements and functional requirements; therefore, security controls, processes, and automation must be incorporated into the workflow. By utilizing the mentioned strategic approach, developers are empowered to proactively address vulnerabilities, thereby guaranteeing the resilience and protection of software systems.

16.1 DevSecOps and Digital Resilience

In our current, rapidly changing digital world, the need to effectively incorporate cybersecurity measures into the software development process goes beyond being a choice and has become an undeniable requirement. DevSecOps, an integration of cultural and technical elements that effectively combines development, security, and operations, represents a notable aspect of this paradigm. DevSecOps assumes a key position in achieving digital resilience, extending beyond its utilitarian function in enhancing cybersecurity. This paradigm enables organizations to effectively defend against cyberattacks and quickly recover from disruptive situations. One of the main principles of DevSecOps is the adoption of the 'shift left' methodology, which involves the proactive integration of security considerations at the initial phases of software development. Instead of considering security as a secondary concern, it is regarded as a fundamental aspect throughout all stages of growth. The adoption of a proactive approach helps to identify weaknesses and vulnerabilities in the codebase, hence reducing the potential for security breaches and minimizing the attack surface.

The DevSecOps approach promotes and advocates for the extensive utilization of automated security testing tools throughout the whole developmental pipeline.

The ongoing assessment and verification of code, configurations, and infrastructure facilitate the quick detection of possible security vulnerabilities, allowing for timely mitigation before they can impact the production environment. The ultimate consequence is a decrease in the duration of development cycles and the provision of software that exhibits exceptional quality, resistant to security vulnerabilities. The creation of a collaborative mindset is a fundamental aspect of DevSecOps. This phenomenon is observed in the coming together of development, security, and operations teams, resulting in the breakdown of organizational barriers. The establishment of this symbiotic relationship promotes a culture where there is a shared sense of responsibility towards cybersecurity. This ultimately leads to swift and efficient actions taken in response to new threats and vulnerabilities.

The utilization of immutable infrastructure represents a significant advantage inside the DevSecOps toolkit. This concept involves the utilization of identical and unchangeable instances of software or infrastructural components. This measure functions as a potent deterrent against malicious actors making it difficult for them to get unauthorized access to systems or carry out unauthorized alterations, hence augmenting the overall security of the system. The significance of ongoing surveillance of systems and apps in the production environment is emphasized by the DevSecOps approach. The consistent state of alertness enables the prompt identification of irregularities, possible risks, or security breaches. By leveraging timely feedback, organizations are well-positioned to swiftly address events, thus reducing their potential for disruption.

Digital resilience comprises a range of factors that extend beyond cybersecurity. However, it is worth noting that DevSecOps, a methodology that integrates security practices into the software development process, can be flexibly adapted to include resilience planning. Organizations enhance their ability to quickly recover from cyberattacks and system failures by effectively integrating disaster recovery protocols, backup processes, and redundancy measures within their developing and deployment pipelines. Meanwhile, DevSecOps assists in the convergence of regulatory compliance and governance obligations. The automation of security checks and documentation allows for the presentation of regulatory compliance, lowering the risk of heavy penalties and legal problems and issues. Added to that, DevSecOps teams may successfully use threat information feeds as a key component of their operational operations. The integration makes proactive mitigation of new cyber threats and vulnerabilities possible, hence increasing an organization's defensive capabilities.

DevSecOps advocates for the preservation of immutable audit trails within post-incident analysis and regulatory compliance reporting. The careful practice of maintaining detailed records guarantees that any modifications made to infrastructure and code are thoroughly recorded. This enables organizations to effectively analyse the underlying factors contributing to incidents and unequivocally showcase their firm dedication to security. All-inclusive, DevSecOps extends beyond the boundaries of a simple development technique, including a strategic framework that enhances cybersecurity and develops digital resilience. By including security as an inherent element of the developmental framework, organizations enhance their defensive capabilities against cyber threats while also maintaining the flexibility required to succeed in the modern digital environment. This denotes a paradigm change from a passive incident response to a proactive approach to mitigating security events.

16.2 Advantages

The adoption of DevSecOps presents a multitude of advantages that hold the capacity to transform the approaches utilized by enterprises in the domains of software development and security. The advantages are diverse, ranging from an understanding of security to the timely integration of security measures into the development process. The application of DevSecOps requires the utilization of automation, the process of a cultural shift towards increased security awareness, and the equipping of developers with the necessary tools and knowledge. This section examines the previously listed benefits, providing a clear explanation of how they contribute to the promotion of software development practices that exhibit improved resilience, security, and efficiency. The attainment of a security perspective is facilitated through the implementation of DevSecOps, which advocates for an approach to security by incorporating security measures throughout all stages of the development process.

- The integration of security measures is initiated at the early stages of development, hence reducing the probability of vulnerabilities being undetected until subsequent phases.
- Continuous testing is a fundamental aspect of the DevSecOps approach, as it emphasizes the ongoing evaluation of security measures across the whole software development lifecycle. This proactive approach ensures that potential security vulnerabilities are identified and resolved in a timely and efficient manner.
- Automation is critical and integral in decreasing the workload of development teams and facilitating the efficient execution of security activities.
- Real-time monitoring and analysis facilitate the swift identification of vulnerabilities and security concerns, enabling appropriate responses.
- DevSecOps promotes a security-oriented culture within the developer community, operational teams, and security professionals, by promoting transparency and collaboration.
- The concept of 'Security-as-Code' involves the seamless integration of security measures into the development process, following agile principles.
- Enhancing Developer Empowerment: DevSecOps facilitates the provision of essential tools, training, and support to developers, hence developing an atmosphere that promotes collaboration and knowledge sharing.

- The concept of infrastructure security includes the implementation of security measures that include all aspects of the infrastructure, such as networks, pipelines, code repositories, and other relevant components.
- The process of evaluating security in DevSecOps involves the use of an assessment methodology that incorporates both quantitative and qualitative indicators to gauge the efficacy of security measures.

16.3 Disadvantages

Although DevSecOps offers significant advantages, its implementation is not without obstacles and limitations. Organizations may encounter challenges when undergoing the move to DevSecOps, necessitating substantial allocation of resources and encountering opposition to cultural shifts. This section examines the drawbacks associated with this approach, offering valuable perspectives on the challenges that organizations may face when implementing this security-focused methodology. The implementation of DevSecOps might present complexities and necessitate substantial modifications to preexisting development and operational procedures.

- This undertaking may necessitate significant allocation of resources, including staff, tools, and training.
- Resistance to change can be observed in teams who are accustomed to traditional development practices, as they may exhibit reluctance in embracing the cultural shift towards DevSecOps.
- The process of acquiring knowledge and skills concerning security practices and tools may require a certain amount of time for developers and operational teams to adjust and become proficient.

16.4 Security Challenges

The process of ensuring the security of software development in DevSecOps poses distinct and specific difficulties. This section describes the complexities associated with managing third-party dependencies, assuring API security, maintaining container security, and implementing continuous testing.

- The management of security for open-source and third-party dependencies is important, yet it can present difficulties.
- The maintenance of API security necessitates the implementation of strong authentication and protection measures to safeguard the exchange of data between apps.
- Container security is a critical aspect that necessitates robust measures in authentication, authorization, and isolation to effectively mitigate vulnerabilities and potential attacks.

- The maintenance of efficient ongoing security testing and protection measures is crucial; yet, it can require a significant allocation of resources.
- The nurturing of secure staff behaviour and awareness are important to mitigate the risk of social engineering attacks, such as spear-phishing.
- To effectively address the security concerns of application environments and infrastructures, it is important that security solutions possess the ability to adapt to their dynamic nature.

16.5 Research Avenues

Given the continuous growth of the DevSecOps culture, it is critical to participate in ongoing research efforts to effectively address new security risks and improve current methodologies. This section will examine critical research topics that may have a substantial impact on the future trajectory of DevSecOps. Among the features included in the previously described categories are automated security solutions, sophisticated analytics, expanded training programmes, and containerization security, among others. The results underline the need for continuous improvement and innovation in the field. Automated security has emerged as a key area of study, with great interest in academic research. The current focus is on the study and evaluation of advanced automated security technologies and methods, intending to increase the efficacy and accuracy of security assessments.

In DevSecOps, timely and dependable execution of security assessments is of first preference, considering the swift and uninterrupted deployment of software. Security analytics is of great significance to developers and engineers. The main aim of the continuing research is to enhance the effectiveness of security analytics in identifying vulnerabilities and promptly monitoring their resolution in real time. The importance of adopting a proactive approach to maintaining a competitive advantage amidst constantly changing challenges is well acknowledged. Similarly, it is worth noting the increasing recognition of the rising importance of security training in research activities. Presently, research is focused on the exploration of diverse methods and efforts aimed at augmenting employee awareness and preparedness in security training. The significance of having sufficiently qualified employees in ensuring security is widely recognized, as the human element plays a substantial role in shaping this aspect. Engineers are now engaged in continuous research on the topic of API security. At the moment, the focus is on looking into and analysing ways to make API security better to protect against new attack methods and vulnerabilities in a digital society that is becoming more connected.

There has been an increasing interest in container security in recent years. A group of researchers is presently involved in the analysis of sophisticated methods to enhance the security of containerization. Our primary focus is on the implementation of more robust isolation mechanisms to augment the overall efficacy of security measures. The primary objective is to guarantee the level of security for applications that are deployed within containers. Interestingly, studies revolve around the advancement of metrics within DevSecOps. This involves the development of standardized metrics that enable the assessment of the effectiveness and influence of DevSecOps implementations, thereby offering useful insights into security posture. We are presently involved in the active assessment of behavioural analytic methodologies. Presently, research is primarily concentrated on the progression and enhancement of behavioural analysis methodologies to identify and mitigate insider threats and social engineering attacks, both of which frequently present notable vulnerabilities.

The present research emphasizes the security of dynamic environments, which arises from the perpetual evolution of IT environments. The awareness of the need to build effective solutions to guarantee the security of applications and infrastructure in contexts characterized by rapid and dynamic change is clear. The emphasis on vulnerability management continues to be significant. The primary aim of this is to enhance vulnerability management approaches, which are necessary in enabling organizations to efficiently prioritize and tackle security issues.

By using this particular method, it is possible to efficiently reduce the probability of prospective security breaches. Again, it is meaningful to acknowledge that threat modelling is currently undergoing continuous advancement and development. Current efforts are dedicated to the development of increasingly advanced threat modelling approaches to achieve a more thorough identification and mitigation of risks. These strategies enable organizations to proactively address vulnerabilities within their software development lifecycle. The concept of DevSecOps signifies a significant shift in software development, with a primary focus on integrating security practices within the initial stages. The need for continuous research and innovation should not be underestimated, as they have a key role in addressing dynamic difficulties and maintaining the long-term viability of DevSecOps methodologies.

GitOps is a prominent trend that has become apparent in recent breakthroughs in the DevOps field. This new phenomenon has received a lot of interest and is currently being extensively encouraged by enterprises because of its capacity to provide them with independence and streamline their infrastructure through automation. The GitOps approach makes use of Kubernetes functionality to make it easier for IT operations managers and developers to use Git as a centralized repository for the aggregation and deployment of multiple applications. Also, there is a growing prominence in the use of microservice architecture. This architectural approach allows for the division of the backend of an application into distinct and specialized components. Independent teams are responsible for developing, testing, and deploying these components, which lowers operational costs. It is advisable to improve the skill sets of existing teams by integrating modern technologies, replacing obsolete platforms with cloud and cloud-native technologies, and complementing these changes with appropriate tools and procedures, such as policy-as-code and continuous integration/continuous deployment (CI/CD).

The challenges encountered during the deployment of DevSecOps can be categorized into four broad domains: individuals, techniques, technology resources, and physical infrastructure. The challenges faced by individuals include issues related to knowledge acquisition and competence, effective collaboration within heterogeneous teams, and organizational culture. The obstacles encountered when attempting to integrate security practices within the DevOps framework serve as a distinguishing feature of the challenges that manifest in practice. On the other hand, difficulties with tools stand out due to their complexity and a lack of clear documentation for both DevOps and security systems. The adoption of DevSecOps in different infrastructure settings, particularly complex cloud environments, presents challenges related to infrastructure. It is expected that shortly there will be significant advancements in the DevOps industry, particularly concerning an increased emphasis on automation and the incorporation of AI to facilitate process automation.

As organizations progressively embrace agile approaches and transition to cloud-based infrastructures, there emerges a demand for solutions that enable prompt and effective answers to customer demands and market dynamics. In conjunction with our research, we have identified security risks and vulnerabilities associated with tools, with a particular emphasis on their implications for containers. Developers must adopt appropriate strategies to lighten the negative consequences of these security issues when incorporating such tools into their development workflow. Shift-left security and continuous security evaluation are commonly recommended as optimal strategies for prioritizing security throughout the deployment cycle. Nevertheless, it is important to acknowledge that there is a requirement for more improvement in tool support within this specific sector. One of the primary obstacles in the industry refers to the inability to fully automate security protocols that have conventionally been executed through manual means.

The incorporation of these methodologies into the continuous integration, continuous delivery (CD), or continuous deployment (CD) practices of DevOps presents an interesting challenge. The role of cultural and human variables in attaining success in DevSecOps is sometimes disregarded in discussions. Therefore, it is crucial to undertake additional socio-technical studies to examine the diverse challenges and possible remedies concerning humans in this domain. Besides, the use of DevSecOps principles and practices in complex, resource-constrained, and highly regulated infrastructures is an emerging area of research. However, it is important to prioritize the adoption of empirically validated solutions to promote greater acceptance and use of these tools and frameworks. It is essential to recognize that there exists a significant interconnectedness between various difficulties and solutions in the implementation of DevSecOps, which may be categorized under distinct themes. When conducting research in this specific sector, researchers must take into account these interrelationships.

17 Concluding Remarks

As the exploration of digital resilience in the context of cybersecurity progresses, it becomes more evident that the ever-changing nature of our digital environment is becoming more prominent. It is important to recognize, from a fundamental perspective, that the concept of digital resilience is closely interconnected with the implementation of strong cybersecurity measures. This chapter describes the key role of cybersecurity in enhancing our capacity to deal with the dynamic setting of digital infrastructures.

It is undeniable that the digital domain is expanding at an unprecedented pace. Because of the introduction of sophisticated technologies such as quantum computing, AI, and 6G networks, the digital infrastructure is experiencing a substantial transition. These technological innovations open up new opportunities while also introducing new challenges and vulnerabilities. Because cybersecurity is at the forefront of protecting digital systems, it requires ongoing adaptation and development in conjunction with the technology it seeks to defend. As we consider the future of digital resilience and cybersecurity, it becomes clear that the present ecosystem is poised for a dramatic transformation.

To accurately forecast the course of cybersecurity, it is crucial to recognize the dynamic nature of the threats. The level of sophistication shown by cybercriminals is on the rise as they deploy evasive and destructive strategies. As we progress, it is essential to anticipate the advent of new cyber dangers and the need for proactive defence methods. The maintenance of a fragile counterbalance between security and convenience will continue to be a prominent and recurring topic. Our aim requires the creation of cybersecurity solutions that place a strong emphasis on user-centricity, prioritizing the dual objectives of providing powerful protection and delivering user-friendly experiences. The process of evolution in this context includes several developments, such as improvements in biometric identification, the implementation of seamless encryption, and the enhancement of user empowerment in the management of their digital identities.

Education and awareness will continue to play an essential role in strengthening our digital resilience. There is a present and predicted trend towards a significant increase in cybersecurity education activities, which include not just academic institutions but also professional settings and daily routines. Individuals and businesses alike agree that cybersecurity is a shared responsibility. All-inclusive, while contemplating the prospects of attaining digital resilience via cybersecurity, we are confronted with a ripe environment favourable to innovation and adaptability. The forthcoming expedition presents a combination of obstacles and prospects, necessitating a firm dedication to safeguarding our digital future with constant attention, collaborative efforts, and an uncompromising commitment to digital resilience.

Acknowledgments We would like to extend our heartfelt gratitude to Aditya Srivastav for his invaluable support throughout the process of writing this chapter. Without his insights and annotations, the chapter's comprehensiveness in the field of digitally resilient systems would not have been feasible. His assistance was of considerable value in this regard. Sincerely, his preliminary lecture notes and observations on DevSecOps and AI, among other topics, were of great help.

Part II
Building Digital Resilience

Chapter 3
Navigating the Regulatory Landscape

1 Introduction

In attempting to address the multiple problems presented by changing global dynamics, many actors, such as governments and international legal systems, are actively looking for remedies to enhance the capabilities of businesses, provide assistance to employees, and enable the empowerment of communities. Simultaneously, corporate firms are experiencing a compelling need to engage in a reassessment of their operational paradigms. This involves the identification of innovative business models, a thorough reconsideration of existing human resource strategies, and a strategic realignment within the evolving digital environment. The current state of affairs on a global scale highlights the ongoing importance of sustainability as the primary factor shaping the direction of progress. This paradigm requires the active and inclusive involvement of several stakeholders, including governments, schools, corporate enterprises, and individual consumers. The objective is to facilitate a shift towards sustainable consumption and production (SCP).

The rapid advancements in technology, along with the growing complexity presented by challenges such as the climate crisis and widespread socio-economic inequality, highlight the mounting complexity and demanding characteristics of these modern-day crises. Therefore, it is expected that the concerns will gradually develop into more complex challenges in the future. The incorporation of sustainable development ideas into the fundamental structure of national legislation, as well as associated institutions and their digital platforms, has considerable importance in this particular context.

The use of this strategy effectively augments stakeholder endorsement and strengthens the resilience of both the country and the organization. The global pandemic has led to a significant advancement of digital infrastructure in modern society. The expansion has prompted a pressing need for the development of legal

S. K. Shandilya et al., *Digital Resilience: Navigating Disruption and Safeguarding Data Privacy*, EAI/Springer Innovations in Communication and Computing,
https://doi.org/10.1007/978-3-031-53290-0_3

frameworks that are specially designed to strengthen cybersecurity measures. The frequent and significant occurrences of cybersecurity breaches have resulted in increased concern, hence instilling fear among individuals and legal systems. This chapter discusses the regulations and laws related to data breaches, as well as the increasing risk matrix associated with them and the potential negative outcomes they may result in. It also investigates the legislative structures implemented to protect against such violations.

The chapter emphasizes the need to adopt a proactive and prepared approach. This requires the establishment of a proficient incident response team, the formulation of response strategies, and the enforcement of periodic breach simulations that are mandatory for all personnel to comply with, with possible legal consequences for negligence in cases of non-compliance. This chapter offers an analysis that explores the core components of laws about data privacy and cybersecurity. The components include the procedures of exploration, investigation, containment, elimination, restoration, and integration of obtained knowledge in incident response, which serve as the principal means of mitigating substantial losses in terms of both information and financial consequences. Also, it is crucial to emphasize the significance of effective communication in the administration and damage control to an organization's image within the framework of a data breach, both in the immediate aftermath and in subsequent stages.

The complex attributes of data breaches need a thorough examination of legal and regulatory considerations. This chapter provides a rationale for the need to gather data from previous security breaches, conduct post-incident evaluations, and consistently enhance cybersecurity tactics. The future that lies ahead will see a process of change marked by the integration of advanced technologies, including AI and ML, which will have a substantial impact on the conduct of nations.

Legal systems face difficulties in developing appropriate regulations and obtaining precise guidance as a result of the existence of ambiguous legislation and fragmented legal provisions about data privacy and cybersecurity. To tackle these concerns, countries must take action to implement more informative legislation about cybersecurity. This should involve the establishment of clear guidelines and necessary adaptations to establish a resilient cybersecurity framework and legislation for safeguarding data, aligning with globally accepted standards.

The legislation in question refers to data that can be used to identify individuals, including but not limited to their names, phone numbers, and addresses. A structured framework is established to ensure the protection of data privacy by defining the rights of individuals, simplifying requests for data erasure, and establishing fines for non-compliance. Although there may be differences in privacy legislation across various locations, their primary goal remains consistent: safeguarding consumers and reducing unjust data-gathering practices. Prominent legislation in this field includes the General Data Protection Regulation (GDPR), the Personal Information Protection and Electronic Documents Act (PIPEDA), the California Consumer Privacy Act (CCPA), and similar regulations. The purpose of this chapter is to provide an analysis of significant data privacy regulations that will be in force in the coming years. It will also examine upcoming legislation in several jurisdictions, including the United States, the European Union, and Australia.

2 Regulations and Impact on Shaping Technology

The concept of a regulatory system refers to a diverse set of regulations, legislation, and procedures that control a wide range of human activities, primarily under the authority and oversight of governmental bodies. The regulatory system functions across various levels, including local, national, and international, and covers a diverse range of sectors, such as banking, healthcare, environmental conservation, and consumer protection. The regulatory system functions as a complete framework of principles and legislation that outlines the parameters within which individuals, organizations, and governments are required to function. The main objective of these regulations is to serve multiple purposes, including promoting equitable economic transactions, ensuring public health protection, preserving the environment, and maintaining social order. The regulatory system is inherently complicated and demonstrates a dynamic nature. Its evolution is influenced by various elements, including legislative determinations, judicial interpretations, and administrative directives.

It is crucial to recognize that this system is susceptible to ongoing modification to respond to evolving societal demands, improvements in technology, and economic circumstances. The regulatory system can be conceptualized as a multidimensional space that is defined by a multitude of qualities. Each dimension within this spatial framework indicates a distinct regulatory feature or requirement. Within the given context, the rules and regulations can be interpreted as limitations that establish acceptable boundaries and behaviours for individuals and entities. Compliance with these limitations is important to prevent legal consequences and maintain the efficient operation of society. The regulatory system assumes a key role in upholding societal structure, safeguarding the welfare of the general public, and promoting ethical conduct. Nevertheless, this initiative is not devoid of obstacles, as effectively operating through this complex structure and obtaining resources occasionally necessitate expertise and ample resources.

Accordingly, it is essential for individuals and organizations to regularly participate in initiatives aimed at comprehending, adjusting to, and complying to the evolving regulatory framework to prosper in society. The relationship between technological breakthroughs and the construction of legislative frameworks is important in shaping the scope of digital resiliency, a concept that is widely acknowledged for its rapid and ongoing evolution. Given the difficulties governments face in overseeing the dynamic digital society, regulatory frameworks have emerged as important tools. These frameworks hold significant relevance in not only avoiding risks but also encouraging innovation, safeguarding user rights, and nurturing a secure and resilient digital ecosystem. This section offers a better examination of the complex relationship between rules and technology, explaining how these two components combined influence the formation of the global regulatory framework. This paradigm possesses the capacity to either facilitate constructive encounters or stimulate disagreements.

- **Developing Innovation While Upholding Safety Protocols:** Regulations can serve as catalysts for the promotion of innovative product development. When carefully established, these norms have the potential to infuse a sense of assurance and generate a degree of confidence that can incentivize the allocation of resources towards emerging technologies. In cybersecurity, rules frequently require the use of particular security protocols. It is essential for organizations and technology developers to actively participate in innovation and design processes to offer solutions that exhibit both durability and efficacy. Considering the constantly evolving set of vulnerabilities and threats, it is crucial to adopt a proactive approach in establishing regulations that promote the advancement of sophisticated cybersecurity protocols.
- **Balancing Privacy and Surveillance:** One of the primary concerns in the fieldwork of digital resilience in current times revolves around the challenge of effectively managing the competing demands of user privacy and security. The establishment and maintenance of equilibrium are reliant upon regulations about data security and monitoring, as they assist in determining an appropriate balance. Inflexible privacy regulations play a crucial role in governing the proper management of user data and ensuring that it is handled responsibly and ethically. Simultaneously, legislation that allows for surveillance grants authorities the capacity to monitor and confront any digital threats. The current international dialogues about encryption, data retention, and legal interception highlight the difficulty of achieving a careful balance between these contradictory elements.
- **Global Harmonization vs. Local Specificities:** The pervasive nature of digital connectivity and technical progress extends beyond the confines of national borders, operating without prejudice or bias. Regulations, on the other hand, occur frequently. The recurring theme of digital resilience includes the conflict that arises when localized particularities collide with global uniformity. Local legislation frequently diverges from international accords and conventions to accommodate unique cultural, political, and safety factors. Despite the presence of international agreements and conventions designed to build a framework, the development of international regulatory frameworks is shaped by the complex task of attaining consensus while recognizing the distinct needs of diverse nations.
- **Adaptability:** The frequent emergence of novel advancements and their associated threats consistently perturb digital technology at a tremendous pace. To effectively achieve its desired objectives, regulation must demonstrate the necessary degree of adaptability. The presence of overly strict legislation can impede innovation, whereas excessively flexible regulations may fail to address previously unnoticed weaknesses. The achievement of equilibrium, characterized by the establishment of a stable framework that accommodates adaptation, is of crucial value in effectively managing changes within the context.

The relationship between technology and regulation within digital resilience is a phenomenon with multiple aspects. The world's growing interconnectedness has increased its reliance on digital infrastructure. As a result, the importance of rules in the advancement of technology and the guarantee of a secure and resilient digital future is gaining prominence.

3 Importance of Legal Frameworks

The construction of appropriate legislative and administrative frameworks holds significant importance in safeguarding the fundamental right to social security. The establishment of a well-defined legal structure is crucial to effectively determine individuals' entitlements, assist eligible individuals in exercising their rights to benefits, seek appropriate remedies in cases of rights violations, and prevent inconsistent findings. The objective of this measure is to guarantee fair and equal access to social protection while also ensuring the distinct obligations of individuals and organizations responsible for overseeing social protection systems. The utilization of this framework is crucial for ensuring the optimal functioning of the system, as it helps to minimize redundancy and flaws. This section delves into the significant role that legal frameworks play in the global regulatory setting and their influence on the promotion of digital resilience and sustainability.

In the current digital revolution, legal frameworks assume increased importance within the global regulatory environment, demanding swift consideration. The legal frameworks in question are not simply administrative documents; rather, they serve as the fundamental underpinnings for the broader notion of digital resilience. They have an important part in ensuring the efficacy of digital tactics, protecting individual liberties, and developing a secure digital age. The implementation of legal frameworks are important in ensuring the rights and obligations of individuals in the digital age. These frameworks establish explicit norms that regulate the conduct of all individuals involved in this field and enable the effective implementation of legal statutes and regulations.

These measures function as protective mechanisms for upholding rights about personal data and privacy. These frameworks provide procedures for the management of data, guaranteeing individuals' autonomy over their digital identities and mitigating the potential for data breaches and illegal access. The facilitation of mitigating cyber-risks is enhanced by the implementation of legislative frameworks that set norms and guidelines to effectively solve difficulties related to cybersecurity. The classification of cybercrimes, the imposition of sanctions, and the promotion of international collaboration are crucial elements in effectively addressing digital risks, deterring hostile actors, and establishing a secure digital infrastructure. The development of trust within the business community and among entrepreneurs can be effectively accomplished by implementing legislative frameworks that relates to digital technology while simultaneously promoting an environment conducive to innovation and investment. Well-defined legislation and strong intellectual property safeguard a sense of assurance, thereby promoting increased investment in initiatives related to research and development.

In the interconnected global society, the attainment of digital resilience necessitates the implementation of collaborative initiatives. The establishment of internationally acknowledged rules and standards through legal frameworks is of utmost importance in facilitating international collaboration. International accords are fundamental in encouraging efficient cooperation among countries in tackling global digital challenges. Legal frameworks are implemented to establish protocols that guarantee responsibility and offer solutions in instances of digital transgressions or conflicts.

They enable individuals and organizations to engage in legal proceedings, thereby promoting confidence in the digital domain and deterring unethical conduct. Legal frameworks play a vital role in the establishment and preservation of digital resilience within the scope of global regulatory systems. These guidelines offer crucial assistance in safeguarding personal freedoms, managing cyber threats, and promoting the development of new ideas. In the current context, where the ability to adapt and anticipate is becoming increasingly important, the use of these frameworks is essential to guaranteeing a resilient and prosperous digital future for both governmental institutions and individuals.

The benefits of cyber-laws are numerous and involve the sheltering of both individuals and entities from the risks associated with cybercrimes, as well as the protection of their online privacy. Similar to traditional legal systems that define acceptable and forbidden behaviour within a society, these rules also include all behaviours occurring on the Internet and inside cyberspace as a whole. In addition to physical transactions, Internet transactions also enjoy the advantages of federal legal protection. The authorities responsible for cybercrime diligently monitor online activity to promptly address any unlawful activities, such as cybercrimes and fraud, thereby upholding the principles of cyber-law enforcement. Also, these laws are integral in formalizing regulations for the sanctioning of individuals involved in cybercrime, thereby promoting the establishment of a safe and protected digital space. Cyber-laws carry a crucial function in the establishment of the fundamental principles of digital governance, mirroring the dynamic nature of the Internet.

The ongoing growth of cyberspace necessitates a constant process of refining and reinforcing cyber-regulations to ensure their full coverage. Several noteworthy emerging developments in cyber-law warrant consideration. The increasing public consciousness over online privacy calls for substantial engagement from the federal government in safeguarding it. The emergence of cloud computing has brought about a substantial exchange of data among computer systems, revealing dangers that necessitate legal consideration. Cryptocurrencies, such as Bitcoin and Ethereum, have gained significant prominence, necessitating the implementation of regulatory measures to safeguard the security of banking systems and transactions. The tendencies highlight the ever-evolving character of cyber-law and the necessity to adjust to technological progress.

Cyber-law incorporates a wide range of legal considerations, recognizing the significant progress made in cybersecurity as a response to the increasing threat posed by malicious cybercrimes. The current rules and regulations are insufficient in addressing the dynamic nature of these emerging risks, necessitating government intervention to establish new, all-including, and efficacious legislation. The successful mitigation of escalating cybercrimes relies on the establishment of collaborative initiatives among governments, regulatory entities, and enterprises on a global scale. The increasing prevalence of cyberspace necessitates the implementation of cybersecurity rules and regulations that include all aspects of online activities. This is important to establish a secure and inclusive environment for all individuals. It is expected that there will be notable advancements in cyber-law shortly. The effectiveness of these laws will depend on the extent to which users comply with them and the level of global collaboration achieved.

4 Understanding Regulatory Authorities

Regulatory authorities are crucial in modern societies by overseeing various aspects of government and industry operations. They are responsible for creating and enforcing rules, regulations, and standards that ensure the safety, fairness, and efficiency of governmental and industrial activities. Regulatory authorities can be broadly categorized into two main types: government agencies and industry-specific bodies. This section aims to provide an understanding of these regulatory entities, their functions, and their significance in modern society.

4.1 Government Agencies

Government agencies are public entities established by governments at various levels to regulate and oversee specific sectors or domains of public life. These agencies derive their authority from legislative acts and are accountable to the respective governments. Government agencies serve as crucial instruments of governance and are entrusted with the responsibility of safeguarding the public interest. They typically operate within a broader legal framework that outlines their powers, duties, and limitations. In this section, we will discuss the key aspects of government agencies as regulatory authorities.

4.1.1 Nature and Purpose of Government Agencies

Government agencies play an important role in facilitating the efficient and regulated operation of a state. The fundamental objective of government is to achieve specified policy objectives. Policy objectives can span a diverse set of domains, such as public health, environmental conservation, financial stability, national security, and other related areas. Government agencies are created to function as specialized entities possessing competence in their designated fields, facilitating the ability to make well-informed judgements and develop legislation that provides advantages for the entirety of society.

4.1.2 Legal Framework and Authority

The legitimacy and power of government agencies are derived from the enactment of legislation by the government or the appropriate legislative body. This legislation confers agencies the authority to formulate and implement regulations within their specified jurisdictions. Also, it establishes the extent of their jurisdiction, outlines the protocols they are obligated to comply to, and stipulates the potential sanctions they are authorized to levy in cases of nonconformity. Government agencies typically possess the jurisdiction to establish regulations that possess the legal weight of statutes, albeit within specific legal limitations and following the principles of due process.

4.1.3 Functions and Responsibilities

Government agencies are generally entrusted with an extensive spectrum of activities and carry substantial obligations. Their duties can be classified into the following general categories:

- **Regulation and Oversight:** Government agencies have responsibility for the establishment and implementation of regulations that oversee a wide range of public and private activities. The primary objective of these regulations is to facilitate compliance to set standards and norms, thereby promoting the principles of safety, justice, and accountability.
- **Licensing and Permitting:** Numerous governmental entities bear the responsibility of granting licences and permits to individuals and enterprises, thereby authorizing their participation in certain activities or sectors. These licences frequently require conditions and prerequisites that must be fulfilled to ensure compliance.
- **Monitoring and Inspection:** Government agencies conduct inspections and audits to assess compliance to regulatory requirements. This involves the examination of records, the conduct of on-site inspections, and the evaluation of procedures to ascertain compliance with established standards.
- **Enforcement:** Government agencies possess the requisite authority to undertake enforcement actions in response to instances of regulatory non-compliance. These measures include a spectrum of consequences, including the imposition of fines and penalties, the revocation of licences or permits, and, in the most severe instances, the initiation of judicial proceedings.
- **Research and Analysis:** Numerous organizations actively participate in research and analytical initiatives to remain adequately informed about emerging develop-

ments and difficulties within their respective fields. The research being conducted provides helpful information for the formulation of new legislation and the revision of current ones.

4.1.4 Accountability and Independence

Government agencies must navigate an equilibrium between being held accountable and maintaining their independence. While they need to be answerable to the government and the constituents they serve, it is equally important for them to possess a certain level of autonomy to provide unbiased judgements and implement legislation without being influenced by political pressures. The preservation of this autonomy is frequently ensured through the implementation of diverse measures, including the establishment of predetermined tenures for agency leaders, the inclusion of non-interference provisions in legislative acts, and the adoption of transparent reporting frameworks.

4.1.5 Examples of Government Agencies

Government agencies are present at multiple tiers of government, including the federal, state, and municipal levels. For instance, government agencies in the United States are demonstrated by the following agencies:

- **Federal Aviation Administration (FAA):** Civil aviation in the United States is regulated and supervised by the FAA, which ensures air travel safety and efficacy.
- **Environmental Protection Agency (EPA):** By regulating environment-impacting activities, such as air and water quality, the EPA is responsible for protecting the environment and public health.
- **Securities and Exchange Commission (SEC):** Surveillance of the securities industry in the United States falls under the purview of the SEC, which ensures investor protection, efficient and equitable markets, and the facilitation of capital formation.
- **Food and Drug Administration (FDA):** The FDA makes sure the safety and effectiveness of medical devices, pharmaceuticals, and the food supply within the United States through regulation.
- **Central Intelligence Agency (CIA):** The CIA is an example of a government agency responsible for national security, intelligence gathering, and analysis.

4.2 Industry-Specific Bodies

In addition to the governmental agencies, industry-specific entities constitute another essential classification of regulatory authorities. These entities are commonly comprised of individuals who possess expertise, have a vested interest, or serve as representatives within a specific industry or area. They bear the responsibility of creating and enforcing regulations and standards within their respective sectors. Industry-specific organizations are established with the primary objective of safeguarding the integrity, safety, and quality of products and services provided within their respective sectors.

4.2.1 Composition and Structure

Sector-specific organizations are often comprised of individuals who possess specialized knowledge and experience in the corresponding subject or sector. The individuals comprising this group may consist of professionals from the industry, representatives from companies operating in the sector, researchers, and specialists in relevant disciplines. The organizational framework of these entities exhibits variability, commonly including boards or committees entrusted with the authority to render judgements and establish benchmarks. Participation in these entities might be either voluntary or obligatory, contingent upon the specific industry and jurisdiction.

4.2.2 Standardization and Regulation

One of the most significant responsibilities of industry-specific organizations is to set and uphold standards and regulations that are tailored to their particular sectors. These standards are established through a collaborative approach that incorporates feedback from industry stakeholders and subject-matter experts. Once defined, these standards function as a set of principles that inform optimal methodologies, ensure product excellence, guarantee safety, and ensure compliance to regulatory requirements within the respective industry.

4.2.3 Self-regulation and Compliance

Industry-specific organizations frequently participate in self-regulatory practices, wherein they establish and enforce their own set of norms and regulations. The use of this self-regulatory strategy enables industries to uphold a certain level of autonomy and skill in effectively handling the distinct difficulties they face. It is anticipated that individuals within the sector will comply to these norms. Nevertheless, failure to comply may lead to the imposition of penalties or sanctions by the relevant industry regulatory authority.

4.2.4 Advantages and Challenges

Industry-specific organizations provide numerous benefits, such as their extensive understanding of the industry, their ability to adapt quickly to changing issues, and their capacity to promote innovation within their respective sectors. Academic institutions have the potential to serve as a helpful intermediary, facilitating the connection between governmental rules and the specific requirements of industries. The industry stakeholders may encounter problems, including the possibility of conflicts of interest, the possibility of self-regulation, and the necessity for robust enforcement mechanisms.

4.2.5 Examples of Industry-Specific Bodies

There is a large number of industry-specific organizations that exist within diverse sectors. The following are some of the examples:

- **International Electrotechnical Commission (IEC):** The IEC is a global organization that develops and publishes international standards for electrical and electronic technologies.
- **Financial Industry Regulatory Authority (FINRA):** FINRA is a self-regulatory organization in the United States responsible for regulating broker-dealers and protecting investors.
- **International Air Transport Association (IATA):** IATA is an industry association representing airlines worldwide, setting standards and best practices for the aviation industry.
- **International Organization for Standardization (ISO):** ISO is a global body that develops and publishes international standards for a wide range of industries, including manufacturing.

5 Adaptations for Legal Frameworks

There is a need for legal frameworks that reflect the dynamic nature of the constantly expanding digital interconnectedness. The ability to effectively handle and respond to the concept of digital resilience requires agility and flexibility. This involves engaging in proactive measures to prepare for and manage any digital risks. The attainment of digital resilience necessitates the adoption of a proactive methodology that foresees and effectively tackles forthcoming issues.

(1) **AI Governance:** With the rapid advancement of AI technology, there is a growing need for governance and ethical guidelines. These guidelines should emphasize principles like openness, accountability, and equity to promote innovation while preventing the misuse of AI for the betterment of society.

(2) **Data Sovereignty:** As data flows across national boundaries, legal frameworks must grapple with international agreements, data localization laws, and the need for seamless data movement in the digital space. Achieving consensus on data governance at a global level is essential.
(3) **Cybersecurity:** To combat digital threats effectively, legislative frameworks for cybersecurity are essential. These frameworks should have strict requirements to mitigate and control cyberattacks, requiring international cooperation due to the interconnected nature of the digital economy.
(4) **Digital Rights and Privacy:** Ensuring the protection of individuals' digital rights and privacy in times of pervasive connectivity is a significant concern. Legal frameworks should be adapted to address emerging technologies like biometrics and facial recognition while mitigating intrusive surveillance practices.
(5) **Blockchain and Cryptocurrencies:** The importance of blockchain technology and cryptocurrencies necessitates regulatory clarification across various domains. Regulatory measures are crucial to preventing financial crimes associated with cryptocurrencies.
(6) **International Collaboration:** Given the interconnectedness of the digital world, international collaboration in the development of legal frameworks is increasingly critical. Nations should work together to establish uniform standards, exchange threat intelligence, and address global issues like cybercrime and digital resource distribution.
(7) **Digital Inclusivity:** Legal frameworks should emphasize equitable access to digital resources and services for all individuals. Regulatory interventions should aim to bridge the digital divide, promote digital literacy, and ensure inclusivity in the digital space.
(8) **Emerging Technologies:** Rapid advancements in emerging technologies such as quantum computing, advanced robotics, and augmented reality pose new challenges. Anticipatory rules may be necessary to guide the ethical development and utilization of these technologies.
(9) **Environmental Considerations:** Environmental concerns, particularly related to data centres and energy-intensive digital operations, are gaining prominence. Implementing sustainability and environmental criteria into regulatory systems can align technology with environmentally advantageous practices.

The progression of AI highlights the necessary need for governance and ethical frameworks. It can be anticipated that there will be substantial modifications to legal frameworks, resulting in the establishment of detailed guidelines for the development and utilization of AI. The rules will emphasize fundamental concepts such as transparency, accountability, and equity to strike a balance between encouraging innovation and safeguarding against potential misuse. An additional growing issue is the concept of data sovereignty and the unrestricted movement of data across international boundaries.

Legal frameworks are confronted with the complex challenges that emerge from international agreements on data sharing, legislation about data localization, and the inherent absence of borders in the digital world. The achievement of global digital resilience and the facilitation of uninterrupted data flows necessitate the establishment of a shared understanding and agreement on the principles and mechanisms of data governance. With the ever-evolving digital threats, it is crucial to establish legal frameworks that address cybersecurity concerns. These frameworks should incorporate strong standards aimed at effectively mitigating and controlling attacks.

The interconnected digital economy also necessitates the importance of international collaboration. The preservation of digital rights and privacy in a corresponding global context is of importance. The existing legal frameworks necessitate amendments to effectively tackle the increasing range of rights associated with data, minimize invasive monitoring practices, and guarantee the ethical utilization of developing technology such as biometrics and facial recognition. The increasing importance of blockchain technology and cryptocurrencies calls for the adaptation of legal systems. Ensuring regulatory transparency about the use of blockchain technology in supply chains and voting systems is of high importance, along with implementing methods to mitigate financial frauds associated with cryptocurrencies.

In the interconnected digital setting, it is essential to enhance international collaboration to develop consistent standards, facilitate the exchange of threat intelligence, and effectively tackle global challenges such as cybercrime and the equitable distribution of digital resources. The need to ensure fair and equal availability of digital resources and services is very important. The effective management of difficulties posed by emerging technologies, including quantum computing, advanced robotics, and augmented reality, may necessitate prompt adaptations to current legal frameworks. The digital infrastructure is currently facing problems, particularly concerning data centres and energy-intensive activities.

To address these concerns, it is necessary to incorporate sustainability and environmental standards into regulatory frameworks. This will help ensure that technology is aligned with eco-friendly practices. The ever-evolving character of the modern digital setting necessitates the continuous revision of international legal frameworks. The adaptability and forward-looking nature of these frameworks are crucial to ensuring global digital resilience. They play a significant role in promoting innovation, protecting human rights, and tackling the complex difficulties brought about by the digital revolution, which continuously alters societies and economies. This section highlights the significance of proactively predicting and embracing transformative developments in digital resilience, specifically within a digital space.

6 Cybersecurity and Privacy Laws, USA

The United States of America is widely recognized as the largest global economy, boasting numerous military installations abroad and occupying prominent leadership roles within major international organizations. Further, the country is home to several prominent technology firms and corporations that have established their headquarters within its borders.

In the absence of federal privacy legislation, numerous organizations enjoy a significant degree of autonomy in their use of data, unless a particular state has enacted its legislation about data privacy. In the majority of states, corporations have the legal authority to utilize, disclose, or trade any data they acquire about individuals without the obligation to provide prior notification of such activities. The absence of uniform national legislation mandates notification requirements for companies in the event of a data breach or unauthorized exposure of personal information.

If a corporation discloses an individual's sensitive data, such as health records or location data, to third-party entities such as data brokers, these intermediaries possess the ability to subsequently vend or disseminate said data without providing any form of notification to the individual concerned. The prevailing belief among individuals is that they possess a sense of protection until this perception is proven false. The majority of laws about privacy are jurisdiction specific, focusing on particular sectors and features of the domain rather than providing a complete framework that addresses all potential issues in most circumstances. Regardless, it is important to note that specific industries, including healthcare, banking, and credit reporting, are subject to a range of regulatory measures.

6.1 FCRA, 1970

The section is dedicated to examining a comparably complex matter that has posed significant difficulties within the domains of the real estate and finance industries. Before going into the particulars, it is essential to clarify the precise definitions and ramifications of the Fair Credit Reporting Act (FCRA) [Uni70]. The FCRA, which was established as a federal statute in 1970, functions as the legal structure that oversees the acquisition of consumer credit data and the authorization for accessing their credit reports. The primary objective of this legislation is to guarantee the equitable, precise, and confidential handling of individuals' data stored in the databases of credit reporting entities. The FCRA contains several significant points.

Firstly, it governs the practices of credit bureaus in the acquisition and distribution of personal consumer information. Businesses use this information for a variety of purposes, such as assessing loan applications and deciding whether to purchase insurance. Employers may also have the potential to access these credit reports. The FCRA bestows upon consumers some entitlements, such as the privilege to acquire their credit reports without incurring any fees, with a minimum frequency of once per year. Also, non-compliance with the FCRA may lead to various consequences, such as the imposition of penalties and the possibility of incurring financial losses. The responsibility for enforcing the FCRA is assigned to two federal entities, namely the Federal Trade Commission (FTC) and the Consumer Financial Protection Bureau (CFPB).

The FCRA serves as the predominant federal legislation that regulates the processes involved in obtaining and reporting consumer credit information. It outlines the specific methods for acquiring, retaining, and disclosing such information while also providing safeguards to ensure consumers have access to their credit reports. The FTC and the Consumer Financial Protection Bureau (CFPB) are responsible for overseeing and enforcing the FCRA. Also, some states have established their credit reporting standards. The FCRA is located within United States Code Title 15, specifically in Section 1681.1. The credit reporting domain includes prominent credit reporting agencies, including Equifax, Experian, and TransUnion, in addition to specialist entities that gather and commercialize information related to the financial backgrounds of individual consumers.

The data is crucial in the computation of individuals' credit scores, exerting an impact on variables such as loan interest rates and loan eligibility. The FCRA, officially known as Public Law No. 91-508, was established by the US Congress in 1970 to promote the precision, equity, and confidentiality of personal data contained in credit reports. Since its inception, the FCRA has undergone several changes. The scope of the FCRA includes the authorized collection of data, describing the specific categories of information that credit bureaus are permitted to gather. These categories contain various aspects such as payment history, past loans, existing debts, employment details, residential addresses, bankruptcy records, child support responsibilities, and records of arrests. Also, the FCRA establishes limitations on the individuals and situations in which credit reports can be accessed.

Lenders use credit reports in credit-related transactions, insurance companies examine them when evaluating policies, and government entities access reports following specific legal requirements. It is important to highlight that, in specific cases, the acquisition of consumer consent or written authorization is mandatory before the disclosure of credit reports. This requirement is particularly relevant when employers are in the process of requesting an applicant's credit report. However, it is important to note that certain lenders and insurers can obtain certain sections of a credit report without obtaining consent from the individual to make unsolicited credit or insurance offers. This process, known as prescreening, is allowed under the FCRA, although consumers have the option to choose not to participate in this practice. The FCRA stipulates that businesses are required to indicate the intended reason when acquiring a credit report, whether it is for loan applications, employment evaluations, or landlord credit assessments. The use of credit reports for unauthorized purposes is considered a breach of the FCRA and can lead to financial penalties.

6.2 FERPA, 1974

The Family Educational Rights and Privacy Act (FERPA) [Edu21] is federal legislation enacted in 1974, aiming primarily to protect the confidentiality of student educational information. The entity in question exercises authority over educational institutions at all levels, including both public and private establishments catering to elementary, secondary, and post-secondary education. This includes state or local education agencies that get financial assistance from programmes administered by the US Department of Education. The FERPA has a twofold objective. Firstly, it grants parents or eligible students increased authority over their educational records. Secondly, it establishes a restriction on educational institutions, preventing them from disclosing 'personally identifiable information in education records' unless they obtain written consent from an eligible student or, in the case of a minor student, the consent of the student's parents. An eligible student is operationally defined as an individual who has attained the age of 18 or is currently enrolled in an educational institution beyond the secondary level.

Parents or qualified students are granted certain rights under the FERPA, which includes the entitlement to see and evaluate the educational records of the student that are maintained by the educational institution. Although educational institutions are not legally required to furnish duplicates of these documents, they may consider making exceptions in cases where geographical constraints render it unfeasible for parents or eligible students to access the original records for examination. Individuals possess the right to seek rectification for documents that they judge to be erroneous or deceptive. If the educational institution refuses to make changes to the academic record, it is possible to initiate a formal hearing by making a request. Following the hearing's completion, if the school maintains its decision to retain the record without any modifications, the parent or qualifying student is entitled to append a statement to the record articulating their perspective on the contested facts.

The FERPA confers individuals with the entitlement to restrict the dissemination of personally identifiable information and to procure a copy of the educational institution's policy that governs the accessibility of educational records. It is important to highlight that educational institutions are required to have written consent from parents or eligible students to disclose any information contained in a student's education record. Failure to comply to the restrictions outlined in the FERPA exposes these institutions to the potential consequence of losing federal funding. It is essential to acknowledge that parochial and private elementary and secondary schools, which typically do not receive federal financing through programmes overseen by the US Department of Education, are excluded from the FERPA. Nevertheless, private post-secondary institutions commonly receive this type of funding and, as a result, are obligated to comply to the requirements outlined in the FERPA.

The FERPA permits some circumstances in which student educational records may be released without obtaining authorization. The allowable disclosures encompass the dissemination of information to school officials with genuine educational interests, educational institutions that a student is transferring to, authorized officials for audit or evaluation purposes, relevant parties involved in matters pertaining to a student's financial aid, organizations conducting specific studies on behalf of the school, accrediting organizations, authorities in situations involving safety and health emergencies, local and state officials within a juvenile justice system in accordance with applicable state legislation, and complying with a court's decree or lawfully issued summons.

6.3 *ECPA, 1986*

The Electronic Communications Privacy Act of 1986 (ECPA) [Uni86] was enacted by the US Congress as federal legislation. The act consists of various significant sections, each meticulously designed to guarantee the privacy of electronic communication through different methods. The Wiretap Act, which is a component of the Electronic Communications Privacy Act (ECPA), establishes the prohibition against purposefully intercepting wire, oral, and electronic communications during their transmission. The act of exploiting or revealing the content of communications acquired in contravention of this rule is deemed illegal. Similarly, the primary objective of the Saved Conversations Act (SCA) is to protect the confidentiality and privacy of wired and electronic conversations that are stored in digital form. The rule includes the permission and accessibility of these communications while concurrently setting protocols for unauthorized entry into systems used for the transmission of said data. The ECPA includes the legislation known as the pen register and trap and trace device statute, which regulates the use of wired or electronic communication devices by governmental authorities.

The ECPA defines the specific protocols that law enforcement entities must follow to legitimately obtain access to stored data. Similarly, it defines rules for service providers concerning the discretionary disclosure of those communications. Under specific circumstances, the ECPA grants individuals the ability to initiate civil legal action to seek financial compensation or court orders to prevent further violations of the provisions outlined in the law. The main purpose of the ECPA is to enhance the scope of legal limitations placed on governmental wiretapping activities. It was originally developed to govern telephonic interactions, but these restrictions have been broadened to include the transfer of digital information facilitated by computerized systems. The legislation currently under examination includes a variety of provisions aimed at prohibiting unauthorized access to electronically stored communications, commonly referred to as the Stored Communications Act. It facilitates the surveillance of telephonic communications by utilizing pen trap functionalities.

The enactment of the ECPA served as an amendment to the wiretap statute found under the Omnibus Crime Control and Safe Streets Act of 1968. The primary objective of this initiative was to establish legal restrictions aimed at mitigating illegal government intervention in private electronic communications. The ECPA has experienced multiple revisions during its history, which have been enacted through the passing of different legislative acts. The legislative measures contain the Communications Assistance for Law Enforcement Act (CALEA) of 1994, the USA Patriot Act of 2001, additional reauthorization acts of the USA Patriot Act in 2006, and the FISA Amendments Act of 2008. The ECPA provides an inclusive definition of 'electronic communications', which includes the transmission of many types of data over different mediums such as wire, radio, electromagnetic, photoelectronic, or photo optic means.

These forms of communication are regarded as having an impact on trade between states or with international entities. However, the law explicitly outlines several exceptions, including wire or oral communication, communications facilitated by tone-only paging devices, communications facilitated by tracking devices, and electronic fund transfer information retained by banking institutions solely for fund transfers. The first provision of the ECPA provides protective measures for wire, oral, and electronic communications during their transmission. In certain instances, it enforces strict criteria for the acquisition of search warrants. Title II, also known as the Stored Communications Act (SCA), provides safeguards for electronic communications stored on computer systems; however, the criteria for obtaining a warrant are slightly less strict compared to those outlined in Title I. The legal framework in the United States, namely Title III of the United States Code and Section 3121, establishes the specific restrictions that regulate the use of pen registers and trap and trace devices for obtaining information linked to the transmission of wire or electronic communications. The limitations dictate that engagement in data-gathering initiatives is contingent upon the acquisition of appropriate permission through a court order.

The principal objective of the ECPA is to protect individuals from illegal interception of electronic communications, irrespective of whether such eavesdropping is conducted by the federal government or private individuals. The actions outlined above include wiretapping, interception of computer transmissions, acquisition of stored electronic communications, and monitoring of telephone calls. Participating in such actions without proper legal authorization or getting consent from the pertinent communication party is regarded as a violation of federal legislation. It is essential to acknowledge that the ECPA does not embrace video recordings that lack an accompanying audio component. Numerous states have the legal authority to enforce supplementary limitations on the act of recording private communications, frequently mandating the specific consent of all those participating in the communication. The ECPA imposes stricter requirements for search warrants in cases involving current conversations, but it does not elevate the threshold for communication data.

6.4 VPPA, 1988

The Video Privacy Protection Act of 1988 (VPPA) [Uni88] is a federal law implemented in the United States. Its main purpose is to govern the dissemination of data about individuals' video usage habits. The legislation imposes strict criteria that must be satisfied before the divulgence of said data, emphasizing the essential nature of getting the consent of consumers. This section will examine the historical context of the VPPA, its ramifications for suppliers of online video content, and the obstacles it poses in the digital age. The beginnings of the VPPA can be attributed to a particular incident that took place more than thirty years ago. In the course of the confirmation hearings for Judge Robert H. Bork, a nominee for the Supreme

Court, a journalist disseminated information about Judge Bork's rental records from a Blockbuster video store. The lack of details within the rental history precipitated quick legislative measures, resulting in the establishment of the VPPA.

The primary component of the VPPA forbids video tape service providers (VTSPs) from intentionally revealing identifiable information about specific video content or service requests made by an individual. Instances of non-compliance with this legislation may lead to substantial statutory penalties amounting to USD 2,500 for each infraction, alongside the possibility of punitive damages, reimbursement of legal expenses, and the provision of equitable remedies. At the outset, there was a scarcity of legal action taken under the VPPA, mostly because traditional video tape service providers (VTSPs) had less motivation to provide the rental histories of their clients unless it served marketing objectives or was prompted by law enforcement inquiries. The legislation included provisions that allowed for exceptions in such situations. The emergence of online video platforms has presented a considerable obstacle for distributors of online video content due to the VPPA. The objective of these providers was to incorporate their services into social media platforms, enabling users to share their experiences of consuming videos and participate in data sharing with third-party technology platforms to enhance user experiences and facilitate targeted advertising.

Before a significant modification in 2013, the VPPA lacked provisions or exemptions that would enable the sharing of data or content on social platforms. This posed a predicament for Internet streaming platforms, which were also classified as VTSPs. In light of the increasing legal disputes and operational difficulties encountered by online video content providers, Congress enacted changes to the VPPA in 2013. The revisions enabled the sharing of information with third parties, contingent upon customers granting 'informed, written consent' that is distinct from any other legal or financial responsibilities. Also, users needed to be provided with the ability to revoke their consent either on a per-instance basis or for a predetermined period. However, the language of the amendment continued to exhibit ambiguity, particularly concerning online platforms. There has been a growing inquiry into the definition of 'informed, written consent' and whether activities such as clicking a 'like' button on social media platforms can be considered as meeting the criteria for such consent. The VPPA does include additional provisions that allow for restricted disclosures to third parties.

Importantly, it facilitated the disclosure of consumers' names and addresses to third parties only to directly sell goods and services to the consumer. Despite its complexity, this clause was designed to facilitate the disclosure of client lists by VTSPs for direct marketing. Despite the enactment of the 2013 Amendment and the subsequent allowance of these authorized disclosures, legal disputes about the VPPA continued to persist within online platforms. The courts have rendered multiple decisions about the extent and implementation of the law. These rulings have dealt with issues such as determining whether individuals who use free mobile applications can be considered 'subscribers' according to the VPPA, as well as establishing the liability of video tape service providers (VTSPs) for personal information that is displayed on devices outside of their jurisdiction.

The VPPA, which was passed into law in 1988, has experienced notable modifications and faced various obstacles in the age of digital technology. The introduction of the 2013 Amendment brought about certain degrees of flexibility for suppliers of online video material. However, the wording of the legislation and the constantly changing nature of the Internet environment continue to give rise to inquiries over its implementation. Legal advancements and judicial decisions continue to play an essential role in influencing the understanding and application of this legislation concerning modern technology and the exchange of data.

6.5 HIPAA, 1996

The Health Insurance Portability and Accountability Act (HIPAA) [Con96], a legislative measure implemented in the United States on August 21, 1996, is designed to govern the privacy and security of medical information to protect its confidentiality. HIPAA supersedes state legislation unless the state laws provide greater protection for medical information. HIPAA, formally referred to as Public Law 104–191, includes two fundamental aims. First and foremost, this policy guarantees uninterrupted access to health insurance for individuals, irrespective of any transitions in employment. Besides, its objective is to mitigate healthcare expenditures through the implementation of standardized electronic protocols for administrative and financial procedures. Similarly, it aims to address the problems of misappropriation, fraud, and inefficiencies within the healthcare and insurance sectors while also enhancing the accessibility of long-term care services and insurance.

HIPAA is federal legislation that creates a set of national standards. These standards aim to ensure the confidentiality and security of patient health information. Precisely, HIPAA focuses on prohibiting the unauthorized publication of such information without the consent of the patient. The HIPAA Privacy Rule was created by the US Department of Health and Human Services (HHS) to effectively execute the provisions. Likewise, the HIPAA Security Rule emphasizes safeguarding electronic protected health information (e-PHI), including health data that is individually identifiable and kept in electronic format.

The Privacy Rule regulates the utilization and dissemination of safeguarded health information (PHI) by entities that fall under its jurisdiction, including both individuals and organizations. This outlines the rights of individuals regarding their health data, to strike a balance between safeguarding privacy and facilitating the essential exchange of data for the provision of high-quality healthcare, public health initiatives, and general welfare. The entities included within the scope of covered entities consist of healthcare providers, health plans, healthcare clearinghouses, and business associates. The utilization and disclosure of protected health information (PHI) for designated reasons, including treatment, payment, and healthcare operations, can be carried out without the need for individual authorization.

The Privacy Rule also permits the utilization and disclosure of protected health information for purposes that serve the public interest or benefit, legal obligations, deceased individuals, organ donations, and research conducted under specific circumstances, among other instances. The primary objective of the HIPAA Security Rule is to ensure the protection of electronic protected health information (e-PHI) by addressing its confidentiality, integrity, and availability. This means that entities subject to HIPAA regulations must actively identify and address potential security threats as well as implement measures to prevent unauthorized access, use, or disclosure of e-PHI.

The oversight of HIPAA laws is conducted by the Office for Civil Rights within the Department of Health and Human Services (HHS), which has the authority to impose civil, monetary, or criminal penalties for any breaches of these regulations. It is worth mentioning that HIPAA predominantly refers to the exchange of information between individuals and entities classified as 'covered entities', such as medical practitioners, healthcare facilities, pharmacies, and insurance providers. The scope of health data covered by the system is limited since it does not include data from sources such as Google Fit and the Wear Operating System, nor does it address inquiries related to COVID-19 vaccination status.

6.6 COPPA, 1998

The Children's Online Privacy Protection Act (COPPA) [Uni98] is federal legislation that was passed in 1998 in the United States. Its main objective is to protect the online privacy of individuals who are under the age of 13. In the dynamic and ever-changing environment of the Internet, characterized by rapid technological advancements and widespread integration into many industries, including business, healthcare, insurance, finance, and law, the COPPA is significant. To effectively mitigate potential concerns stemming from these expeditious transformations, COPPA expands its jurisdiction beyond mere websites, embracing various online services. The essence of the COPPA is the implementation of strict laws about the gathering and utilization of data associated with individuals under the age of 13. These regulations grant parents, including legal guardians, the power and legal responsibility to supervise and give consent for the disclosure of their children's personal information on digital platforms.

The FTC serves as the regulatory agency tasked with overseeing and implementing the COPPA, thereby ensuring compliance with its provision. One of the fundamental principles of the COPPA is the mandate for websites and online services to acquire verifiable parental consent before gathering or utilizing any personal data from young users. Along with that, it requires the incorporation of precise components in privacy policies and compels the prominent exhibition of those policies in all locations where data is collected. The COPPA also describes the specific procedure and time frame for acquiring verified consent from a parent or legal guardian. The legislation clearly defines the legal obligations of website operators concerning the protection of children's privacy and ensuring their safety in the online environment. These obligations include imposing limitations on targeted marketing activities aimed at children below the age of 13. The origins of COPPA may be traced back to the need to address the significant increase in online marketing aimed at children throughout the 1990s. This age witnessed a substantial number of websites engaging in the covert collection of personal information from children without obtaining the necessary awareness or agreement from their parents. The research provided evidence supporting the assertion that children had a limited comprehension of the possible threats linked to the disclosure of personal information on the Internet.

The inclusion of media exposés has served to highlight the inherent simplicity with which private data may be obtained from minors, thereby intensifying the public's call for legal measures to be taken. In terms of its scope, the COPPA includes all websites and online platforms that are involved in the gathering of data from individuals under the age of 13, including social media platforms. Although the legislation does not explicitly prescribe a specific approach for gaining parental agreement, the FTC has provided suggestions to aid website operators in ensuring compliance with the COPPA. The rules comprise the clear and transparent display of consent forms that can be downloaded and delivered through mail or fax to the operator. They also involve the transparent disclosure of privacy policies that provide detailed information about data practices. Besides, conscientious efforts are made to enlighten parents about the practices of data collection.

Correspondingly, it is important to emphasize the importance of implementing appropriate measures to safeguard the privacy of children. This includes the responsible handling of personal information, limiting its use to its intended purpose, and ensuring its secure destruction once it is no longer required. Another aspect to consider is the age and identity verification process, which involves credit card validation as mandated by the COPPA. Similarly, in compliance with COPPA regulations, website owners are required to provide parents with access to their child's personal information upon request, with the exception that updates to this information are not allowed. The introduction of COPPA also implemented the Safe Harbour programme, which grants organizations the authority to develop self-regulatory principles to safeguard children. These organizations have the option to submit their guidelines to the Federal Trade Commission (FTC) to obtain certification. This certification then grants them the authority to certify other commercial websites for their compliance with the COPPA.

6.7 GLBA, 1999

The Gramm-Leach-Bliley Act (GLBA) [Uni99], also referred to as the Financial Services Modernization Act of 1999, is federal legislation in the United States that establishes regulations and guidelines about the handling and safeguarding of personal information by financial institutions. The legislation consists of three primary components: the Financial Privacy Rule, which governs the acquisition and distribution of sensitive financial data; the Safeguards Rule, which mandates that financial institutions establish protective measures to ensure the security of this information; and the Pretexting Provisions, which expressly forbid deceptive tactics used to obtain private data. According to the provisions outlined in the GLBA, financial institutions are required to furnish written notifications to their clients that explicitly outline their policies and procedures about the disclosure and exchange of information.

The legislation, enacted in 1999, was critical in the demolition of substantial regulations outlined in both the Glass-Steagall Banking Act of 1933 and the Bank Holding Company Act of 1956. The enactment of this legislative measure enabled the amalgamation of banks, brokerage firms, and insurance companies via mergers, leading to the establishment of a novel institutional structure that allowed bank-holding corporations to obtain full-service investment banks and insurance enterprises and vice versa. The GLBA also granted the US Federal Reserve increased supervisory power in overseeing the development of financial arrangements while also augmenting the data security obligations of the Federal Deposit Insurance Corporation (FDIC). The main goal of the GLBA is to implement regulations that mandate financial institutions and their affiliated organizations to safeguard personally identifiable information (PII) acquired from customer records, irrespective of the medium in which the data is held, including but not limited to physical documents, electronic files, or any other form. Businesses that have been impacted are required to comply with strict data security rules as stipulated by the Act. They have a legal responsibility to protect client confidentiality and prevent unauthorized access.

To comply to the requirements set forth by the GLBA, organizations must develop privacy practices and policies that clearly outline procedures related to data collection, sharing, distribution, and utilization. These practices and policies should provide consumers with the opportunity to exercise their autonomy by deciding the degree to which a company can disclose or retain their personal information. The provisions of the GLBA about personally identifiable information (PII) apply to all non-public personal information obtained through different channels or supplied willingly by consumers in the course of transactions. The holistic strategy for safeguarding privacy includes the incorporation of directives about data storage and security within information security rules, intending to minimize any risks or dangers that may lead to unfavourable outcomes for users. The Act also implemented restrictions on the exchange of consumer information and sought to mitigate the practice of pretexting, which involves the acquisition of personal information using deceitful methods.

The Gramm-Leach-Bliley Act of 1999, sometimes known as GLBA, is a notable legislative initiative aimed at the modernization of the financial sector. The Glass-Steagall Act of 1933, which prohibited commercial banks from participating in activities such as investments and insurance, has formerly constrained the capacity of banks to provide a broader spectrum of financial services. According to the provisions of the GLBA, individuals who have responsibility for providing financial services are subject to a restriction that prohibits them from engaging in the sale or management of many types of financial products or instruments. Also, financial institutions are obligated to provide customers with clear and transparent information regarding their practices related to the sharing of customer information.

The source of the GLBA can be attributed to the merger of Citigroup, a financial institution specializing in commercial banking, and Travellers Group, an insurance firm, resulting in the establishment of Citigroup. The merger in question was found to violate both the Glass-Steagall Act and the Bank Holding Company Act of 1956. Therefore, the enactment of the Gramm-Leach-Bliley Financial Services Modernization Act (GLBA) was intended to streamline mergers and acquisitions, resulting in the removal of the Glass-Steagall Act and granting financial institutions the ability to expand their service offerings. The GLBA requires financial institutions that offer lending services, financial or investment advice, and insurance to provide disclosure of their practices related to sharing client information, thereby safeguarding consumer privacy. Customers are provided with the opportunity to 'opt out' if they do not wish to have their sensitive information shared. The primary objective of this Act was to establish a framework for safeguarding privacy by imposing certain limitations on the sale of personal data and addressing the misleading acquisition of personal information, commonly known as pretexting.

6.8 CCPA, 2018

In the latter half of June 2018, the state of California approved Assembly Bill 375 (AB 375), a legislative bill about consumer privacy. The prospective impact of this legislation on companies in the United States is expected to be stronger in comparison to the GDPR of the European Union, which was implemented in May 2018. While the California law lacks certain strict standards present in the GDPR, such as the strict 72-hour breach reporting deadline, it surpasses the GDPR in several aspects. The California Consumer Privacy Act (CCPA) [Sta18] adopts a broader interpretation of the extent of personal data. Hence, the foremost focus on security revolves around the recognition and safeguarding of this vast assortment of sensitive data. The implementation of the CCPA has garnered widespread recognition as a robust data privacy policy in the United States.

The legislation in question relates to enterprises that engage in the acquisition of personal data from individuals. It defines the specific entitlements that are granted to these individuals. As per the provisions of the CCPA, individuals possess the right to get information about the personal data that organizations gather as well as the intended recipients of that data. Individuals also have the right to legally request the deletion of their collected personal data, the ability to refuse the sale of their personal information, and the assurance of fair treatment when exercising their privacy rights.

The CCPA had a subsequent modification known as the California Privacy Rights Act (CPRA), which was officially ratified in 2020 and subsequently became enforceable in 2023. The alteration expanded the range of consumer rights by including the correction of inaccurate data obtained by commercial entities as well as providing consumers with control over the use and distribution of sensitive information. The CCPA confers upon California customers the power to solicit access to all data retained by a firm about their personal information, together with a list of third-party entities with whom their data is shared. It grants customers the ability to commence legal actions against corporations for violations of privacy standards, irrespective of the existence of a data breach.

The CCPA applies to organizations that cater to individuals residing in California and satisfy specific eligibility requirements. The conditions include the requirement of annual revenues exceeding USD 25 million, the retention of personal data for a minimum of 50,000 individuals, or the generation of over 50% of their revenue through the sale of personal information, irrespective of the company's size. It is important to highlight that firms are not obligated to establish their headquarters in California or within the United States to fall under the purview of this Act. However, in April, an amendment was ratified that explicitly exempts 'insurance institutions, agents, and support organizations' from the provisions of the legislation. The rationale for this exclusion is supported by the fact that these businesses are already subject to similar requirements outlined in the Insurance Information and Privacy Protection Act (IIPPA) of California.

The scope of the CCPA includes a wide set of data categories that extend beyond the definition of sensitive data as specified by the GDPR. This includes olfactory data, browsing history, and records of user interactions with websites or applications. The legal definition of 'personal information' includes various elements, such as identifiers, attributes associated with protected groups, commercial data, biometric data, Internet usage details, geolocation data, audiovisual data, professional particulars, educational particulars, inferences derived from the data, and publicly accessible information, subject to certain exemptions. The primary objective of AB 874, the legislative proposal under consideration, is to establish a distinct classification for publicly accessible consumer data that has been deidentified and aggregated to differentiate it from personally identifiable information (PII).

The CCPA initially included employee data within its purview, but it was subsequently exempted by an amendment that received approval in April. AB 25 introduces an amendment that grants a restricted exemption for the gathering of personal data from various individuals, including job seekers, proprietors, directors, executives, medical staff, and contractors. It is significant to acknowledge that the exemption is scheduled to conclude on January 1, 2021. The implementation of the CCPA enforces substantial privacy regulations, which require enterprises to offer customers the choice to abstain from sharing their data with other parties. The capacity to classify collected data based on user privacy preferences is necessary for this task. Organizations can offer incentives to individuals who want to reveal their data, perhaps leading to adjustments in price structures based on privacy preferences.

The CCPA provides California customers with a wider range of opportunities to view their data records, in contrast to the GDPR. Organizations may face challenges when attempting to integrate a significant amount of data, primarily due to its dispersion across multiple platforms and file types. Compliance with the 45-day time frame stipulated by the CCPA for the provision of thorough data reports poses notable difficulties, especially when it comes to the disclosure of data sales to external entities. The enforcement of security and breach response criteria under the CCPA is not as that under the GDPR. However, this legislation imposes penalties on firms that are responsible for the disclosure or negligence leading to the compromise of consumer data. The courts have the authority to grant injunctive or declaratory remedies at their discretion.

In contrast to the GDPR, the California Customer Privacy Act (CCPA) does not provide a legal obligation for firms to disclose breaches. The levying of penalties by the CCPA is contingent upon the presence of customer grievances. Hence, it is important to comprehend the precise definition of private data as per the CCPA and use suitable security protocols to safeguard it. It is commonly understood that organizations that have successfully attained compliance with the GDPR are generally not required to conduct further actions to align with the data security standards specified in the CCPA.

6.9 ADPPA, 2022

The American Data Privacy and Protection Act (ADPPA) [Uni22] is federal legislation enacted in the United States with the primary objective of safeguarding the personal data of individuals. The fundamental objective of this initiative is to guarantee the protection and privacy of consumer data. The legislation grants privacy protections to consumers in the United States concerning the use of their data while also instituting a mechanism for ensuring compliance. If the ADPPA is enacted, it will supersede analogous laws at the state level. This implies that organizations that contravene a regulation at both the state and ADPPA levels may face sanctions imposed by the FTC, the entity entrusted with the enforcement of the ADPPA. As an illustration, the California Consumer Privacy Act (CCPA) mandates that businesses apprise customers of their privacy rights through the provision of a privacy policy or the display of pertinent information on their websites.

On the other hand, the ADPPA stipulates that enterprises are required to uphold a transparent and readily accessible privacy policy on their online platforms. The responsibility for enforcing this provision for companies that provide services to citizens of California would be under the purview of the FTC rather than the California Attorney General. Numerous state laws within the United States provide privacy protection to their respective residents. However, the ADPPA, a bipartisan federal initiative, seeks to establish itself as the inaugural all-including federal privacy legislation, with the primary objective of defending the privacy of all customers in the country.

The objective of this section is to describe the characteristics of the ADPPA, including its extent, prerequisites, measures for compliance, recommendations for formulating a privacy policy that aligns with ADPPA standards, and the possible repercussions for failure to comply. In the case of enterprises catering to the demographic of California residents, the absence of a privacy policy would result in a transfer of the obligation to enforce the relevant legislation from the California Attorney General to the FTC. The Administrative Procedure and Practice Act (ADPPA) relates to entities that are specifically classified as covered entities. These include any institution or individual engaged in the collection and processing of data that falls under the purview of regulation, either through direct involvement or as a representative of another organization.

The term 'covered data', as stipulated by the American Data Privacy and Protection Act (ADPPA), includes both information and technologies that can identify an individual, either on their own or when combined with other data. This definition does not include de-identified data (data from which personal information has been removed), employee data, or publicly available information. Sensitive covered data refers to a distinct classification of covered data, including non-public personal information such as social security, driver's licence, and passport numbers, as well as personal health or financial data. It includes biometric data like fingerprints or retinal scans, precise geolocation information, and private communication data such as emails, texts, phone calls, and voicemails. The ADPPA requires that covered organizations comply with its standards in four main domains:

(1) **Data Minimization:** Businesses must only collect essential information for service provision, communication, or permissible purposes.
(2) **Loyalty Duties:** The ADPPA prohibits specific data practices unless certain circumstances apply, including the transfer of sensitive personal information to third parties and the collection, processing, or transfer of users' personal Internet browsing history.
(3) **Privacy by Design:** Covered entities must prioritize minors' privacy, mitigate general privacy risks, and comply with all relevant privacy laws, following the concept of privacy by design.
(4) **Loyalty to Individuals Concerning Pricing:** Covered entities cannot condition services or prices based on consumers' data.

To comply with the ADPPA, businesses should:

- Publish a publicly accessible privacy policy on their websites.
- Maintain a concise, clear short-form privacy policy (under 500 words) if they are significant data holders.
- Enable users to access, modify, or delete their personal information and inform users about opting out of data processing, transfer, or sharing.
- Establish measures to prevent the collection of covered data from individuals under 17 years of age.

The privacy policy should include details about the data collected or processed, the reasons behind it, data sharing practices, and data retention durations. To be compliant with the ADPPA, a privacy policy must include several critical elements:

- **Exercising User Rights:** The privacy policy should explain how users can exercise their rights as stipulated in the ADPPA.
- **Security Practices:** Details regarding the security practices used to safeguard user data must be included.
- **Effective Date:** The privacy policy must specify its effective date, indicating when it became applicable.
- **Data Transfer to Restricted Countries:** If the organization transfers, processes, or shares data in countries such as China, Iran, North Korea, or Russia, this must be disclosed.
- **Contact Information:** Information on how users can contact the organization for privacy-related inquiries or concerns should be provided.

For large data holders, which are entities or service providers generating annual revenue exceeding USD 250 million, collecting, processing, or transferring personal data from more than five million individuals or devices and sensitive covered data from more than 200,000 individuals or devices, additional requirements apply:

- **Archiving Privacy Policies:** Large data holders must maintain easily accessible copies of their privacy policies for the preceding 10 years on their websites.
- **Short-Form Notices:** These entities should provide concise, user-friendly notices to consumers. These notices must be clear and easily accessible, include information on unexpected data practices, cover sensitive data practices, and be less than 500 words in length.

Following privacy best practices, organizations should implement the following measures:

- **Regular Privacy Policy Updates:** Ensuring that the privacy policy is periodically reviewed and updated is crucial. This practice ensures that the policy aligns with the latest privacy laws and any modifications to existing regulations.
- **Data Access:** Organizations should provide users with a straightforward and convenient mechanism to access, modify, or delete their personal information. This can be achieved by including clear instructions within the privacy policy.
- **Opt-Out Information:** Users must be informed about how they can opt out of data processing and data transfer. Also, they should have the option to decline targeted advertising.
- **Child Data Protection:** Organizations must strictly follow the Children's Online Privacy Protection Rule (COPPA) when collecting personal information from children. This regulation is essential for safeguarding the privacy of minors.
- **Privacy and Data Security Officers:** Covered entities are required to designate privacy and data security officers who are responsible for the development and implementation of data and privacy protection programs to ensure compliance with relevant regulations.

To create an ADPPA-compliant privacy policy, organizations should focus on the following key components:

- **Content:** The privacy policy should be clearly written and regularly updated on the website.
- **Effective Date:** The policy should specify its effective date, be presented in languages spoken by consumers, and be easily accessible to users with disabilities.
- **Data Collection Details:** Provide information about the types of data collected. Consider breaking this section down further, addressing data collected automatically, voluntarily provided by users, and collected via cookies. Also, specify data retention durations and the actions taken with data once its purposes are fulfilled.
- **Why Data Collection Occurs:** The privacy policy should explain the reasons behind the collection of users' data and how it is used.
- **Whom Data Is Shared with:** Transparently identify any third parties or service providers with whom the collected data is shared.
- **International Data Sharing Disclosure:** Disclose whether data is shared with countries like China, Iran, North Korea, or Russia.
- **User Rights:** Inform users of their privacy rights as per the ADPPA, including the right to access, correct inaccuracies, and opt out of data selling.
- **Data Security Measures:** Include details about the security measures implemented to protect collected data.
- **Contact Information:** Provide means for individuals to contact the organization for privacy-related inquiries or concerns.
- **Placement and Consent:** Display the privacy policy prominently and consider using an 'I Agree' checkbox to ensure users explicitly agree to the privacy terms.

Penalties for Non-compliance: Non-compliant businesses can face fines under the Federal Trade Commission Act, as the enforcement of the ADPPA falls under the jurisdiction of the Federal Trade Commission (FTC) through its Bureau of Privacy.

The American Data Privacy and Protection Act (ADPPA) is federal legislation in the United States that is designed to safeguard individuals' personal information, ensure the preservation of privacy rights, and develop mechanisms for effective enforcement. The term covered entities include organizations, individuals, or third parties that are responsible for managing the personal information of US customers. Covered data refers to a collection of information that can uniquely identify an individual, while sensitive covered data specifically refers to confidential personal information. The ADPPA requires compliance to regulations about the principles of data minimization, loyalty obligations, and privacy by design.

To follow the regulations, organizations must uphold a transparent privacy policy that is consistently revised while also ensuring that users have convenient access to their data and are provided with details regarding data processing, transfer, and the option to opt out of data sharing. Also, it is advisable for entities to refrain from purposefully gathering data from individuals who are under the age of majority and to designate specific officers responsible for overseeing privacy and data security matters. A privacy policy that meets the requirements of the American Data Privacy and Protection Act (ADPPA) should include various aspects, including the types of data collected, the purposes for which it is used, any sharing of data with third parties, the security measures in place to protect the data, the rights of the users about their data, and the contact details for any inquiries or concerns.

7 Cybersecurity and Privacy Laws, UK

The United Kingdom occupies a prominent position on the world stage, characterized by its historical legacy, and a substantial engagement in international affairs. In data privacy legislation, the United Kingdom has adopted a distinct strategy in contrast to its American counterpart.

In contrast to the lack of federal privacy legislation in the United States, the United Kingdom has enacted strict data protection rules that are uniformly applicable throughout the nation. The GDPR, which was implemented in May 2018, is considered a significant regulatory framework in the United Kingdom, and it imposes strict regulations on the collection, processing, and safeguarding of personal data. It includes all individuals and entities, regardless of their public or private nature, which operate within the United Kingdom.

The United Kingdom's approach to data privacy prioritizes the protection of individuals' sensitive information, such as health records and location data. This implies that when organizations gather such data, they have a responsibility to handle it with the highest level of caution and openness. The United Kingdom has facilitated the creation of a data protection authority in the country, a regulatory body that is responsible for monitoring compliance to data protection legislation and safeguarding the rights of individuals.

The United States utilizes a decentralized approach with a collection of rules and regulations that vary by state, but the United Kingdom has centrally implemented regulations which establish a unified framework for safeguarding data privacy throughout the entire country. These strategies not only provide individuals with enhanced safety but also establish a more foreseeable and controllable setting for enterprises operating within the United Kingdom.

7.1 Computer Misuse Act, 1990

The Computer Misuse Act [Par90] of 1990 was a legislative milestone in response to the burgeoning domain of cybercrime. The establishment of a legislative framework was undertaken to address the threats connected with unauthorized access to computer systems, acknowledging the dynamic nature of technology and its continuous evolution. The action played a significant part in ensuring the security of digital systems and preserving the integrity of the data they housed.

The fundamental purpose of the Computer Misuse Act is to establish the illegality of unauthorized access to computer systems. This provision served as a fundamental component in safeguarding the authenticity and confidentiality of electronic data. The criminalization of unauthorized access serves as a protective measure against several manifestations of hacking and unauthorized intrusion into computer networks.

The legislation took an additional measure by making unauthorized entry to conduct subsequent offences a criminal act. This acknowledgement acknowledges that instances of unauthorized access frequently operate as a precursor to more severe criminal activity. Cybercriminals possess the potential to exploit existing vulnerabilities to illicitly acquire sensitive data, perpetrate financial fraud, or even compromise the security of a nation. Through the imposition of penalties on these intentions, the legislation effectively served as an effective approach against individuals with evil motives, dissuading them from participating in cybercriminal initiatives.

The Computer Misuse Act also has provisions about the production and dissemination of malicious software, recognizing the possible detrimental consequences that may arise from the creation and propagation of destructive code. The use of a proactive approach was crucial in effectively addressing a diverse set of cyber threats, including viruses, malware, and ransomware. The legislation sought to safeguard digital networks and individuals' privacy by implementing regulations and criminalizing the production and dissemination of dangerous software.

The Computer Misuse Act of 1990 emerged as a crucial legislative enactment that established clear parameters and repercussions for illicit actions within digital technology. The statement acknowledges the evolving nature of cybercrime and sets the foundation for the establishment of legal safeguards for digital infrastructure, data integrity, and computer system security.

7.2 *Privacy and Electronic Communications Regulations, 2003*

The Privacy and Electronic Communications Regulations [Par03] enacted in 2003 was helpful in effectively dealing with the dynamic environment of electronic communications and digital marketing. During a period characterized by the growing adoption of electronic marketing strategies and the prevalent utilization of cookies on online platforms, these policies have assumed a pivotal role in upholding individuals' privacy and influencing conscientious digital conduct.

The legislation has implemented precise guidelines that govern electronic marketing practices, including email marketing and SMS marketing. One of the primary stipulations required that organizations secure individuals' agreements before transmitting commercial materials. The consent-based strategy included a diverse range of objectives.

The principal aims of this initiative were to address the prevalent problem of unsolicited spam, protect personal privacy, and encourage ethical marketing strategies. In response to the first concern, the regulations stipulated that marketing messages were permissible solely for individuals who had provided explicit consent, effectively mitigating the prevalence of intrusive and unwanted commercial communications that had emerged as a substantial inconvenience for individuals.

Subsequently, this legislation also exhibited a significant focus on safeguarding the privacy of individuals, thereby empowering them with the autonomy to determine their preference regarding the receipt of marketing communications, consequently affording them enhanced dominion over their personal information.

Ultimately, with the implementation of consent-based practices, organizations were prompted to adopt more responsible and focused marketing tactics. This transition encouraged firms to actively interact with individuals who possess a genuine interest in their products or services, thereby supporting a marketing technique that is both ethical and efficient.

The scope of the legislation was expanded to include the governance of cookies on websites, acknowledging the critical function that cookies fulfil in the domains of online tracking and personalization. Websites were obligated to conform to a prescribed set of explicit criteria to improve user privacy and grant greater control. Transparency concerning the utilization of cookies was a pivotal aspect of these regulations. Websites need to furnish users with transparent information regarding the use of cookies, thereby allowing individuals to acquire knowledge about the tracking of their online activity and the potential utilization of their data.

On top of that, the legislation also mandated that websites must provide users with the option to either provide or withhold consent for the utilization of cookies. This crucial component offered individuals the ability to make informed decisions regarding their acceptance of data gathered through cookies throughout their online activities, so strengthening their autonomy over their online privacy.

7.3 Investigatory Powers Act, 2016

The Investigatory Powers Act of 2016 [Par16], commonly known as the 'Snooper's Charter', constituted a measure aimed at addressing the various difficulties arising from the digital settings regarding national security and law enforcement. Nevertheless, it has also sparked substantial discussions regarding the complex equilibrium between safeguarding security and upholding individual privacy.

The permission for mass data collecting was a highly disputed component of the act. This clause enabled law enforcement and intelligence organizations to gather enormous quantities of communication data for national security and criminal investigations. Although the primary objective was to ascertain potential risks and mitigate criminal activities, this approach has drawn out significant concerns. The concerns around mass surveillance have been brought to light due to the provision within the act that allows for bulk data collection. This provision has ignited debates over the potential implications of gathering and analysing large volumes of data. Critics contended that this methodology had the potential to encroach upon the private rights of individuals, as it permits the indiscriminate gathering of data.

A noteworthy element of the legislation was the provision for the sanctioning of communication interception, including electronic mail and messages. This ability bestowed law enforcement and security services with indispensable resources for carrying out investigations about matters of national security and grave criminal offences. The necessity of intercepting communications for investigative purposes necessitated the implementation of strict supervision measures to mitigate the risk of potential abuses and safeguard the privacy of individuals. The legislation sought to achieve an equilibrium between the need to conduct efficient investigations and upholding the fundamental rights to personal privacy.

The legislation implemented a requirement mandating that Internet service providers (ISPs) maintain records of Internet connection activity (ICRs) for a designated duration. The information collection requests (ICRs) included documentation about the browsing history of individuals and the communication platforms they utilized. The purpose of this provision was to furnish law enforcement agencies with significant instruments for investigation. However, it also sparked substantial discussions and deliberations. The necessity for ISPs to preserve ICRs has sparked debates on data retention and its potential consequences on personal privacy. Critics expressed concerns regarding the possibility of misuse or unauthorized access to the data, highlighting the significance of implementing strong security.

7.4 Payment Services Regulations, 2017

The Payment Services Regulations [Par17] enacted in 2017 have had a substantial impact on the regulation of the payment sector in the United Kingdom, facilitating notable transformations and advancements in digital finance.

The incorporation of strong customer authentication (SCA) was a fundamental component of these requirements. This security precaution required payment service providers to incorporate supplementary authentication measures for specific high-risk transactions. The main aim of the initiative was to enhance security measures and safeguard consumers against instances of fraudulent behaviour, particularly within the rapidly evolving domain of digital finance.

The implementation of SCA has effectively enhanced security measures by introducing supplementary tiers of authentication, hence increasing the difficulty for fraudulent individuals to carry out unauthorized activities. As a result, this phenomenon enhanced consumer trust in digital payment systems, as individuals became more assured of the better security measures implemented to safeguard their financial transactions.

The regulations were designed with the objective of not just enhancing security measures but also encouraging competition and promoting innovation within the payment industry. The policies effectively promoted the establishment of novel payment service providers (PSPs) and aimed to streamline their entry into payment networks.

The legislation facilitated a competitive spirit among payment service providers (PSPs) by creating an atmosphere that encouraged the entry of new participants. The competition ultimately conferred benefits to both consumers and businesses through the provision of increased options and competitive pricing within the payment industry.

The promotion of competition raised a climate that was favourable to the flourishing of invention. Start-ups and enterprises powered by technology have identified favourable conditions to prosper and create inventive payment solutions, thereby making significant contributions to a dynamic and ever-changing payment ecosystem.

The Payment Services Regulations exerted a significant influence on the payment sector in the United Kingdom. The implementation of SCA not only improved security measures but also raised a climate of competition and creativity within the digital payment industry, ultimately leading to increased security, competitiveness, and innovation.

7.5 *Network and Information Systems Regulations, 2018*

The Network and Information Systems Regulations [Par18b] of 2018 was important in enhancing the cybersecurity resilience of critical services and digital service providers inside the United Kingdom. The implementation of these regulations was undertaken following the EU Directive on Security of Network and Information Systems (NIS Directive), to introduce significant steps to enhance the safeguarding of critical infrastructure and digital services from cyber threats.

One of the important elements of these laws involved the categorization of critical service operators and digital service providers. The classification of these companies was of importance as it enforced distinct cybersecurity rules and mandated incident reporting obligations. The primary objective was to guarantee the security of critical infrastructure and digital services, with a special focus on sectors such as electricity, healthcare, and transportation.

The regulations set responsibilities regarding incident reporting, necessitating these organizations to quickly report cybersecurity incidents. The action enabled a synchronized reaction to cyberattacks, reducing possible harm and guaranteeing the uninterrupted provision of essential services.

The significance of risk management in cybersecurity is emphasized by the Network and Information Systems Regulations. Organizations that were bound by these requirements were obligated to aggressively evaluate and address cybersecurity vulnerabilities. The methodology aids in safeguarding the integrity and availability of their services, especially when confronted with ever-changing cyber threats.

Organizations were required to perform risk assessments, including the identification of vulnerabilities and potential threats, to adopt a proactive approach towards risk management. This procedural measure facilitated the formulation of focused approaches for cybersecurity.

The regulations push organizations to adopt and execute mitigation strategies to efficiently manage identified risks. The implementation of this proactive strategy ensured that potential vulnerabilities were promptly identified and mitigated, hence minimizing the risk of exploitation by malicious entities.

7.6 Data Protection Act, 2018

The Data Protection Act [Par18a] of 2018 aided in connecting the European Union's GDPR with the United Kingdom's data protection framework following its departure from the EU. The legislation effectively tackled multiple crucial aspects of data protection, facilitating a smooth and uninterrupted transition while maintaining strict levels of privacy safeguarding.

The Data Protection Act placed significant emphasis on the enhancement of the rights of individuals concerning their data. The rights, including the entitlement to access one's data, rectify inaccuracies, and raise objections to processing for specified reasons, were consistent with the concepts described in the GDPR. The implementation of this reinforcement provided individuals with the ability to exercise authority over their data, guaranteeing the preservation of their privacy.

The act prioritized the reinforcement of data subject rights, highlighting the significance of individuals possessing agency and control over their personal information. This aligns with the primary objective of the GDPR to safeguard individual privacy.

The Data Protection Act implemented regulations about the processing of data by law enforcement agencies and national security organizations. The legislation's crucial dimension was to achieve a nuanced equilibrium between the need for security and the safeguarding of privacy rights. To ensure the protection of data subjects' interests, specific measures were implemented.

The act recognizes the essentiality of data processing for the objectives of law enforcement and national security while placing significant emphasis on the significance of implementing mechanisms to ensure accountability and oversight. The implementation of protective measures guaranteed the preservation of the rights of individuals whose data was being processed, even in situations when the prioritization of security was of importance.

The Data Protection Act was necessary in ensuring that data protection regulations in the United Kingdom remained in alignment with the GDPR. The establishment of this alignment has furnished a unified and all-including structure for safeguarding data in the times following Brexit, thereby ensuring the dedication to upholding elevated levels of privacy protection for both individuals and organizations.

Through its alignment with the principles outlined in the GDPR, the act provided a sense of continuity in data protection practices inside the United Kingdom. Organizations can effectively adjust to the evolving situation with a sense of assurance, as they are aware that the standards for data protection have remained resilient.

The Data Protection Act played a crucial part in modifying UK data protection legislation to align with the changing post-Brexit environment. The measures served to strengthen the rights of data subjects, establish a framework for the handling of data by authorities, and guarantee compliance with the principles outlined in the GDPR. The legislation laid the groundwork for ethical data processing practices while also maintaining strict privacy safeguards.

7.7 UK-GDPR, 2021

The UK General Data Protection Regulation (UK-GDPR) [Par18c] enacted in 2021 exemplifies the United Kingdom's firm dedication to upholding strict data protection measures following its withdrawal from the European Union. The act includes the following fundamental elements:

The UK-GDPR exhibits significant similarities to the European Union's General Data Protection Regulation (EU GDPR) in various aspects. The preservation of important principles and rights established by the EU GDPR is maintained by this alignment, thereby guaranteeing uniformity in the implementation of data protection measures. Significantly, the preservation of data subject rights and the overarching significance of data protection are upheld, thereby reaffirming the United Kingdom's commitment to ensuring the protection of personal privacy.

The UK-GDPR demonstrates its dedication to safeguarding the fundamental principles of data privacy by embracing the concepts and rights outlined in the EU GDPR.

An essential aspect of the UK-GDPR relates to the transmission of personal data between the United Kingdom and the European Union. The measures have been strategically formulated to effectively support the uninterrupted transmission of data post-Brexit, thereby ensuring that entities can persistently move personal data across international boundaries without encountering any hindrances.

The rules related to data transfer in the act contribute to maintaining the uninterrupted data flow between the United Kingdom and the European Union, thereby safeguarding the seamless operation of digital and economic operations.

The UK-GDPR designates the Information Commissioner's Office (ICO) as the primary regulatory body responsible for overseeing data protection in the United Kingdom. The ICO is tasked with the duty of implementing data privacy laws and regulations, thereby ensuring that organizations comply with the prescribed standards.

The designation of the ICO as the data protection authority in the United Kingdom, as stipulated by the act, serves to strengthen the country's dedication to ensuring robust supervision and enforcement of data protection protocols.

The UK-GDPR of 2021 reflects the commitment of the United Kingdom to maintaining data protection consistency while accommodating the distinct legal and regulatory environment resulting from the Brexit process. This act ensures the maintenance of elevated levels of data protection and privacy within the United

Kingdom by closely adhering to the European Union's GDPR, regulating the movement of data, and designating the ICO as the primary supervisory authority.

8 Cybersecurity and Privacy Laws, Australia

Australia, like the United States and the United Kingdom, is a prominent player in the global economy and holds a significant presence on the international stage. The country is renowned for its rich natural resources, strong financial sector, and well-developed technology industry. Despite its global influence, Australia, similar to the United States, lacks federal legislation about data privacy that would be consistently applicable across all its states and territories.

In the absence of federal privacy legislation, organizations in Australia enjoy a considerable degree of autonomy in handling data, unless specific states or territories have enacted their privacy laws. This fragmentation in regulations leads to varying privacy standards across the nation. In many instances, corporations have the legal authority to collect, use, disclose, or trade data acquired from individuals without being obliged to provide prior notice regarding such activities.

When a corporation shares an individual's sensitive data, such as health records or location information, with third-party entities like data brokers, these intermediaries often possess the freedom to further distribute or sell this data without any obligation to notify the concerned individual. Many Australians might assume they have a greater degree of data protection than they do.

Privacy laws in Australia primarily remain jurisdiction specific, concentrating on particular industries and aspects of data privacy rather than offering a framework addressing all potential privacy concerns across various contexts. Nonetheless, certain sectors such as healthcare, banking, and credit reporting are subject to specific regulatory measures to safeguard personal information, reflecting the country's approach to data protection.

However, it's also worth mentioning that Australia takes cybersecurity very seriously and has implemented a range of measures to defend against cyber threats. The Australian Cyber Security Centre (ACSC), established by the federal government, serves as the nation's primary cybersecurity agency. The ACSC collaborates with international partners to share information and coordinate responses to cyber-incidents. It provides valuable guidance and advice to government agencies, businesses, and individuals on how to protect themselves from cyber-attacks. Australia has enacted several cybersecurity-related laws, including the Privacy Act, the Cybercrime Act, and the Telecommunications (Interception and Access) Act, to control and safeguard against cyber threats. This demonstrates the country's commitment to enhancing cybersecurity and combating cyberattacks through government and business sector collaboration.

8.1 The Privacy Act, 1988

In a time marked by the abundant production and dissemination of data, safeguarding personal privacy emerges as a crucial matter. The foundation of Australia's dedication to protecting private rights lies in the Private Act of 1988. This significant legislation, implemented by the federal government, bears the crucial duty of setting and implementing strict requirements to safeguard personal information. The authority of this act includes both public sector agencies and private business companies that operate within the limits of the nation. The carefully crafted legal framework in question aims to guarantee responsible, ethical, and transparent handling of personal data. The act successfully achieves a nuanced and crucial equilibrium between the protection of individual privacy rights and the operational needs of organizations.

The Australian Privacy Act [Gov88] is centred around the Australian Privacy Principles (APPs), which consist of a collection of 13 guiding principles. These principles serve as the fundamental basis for the responsible management of data:

(1) **Open and Transparent Management of Personal Information:** The first principle emphasizes the need for organizations to be open and transparent about how they handle personal information. This includes having a clear and readily accessible privacy policy that explains the organization's practices.
(2) **Anonymity and Pseudonymity:** APP 2 allows individuals the choice of not identifying themselves or using an alias when interacting with organizations, where it is practical to do so. This offers individuals a level of privacy and control over their information.
(3) **Collection of Solicited Personal Information:** This principle describes the criteria for the legal and equitable gathering of personal data. The requirement states that organizations are obligated to gather only the information that is substantially essential for their tasks or activities.
(4) **Dealing with Unsolicited Personal Information:** APP 4 guides organizations on how to handle unsolicited personal information. It requires organizations to estimate whether they could have collected the information under APP 3 and, if not, to destroy or de-identify it.
(5) **Notification of the Collection of Personal Information:** Organizations have a duty to tell individuals about the gathering of their personal data, including the intended use, potential recipients, and other pertinent information.
(6) **Use or Disclosure of Personal Information:** APP 6 defines when personal information can be used or disclosed. Unless specific exceptions apply, it specifies that personal information may generally only be utilized for the purpose for which it was initially collected.
(7) **Direct Marketing:** APP 7 regulates the use of personal information for direct marketing purposes. It requires organizations to provide a simple way for individuals to opt out of receiving direct marketing materials.
(8) **Cross-Border Disclosure of Personal Information:** APP 8 addresses the cross-border disclosure of personal information. It requires enterprises to

ensure that the recipient of the information provides similar privacy protections or is subject to privacy laws that are comparable to Australia's.

(9) **Adoption, Use, or Disclosure of Government-Related Identifiers:** This principle limits the usage of government-related identifiers, such as driver's licence numbers, for most purposes. Exceptions are applicable under specific conditions.

(10) **Quality of Personal Information:** Organizations must take suitable measures to ensure that the personal information they collect, use, or disclose is accurate, complete, and up to date.

(11) **Security of Personal Information:** APP 11 mandates organizations to take appropriate measures to safeguard personal information from unauthorized access, disclosure, loss, or interference. It highlights the importance of data security.

(12) **Access to Personal Information:** Individuals have the right to access their personal information held by an organization and to request corrections if needed. APP 12 outlines how organizations should handle such requests.

(13) **Correction of Personal Information:** This final principle sets out the process for correcting personal information when an individual requests it. Organizations must ensure that corrections are made promptly.

These principles define the appropriate protocols that govern the acquisition, utilization, retention, and disclosure of personal data. Encompassing a range of fundamental elements about the management of personal data, including the requirement for informed permission and the importance of data security, the APPs offer organizations a guide to effectively handle privacy management.

The Privacy Act serves the purpose of establishing privacy standards and confers upon an entity the power to enforce compliance with these requirements. The entity refers to the Office of the Australian Information Commissioner (OAIC) and undertakes the responsibility of protecting privacy rights in Australia, fulfilling this objective through two main functions: promotion and enforcement. The OAIC actively strives to enhance public consciousness regarding privacy rights and obligations, as it serves as an advocate for privacy protections. Concurrently, it serves as a diligent overseer, ensuring that organizations comply with the prescribed privacy regulations. The dual role mentioned above serves a crucial role in safeguarding the rights of individuals and ensuring that organizations are held responsible for their actions.

In response to the increasing vulnerability of data breaches in the current digital, the Privacy Act implemented the Notifiable Data Breaches (NDB) scheme. This policy establishes a requirement for organizations to quickly inform both impacted individuals and the OAIC in the occurrence of a qualifying data breach. The implementation of this proactive action by the Privacy Act serves to greatly augment transparency and accountability. This technology enables individuals to proactively engage in safeguarding their personal information, transforming them from passive recipients of data to active contributors in the domain of data security.

In addition to surpassing the notification system, the Privacy Act grants individuals several essential rights. One of the primary rights individuals possess is the entitlement to obtain the personal information that is being retained by an entity. This access privilege not only ensures the visibility of information but also enables individuals to assume control over their data. Similarly, individuals possess the entitlement to seek rectifications to their data if inaccuracies are detected. These features serve to enhance individuals' authority over their information and develop a deep sense of empowerment and trust.

To guarantee the strict implementation of the APPs and the Privacy Act, a framework of enforcement procedures has been firmly established. Organizations that do not fulfil their privacy duties are subject to various potential repercussions, such as monetary fines and penalties. The implementation of a strict framework effectively develops a culture of responsibility in the management of data. Also, the OAIC, which possesses the authority to conduct investigations and pursue legal action, is prepared to undertake appropriate measures to rectify violations and uphold the integrity of privacy regulations.

The Privacy Act serves as a strong witness to Australia's firm dedication to safeguarding individual privacy rights. In the current world, characterized by the increasing significance of data as a very valuable commodity, the relevance of this regulation persists without diminishing. Not only does it serve to protect individual privacy, but it also promotes transparency and accountability in the handling of data.

8.2 Telecommunications Act, 1997

The Telecommunications Act [Gov97] of 1997 is a federal legislation in Australia that serves as the regulatory framework for the telecommunications sector. The enactment was implemented to govern multiple aspects of telecommunications services, including the licencing of carriers and service providers, the allocation of telecoms numbering, and the administration of the telecommunications infrastructure.

The legislation sets a regulatory framework for carriers, which are corporate companies engaged in the provision of telecommunications services. To conduct operations within Australia, carriers are required to acquire a licence from the Australian Communications and Media Authority (ACMA). Further, the Act enforces a universal service obligation (USO) upon carriers. This implies that telecommunications providers are obligated to guarantee that standard telephone services are accessible to all individuals in Australia fairly and impartially, regardless of their geographical location or place of employment.

The Act has provisions about the distribution and administration of telecommunications numbers, including phone numbers and other distinctive identifiers. The administration of the numbering plan falls within the purview of ACMA. Consumer protections are a crucial component of the Act, containing laws designed to safeguard the welfare and rights of consumers. The provisions include several aspects like billing procedures, mechanisms for resolving complaints, and protocols for managing client information.

The legislation sets a system for the provision of telecommunications infrastructure, guaranteeing that alternative telecommunications service providers are permitted to utilize crucial facilities owned or managed by carriers under fair and equitable conditions. The Act also incorporates provisions that address privacy considerations. The document has clauses about the safeguarding of consumer privacy and the management of customer data by carriers and service providers.

The primary objective of the Act was to promote competition within the telecommunications market by prohibiting anti-competitive conduct and establishing regulatory frameworks for the interactions among carriers, service providers, and other entities involved in the industry. It is worth noting that although the Telecommunications Act primarily centres its attention on fixed and mobile telecommunications services, the Act also recognized the significance of spectrum management, specifically about radio frequencies. Nevertheless, the regulation of this element is conducted in conjunction with other legislative measures, like the Radiocommunications Act of 1992.

8.3 *Electronic Transactions Act, 1999*

The Electronic Transactions Act [Gov99] 1999 (ETA) is another federal legislation that provides a legal structure for electronic transactions and electronic commerce within the nation of Australia. The fundamental objective of this system is to streamline and oversee the utilization of electronic communication and digital technologies across several domains of business and government activities. The Electronic Transactions Act (ETA) achieves this objective by acknowledging the legal legitimacy and enforceability of electronic transactions and documents.

The ETA includes a key provision that grants legal recognition to electronic transactions, thereby confirming their equivalence to conventional paper-based transactions in terms of legal validity. As a result, this Act deems contracts, agreements, and other documents that are formed and signed electronically to be legally enforceable. Also, the ETA recognizes the utilization of digital signatures as a valid method for authenticating electronic documents. The statement outlines the establishment of precise criteria aimed at guaranteeing the genuineness and reliability of digital signatures, hence augmenting the trustworthiness of electronic transactions.

The Act also includes provisions for the preservation of electronic records. It defines the necessary criteria for the preservation and management of electronic documents and records, including aspects such as accessibility and the preservation of data integrity over time. These requirements contribute to the establishment of a robust and dependable system for maintaining records of electronic transactions. The ETA not only applies to the commercial sector but also includes transactions that include government entities. This facilitates electronic interactions between individuals, corporations, and government bodies, streamlining procedures such as the submission of applications and tax forms.

Consumer protection is another significant aspect of the ETA. The Act aims to facilitate electronic transactions while also guaranteeing that customers are provided with information that is accessible and comprehensible to them. It provides customers with the option to engage in electronic transactions or stick to traditional paper-based methods, thereby protecting their rights and accommodating their preferences. The ETA is intentionally designed to maintain technological neutrality, allowing for the inclusion of various electronic methods and technology utilized in electronic transactions. The inherent adaptability of this system enables the seamless incorporation of emerging digital technologies into the domain of electronic commerce methodologies.

The Act further recognizes its reciprocity with other Australian laws to ensure its harmonization and compatibility with the prevailing legal structures. This practice guarantees uniformity and mitigates any clashes among various legislative measures. The Electronic Transactions Act 1999 is significant in the modernization and facilitation of electronic commerce and commercial operations in Australia as the acknowledgement of the legal legitimacy of electronic transactions, digital signatures, and electronic records establishes a robust legal framework that ensures the secure facilitation of commercial activities in a digital space.

8.4 Spam Act, 2003

The Spam Act [Gov03] 2003 is a legislative measure in Australia that aims to govern the transmission of unwanted commercial electronic communications, colloquially known as 'spam'. The main objective of this legislation is to mitigate the occurrence of unsolicited and intrusive electronic communications, protect the privacy and interests of individuals and organizations, and promote appropriate electronic marketing practices within the jurisdiction of Australia.

According to the provisions outlined in the Spam Act, unsolicited commercial electronic messages and communications transmitted over diverse electronic channels, including email, SMS, MMS, or instant messaging platforms. These messages are primarily intended to promote goods, services, or business-related initiatives. The regulation places significant emphasis on the necessity for senders to acquire the authorization of recipients before transmitting commercial electronic messages. Consent may be explicitly given, often acquired through an opt-in process, or implicitly understood in particular circumstances. Senders must have a distinct and operational unsubscribe option in their messages, which allows receivers to voluntarily discontinue receiving any future communications.

The act of identifying the sender is an essential element of the Spam Act. The requirement is imposed on individuals sending messages to furnish precise and transparent information regarding their identity and contact particulars in every communication, thereby enabling receivers to promptly ascertain the sender's identity and undertake suitable measures, such as opting out. To provide recipients with greater autonomy, the legislation mandates the incorporation of an opt-out feature within commercial messages. The implementation of this method should be designed to be uncomplicated and provided to receivers without any cost. After a recipient chooses to opt out, senders are required to quickly discontinue sending commercial electronic messages to that particular individual.

Failure to comply with the Spam Act can result in significant penalties for both individuals and businesses. Violators may be subject to civil penalties, with the ACMA being entrusted with the task of enforcement. ACMA possesses the authority to issue infringement notices, provide formal warnings, and initiate legal proceedings against those found to be in violation. The Spam Act governs the majority of commercial electronic messages; however, it does include some exemptions. The exemptions pertain to various categories such as messages associated with established business ties, government communications, charitable organizations, and other particular categories. Plus, specific communications that are associated with legal responsibilities, such as financial statements, are granted an exception.

The efficacy of the Spam Act in mitigating the influx of unsolicited commercial electronic messages in the email accounts of Australian recipients and nurturing ethical marketing strategies has been substantiated. Compliance with this Act holds importance for organizations and individuals involved in electronic marketing activities within Australia, as non-compliance can lead to substantial penalties.

8.5 DNCR Act, 2006

The legislative basis for the implementation and functioning of the 'Do Not Call Register' (DNCR) [Gov06] in Australia is provided by the Do Not Call Register Act 2006. The DNCR, also known as the Do Not Call Registry, is a programme that grants individuals and businesses the option to decline unsolicited telemarketing calls and marketing faxes. This service empowers them with enhanced authority over the marketing communications they receive.

The establishment of the DNCR is a fundamental aspect of this legislation. The DNCR functions as a medium through which individuals and organizations can formally register their telephone numbers and facsimile numbers, thereby exercising their choice to abstain from receiving unsolicited commercial calls and facsimiles. The legislation provides a clear and uncomplicated procedure for individuals and organizations to exercise their right to exclude their phone numbers and fax numbers from receiving unsolicited communications by registering them on the DNCR. The process of registration is commonly free of charge and can be simply accomplished using Internet platforms, providing a handy and easily accessible method for individuals to decline unsolicited commercial communications.

Telemarketers and fax marketers are legally obligated to comply to the regulations outlined in the Act, which explicitly forbid them from engaging in unsolicited marketing calls or transmitting marketing faxes to phone lines that are registered on the DNCR. Nevertheless, there are several instances where exclusions and exemptions apply, such as receiving calls or faxes from registered charitable organizations, governmental entities, and political parties, among other entities.

The Act also governs the telemarketing sector by putting obligations on telemarketers to authenticate phone numbers against the DNCR before conducting marketing calls, as well as to uphold their roster of numbers to circumvent. This initiative aims to promote responsible telemarketing practices and ensure compliance to the preferences of registered individuals and organizations. To guarantee compliance to its stipulations, the Act has mechanisms for surveillance and implementation. The enforcement of the Act is the responsibility of the ACMA, which is empowered to levy penalties on telemarketers and fax marketers who fail to comply with its stipulations.

The Act incorporates periodic evaluations and prospective modifications as an integral component of its structure, allowing for necessary improvements to be implemented to effectively tackle concerns about unsolicited commercial calls and faxes. The DNCR Act of 2006 holds substantial importance within the Australian legal framework as it aims to mitigate the occurrence of unwelcome and intrusive telemarketing calls and marketing faxes. This framework grants consumers and organizations the authority to manage the marketing communications they receive while simultaneously imposing responsibilities on marketers to honour their expressed preferences.

8.6 HI Act, 2010

The Healthcare Identifiers Act [Gov10] 2010 is a legislative statute at the federal level in Australia that provides a structure for the implementation and use of healthcare identifiers within the healthcare system of the country. Identifiers are numerical codes that are assigned to individuals and healthcare providers to improve the precision and effectiveness of health information management. According to the legislation, healthcare identifiers are specifically designated as distinct numerical values that function to differentiate individuals and healthcare providers inside the healthcare system. Health information exchanges is key in the digital health infrastructure by facilitating the secure and precise matching of health records and information.

One of the key purposes of the Act is the establishment of individual healthcare IDs (IHIs), which serve as distinctive IDs allocated to individuals. Integrated Health Information Systems (IHIs) facilitate the precise integration of an individual's health information across diverse healthcare providers and environments, hence enhancing the coordination of patient care and the overall standard of healthcare. In addition to personal IDs, the legislation introduces healthcare provider IDs (HPIs). Healthcare provider identifiers (HPIs) are distinctive designations allocated to individuals and entities involved in the provision of healthcare services, including physicians, nurses, and healthcare institutions. Health information interchange among providers is facilitated by them, playing a crucial role in ensuring security.

The legislation lays a significant emphasis on the protection of privacy and the maintenance of security. The regulations and safeguards are designed to ensure the privacy and security of healthcare identifiers and the corresponding health information. These safeguards have been specifically devised to mitigate the risk of unauthorized access or disclosure of sensitive health data. The utilization of healthcare identifiers is intended to improve the precision and effectiveness of health information administration. The Act plays a major role in enhancing patient care coordination and healthcare outcomes by facilitating healthcare professionals' access to relevant patient records and mitigating errors.

The act of registering for healthcare IDs is often discretionary, although certain healthcare providers and organizations may have a legal obligation to register and utilize these identifiers in the course of their operations. The Healthcare Identifiers Service (HI Service) is created by legislation as the governing body responsible for the issuance and administration of healthcare identifiers across Australia. The HI Service is necessary in upholding the integrity and security of healthcare identifiers.

Healthcare IDs are of significant importance in facilitating the integration of electronic health records (EHRs) and other health information systems. These systems are required in facilitating accurate linkage and sharing of patient information across various healthcare providers and organizations, hence promoting a more integrated and complete healthcare system. To ensure compliance and protect the confidentiality and integrity of data, the legislation has provisions for sanctions and mechanisms for implementation. Penalties may be imposed on individuals or organizations that are found to violate the Act. The Healthcare Identifiers Act 2010 in Australia is a measure that is important in enhancing the precision, compatibility, and protection of health data within the nation's healthcare system. The objective of assigning distinct IDs to individuals and healthcare professionals is to improve the quality of patient care while also upholding the confidentiality and protection of health data.

8.7 MHR Act, 2012

The My Health Records Act [Gov12] 2012 is a legislative measure that serves as the fundamental basis for the establishment and operation of the MHR system in Australia. This system is a digital platform for health records that has been specifically developed to offer individuals a safe means of accessing their health information online, hence facilitating improved coordination of healthcare services. It allows authorized healthcare practitioners to get access to patient records to enhance the standard of healthcare provided.

The establishment of the MHR system itself is a key aspect of the Act. The digital health record system provides individuals with the ability to create and oversee their electronic health records, commonly referred to as MHRs. These records contain crucial health-related data, including an individual's medical history, allergies, prescribed prescriptions, and outcomes of diagnostic tests.

The Act incorporates provisions for both voluntary and involuntary methods of engagement with the MHR system. In the beginning, individuals were mandated to actively choose to participate in the establishment of an MHR. Nevertheless, the Act underwent changes in 2018 which implemented an opt-out strategy, implying that unless individuals explicitly want to opt out, an MHR will be issued for them by default. The objective of this modification was to promote wider engagement inside the system.

The Act highlights the significance of access control methods within the MHR system. Individuals can ascertain the parties authorized to access their MHR, thereby establishing access controls that either permit or limit the access of certain healthcare professionals. This measure guarantees the confidentiality and protection of individuals' health data. The legislation also includes the responsibilities of healthcare providers, conferring authorized professionals, including physicians, nurses, and chemists, with the privilege of accessing a patient's MHR. The utilization of this access has the potential to improve the coordination of patient care and offer critical information during emergency scenarios.

The Act places great importance on security and privacy and includes strong measures to protect the confidentiality and integrity of health information kept in MHRs. The inclusion of these provisions is crucial in upholding patient confidence and guaranteeing the safe management of confidential health information. Along with that, the legislation requires healthcare professionals to maintain precise and current documentation within a patient's MHR, contingent upon the patient's explicit agreement and compliance to access restrictions. This practice guarantees that the record includes thorough and up-to-date health information.

To promote the awareness and comprehension of individuals regarding their MHR and the associated rights and duties, the Act incorporates provisions for the dissemination of notifications. The alerts serve to provide clarification regarding the management of individuals' health information within the system. The legislation also has provisions about the consequences of privacy and security violations of MHRs. It outlines enforcement procedures aimed at addressing instances of non-compliance exhibited by healthcare practitioners and organizations.

The MHR system has been specifically built to ensure accessibility and interoperability with diverse healthcare providers' systems. This facilitates the smooth incorporation of electronic health data and improved coordination of patient care. In brief, the MHRs Act of 2012 offers a complete legal framework governing the implementation and operation of the MHR system within the Australian context. The fundamental objectives of this system are to facilitate individuals in effectively managing their health information, facilitate healthcare practitioners in securely accessing this information, and guarantee the confidentiality and security of sensitive health data.

8.8 TOLA Act, 2018

The Assistance and Access Act, officially known as the Telecommunications and Other Legislation Amendment (Assistance and Access) Act [Gov18a] 2018, is a prominent and frequently debated Australian federal law.

The fundamental aim of the Assistance and Access Act is to provide law enforcement and intelligence services in Australia with the necessary tools and methods to proficiently obtain access to and intercept electronic communications. The capabilities are considered indispensable in safeguarding national security, conducting inquiries into illegal acts, and operating within the wider context of law enforcement.

The requirements for Technical Assistance Notices (TANs) are crucial for the operation of the Assistance and Access Act. The notifications, which are issued by duly authorized entities, possess the authority to enforce selected communication providers to give technical support. The provision of help might manifest in diverse ways, including the dissemination of critical information, the provision of authorized entry to particular systems or data, or the creation of decryption skills. The primary objective is to enhance the ease of accessing encrypted communications in situations where it is crucial for current investigations or national security concerns.

The Assistance and Access Act incorporates the notion of Technical Capability Notices under specific conditions. Nevertheless, the issuance of Temporary Custody Notices is not a decision made unilaterally, as it necessitates the permission of the attorney general. The mandate imposed on communication providers by TCNs requires them to develop novel technical capabilities or make alterations to their current systems to facilitate access to encrypted communications. Significant debate arises at this point. There have been expressed concerns, both domestically and internationally, regarding the potential unintended consequences of the constraints imposed on third-country nationals, which could result in the emergence of systemic flaws or backdoors inside encryption systems. Theoretical exploitation of these vulnerabilities has the potential to undermine the digital security and privacy of both individuals and organizations.

To address these concerns and uphold the rights to individual privacy, the Assistance and Access Act provides a series of safeguards and control procedures. These procedures guarantee that the powers bestowed by the Act are utilized prudently and by established legal norms. To ensure the proper implementation of the Act's provisions, mechanisms for oversight and reporting have been put in place. These mechanisms serve to monitor and scrutinize the application of the Act, intending to strike a careful equilibrium between the need for security and the protection of privacy.

However, the Assistance and Access Act continues to be a subject of significant controversy, giving rise to vigorous discussions both within Australia and on a global scale. Detractors of the Act contend that it has the potential to compromise the integrity of digital security and encryption measures. The individuals articulate concerns over the gradual erosion of privacy rights and the potential compromise of encrypted communications. On the contrary, advocates argue that the Act is indispensable for the sake of law enforcement and national security initiatives in current society filled with digital technology and encrypted communications.

The ramifications of the Assistance and Access Act transcend national boundaries since it has the potential to affect multinational technology corporations that offer services to customers in Australia. This issue has sparked more extensive debates regarding the complex and dynamic task of achieving a suitable equilibrium between security priorities and the rights to personal privacy at a time when encryption is vital in protecting digital communication and sensitive data.

In simple terms, the Assistance and Access Act encapsulates the complex and disputed terrain of digital security, privacy, and law enforcement in Australia and other jurisdictions. It highlights the continuous effort to address the complex and changing obstacles presented by encryption technology, to balance national security concerns with the fundamental values of privacy and individual rights. The constant and expanding worry in our more digital and networked society lies in maintaining a delicate equilibrium between these essentials.

8.9 NDB Scheme, 2018

The Notifiable Data Breaches (NDB) scheme [Aus18], which became operative in Australia on February 22, 2018, represents a watershed moment in the nation's approach to data protection and privacy. Rooted in the Privacy Act 1988, this regulatory framework was instituted to address the escalating challenges posed by data breaches and the need to enhance the safeguarding of individuals' personal information.

The fundamental essence of the NDB plan revolves around a particular goal, which is to enhance the safeguarding of personal information through the imposition of certain obligations on organizations. These requirements refer to the requirement of notifying both individuals and the OAIC following data breaches that are anticipated to cause significant harm to the affected parties. This event marked a notable transformation in the domain of data privacy and security within the Australian context.

The NDB scheme demonstrates a wide-ranging applicability, spanning a diverse range of organizations. These entities include not just governmental institutions in Australia but also corporations, not-for-profit organizations that surpass a specified yearly turnover threshold, credit reporting bodies, and health service providers. The incorporation of these categories highlights the scheme's ambition to include a wide range of sectors and companies, emphasizing the widespread presence and significance of data protection.

The fundamental principle behind the NDB scheme is around the notion of a 'notifiable data breach'. Instances of such breaches arise when an entity's data is accessed or disclosed without proper authorization or when it is lost in a manner that presents a substantial risk of serious harm to the individuals whose information is involved. The term 'harm' is interpreted broadly, embracing a range of aspects, including but not limited to pecuniary and physical harm, as well as emotional and reputational harm. During such occurrences, organizations are obligated to initiate the notification process.

The fulfilment of notification responsibilities is a fundamental aspect of the Notifiable Data Breaches (NDB) scheme. Organizations that fall under the purview of its jurisdiction are obligated to promptly respond by notifying affected individuals promptly after the identification of a data breach. Concurrently, it is essential for them to promptly inform the OAIC regarding the occurrence of the breach. Notifications intended for individuals should contain accurate and detailed information regarding the breach, including data about the nature of the breach, the exact type of compromised information recommended actions for impacted individuals, and contact details for the organization. The process of notifying the OAIC involves the filing of a complete report that provides detailed information on the specific aspects of the data breach.

The potential ramifications of failing to comply to the NDB plan should not be underestimated. The penalties, fines, and regulatory measures issued by the OAIC are included within this framework. These consequences highlight the significance of the scheme's responsibilities and the necessity for organizations to prioritize data privacy and security with the highest level of importance.

Although notification is an important aspect in the system, it should be noted that it is not the only dimension involved. The NDB system functions as a proponent for proactive steps in the domain of data protection. It is highly recommended that organizations implement strong data security protocols, perform thorough risk evaluations, and develop strategies for responding to data breaches. The prioritization of prevention reflects the well-recognized maxim that 'prevention is superior to treatment', hence emphasizing the significance of proactive measures in protecting personal data.

In addition to the technical requirements of compliance, the NDB system aligns with overarching ideals of transparency and responsibility. This highlights the importance of organizations fulfilling their obligations diligently when they are entrusted with personal data. It emphasizes the importance of promptly and openly informing individuals who have been impacted by data breaches. This develops a climate of trust and emphasizes the significance of swiftly notifying individuals when their data is in jeopardy.

The implementation of the Notifiable Data Breaches (NDB) programme in Australia's legal framework in 2018 marks a significant milestone in the nation's efforts to strengthen data security and safeguard privacy. The statement demonstrates a strong dedication to safeguarding the privacy rights of individuals and developing an environment that prioritizes data security. The NDB scheme plays a vital role in safeguarding data protection by incorporating measures such as prevention, notification, transparency, and responsibility. Broadly speaking, these legislations function as a guiding principle for organizations, ensuring the protection of individuals' personal information, thereby demonstrating Australia's commitment to preserving the security of the digital space.

9 Cybersecurity and Privacy Laws, EU

This section covers the extensive legislation that governs data privacy in Europe, with a particular emphasis on the GDPR. The GDPR represents an innovative legal initiative to develop a structure to protect personal data, specifically in its management by commercial entities. The statement signifies the establishment of a legal framework aimed at protecting consumer privacy.

The GDPR, which was formulated and implemented by the European Union (EU), is widely recognized as the most rigorous legislation about privacy and security on a worldwide scale. The authority of the EU includes entities on a global scale, provided that they engage with or collect data about individuals within the EU. The regulation was implemented on May 25, 2018. It is worth mentioning that the GDPR imposes significant financial penalties on individuals or organizations that fail to comply with its privacy and security requirements. These penalties can amount to millions of euros.

The implementation of the GDPR by Europe serves as a clear indication of its dedication to safeguarding data privacy and security. This commitment is especially crucial at a time when individuals are increasingly relying on cloud services to store their personal data, and data breaches are occurring with alarming frequency. The task of attaining compliance with the GDPR can be a significant difficulty, especially for small- and medium-sized businesses (SMBs), due to its broad reach and limited set of explicit instructions.

9.1 GDPR, 2018

The General Data Protection Regulation (GDPR) [Com16] is widely determined as the principal legislation governing data privacy within the EU. The legislative framework under consideration includes the implementation of financial penalties on businesses that provide customers with inaccurate personal information, including incorrect bank account details. As per the provisions outlined in the GDPR, individuals who possess citizenship inside the EU are officially classified as data subjects. Consequently, the collection and sharing of personal data within the European Union is prohibited for all organizations, irrespective of their geographical location, even if they operate in nations like Canada that possess data privacy laws.

The primary objective of the GDPR is to improve the rights and autonomy of individuals inside the EU by granting them greater control and authority over their data. The aim is to establish restrictions on the extent of data availability for service providers and reduce the impact imposed by data brokers, including marketing aggregation firms and certain financial organizations. The jurisdictional reach of the GDPR surpasses that of analogous regulations in the United States, namely the California Consumer Privacy Act (CCPA) and the California Privacy Rights Act (CPRA), which predominantly operate within the confines of state-level governance. Business entities are bound by restrictions on their ability to obtain data, and corporate entities that function as intermediaries for data are susceptible to substantial penalties.

The GDPR places significant emphasis on the role of a data protection officer (DPO), whose responsibility is to supervise the lawful processing of personal data for individual resident within the European Economic Area (EEA). Cloud service providers and other entities involved in data processing bear the responsibility of managing and processing data on behalf of data controllers. The controllers are responsible for gathering data from individuals who reside within the EEA. The GDPR outlines a framework of eight essential rights that individuals possess concerning their data.

These rights include the entitlement to be informed about the handling of their data, the entitlement to access their data, the entitlement to correct any inaccuracies in their data, the entitlement to have their data deleted, the entitlement to transfer their data to another service provider, the entitlement to object to specific data processing activities, the entitlement to limit data processing, and the entitlement to be exempt from decisions made solely through automated processing. The legislation prohibits entities from participating in the processing of personal data without appropriate legal authorization or explicit consent. To give permission, it has to be done voluntarily, clearly, and with full knowledge of all the important details, without any kind of manipulation.

- The entitlement to receive information. It is essential to provide data subjects with adequate notification regarding the acquisition and utilization of their data at the time of its collection.
- Individuals have the right to view their data. A data subject has the right to submit a data subject request to obtain a copy of their data. Data controllers are obligated to explain the methods used for data collection, the specific data being processed, and the parties with whom it is shared.
- The concept of the right of rectification. If the data about an individual is found to be erroneous or incomplete, the individual has the right to request rectification from the data controller.
- The concept of the right to erasure. Individuals have the right to seek the deletion of personal data about them based on specific criteria within a stipulated time frame of 30 days.
- The right to impose limitations on the processing of personal data. Individuals possess the right to request the imposition of limitations or the withholding of their data while allowing for its retention.
- The concept of data portability refers to the legal entitlement of individuals to get and transfer their data from one organization to another in a structured, commonly used, and machine-readable format without hindrance or restriction. Data subjects can securely and reliably transfer their data from one electronic system to another without causing any disruption to its usability.
- The entitlement to object. Individuals have the right to express their objection to the use of their data for marketing, sales, or any other purposes unrelated to

the provision of services. The right to object is not applicable in situations where legal or official power is being exercised, tasks are being performed in the public interest, or data processing is necessary for the provision of a service to which the individual has consented.

Organizations that are involved in significant data processing activities about criminal offences or sensitive categories, such as health, sexual orientation, race, or religion, are required to designate a data protection officer (DPO). This need also includes public bodies, authorities, and businesses engaged in the systematic and consistent surveillance of individuals' data. The GDPR is enforced by independent data protection authorities (DPAs) that act on behalf of the 27 member states of the EU.

Data protection authorities (DPAs) are responsible for the examination of complaints, the identification of violations of the GDPR, and the provision of advice on data protection. Failure to comply to EU data protection regulations, such as the GDPR, can lead to financial consequences in the form of penalties. These penalties may amount to 4 per cent of a company's worldwide annual revenue or USD 20 million, whichever is greater. When comparing the penalties, it is often observed that they are more strict regulations at the state level in the United States.

The GDPR is a pivotal piece of legislation that regulates the various aspects of data collection, utilization, transmission, and security of individuals residing in the 28 member countries of the EU. The scope of the application extends to all European Union residents, irrespective of the geographical location of the organization responsible for the collection of their data. One of the primary requirements of the GDPR is the necessity to get clear and unambiguous consent from individuals whose personal data is being gathered, including data that is collected via the use of cookies. It is essential for organizations to quickly inform both regulatory authorities and data subjects within a 72-hour time frame in the event of a data breach that compromises the personal information of users.

Data subjects own distinct entitlements, including the entitlement to receive information, gain access to their data, rectify any inaccuracies, request the deletion of their data, impose limitations on data processing, exercise data portability, and voice objections to specific uses of their information. The GDPR is an important data privacy policy that holds significant importance in the EU. The primary objective of this initiative is to enhance the autonomy and agency of individuals, limit the ability of service providers to access data, and establish strict consequences for failure to meet the prescribed regulations. The provisions of the GDPR include several aspects, such as consent, data breach notifications, and data subjects' rights. This framework plays a critical role in safeguarding data protection throughout the European Union.

9.2 DSA, 2022

The Council of the European Union introduced the Digital Services Act (DSA) [Com22b, Com22c] on November 16, 2022. The primary aim of this regulation is to address the problem of unauthorized and harmful content present on digital platforms such as Google and Facebook. The assertion suggests a core principle: Behaviours considered illegal in physical environments should also be considered illegal in the virtual domain. The EEA contains four main business sectors within its ambit. These include intermediate services involved in the supply of network infrastructure, hosting services that include cloud and web-hosting services, online platforms that promote interactions between merchants and consumers, and highly extensive online platforms. The latter category refers to platforms that can engage with over 10 per cent of the total consumer base in Europe, which amounts to over 450 million individuals.

The DSA applies certain criteria to each of these categories of companies. Universal obligations that apply to all categories include the following: engaging in transparency reporting regarding court orders, efforts related to content moderation, and other relevant actions; revising their terms of service to align with fundamental rights; cooperating with national authorities; and establishing communication channels for authorities and legal representatives. Further, there is a necessity for hosting providers, online platforms, and specifically large online platforms to have a notice-and-action mechanism. This system facilitates the reporting of instances of potentially illegal content for removal, as well as the reporting of criminal acts.

It is important for online platforms, especially those with substantial reach, to establish a formalized framework for resolving grievances and offering redress. Also, it is recommended that these platforms establish a system wherein individuals or entities possessing expertise in content identification are designated as trusted authorities responsible for sending content notifications. Similarly, it is essential to implement suitable actions to effectively handle cases involving abusive notices and counter-notices. To successfully incorporate a marketplace function, it is crucial to prioritize compliance-by-design principles and conduct an assessment of the qualifications of third-party vendors.

Large platforms are obligated to undertake additional obligations, including the execution of risk management strategies, the formulation of protocols for crisis management, the involvement of external independent auditors, the creation of an internal compliance department, and the demonstration of public accountability. These platforms must offer users the option to refuse content recommendations that are generated through profiling, share data with both authorities and researchers, meet self-imposed codes of conduct, and cooperate with authorities during crisis response efforts.

It is necessary to remark that European Union data protection authorities have the authority to exercise their power by accessing, obtaining information from, and conducting inspections of service providers to collect pertinent data to issue directives and enforce penalties. Failure to comply with the EEA can result in significant financial penalties, with businesses facing fines of up to 6 per cent of their annual global revenue for serious offences and up to 1 per cent for breaches related to the information obligations outlined in the DSA.

9.3 DMA, 2022

The Digital Markets Act (DMA) [Com22a] comprises a regulatory framework that mainly focuses on significant digital platforms, often known as gatekeepers. Prominent corporate corporations such as Facebook, Apple, Microsoft, and Google serve as gatekeepers in their domains. The primary objective of the DMA is to promote equitable and unbiased circumstances for digital enterprises while concurrently forbidding dominant firms from imposing unjust conditions on their competitors. An example of a regulatory measure imposed on companies, such as Amazon, is the prohibition of offering preferential placement to their products and services on their platform to obtain a competitive advantage.

A company can be classified as a gatekeeper if it meets certain requirements, such as having a significant economic presence, exerting a significant impact on the EU market by operating in multiple EU member states, and playing a prominent intermediary role by facilitating connections between a large user base and multiple businesses. Further, a firm might be classified as a gatekeeper if it has or is expected to attain a commanding market position, as assessed by its compliance with the criteria throughout the preceding three fiscal years.

Organizations that meet the criteria specified in the DMA concerning gatekeeper standards are obliged to comply with a variety of duties. These obligations include refraining from self-promotion, where the gatekeeper prioritizes its own products and services over equivalent third-party offerings on its platform, abstaining from repurposing users' data outside the original context without consent, refraining from tracking users outside the platform for targeted advertising purposes without consent, ensuring the accessibility of communication and content between businesses and end users, maintaining transparency regarding prices and fees in advertising intermediation services, offering access to marketing and advertising performance data to platform users, facilitating user modification of default settings and software uninstallation, ensuring compatibility between third-party technologies and the gatekeeper's systems, enabling user data portability to other platforms, granting businesses real-time access to their data on the gatekeeper's platform, refraining from hindering user complaints to authorities, not making user registration for additional services a prerequisite for accessing a given service, and abstaining from leveraging non-public data from businesses to engage in competitive activities.

Non-compliance with the DMA may result in fines of up to 10 per cent of the gatekeeper's annual global sales. In cases of recurring violations, the possible penalty may increase by a maximum of 20 per cent. In cases where violations occur repeatedly, other remedies, such as requiring the disposal of certain assets or businesses, may be used as alternatives to financial sanctions.

9.4 DPF, 2023

The EU-US Data Privacy Framework (DPF) [Com23d] refers to a set of policies that have been developed to control the transfer of personally identifiable information from the EU to the United States. The framework in question, which was granted approval by the European Commission on July 10, 2023, may be seen as the further evolution of the EU-US Privacy Shield. The Data Protection Framework (DPF) has been enhanced with further protective measures in response to concerns raised by the European Court of Justice.

The recommended protective measures include the implementation of limitations on the access of US intelligence agencies to European Union data, allowing access only to information that is deemed necessary and proportional. More importantly, within the scope of the proposal now under examination, there is the potential to create a legal institution known as the Data Protection Review Court (DPRC). The EU has recently implemented a set of additional rights for individuals, which include the right to get their data and the right to correct or delete any data that is shown to be incorrect or improperly handled. In a specific and defined setting, the diesel particulate filter (DPF) has significant importance for small- and medium-sized enterprises (SMEs). These firms now possess a cost-effective and efficient method to facilitate the transfer of personal data from the European Economic Area.

The EU-US Data Privacy Framework has significant importance on a global scale, yet it is crucial to acknowledge that it lacks the status of a formal legislative act. The relevance of this framework lies in its influence on the transfer of data belonging to citizens of the EU to the United States (USA). To fully grasp its significance, it is essential to examine the historical circumstances that precipitated its inception and gain insight into its subsequent development. The Privacy Shield framework was once seen as a fundamental element for enterprises involved in the transfer of data over the Atlantic, as it offered reassurance that the data of European Union individuals would be appropriately protected throughout its transfer to the United States. However, the framework faced a substantial defeat in the Schrems II legal proceeding, resulting in its invalidation. As a result, enterprises pursued different methods to guarantee the legality of data transfers.

The standard contractual clauses refer to a collection of contractual terms that have been officially endorsed by the European Commission. These provisions are designed to function as legal protections to facilitate data transfers. Nevertheless, the stability of self-compacting concretes might be considered relatively fragile. One notable constraint emerges when American enterprises are subjected to the provisions of Section 702 of the Foreign Intelligence Surveillance Act (FISA). This particular section grants authority to spy services in the United States to carry out searches on overseas communications, which include the data of individuals who are citizens of the European Union.

The complexities of Section 702 extend beyond the scope of this discussion; nevertheless, it is crucial to understand the uncertainty surrounding its relevance to companies. Therefore, the use of SCCs for data transfers presents a venture fraught with risks. The situation endured until October 7, 2022, when President Biden issued an Executive Order entitled 'Enhancing Safeguards for United States Signals Intelligence Activities'. The implementation of this executive action has led to the establishment of the EU-US Data Privacy Framework, representing a noteworthy advancement in transatlantic data privacy.

The recently implemented system, as outlined in President Biden's Executive Order, has several significant aspects. First and foremost, it integrates advanced security procedures to strengthen the safeguarding of sent data. Besides, it introduces a framework for addressing grievances, providing EU and US citizens with the opportunity to seek redress in cases where they believe their data privacy rights have been violated. Likewise, it enhances the protective measures about the information of non-domestic individuals that is being transmitted to the United States. Again, the framework mandates that intelligence agencies implement modifications to their rules and procedures about surveillance.

These alterations will then undergo scrutiny by the Privacy and Civil Liberties Oversight Board. At present, the framework is undergoing assessment by the European Commission, and the European Data Protection Board has been invited to provide comments. Anticipating possible critiques from European privacy advocacy organizations is of importance since these organizations may thoroughly examine the effectiveness and sufficiency of the framework in safeguarding data privacy. If the framework effectively completes this examination process and obtains support, it has the potential to establish itself as the authoritative mechanism regulating data transfers between the European Union and the United States. This would provide companies with a necessary and legally valid resolution for effectively handling foreign data privacy requirements.

The EU-US Data Privacy Framework has significant importance within the complex network of global data privacy rules. The progression of its development, starting with the invalidated Privacy Shield, followed by the Standard Contractual Clauses, and now the proposed framework presented in President Biden's Executive Order, demonstrates the continuous initiative to achieve a nuanced equilibrium between safeguarding data privacy and upholding national security concerns. The review conducted by the European Commission will have significant implications for the future direction of data transfers between the European Union and the United States. As a result, it has a central position in the constant shift towards global data privacy regulation.

9.5 AI Act, 2023

The Artificial Intelligence Act (AI Act) [Com23b] of the EU is a regulatory measure that aims to provide a legislative structure governing the advancement, implementation, and use of AI systems. In June 2023, the AI Act was approved by the European Parliament. This legislation is a pioneering milestone, being the first law of its kind globally. The primary objective of the AI Act is to establish measures that guarantee the safety and compliance to basic rights and values of AI systems inside the European Union. Also, its objective is to guarantee the unrestricted mobility of products and services that are based on AI across international boundaries. The AI Act has provisions that restrict the deployment of apps that offer a degree of danger deemed unacceptable to the safety and privacy of individuals. The regulations also include the prohibition of AI systems that are specifically developed to influence those who are considered vulnerable, such as minors or individuals with impairments.

The AI Act was endorsed by the European Parliament with a significant majority of 499 votes in favour, 28 votes against, and 93 abstentions. The EU AI Act includes individuals and entities engaged in the development and deployment of AI systems inside the European Union, irrespective of their geographical location, thereby extending its jurisdiction beyond EU borders. The scope of the Act extends to include suppliers and users of AI systems situated outside the borders of the European Union, provided that the output generated by those systems is intended for use inside the European Union. The degree of regulation is contingent upon the level of threats associated with a certain application, ranging from minor to those deemed unacceptable. The legislation prohibits the use of apps that pose an intolerable degree of risk to individuals' well-being and confidentiality.

On June 14, 2023, the European Parliament approved the drafted EU AI Act. The European institutions will commence talks to achieve consensus on the final document. The proposed legislation aims to establish laws for AI systems, focusing on those deemed high risk. Its scope extends to firms operating inside the EU as well as those outside of it but serving EU clients or users. This paraphrase will explore the fundamental provisions and consequences of the AI Act in a way that is academically straightforward and mathematically rigorous. The major emphasis of the AI Act is on AI systems that are classified as high risk. These systems with high levels of risk contain applications in several areas, including transportation, education, employment, and welfare. These sectors have a significant effect on critical domains such as employment, credit, and healthcare.

Organizations that have the intention of implementing high-risk AI systems inside the EU are obligated to meet the strict compliance standards defined in the Act. These procedures include the implementation of impact assessments, the maintenance of records, and the fulfilment of transparency standards to guarantee the responsible utilization of AI. A key element of the AI Act is the banning of AI systems capable of inducing mental or bodily damage via the manipulation of an individual's conduct. The policy explicitly prohibits the creation of systems that capitalize on weaknesses within certain demographics, such as age, physical or mental disability, or behaviours.

This is because such exploitation has the potential to cause both psychological and physical damage. The Act places considerable emphasis on the protection of vulnerable populations, therefore confirming the European Union's commitment to the ethical and responsible use of AI. Also, the Act exhibits extraterritorial application by extending its jurisdiction outside the confines of the European Union. Consequently, non-EU enterprises that possess a client base or user population inside the EU are also bound by these restrictions. Therefore, the AI Act establishes an international legal framework that has wide-ranging implications for AI developers and adopters on a global scale.

To classify AI systems and effectively mitigate any biases and threats connected with AI, the AI Act establishes a framework that categorizes AI systems into four distinct degrees of risk: unacceptable, high risk, restricted risk, and minimum risk. The category of unacceptable risk explicitly precludes the use of subliminal tactics, exploitative systems, or social scoring systems used by public agencies in applications. Also, it enforces a prohibition on the use of real-time remote biometric identification technology by law enforcement agencies inside publicly accessible areas.

As previously said, high-risk AI systems are used in areas that are of importance. To assure compliance with specific safety criteria, companies intending to implement high-risk AI systems inside the EU are obligated to carry out a conformity assessment. To maintain openness, these corporations need to provide details about their high-risk AI systems in a database that is readily available to the public. AI systems with less risk are accompanied by distinct duties regarding transparency.

When users engage with a chatbot, it is essential that they be provided with explicit notification of the automated nature of the interaction and are offered the alternative of conversing with a human agent. AI systems with less risk, such as commonly used applications like spam filters and inventory management systems, possess lower inherent threats and are subject to less severe standards. The primary obligation for ensuring compliance with the AI Act lies with the vendors of AI systems. Nevertheless, the legislation also imposes specific obligations on distributors, importers, users, and other relevant third parties participating in the AI ecosystem. The primary objective of this collective duty is to provide a regulatory framework for AI systems, with a particular focus on promoting ethical and responsible use.

To ensure compliance with the requirements, it is essential for businesses engaged in the development and deployment of AI to carefully contemplate several crucial actions. It is essential to include an AI Risk Assessment Framework that effectively mitigates bias concerns at every stage of an AI project's lifespan. This process includes comprehending and recording the attributes of data, establishing objectives for algorithms, using suitable training data, and keeping thorough records of model parameters and performance indicators. It is essential for enterprises to build a governance architecture that aligns with existing rules and industry best practices about technological resilience, thorough testing protocols, effective data management, and strong cybersecurity measures.

The inclusion of human supervision at every stage of an AI system's lifespan is an obligatory prerequisite, highlighting the significance of transparency in the outputs generated by AI systems. Also, enterprises need to assess and improve their privacy initiatives to conform to the regulations outlined in the AI Act. These regulations permit the use of certain personal data for monitoring, detection, and rectification of bias. The preservation of individuals' rights and the development of public trust are of importance. Finally, it is important to boost AI literacy across all teams within a company, including those in legal departments. This measure guarantees that all relevant parties possess an understanding of the operational mechanisms of their organization's AI systems, thereby promoting compliance to the forthcoming legislation.

In brief, the 'Artificial Intelligence Act' of the European Union aims to address the potential problems linked to AI, specifically focusing on prejudice. This is achieved by the classification of AI systems into various categories of risk and the imposition of extensive compliance obligations on providers. To effectively manage these requirements, enterprises must take proactive steps to align their operations with the standards outlined in the Act and reduce any risks associated with AI.

9.6 ePR, Expected 2023

The ePrivacy Regulation (ePR) [Gov23b] is an upcoming legislative measure of the EU that is intended to replace the existing ePrivacy Directive. The primary aim of this initiative is to establish regulations about the protection of privacy and confidentiality in electronic communications within the European Union. The ePrivacy Regulation holds significant significance as it will establish a framework that outlines data protection standards for various electronic communication modalities. These modalities include a range of methods such as emails, text messages, messages sent through WhatsApp or Telegram, call contents, file metadata (which includes GPS tracking data), and cookie information. Simultaneously, the implementation of the ePrivacy Regulation will provide users with an increased level of authority over their data, thereby demonstrating a heightened focus on safeguarding individual privacy rights.

The legal framework in question also includes a series of stricter regulations targeting companies involved in the acquisition or processing of personal data. One important clause within the proposal put out by the Commission is that cookies used only to process data in an anonymized manner may no longer need agreement from end users. Further, it is worth noting that the ePrivacy Regulation will broaden its coverage to include rising entities in electronic communication services, which may include but are not limited to platforms like Facebook Messenger, WhatsApp, and Skype for Business. The primary objective of this expansion is to guarantee that these widely used service providers maintain the same rigorous requirements of communication confidentiality as conventional telecommunications companies. Within the wider framework of data protection at the EU level, the ePrivacy Regulation serves as an essential component that works in conjunction with the GDPR.

It is worth mentioning that the ePrivacy Regulation, often known as the ePR, has undergone a prolonged period of development and was initially intended to come into effect simultaneously with the EU's GDPR. If the anticipated ePR is approved, it would act as a precursor to a new set of privacy laws that would include both traditional electronic communication services and previously unregulated businesses, such as messages on Facebook, WhatsApp, and the messaging service Skype, which were not covered by the earlier ePrivacy Directive. The ePrivacy Regulation (ePR) aims to establish a regulatory framework that enhances the protection of privacy in electronic communications.

This framework includes both the substantive content of communications and their associated metadata, which refers to data that offers descriptive information about other data. Significantly, the ePR necessitates that service providers and electronic communication networks get explicit authorization from consumers before participating in the processing of metadata related to their electronic interactions. One notable aspect of this legislative proposal concerns cookies since the ePR implements more efficient protocols. Individuals would be provided with the opportunity to either provide or refuse permission for the use of tracking cookies at the browser level. It is important to note that websites are not required to seek approval for the use of non-privacy intrusive cookies. These specific cookies enable features such as shopping carts, which monitor user orders. Significantly, the ePR also includes a provision that mandates organizations to facilitate the ability of end users to withdraw permission that was previously given yearly.

10 Cybersecurity and Privacy Laws, India

India, a vast and diverse nation, holds a significant position in the world due to its burgeoning population, thriving technology sector, and increasing participation in international affairs. When it comes to federal legislation on data privacy.

India has been actively engaged in the development of data protection laws. The enactment of the Personal Data Protection Act signifies a substantial advancement in establishing a resilient framework for safeguarding data privacy.

The primary objective of the Personal Data Protection Act is to develop a framework that regulates the processing of personal data within the jurisdiction of India. The document defines strict protocols governing the acquisition, manipulation, and safeguarding of individualized information while also introducing the notion of data fiduciaries and data principals. Data fiduciaries refer to entities that are responsible for the collection and management of personal data, whereas data principals are the individuals who own the data. The legislation prioritizes the concept of informed consent, granting individuals increased authority over the management of their data.

India's approach is characterized by the establishment of the Data Protection Authority of India (DPAI), which serves as a regulatory entity responsible for the oversight and enforcement of data protection regulations. The responsibility of monitoring compliance to the terms of the Data Protection Act (DPA) and protecting the rights of individuals is bestowed upon the DPA. In addition to the data protection authority (DPAI), the country possesses various other regulating entities, such as the following:

- **CERT-In**: The Computer Emergency Response Team-India (CERT-In), which was created in 2004, holds a significant position in the cybersecurity domain of India. The major national authority responsible for the collection, analysis, and dissemination of information about cybersecurity incidents is identified. The CERT-In offers recommendations to Indian organizations regarding optimal cybersecurity practices and emphasizes the importance of promptly reporting any cybersecurity concerns. These tasks are of utmost importance in addressing cyber threats, effectively handling incidents, and advocating for the implementation of cybersecurity measures.
- **NCIIPC**: NCIIPC (National Critical Information Infrastructure Protection Centre), founded in 2014, is tasked with safeguarding India's critical information infrastructure. It distinguishes between Critical Information Infrastructure (CII) and Non-Critical Infrastructure (NCI) and focuses on sectors like energy, finance, transportation, telecommunications, and government enterprises. NCIIPC establishes policies, facilitates information exchange, and enforces cybersecurity regulations for critical industries, enhancing national cybersecurity resilience.
- **CRAT**: CRAT (Cyber Regulations Appellate Tribunal), established under Section 62 of the IT Act, serves as a tribunal for accepting cyber-evidence and examining witnesses in cyber-related cases. While its jurisdiction in cybersecurity reporting is limited compared to CERT-In, CRAT is significant in addressing cybersecurity incidents and maintaining the integrity of legal proceedings related to cybercrimes.
- **SEBI**: SEBI (Securities and Exchange Board of India), founded in 1988, regulates India's stock and commodity markets. It enforces regulations on market intermediaries, ensuring data confidentiality and integrity. SEBI's recent focus on cybersecurity initiatives includes the formation of a committee to develop and implement cybersecurity laws aligned with global standards, strengthening data protection in financial markets.
- **IRDAI**: IRDAI (Insurance Regulatory and Development Authority of India) oversees the insurance industry and has introduced information and cybersecurity guidelines. These guidelines mandate the appointment of chief information security officers (CISOs), the establishment of information security committees, and the development of strategies to handle cyber-catastrophes. IRDAI aims to protect policyholders' data and enhance cybersecurity in the insurance sector.
- **TRAI and DoT**: TRAI is India's regulatory body for the telecommunications sector, established in 1997. It oversees licensing, tariffs, quality of service, and consumer protection, ensuring fair competition among telecom service providers. DoT, a governmental department under India's Ministry of Communications, formulates policies, allocates spectrum, and regulates telecom services. TRAI and DoT collaborate on policy development and implementation, addressing data privacy and security concerns in India's telecommunications sector. Their joint efforts aim to promote efficient spectrum utilization, protect consumer rights, and ensure the integrity of telecom services.

The data privacy regulations implemented in India reflect a strong dedication to safeguarding sensitive data, such as health records and location information. When organizations collect such data, they will be obligated to treat it with great care, adhering to strict standards of transparency and accountability.

The data privacy framework in India is currently undergoing development and has not yet reached a finalized state at the time of writing this section. However, it reflects the nation's commitment to establishing a strong and all-including data protection ecosystem. This development positions India as an active participant in global conversations surrounding digital resilience and data privacy, emphasizing its commitment to establishing uniform privacy regulations within a country undergoing rapid digitization.

10.1 The Information Technology Act, 2000

The Information Technology Act [Ind00], 2000, often referred to as the IT Act 2000, stands as a significant piece of legislation in India, governing various aspects of electronic commerce, cybersecurity, and the use of electronic records and digital signatures. It includes various diverse set of provisions and objectives, fundamentally shaping the digital setting of the country.

One of the provisions of this Act is the legal recognition it bestows upon electronic records, elevating them to the status of legal validity equivalent to their traditional paper-based counterparts. This pivotal recognition has streamlined electronic transactions and substantially reduced reliance on conventional paper-based processes, promoting digital innovation and efficiency.

The Act establishes a robust framework for certifying authorities (CAs) tasked with issuing digital certificates. These certificates play a pivotal role in verifying the authenticity of digital signatures, thereby ensuring the security and integrity of electronic transactions. This provision has been important in creating a secure setting for online interactions, catalysing the growth of e-commerce and digital communication. Digital signatures, another focal point of the IT Act, are accorded a clear legal framework, legitimizing them as a valid means of signing electronic documents. This recognition not only improves the security of electronic communications but also enhances their legal standing by authenticating the identity of the sender and safeguarding the integrity of the transmitted message.

The Act also confronts the pressing issue of cybersecurity by defining various cybercrimes and offences, ranging from unauthorized access to computer systems to data theft and the introduction of computer viruses. To deter such malicious activities, it prescribes penalties, including imprisonment and fines, emphasizing the importance of safeguarding the digital infrastructure. While the Act doesn't include data protection and privacy regulations, it does incorporate provisions related to the preservation and retention of electronic records. It empowers the government to issue guidelines on data protection, a growing concern in the age of digital data where the need to safeguard personal information is essential.

Addressing the liability of intermediaries, such as Internet service providers and social media platforms, is another aspect of the Act. These intermediaries are obligated to follow specific due diligence practices to ensure compliance with legal standards. This balanced approach encourages digital innovation while holding intermediaries accountable for the content they host or transmit. To ensure a structured and accountable legal process for addressing cybercrimes, the Act establishes adjudication authorities, complemented by appellate authorities to review their decisions. This act enhances the efficiency of handling disputes and offences related to electronic transactions.

The Information Technology Act, of 2000, has evolved to address emerging cybersecurity challenges and align with international standards. The Information Technology (Amendment) Act, 2008, stands as a notable example, introducing changes related to data protection, intermediary liability, and increased penalties for cybercrime; the subsequent sections will provide a better analysis of this amendment. These amendments reflect the dynamic nature of technology and its evolving legal implications in India.

India's approach to cybersecurity is not solely reliant on the IT Act. Instead, it combines various sector-specific regulations and standards to nurture robust cybersecurity practices. The IT Act, coupled with these sector-specific regulations, serves as a legal framework for safeguarding India's critical information infrastructure. For instance, Section 43A of the Information Technology Act mandates that Indian enterprises and institutions implement 'reasonable security processes and procedures' to protect sensitive information from unauthorized access, harm, disclosure, or misuse. Section 72A of the Act further defines consequences for individuals or intermediaries intentionally disclosing personal data without the owner's consent to cause harm, including potential imprisonment of up to three years, a fine of up to 500,000 INR or both.

In essence, India's cybersecurity legal framework, anchored by the Information Technology Act, of 2000, remains a dynamic and evolving structure that responds to the ever-changing digital setting, promoting both innovation and security in the digital world.

10.2 Know Your Customer, 2002

The 'Know Your Customer' (KYC) [Ind02] guidelines in India were established under the Prevention of Money Laundering Act (PMLA) of 2002 and the accompanying rules and regulations. These regulations were introduced to combat money laundering and the financing of terrorism by ensuring that financial institutions and other entities involved in financial transactions have a clear understanding of their customers' identities, financial activities, and risk profiles.

KYC requirements mandate that financial institutions, including banks, non-banking financial companies (NBFCs), and other entities like payment system operators, maintain customer records. This includes verifying the identity and address of customers and keeping records of their financial transactions. The key components of KYC include the following:

(1) **Customer Identification:** Financial institutions are required to verify the identity of their customers using reliable and independent documents, data, or information. This includes obtaining proof of identity and address, such as Aadhaar cards, passports, or utility bills.
(2) **Risk Profiling:** Institutions must assess the risk associated with each customer based on factors like their business activities, financial background, and transaction history. Higher-risk clients must undergo additional due diligence.
(3) **Monitoring Transactions:** Continuous monitoring of customer transactions is essential to identify suspicious or unusual activities that may indicate money laundering or terrorist financing.
(4) **Reporting Suspicious Transactions:** As the central agency tasked with receiving, processing, analysing, and disseminating information pertaining to suspicious financial transactions, the Financial Intelligence Unit-India (FIU-IND) requires financial institutions to promptly notify it of any suspicious transactions.
(5) **Record Keeping:** Records of customer transactions and identity verification documents must be maintained for a specified period, as mandated by the PMLA rules.

The KYC framework is an essential tool in India's efforts to combat financial crimes and maintain the integrity of its financial system. It helps in tracking and preventing illicit financial activities and ensures that financial institutions play a proactive role in safeguarding against money laundering and terrorist financing.

It's worth noting that KYC requirements have evolved over the years, with the introduction of digital KYC options and Aadhaar-based verification methods, making the process more convenient and efficient for both customers and financial institutions. However, compliance with KYC regulations remains a critical aspect of financial services in India.

All-inclusive, KYC plays a pivotal role in enhancing digital resilience by ensuring the authenticity of individuals and entities participating in digital transactions. By verifying identities, assessing risk profiles, and monitoring financial activities, KYC helps prevent fraudulent and illicit transactions, strengthening the overall security of digital ecosystems. This proactive approach not only safeguards against money laundering and terrorist financing but also builds trust among users, promoting a more secure and resilient digital environment where legitimate transactions can thrive while minimizing the risks associated with cybercrime and financial fraud.

10.3 Information Technology (Amendment) Act, 2008

The Information Technology (Amendment) Act [Ind08] of 2008 is a significant landmark within India's legal system, as it addresses the emerging difficulties brought about by the fast growth of digital technology and cyberspace. The amendment included several significant modifications and enhancements to the first Information Technology Act of 2000, therefore restructuring India's legal framework about several aspect of the digital society.

The heightened focus on data protection and privacy was a notable component of the 2008 Amendment Act. In light of the growing significance of personal data in the age of digitalization, the legislation included measures to protect the privacy of individuals' personal information. The legislation rendered the act of divulging personal information without explicit authorization unlawful, therefore strengthening the concept of data privacy as an inherent entitlement. Not only did this resolve issues about unauthorized data exchange, but it also laid the groundwork for the development of India's expanding data privacy legislation.

The Amendment was enacted in reaction to the increasing severity of cybercrimes, and it notably broadened the existing legal framework to effectively combat these offences. The expansion of the cybercrime definition acknowledges the emergence of new risks and various manifestations of criminal behaviour in the digital domain. Through this move, the Act granted law enforcement authorities the authority to take more efficient measures against those involved in cybercrime. It implemented more strict sanctions for a range of cybercrimes, thereby conveying the stance that unlawful actions conducted in the digital space will be met with zero tolerance.

The Amendment also addressed the crucial issue of intermediary liability. The legal status of intermediaries such as Internet service providers, social media platforms, and online forums was described. Although intermediaries were not directly held accountable for user-generated material, they were obligated to comply with particular due diligence responsibilities. This approach achieved equilibrium by both enhancing digital innovation and ensuring that platforms were held responsible for any material that infringed against legal regulations.

With almost nine chapters and 117 parts covering critical information security measures for cybercrime and data protection, the new Information Technology Amendment Act of 2008 covers the following responsibilities:

(1) Enhancing cybersecurity and forensics.
(2) Requiring intermediaries and corporations to notify CERT-In of cybersecurity incidents.
(3) Avoiding unauthorized/illegal computer system usage.
(4) Protecting personal information and data from identity theft, phishing, malware, and cyberterrorism.
(5) Organizational cybersecurity is legally recognized.
(6) E-payment and electronic transaction security, as well as electronic record monitoring and decryption.
(7) Creating a legislative structure for electronic signatures.
(8) Identifying and monitoring intermediaries.

The Act served to strengthen the authority of law enforcement agencies in their ability to intercept and surveil electronic communications, but only in cases directly linked to matters of national security or public order. The objective of this provision was to guarantee that authorities could adequately address emerging risks and acquire crucial information, all while upholding due process and safeguarding individual rights.

The enactment of the Information Technology (Amendment) Act of 2008 marked a significant milestone in the country's progression towards a digital India. The measures implemented enhanced the safeguarding of data, broadened the range of legal tools available to combat cybercrimes, provided clarity about the obligations of intermediaries, and gave law enforcement authorities the authority to effectively tackle growing issues in the domain of cyberspace. The Act retains its significance as a fundamental legislative measure in the country's initiative to handle the complexities of a digital India, all the while ensuring the protection of individual rights and national security.

10.4 Information Technology Rules, 2011

The Information Technology (Intermediaries Guidelines) Rules [Ind11a], 2011, known as the IT Rules 2011, are a collection of rules established within the larger context of the Information Technology Act, 2000, in India. The regulations were implemented to oversee the operations of intermediaries within the domain of digital technology and establish a foundation for the management of content and user engagements on digital platforms.

One of the key aims of the IT Rules 2011 is to establish a set of norms for intermediaries, including Internet service providers, social networking platforms, e-commerce websites, and online forums. These intermediaries are necessary in enabling digital connections while also bearing responsibility for the material that is housed or transmitted on their platforms. Key provisions of the IT Rules 2011 include the following:

- **Due Diligence by Intermediaries:** The rules mandate that intermediaries exercise due diligence while discharging their duties. This involves the adoption of reasonable security practices and procedures to protect sensitive user information and prevent unlawful content from being disseminated through their platforms.
- **Content Removal:** Intermediaries are required to remove or disable access to content that is considered objectionable, harmful, or defamatory within 36 hours of receiving a complaint. This provision aims to curb the spread of offensive or illegal content online.
- **User Agreement:** Platforms are required to establish user agreements or terms of service that outline the rules and regulations governing user behaviour on their behaviour. Users are expected to follow these guidelines.
- **User Information:** Intermediaries are mandated to maintain records of user information for a specified period, making it accessible to law enforcement agencies when required for investigation or prevention of cybercrimes.

- **Notification to Users:** Platforms must inform users that their content has been removed or access to it has been disabled due to a violation of the platform's rules or legal provisions.
- **Compliance Officer:** Large intermediaries are required to appoint a compliance officer responsible for ensuring compliance to these rules and cooperating with law enforcement agencies.
- **Grievance Redressal:** Intermediaries are expected to establish mechanisms for addressing user complaints and grievances promptly.
- **No Monitoring Obligation:** The rules clarify that intermediaries are not obligated to actively monitor the content on their platforms, but they must act promptly upon receiving complaints or knowledge of illegal activities.
- **Preservation of Information:** Platforms are required to preserve user information and content for a specified period as may be required by the government.

The IT Rules of 2011 have considerable importance in the regulation of online material and interactions inside India's digital environment. The objective is to achieve an equilibrium by both upholding the principles of freedom of speech and ensuring that platforms bear responsibility for any material that contravenes the statutes. The regulations have been a topic of extensive deliberation and analysis, especially concerning matters about the preservation of freedom of expression and the safeguarding of personal privacy.

10.5 Indian SPDI Rules for Reasonable Security Practices, 2011

The regulations known as the Indian Sensitive Personal Data and Information (SPDI) [Ind11b] Rules for Reasonable Security Practices and Procedures and Sensitive Personal Data or Information, 2011, are a crucial element of India's extensive framework for data security and privacy. The standards and requirements outlined in this set of regulations are created by Section 43A of the Information Technology Act, of 2000. They pertain to organizations that are responsible for managing SPDI.

At the core of these regulations lies a precisely defined and all-including clarification of SPDI. It includes a diverse set of personal information, including passwords, financial particulars, health records, sexual orientation, and biometric data. The provision of a precise definition by the regulations provides a robust basis for the identification and classification of data that needs enhanced safeguarding and attention.

The SPDI Rules, 2011, place significant emphasis on the implementation of acceptable security practices and procedures by organizations to protect SPDI against unauthorized access, disclosure, change, or destruction. These security practices must be by established international standards and recognized best practices, therefore guaranteeing the efficient and consistent protection of data.

Also, it is emphasized by the regulations and the significance of obtaining express and informed agreement from individuals before the collection of their SPDI. Organizations need to exhibit transparency about the objectives of data gathering and its planned use, to guarantee that individuals possess knowledge regarding how their data will be used.

In scenarios where organizations want to transmit sensitive personally identifiable information (SPDI) to another entity, severe precautions must be implemented by the established standards. These measures include the implementation of security practices and procedures that are consistent with those used by the receiving organization. Any transfer of SPDI must be conducted by a well-defined contractual arrangement that defines the respective rights and obligations of all involved parties.

The regulations governing the disclosure of SPDI to third parties are strict, primarily emphasizing the acquisition of explicit agreement from the data subject before the dissemination of their sensitive information. Instances, when the need for permission is waived, occur in situations where law enforcement agencies or government bodies impose such requirements, therefore emphasizing the delicate equilibrium between safeguarding the privacy and adhering to legal obligations.

The SPDI Rules of 2011 also provide a framework for effectively addressing incidents of data breaches. In the case of a data breach that has the potential to cause damage to the individuals whose data is compromised, organizations have a legal obligation to swiftly inform both the affected individuals and the Indian Computer Emergency Response Team (CERT-In). The prompt dissemination of this timely notice serves to facilitate the implementation of appropriate measures by affected individuals while also providing government authorities with the means to evaluate and address the breach efficiently.

Failure to comply with the SPDI Rules may lead to substantial consequences, such as the need to provide restitution to impacted individuals and monetary sanctions. The imposition of these fines serves as a compelling motivator for organizations to give importance to data security and comply with the restrictions outlined in the guidelines.

The regulations promote the practice of organizations conducting regular audits of their data handling procedures. The audits serve the purpose of ensuring continuous compliance to regulations and provide organizations with a crucial means of evaluating and improving their data security protocols.

The SPDI Rules, 2011, have become essential in defining India's data protection and privacy environment. These policies are primarily designed to safeguard individuals' sensitive personal data while also protecting their privacy and security. They offer a complete legal framework that allows organizations to manage and preserve such data in an increasingly digitalized world efficiently and responsibly.

10.6 National Cyber Security Policy, 2013

The National Cyber Security Policy [Ind13] of India, established in 2013, is a framework that outlines the strategic vision and objectives for safeguarding India's cyberspace. This policy is a crucial component of the country's efforts to address the evolving and complex challenges in the digital society.

One of the central elements of the National Cyber Security Policy, 2013, is its vision and mission. The policy sets forth a clear vision to build a secure and resilient cyberspace ecosystem that promotes economic growth, protects critical infrastructure, and ensures the safety and privacy of citizens. It aspires to establish India as a global leader in the field of cybersecurity, recognizing the role of cybersecurity in the nation's progress.

A significant focus of the policy is on the legal framework governing cyberspace. It emphasizes the need to review and strengthen the existing legal framework, including the Information Technology Act, of 2000, and its amendments, to ensure that they are robust and effective in addressing emerging cyber threats. This highlights the importance of a legal framework that can adapt to the rapidly evolving digital setting.

The National Cyber Security Policy, 2013, also outlines the development of a cyber-crisis management plan. It envisions the establishment of a National Critical Information Infrastructure Protection Centre (NCIIPC) and sectoral Computer Emergency Response Teams (CERTs). These entities help in facilitating a coordinated response to cyber-incidents and crises, ensuring that the nation can respond swiftly and effectively to cyber threats.

The policy places a strong emphasis on promoting research and development in the field of cybersecurity. It recognizes that innovation and indigenous solutions are essential in addressing cyber threats. Therefore, it encourages the creation of an environment that encourages innovation and supports the development of cybersecurity technologies and solutions within the country.

Cybersecurity awareness is another aspect of the policy. It emphasizes the significance of raising cybersecurity awareness among various stakeholders, including government agencies, businesses, academia, and the general public. Promoting a culture of cybersecurity and responsible digital behaviour is seen as crucial in enhancing overall cybersecurity and resilience.

Recognizing the collaborative nature of cybersecurity efforts, the policy encourages public-private partnerships. These partnerships aim to enhance information sharing, threat intelligence sharing, and joint cybersecurity initiatives. By leveraging the strengths of both the public and private sectors, the policy seeks to create a more robust cybersecurity system.

The policy emphasizes the importance of international cooperation in combating cyber threats. India seeks to engage with other nations and international organizations to address global cybersecurity challenges effectively. This highlights the interconnected nature of cybersecurity and the need for a coordinated global response to cyber threats.

In terms of practical measures, the policy encourages organizations to establish mechanisms for reporting and responding to cybersecurity incidents promptly. This includes sharing incident information with relevant authorities and agencies and contributing to a more coordinated and proactive approach to incident response.

The National Cyber Security Policy acknowledges the shortage of skilled cybersecurity professionals and promotes capacity building through training programs, certifications, and educational initiatives. This human resource development aspect is crucial in building a strong cybersecurity workforce to meet the growing demands of the digital infrastructure.

Lastly, the policy highlights the critical need to secure critical information infrastructure, including sectors such as energy, finance, and telecommunications. It mandates the protection of its information infrastructure through strict cybersecurity measures, recognizing the potential impact of cyberattacks on these critical sectors.

10.7 National Cyber Security Strategy, 2020

The National Cyber Security Strategy [Ind20] of 2020 represents a significant and eagerly anticipated initiative by the Indian government. This strategic framework holds the promise of enhancing the nation's overall cybersecurity posture and is currently undergoing evaluation by the National Security Council Secretariat. The primary objective of this initiative is to provide authoritative guidance and support to a diverse range of stakeholders, including policymakers and corporate leaders, in their efforts to mitigate cyber threats, including incidents, cyberterrorism, and cyber-espionage, within the rapidly evolving digital India.

At its core, the proposed strategy seeks to elevate the quality and effectiveness of cybersecurity audits. This is a pivotal aspect as it enables organizations to conduct more robust assessments of their cybersecurity infrastructure and capabilities. By optimizing the standards and procedures for cybersecurity audits, the strategy aims to empower organizations to identify vulnerabilities and weaknesses in their security posture more effectively. It sets the stage for a proactive approach to cybersecurity, encouraging organizations to enhance their security programs and adapt to emerging threats.

One of the anticipated outcomes of implementing this regulation is the enhancement of cybersecurity standards among auditors themselves. By setting higher benchmarks for cybersecurity auditors, the strategy promotes a culture of continuous improvement and professional development within the cybersecurity field. This, in turn, benefits organizations as they can engage with auditors who possess advanced skills and knowledge, ensuring a more thorough evaluation of their cybersecurity measures.

Also, the strategy's emphasis on cybersecurity audits is important for promoting transparency and accountability. Organizations will be better equipped to assess their cybersecurity readiness, identify areas that require attention, and take remedial actions promptly. This proactive approach reduces the likelihood of cybersecurity incidents and strengthens the overall resilience of critical infrastructure and digital assets.

In the broader context of India's cybersecurity setting, the National Cyber Security Strategy signifies a commitment to staying ahead of evolving cyber threats. It acknowledges the dynamic nature of the digital environment and the need for an adaptable strategy to safeguard the nation's interests. By offering guidance and support to policymakers, businesses, and other stakeholders, this initiative seeks to create a collaborative ecosystem that can effectively respond to the challenges posed by cyber-adversaries.

10.8 IT Rules, 2021

The 2021 IT rules [Ind21] represent a significant development in India's regulatory framework for the digital space. These rules have their origins in Section 87 of the Information Technology Act, of 2000, which provides the legal basis for the government to frame guidelines and regulations concerning various aspects of digital communication and online platforms.

What makes the 2021 IT rules noteworthy is that they consolidate and address a range of issues that have become increasingly relevant in the digital age. They are essentially a merger of two key regulatory initiatives: the draft Intermediaries Rules, 2018, and the OTT (Over-The-Top) Regulation and Code of Ethics for Digital Media.

The draft Intermediaries Rules, 2018, were initially proposed to regulate online intermediaries, which include social media platforms, messaging apps, and other Internet-based services. These rules sought to establish a framework for these intermediaries to follow, addressing issues like content moderation, user data protection, and the identification of the 'first originator' of certain content.

On the other hand, the OTT Regulation and Code of Ethics for Digital Media were designed to bring digital streaming platforms and content providers under a regulatory framework. These regulations aim to ensure that digital media content follows specific standards and ethical guidelines, particularly concerning issues like hate speech, violence, and explicit content.

The development of the 2021 IT rules involved close coordination between multiple government bodies, including the Central Government of India, the Ministry of Electronics and Information Technology (MeitY), and the Ministry of Information and Broadcasting (MIB). This collaborative effort reflects the recognition of the dynamic nature of digital technology and its impact on various aspects of society, including media, information dissemination, and online communication.

In essence, the IT rules mark a significant step towards bringing digital platforms, intermediaries, and content providers under a more regulatory framework. They aim to strike a balance between safeguarding individual rights, ensuring the responsible use of digital media, and maintaining the integrity and security of online platforms. These rules reflect the government's commitment to adapting to the challenges and opportunities presented by a digital India.

10.9 The Digital Personal Data Protection Act, 2023

The Digital Personal Data Protection Act [Ind23], 2023 (DPDP Act) is a landmark piece of legislation passed by the Parliament of India. It aims to tackle the pressing concerns related to the handling of digital personal data, ensuring the protection of individuals' rights to safeguard their personal information. At the same time, the act recognizes the legitimate requirements for data processing.

The scope of the DPDP Act is extensive, including the processing of digital personal data within India, regardless of whether it was collected through online or offline means and subsequently converted into digital format. The Act extends its jurisdiction to situations where digital personal data is processed outside of India, particularly if it involves the provision of goods or services to individuals within the Indian territory.

The DPDP Act has a crucial element, which is the formulation of a set of standards that necessitate data fiduciaries (entities entrusted with the responsibility of handling personal data) to comply with. Interestingly, these principles include the significance of conducting data processing activities in a legitimate, equitable, and transparent manner, with a notable focus on obtaining informed permission and ensuring transparency in both data collection and use. The Act emphasizes the concept of purpose restriction, which mandates that data fiduciaries must gather personal data only for explicitly stated and revealed reasons. This ensures that data is not used beyond the scope of individuals' permission.

The principles of data minimization, accuracy, and responsible data retention practices are also required, highlighting the need for organizations to gather just the essential data required for their stated objectives, ensure data precision, and store data only for the duration that it fulfils its intended goals. The Act highlights the importance of data security by assigning data fiduciaries the duty to adopt measures aimed at protecting personal data against unauthorized access, disclosure, change, or destruction. Accountability is a fundamental principle that requires data fiduciaries to assume responsibility for their compliance to the DPDP Act.

The DPDP Act is of importance as it offers specific details by categorizing personal data as any data that connects to an identifiable individual. The definition incorporates several data pieces, including but not limited to names, addresses, contact information, biometric data, and financial details. The Act incorporates the notion of 'sensitive personal data', including personal data that discloses an individual's caste, religious views, political viewpoints, or medical history. The establishment of these specific definitions serves as the fundamental basis for the implementation and enforcement of the Act's provisions and rules.

The DPDP Act is supported by a core set of principles that data fiduciaries, referring to entities responsible for the processing of personal data, are required to follow with care and diligence. The principles include the following:

(1) **Lawfulness, Fairness, and Transparency:** Data fiduciaries must process personal data exclusively for lawful purposes and maintain transparency in their data collection, usage, and sharing practices.
(2) **Purpose Limitation:** Data fiduciaries can only collect personal data for specific, pre-defined purposes that they have disclosed to individuals.
(3) **Data Minimization:** Data fiduciaries should limit data collection to what is strictly necessary for the stated purposes.
(4) **Accuracy:** Data fiduciaries are responsible for ensuring the accuracy and currency of the personal data they collect.
(5) **Storage Limitation:** Personal data should not be retained longer than necessary for the purposes it was collected.
(6) **Integrity and Confidentiality:** Data fiduciaries are required to implement robust security measures, including encryption and regular security assessments, to protect personal data from unauthorized access, alteration, or destruction.
(7) **Accountability:** Data fiduciaries are held accountable for their compliance with the DPDP Act's provisions.

The DPDP Act grants individuals a suite of rights regarding their data, significantly enhancing their control over its processing. These rights include the following:

- **Right to Information:** Individuals have the right to be informed about how their data is collected, used, and shared.
- **Right to Access:** Individuals can request access to their data held by data fiduciaries.
- **Right to Correction:** If personal data is inaccurate or incomplete, individuals have the right to correct it.
- **Right to Object:** Individuals can object to the processing of their data for specific reasons.
- **Right to Erasure:** Also known as the 'right to be forgotten', this empowers individuals to request the removal of their data once its intended purpose is fulfilled or consent is withdrawn.
- **Right to Restrict Processing:** Individuals can limit the processing of their data under certain conditions.
- **Right to Data Portability:** Individuals can request their data in a commonly used format for porting to another service.
- **Right to File Complaints:** Individuals have the right to file complaints with the data protection authority regarding data protection violations.

To oversee and enforce compliance with the DPDP Act, it establishes the Data Protection Authority of India (DPAI), also known as Data Protection Board, as an independent regulatory body with significant powers. The DPAI has the authority to investigate complaints of non-compliance, impose penalties on data fiduciaries found in violation of the Act, and ensure compliance to data protection standards.

The DPDP Act is a complex piece of legislation, and as of the time of writing, there are still several unanswered questions about how it will be implemented. However, the Act is a significant step forward for data protection in India, and it is expected to have a major impact on the way that personal data is processed in the country. The Digital Personal Data Protection Act represents a pivotal moment in India's data protection. It strikes a delicate balance between individual privacy rights and the legitimate needs of organizations to process data. By providing a robust legal framework, empowering individuals with data rights, and establishing a vigilant regulatory authority, it aims to shape a future where data protection and privacy are the priority.

11 Cybersecurity Regulations

Cyber-regulations, commonly known as Internet rules, comprise the legal framework that governs several aspects of the subject of legal informatics. These entities are specifically created to supervise and regulate multiple aspects of the digital environment, including the distribution of information, electronic commercial activities, software functioning, and the safeguarding of information. Cyber-regulations comprise various interconnected sectors, including but not limited to Internet usage and accessibility, the preservation of freedom of expression, and the safeguarding of privacy.

The justification for the implementation of cybercrime rules is rooted in the multitude of security and privacy issues that have emerged as a result of the extensive use of the Internet. Malicious entities have been observed utilizing advanced strategies to partake in unauthorized actions and probable occurrences of deceit. Therefore, it is important to protect against these potential dangers. The enforcement of robust cybersecurity policies is widely regarded as the most efficacious strategy for attaining this objective.

The main objective of formulating these rules and their accompanying regulations is to safeguard the interests of individuals and enterprises involved in online activities. The implementation of measures intended to hold individuals accountable for engaging in hostile online behaviour and subject them to appropriate penalties as determined by the federal government achieves this objective.

The significance of cybersecurity legislation within regulatory frameworks includes various dimensions. The existing regulations under examination are of great significance in promoting the effective utilization of the Internet, serving various objectives. The principal aim of their existence is to safeguard consumers from falling prey to cybercrimes. Concurrently, these guidelines function as regulations that govern the use of the Internet and computers within a wider framework.

The subject matter under consideration is related to instances of fraudulent behaviour. It is important to acknowledge that these effects and consequences are mostly observed, yet it should be recognized that there may exist other detrimental and damaging tertiary consequences in the absence of adequate and standardized legislation concerning cybersecurity. Hence, the primary areas covered by cyber-regulations can be distilled into three domains:

- **Fraud:** The implementation of cyber-rules serves the objective of safeguarding individuals from fraudulent acts that occur in the online domain. These rules operate as a deterrent against offences such as credit card fraud and identity theft. Along with that, these regulations establish the parameters for the use of federal and state criminal sanctions against individuals involved in deceitful behaviours.
- **Copyrights:** The notion of copyright refers to the legal safeguard provided to creative expressions of authorship, ensuring exclusive rights to their creators. The implementation of cyber-limitations has been extended to broaden their reach and address the issue of copyright infringement, ensuring the preservation of copyright protection measures. Copyright laws grant individuals and businesses the legal ability to protect their artistic creations and derive economic benefits from them.
- **Calumniation:** Calumniation or defamation refers to the act of disseminating false statements that detrimentally impact the standing or reputation of an individual or entity. In digital communication, it is crucial to acknowledge that cyber-laws also involve cases of defamation. This service provides individuals and companies with protection against baseless and detrimental allegations disseminated on the Internet, which have the potential to inflict harm on their reputations.

Cybersecurity regulations, referred to as cybercrime statutes, comprise a diverse set of directives that are designed to enhance the resilience and effectiveness of information technology systems. The primary objective of these individuals is to motivate companies and organizations to improve their systems and safeguard their data against cyberattacks through the implementation of various strategies. In this discussion, we have provided a concise overview of different classifications of global cyber-regulations, including those about India, the United States, the United Kingdom, Australia, and the European Union. Therefore, in this section, we will address the standards that are relevant to cybersecurity and its associated technological advances.

11.1 NIST

The National Institute of Standards and Technology (NIST) is an autonomous governmental entity bereft of regulatory authority. Its primary mandate resides in the development of standardized protocols for both governmental bodies and private organizations. The primary operational objective of the NIST is focused on developing and establishing best practices that are relevant to cybersecurity. The

NIST Cybersecurity Framework is a prominent tool that assists in the recognition and management of cyber threats faced by organizations, irrespective of their scale.

The guidelines and benchmarks established by the NIST are widely acknowledged and implemented worldwide. Compliance to these standards endorsed by the NIST promotes an atmosphere of confidence and reliability among organizations, their customers, collaborative partners, and interested stakeholders. Developed with a specific emphasis on the critical infrastructure sector, the NIST Cybersecurity Framework aims to provide a cost-effective approach to the management of cybersecurity risks.

The collection of security foundations, directives, and protocols provided by NIST enables organizations to effectively manage and mitigate the fluctuations in cybersecurity vulnerabilities. Organizations that undertake the implementation of the NIST Framework are more likely to be well-prepared to meet future requirements for regulatory compliance. Established with the primary objective of encouraging innovation and enhancing economic competitiveness, NIST, which was founded in 1901, operates as an essential component of the US Department of Commerce. The primary objective of this organization revolves around promoting innovation within the United States and enhancing industrial competitiveness.

The scope of NIST's activities involves the creation, promotion, and maintenance of standards and measurements for various industries. NIST promotes measurement science, standards, and technology through its wide range of initiatives. The NIST is committed to enhancing economic security and improving the overall quality of life. NIST is actively engaged in the promotion of the NIST Cybersecurity Framework, which is a voluntary framework designed to enhance businesses' comprehension of cybersecurity risk, its governance, and potential mitigation strategies across different organizational sizes.

As of the time of writing, the NIST is actively involved in an ongoing initiative to enhance the NIST Cybersecurity Framework, which is referred to as CSF 2.0. In summary, NIST is located within the US Department of Commerce. The NIST Cybersecurity Framework serves as a fundamental tool that effectively empowers businesses of varying sizes to develop a deeper understanding of cybersecurity risk and implement strategies to manage and mitigate it. This framework has been the de facto standard in safeguarding networks and data resources.

11.2 ISO

The International Organization for Standardization (ISO) is and has been a significant and influential figure at a global level in standardization. The association acts independently and functions as a non-governmental institution, with a membership of 165 national standard bodies. The primary objective of the ISO is to bring together a wide range of specialists from many disciplines, encourage the sharing of knowledge, and advance the establishment of universally recognized standards through the process of consensus building. The formulation of these standards has been pursued to harmonize them with market dynamics, promote innovation, and offer resolutions to developing global concerns. ISO standards perform an essential function by offering organizations a structured framework that enables guidance, coordination, simplification, and harmonization.

The primary objective is to decrease operating expenses while simultaneously improving overall effectiveness. Again, it may be argued that they play a crucial and essential role in the formation of generally recognized benchmarks for goods and services on a global level. The objective of this initiative is to enhance consistency and compatibility among various entities operating in geographically distinct areas. The worldwide use of ISO standards is extensive, including a wide range of countries and organizations. The reason behind its adoption stems from a mutual dedication to creating defined criteria for quality. The objective of the ISO is fulfilled by the creation of reports that cover essential standards, specifications, guidelines, and characteristics. The purpose of these publications is to assist businesses in effectively aligning their materials, products, processes, and services with their intended aims.

The ISO is substantial in promoting an environment in which member nations fully embrace these standards, thus building a framework for standardization. The numerous advantages of international standards are easily evident. The main goal is to establish a conducive environment that promotes enhanced standards of quality and safety, ensuring the safeguarding of both consumers and workers. Along with that, the implementation of international standards enhances the efficiency of company operations, resulting in decreased costs and subsequent financial gains. It is necessary to recognize that they possess the capacity to improve global competitiveness through the promotion of inter-industry trade and the facilitation of technology dissemination. Similarly, these entities serve as catalysts for promoting improved coherence in IT artefacts and can successfully support ethical compliance in domains such as accounting.

The ISO standards also strive to create a fair and equal economic environment on a worldwide scale by advocating for consistent labour standards among states. These entities are important in safeguarding the environment by monitoring and controlling key factors, including atmospheric and aquatic conditions, soil composition, and the release of different gases and radiation. ISO standards are of the greatest significance in ensuring the protection of the environment and advancing the well-being of individuals. The adoption of international standards is necessary in establishing a common language among trading partners, leading to the standardization of various products and services. This, in turn, enhances the efficiency of international commerce.

11.3 NIST Cybersecurity Framework

The NIST Cybersecurity Framework (NIST CSF) is a compendium of guidelines and methodologies developed by the NIST to effectively govern and mitigate security vulnerabilities in information technology (IT) systems. The presented paradigm offers a systematic approach to mitigating, identifying, and addressing cyberattacks. The primary components of this framework include standards, suggestions, and methodologies designed to enhance the cybersecurity readiness of organizations, particularly those operating within the private sector in the United States. Other nations, such as Japan and Israel, have also adopted this approach.

The main purpose of the NIST CSF is to provide a framework for protecting critical infrastructure from cybersecurity risks. Enhancing cybersecurity knowledge can be particularly advantageous for smaller organizations or those with less regulatory control. However, organizations of greater scale that have well-established information technology security programmes may perceive it as less instructive. The development of this framework was the result of a collaborative effort between private enterprises and government agencies, undertaken as a voluntary initiative. The design of this system aims to possess adaptability, cost-effectiveness, and flexibility, allowing organizations to prioritize its components according to their specific needs.

The NIST Cybersecurity Framework is accessible in multiple formats, including spreadsheets, PDFs, and reference tools. The aims of this initiative include the assurance of security for critical IT infrastructure, the promotion of knowledge regarding cybersecurity threats, and the incorporation of cybersecurity issues into routine organizational deliberations. Organizations can utilize the NIST CSF in diverse manners, including the evaluation of their existing cybersecurity measures, identification of prospective standards and policies, dissemination of novel requirements, and establishment of fresh cybersecurity programmes.

It is urgent to recognize that compliance to the NIST CSF is a voluntary initiative that emphasizes performance. The implementation of this policy was first initiated by executive orders issued by former US President Barack H. Obama and has since been consistently carried out by various government agencies, even throughout the administration of former US President Donald J. Trump. Similarly, it is worth noting that the adoption of this framework is not limited to government and private sector institutions, as public enterprises also have the option to implement it. Both the US government and the NIST have developed and made available various tools aimed at aiding organizations in the establishment and implementation of cybersecurity programmes and assessments. The framework's subsequent iteration, denoted as Version 1.1, incorporated a dedicated segment entitled 'Self-Assessing Cybersecurity Risk with the Framework'. This addition aims to provide organizations with guidance on how to effectively carry out their cybersecurity risk assessments.

The NIST Cybersecurity Framework finds its roots in Executive Order (EO) 13636, entitled 'Improving Critical Infrastructure Cybersecurity', which was promulgated on February 12, 2013. The executive order catalysed a cooperative initiative between the NIST and the private sector in the United States. The objective was to ascertain and integrate prevailing voluntary consensus standards and industry best practices into a cohesive framework for cybersecurity. The outcome of this collaborative effort led to the creation of NIST Cybersecurity Framework Version 1.0.

The Cybersecurity Enhancement Act (CEA) of 2014 augmented the responsibilities of the NIST in advancing the development of the framework.

The NIST Cybersecurity Framework has experienced extensive acceptance across several industries in the United States with time, demonstrating its efficacy in aiding organizations in the mitigation of cybersecurity threats. Numerous organizations have attested to the efficacy of version 1.1 of the framework in mitigating cybersecurity risks. In light of the dynamic nature of the cybersecurity domain, the NIST regularly collaborates with stakeholders to ensure the ongoing efficacy and alignment of the upcoming CSF 2.0 with its initial objectives.

The NIST Cybersecurity Framework is composed of several essential elements, which include functions, categories, subcategories, and informational references. Functions offer a perspective on security procedures and optimal strategies, highlighting their ongoing nature as activities that contribute to the development of a cybersecurity-oriented organizational culture rather than mere procedural measures.

Categories and subcategories provide a set of action plans that are customized to address the distinctive requirements of each department or operation inside an organization. Several instances of these functions and categories can be observed:

(1) **Identify:** This function involves comprehending an organization's critical assets and resources, including categories such as asset management, business environment, governance, risk assessment, risk management strategy, and supply chain risk management.
(2) **Protect:** The protect function addresses technical and physical security controls to safeguard critical infrastructure. It includes categories such as identity management and access control, awareness and training, data security, information protection processes and procedures, maintenance, and protective technology.
(3) **Detect:** The detect function's objective is to promptly alert organizations to cyberattacks. Categories under this function include anomalies and events, security continuous monitoring, and detection processes.
(4) **Respond:** The respond function is centred on effectively responding to cyberattacks. Categories within this function cover response planning, communications, analysis, mitigation, and improvements.
(5) **Recover:** Recovery activities include the implementation of plans for cyber resilience and ensuring business continuity following a cyberattack or security breach. The recovery function includes recovery planning improvements and communications.

The NIST Cybersecurity Framework incorporates instructive references that provide linkages between its functions, categories, subcategories, and specific security measures present in alternative frameworks. The document does not provide prescriptive guidelines for the precise methodologies that organizations should use in doing physical or software inventory. Rather, it presents a checklist of tasks that need to be completed. Organizations have the autonomy to select their preferred methodologies for inventory management. If organizations find themselves in need of further guidance, they have the option to consult the instructive references that pertain to the relevant controls included in complementary standards. The ability to adapt and tailor cybersecurity risk management strategies according to specific organizational needs provides organizations with enhanced autonomy and control.

The main purpose of these roles is to provide a strategic outlook on an organization's cybersecurity threats. The application of the framework involves a hierarchical methodology that offers an understanding of an organization's cybersecurity vulnerabilities and the strategies utilized to mitigate these vulnerabilities. The tiers explain the degree to which an organization's cybersecurity risk management practices conform to the characteristics defined in the framework. A Tier 1 organization is defined by its limited level of understanding and its incomplete execution of the framework. The tiering system advances to Tier 4, denoting an adaptive organization that demonstrates proficiency in efficiently addressing cybersecurity risks. The framework profile functions as a method for describing the present security programme status of an organization and evaluating it against the intended state. This procedure has the potential to identify any pre-existing deficiencies that necessitate remediation. The principal objective of a profile is to aid organizations in formulating a strategic plan for mitigating cybersecurity risk.

The NIST Cybersecurity Framework 2.0 (CSF 2.0) represents a methodical approach to the governance of cybersecurity vulnerabilities. The provided framework offers precise guidelines for a range of institutions, including governmental authorities and private companies, intending to reduce cybersecurity threats. The CSF 2.0 includes six primary roles, namely, governance, identification, protection, detection, response, and recovery. The incorporation of the governance function is a noteworthy addition. The existing version of the CSF, as of writing, referred to as 2.0, was officially released on August 8, 2023, as a draft open for comments. The proposed changes include significant alterations, such as broadening the scope, integrating the governance role stated earlier, improving internal communication inside cybersecurity organizations, and placing greater importance on issues related to cyber-risk. It is crucial to acknowledge that the adjustments made to CSF 2.0 were based on feedback obtained from stakeholders, ensuring its congruence with the ever-changing field of cybersecurity.

11.4 ISO Standards

ISO standards are a collection of globally recognized standards that have been formulated by the ISO. These standards are formulated to offer recommendations, specifications, and optimal approaches to guarantee the excellence, security, effectiveness, and compatibility of goods, services, and systems across many industries and sectors. ISO standards include a diverse set of domains, which include, but are not restricted to, the following:

- **Quality Management:** ISO 9001 sets the standards for quality management systems, helping organizations improve their processes and meet customer expectations.

- **Environmental Management:** ISO 14001 provides guidelines for environmental management systems to reduce the environmental impact of organizations' activities.
- **Food Safety:** ISO 22000 focuses on food safety management systems, ensuring the safety of food products throughout the supply chain.
- **Occupational Health and Safety:** ISO 45001 provides guidelines for occupational health and safety management systems to protect employees and prevent accidents.
- **Energy Management:** ISO 50001 sets the standards for energy management systems, helping organizations reduce energy consumption and improve efficiency.
- **Risk Management:** ISO 31000 offers principles and guidelines for effective risk management in organizations.
- **Quality of Healthcare:** ISO 9001 and ISO 15189 are used in healthcare settings to improve the quality of patient care and laboratory services, respectively.
- **Aerospace:** ISO 9001 and AS9100 are used in the aerospace industry to ensure quality and safety in aircraft manufacturing.
- **Automotive:** ISO/TS 16949 outlines quality management requirements specific to the automotive industry.
- **Social Responsibility:** ISO 26000 guides social responsibility, encouraging organizations to operate ethically and contribute to societal well-being.
- **Information Security:** ISO 27001 addresses information security management systems to safeguard sensitive information.

Organizations globally generally embrace ISO standards as a means to showcase their dedication to quality, safety, and optimal operational procedures. Compliance to these standards has the potential to result in enhanced operational procedures, heightened levels of client contentment, and greater competitive advantage within the global marketplace.

This section will explore ISO standards that are relevant to the fundamental concept of digital resilience. Standards have been developed to guide organizations in the construction and maintenance of resilient digital systems and processes capable of withstanding disruptions, cyberattacks, and the dynamic nature of technology. ISO standards about digital resilience offer crucial foundations for organizations to proactively tackle the difficulties posed by the digital infrastructure, thereby guaranteeing the uninterrupted functioning and safeguarding of their digital activities.

In this discussion, we will examine the significant ISO standards that have a crucial impact on augmenting an organization's capacity to adjust, recuperate, and prosper amidst digital disruptions and cyber threats. From the domain of information security to business continuity and risk management, these standards serve as essential instruments for effectively handling the complex and ever-evolving digital world.

11.4.1 ISO 22301

ISO 22301 is a highly recognized standard that connects to Business Continuity Management (BCM). This standard offers organizations a complete framework that enables them to effectively and pre-emptively prepare for, as well as respond to, various interruptions and crises. ISO 22301 provides crucial guidance in traversing the complex corporate environment, which is fraught with many possible risks such as natural calamities and cyber-intrusions.

The fundamental components of ISO 22301 include risk assessment and management, as well as overall resilience, with a particular focus on the identification and evaluation of potential risks and vulnerabilities. Organizations can obtain valuable information regarding the importance of different business activities and efficiently prioritize recovery efforts by conducting a Business Impact Analysis (BIA). The core of ISO 22301 centres around the formulation of a Business Continuity Plan (BCP), which defines specific measures to sustain essential operations in the event of disruptions. The use of response and recovery methods, as outlined in ISO 22301, enables organizations to effectively manage crises, thereby mitigating their adverse consequences.

The continuous evaluation and assessment of the BCP through testing, exercising, continuous monitoring, and performance evaluations are essential to maintain the sustained efficacy of the system. The precise recording of policies, procedures, and records is deemed critical under ISO 22301, emphasizing the significance of documentation for organizations. The ISO 22301 standard possesses worldwide applicability, rendering it advantageous for organizations across various sizes and industries. Obtaining accreditation signifies a dedication to implementing business continuity management practices, which serve to strengthen resilience, protect reputation, and maintain confidence among customers, partners, and stakeholders in an environment characterized by uncertainty.

11.4.2 ISO/IEC 27001

The ISO/IEC 27001 standard centres around the establishment and maintenance of information security management systems (ISMS). The framework offers a systematic and all-including structure for organizations to protect their information assets, efficiently manage risks, and guarantee the confidentiality, integrity, and availability of sensitive information. The fundamental focus of ISO/IEC 27001 lies in highlighting the significance of risk management. Organizations must conduct a complete identification and assessment of information security risks, including several aspects including but not limited to data breaches, cyber threats, and unauthorized access. By comprehending these potential threats, organizations can deploy suitable safeguards and strategies to control their impact.

The implementation of the standard requires the establishment of an information security management system (ISMS), which is a methodical approach to the governance and safeguarding of confidential information. This includes the development of policies, procedures, and protocols that effectively tackle challenges related to information security. The ISO/IEC 27001 standard also places significant emphasis on the significance of ongoing improvement. Regular evaluations, audits, and assessments are essential to maintain the ongoing effectiveness and currency of the ISMS in light of security threats and technological advancements.

The attainment of ISO/IEC 27001 accreditation not only signifies a dedication to safeguarding information security but also promotes trust and assurance among many stakeholders, such as customers and partners. This solution aids organizations in adhering to legal and regulatory obligations about information security, hence enabling them to efficiently address issues. In the current setting characterized by a high occurrence of data breaches and cyberattacks, ISO/IEC 27001 offers a framework for organizations to safeguard sensitive information, mitigate security risks, and improve their overall security stance.

11.4.3 ISO/IEC 27002

The ISO/IEC 27002 standard offers extensive guidance on the implementation of efficient information security management practices in organizational settings. The standard in question is critical as a necessary counterpart to ISO/IEC 27001, which outlines the prerequisites for the establishment and upkeep of an ISMS. The ISO/IEC 27002 standard provides a well-organized framework for protecting information assets and effectively and systematically managing security risks.

The standard includes a variety of crucial elements. First and foremost, this emphasizes the significance of formulating customized information security policies and procedures to establish a strong basis for information security within an organization. Also, ISO/IEC 27002 lays significant importance on the practice of risk management. This resource provides essential recommendations about the identification, assessment, and mitigation of information security threats, to properly safeguard sensitive data and key systems.

The topic of access control is also examined, whereby guidelines are presented to guarantee that solely authorized individuals can access resources, hence mitigating the potential for unauthorized access and data breaches. The topic of cryptography is thoroughly examined to uphold the principles of secrecy, integrity, and authenticity in data management. It provides valuable insights into the adoption of cryptographic approaches and the implementation of best practices.

Security awareness and training are necessary components in promoting a culture of security consciousness among employees and providing them with the necessary abilities to identify and address security risks. Efficient incident management is advocated, defining protocols for quickly identifying, documenting, and addressing security incidents. It is advisable to implement physical security measures, such as the establishment of secure data centres and the enforcement of restricted access protocols in critical areas, to safeguard information assets from potential physical risks.

The importance of business continuity planning and disaster recovery is emphasized, as it offers direction on guaranteeing the availability of essential systems and data in times of crisis. Last but not least, ISO/IEC 27002 places significant emphasis on the necessity of adhering to legal, regulatory, and contractual obligations in the field of information security. Organizations can enhance their information security posture, eliminate risks, and demonstrate their commitment to best practices by implementing the full guidelines outlined in ISO/IEC 27002 about these crucial areas. Not only does this practice ensure the protection of data, but it also boosts confidence among stakeholders and assists in meeting security-related duties.

11.4.4 ISO/IEC 27031

The ISO/IEC 27031 standard is a framework that specifically addresses the preparedness of information and communication technology (ICT) systems concerning business continuity. The importance of this standard is particularly notable in the current digital age, where ICT systems and services serve as the foundation for numerous critical corporate activities. The ISO/IEC 27031 standard offers organizations a systematic method for evaluating and improving the preparedness of their ICT infrastructure to effectively meet business continuity goals.

One of the key factors emphasized in the standard is the acquisition of a thorough comprehension of an organization's ICT needs for the continuity of business operations. This involves the identification of crucial ICT systems and services, along with their interconnections with other critical business processes. The need for risk assessment and management is also emphasized, urging organizations to proactively identify and mitigate ICT-related threats that have the potential to disrupt their operations.

The ISO/IEC 27031 standard guides organizations on the development of a specialized ICT continuity strategy that is in line with their overall business continuity objectives. This particular approach establishes explicit goals, preferences, and resource demands for the continuity of ICT. Proponents of the standard argue in favour of developing ICT continuity plans that outline the strategies and procedures for sustaining or restoring ICT systems and services during times of disruption.

The regular assessment, implementation, surveillance, and evaluation of ICT continuity plans are integral elements within the framework of ISO/IEC 27031. These actions serve to enhance the efficacy of recovery procedures, facilitate staff training for incidents connected to ICT, and enable organizations to respond to changing technological advancements and business requirements. ISO/IEC 27031 is a standard that holds significant value for organizations significantly dependent on ICT. It serves to augment their capacity to preserve the accessibility, reliability, and secrecy of crucial data and services during disruptive occurrences.

11.4.5 ISO/IEC 27032

The ISO/IEC 27032 standard addresses the subject of cybersecurity and the safeguarding of information within the domain of cyberspace. The principles and best practices for managing and enhancing the security of ICT within the digital domain are established by the ISO and the International Electrotechnical Commission (IEC).

The fundamental objective of ISO/IEC 27032 is to advocate for the establishment and execution of a well-defined cybersecurity strategy. An organization's strategy must be in congruence with its overarching aims and objectives, taking into account the distinctive difficulties and risks that arise from the dynamic and ever-changing cyberspace environment. The standard lays considerable emphasis on the governance and management systems of cybersecurity. This emphasizes the significance of clearly defining roles and duties, establishing robust risk management systems, and implementing efficient monitoring measures.

The focal point of this discussion is cyber-risk management, which includes the processes of identifying, evaluating, and effectively handling cyber-risks. The ISO/IEC 27032 standard provides guidelines on the implementation of methods aimed at the prevention, detection, and response to cyber threats and incidents. The principles of collaboration and information sharing aid in encouraging cooperation among many stakeholders, such as governments, enterprises, and organizations, to exchange cybersecurity intelligence and insights. The utilization of a collaborative method significantly improves the capabilities of early threat identification and response.

Also, the established guidelines emphasize the importance of promoting cybersecurity awareness and providing training programmes to effectively mitigate the impact of human factors on cybersecurity. The text emphasizes the necessity of implementing incident response and recovery strategies to mitigate the consequences of cyber-incidents. It highlights the significance of adhering to legal and regulatory requirements within the field of cybersecurity.

ISO/IEC 27032 is an important resource for organizations and governments aiming to improve their cybersecurity stance within the context of an ever more interconnected digital environment. Compliance to this standard facilitates the safeguarding of critical information and infrastructure, promotes international collaboration in the domain of cyberspace, and contributes to the establishment of a more secure digital infrastructure on a worldwide scale.

11.4.6 ISO/IEC 27701

The ISO/IEC 27701 standard focuses on the safeguarding of personally identifiable information (PII) and the preservation of data privacy. The statement serves as a foundational structure for organizations that seek to design, operate, maintain, and improve a Privacy Information Management System (PIMS). ISO/IEC 27701 serves as an extension to ISO/IEC 27001, with a specific emphasis on addressing privacy considerations within the domain of information security.

The standard incorporates a range of essential components. The provided statement defines a thorough framework for privacy management, including the establishment of roles, duties, and processes for the handling of personally identifiable information (PII) and the mitigation of privacy threats. Risk assessment and management are important in this particular setting, as organizations are guided by a standard that provides instructions on how to apply controls to effectively reduce these risks.

One notable characteristic of this feature is in its prioritization of adherence to privacy standards, which have become progressively more rigorous on a global scale. The ISO/IEC 27701 standard offers guidelines to assist organizations in effectively navigating the complex data protection legislation, facilitating the alignment of their practices with the necessary legal requirements. The ideals of transparency and accountability are essential components. It is recommended that organizations prioritize transparency about the handling of personally identifiable information (PII) and strive to conform to privacy guidelines.

The established framework includes provisions that pertain to the rights of data subjects, thereby streamlining procedures that enable individuals to assert their rights concerning personally identifiable information (PII). The document expands the scope of privacy considerations to include connections with third parties and defines protocols for incident response and notification, which are especially critical in cases of data breaches. ISO/IEC 27701 serves as a significant asset for organizations that are dedicated to the responsible management of personally identifiable information (PII) and compliance to data privacy rules. Compliance with this standard not only demonstrates conscientious data management practices but also develops confidence among stakeholders in a time characterized by escalating concerns around data privacy and regulatory oversight.

11.5 UKCIS Digital Resilience Framework

The digital resilience framework [UKC19], developed by the Digital Resilience Working Group of the UK Council for Internet Safety (UKCIS), offers an indispensable strategy for improving individuals' capacity to navigate the digital environment securely and efficiently. Today, there exists a pressing need to provide individuals with the necessary competencies and understanding to effectively handle and excel in the digital setting.

Digital literacy is fundamentally essential in establishing the fundamental basis for digital resilience. The purpose of this particular element within the framework is to provide individuals with a diverse range of digital skills, enabling them to enhance their capabilities and autonomy. The scope of digital literacy extends beyond fundamental skills to embrace more advanced proficiencies, such as comprehension of online content, identification of misinformation, and effective utilization of digital resources. Essentially, it empowers individuals to acquire expertise as digital citizens who are capable of utilizing technology to their advantage.

The compliance to responsible and ethical behaviour in online environments constitutes a fundamental component of digital resilience. The subject matter comprises a diverse set of elements, such as the prevention of cyberbullying, the promotion of respectful dialogue within digital environments, and the development of responsible practices in information sharing. The framework promotes the active consideration and resolution of these concerns, motivating individuals to interact with the digital domain in a manner that upholds personal integrity and mutual respect, ultimately developing an atmosphere of civility in online interactions.

In current times, the protection of personal information and the assurance of online security have become of importance, given the significant value attributed to data. The framework addresses a crucial component by providing education on effective password management, enhancing awareness of phishing efforts, and implementing steps to safeguard personal data from unauthorized access. The concept of empowering individuals to take proactive measures in safeguarding their digital identities is of great importance.

In the context of the abundant wealth of information accessible through online platforms, the capacity to engage in critical thinking assumes a critical role. This particular element promotes the development of critical thinking skills in individuals, urging them to exercise caution and discernment while engaging with online content. It aims to equip individuals with the ability to analyse and evaluate information sources, ascertain their authenticity, and effectively differentiate between factual information and subjective viewpoints. In a time characterized by the increase in spread of disinformation, the development of critical thinking skills emerges as a potent safeguard.

The concept of digital resilience includes not just technical competencies but also emotional well-being. This programme provides individuals with effective techniques for coping with online stress, addressing incidents of harassment or cyberbullying, and developing a constructive online image. By acknowledging the potential emotional burden that online interactions may impose, this component of the framework emphasizes the importance of promoting mental and emotional wellness in the times of digital communication.

In addition to the fundamental components, the digital resilience framework has several other essential factors. Assessment tools help in evaluating an individual's digital resilience level. The utilization of structured assessments allows individuals to effectively identify their strengths and weaknesses, helping them to concentrate their efforts on areas that necessitate improvement. The framework also provides access to a wide range of educational materials, guidance, and best practices to support the development of digital resilience skills.

Acknowledging the crucial significance of education in moulding individuals into competent digital citizens, the framework has been specifically developed to seamlessly include it in educational curricula and programs. The incorporation of digital resilience into education serves to establish it as a core component, equipping students with the necessary skills to confront the complexities of the digital world from an early stage. The digital environment is characterized by its dynamic nature, wherein new risks and difficulties arise regularly. To maintain its efficacy, the framework undergoes ongoing evaluation and revisions, hence assuring its modernity and importance.

Collaboration is crucial in addressing the complex and diverse difficulties presented by the digital setting, necessitating the collaborative initiatives of multiple stakeholders. The framework promotes collaborative efforts among many stakeholders, including educators, parents, policymakers, and technology corporations. The adoption of a collaborative strategy is of importance in the establishment of a safer and more resilient digital society. The digital resilience framework possesses a relevance beyond its function as a tool for personal empowerment within the context of the digital age. The subject matter holds significant consequences for several aspects of society, education, and the collective approach towards addressing the ever-changing complexities of the digital world.

In the current period, characterized by the pervasive integration of technology into our everyday existence, the framework initiatives to respond to an urgent requirement of endowing individuals with the requisite competencies and understanding to proficiently use the digital platforms. The need for digital literacy, which includes fundamental abilities and the capacity to differentiate reliable information from misleading content, has become increasingly crucial. The framework addresses this educational deficiency by developing a cohort of knowledgeable digital citizens who possess the ability to make conscientious decisions.

Through the prioritization of responsible online conduct and ethical practices, it actively contributes to the raising of a culture characterized by online civility and mutual respect. The prevalence of cyberbullying and harassment necessitates a framework that empowers individuals with the necessary resources to effectively address and mitigate these behaviours. By implementing such a framework, online interactions can be made to be safer and characterized by increased levels of respect.

The primary emphasis of the framework lies in the domain of privacy and security, effectively tackling concerns about data breaches, identity theft, and phishing attempts. Providing individuals with education on the need to preserve personal information and enhancing their ability to identify online risks not only serve to protect individuals but also play a role in promoting a more secure digital space.

Significantly, the incorporation of the framework within the educational system guarantees that digital resilience becomes an essential component of contemporary pedagogy. This early preparation equips students with the necessary skills to handle and adapt to the complexities and possibilities of the digital world, thereby augmenting their preparedness for a workforce that is progressively reliant on digital technologies.

The assessment tools, coaching, and continuous improvement methods provided by this system allow individuals to assess their levels of digital resilience, monitor their progress, and remain informed about emerging digital risks. The framework promotes the facilitation of collaboration among various stakeholders, including educators, parents, policymakers, and technology businesses. The recognition of the collaborative approach in addressing the various difficulties presented by the digital setting emphasizes the necessity of communal initiatives in establishing a safer and more accountable online society.

At a higher level, the framework possesses the capacity to exert an impact on national policies and practices about digital education and safety. The ideas of this model have the potential to serve as a paradigm for other nations, thereby making a valuable contribution to the global initiative towards digital resilience. The digital resilience framework signifies a pivotal advancement in the pursuit of establishing a digital space wherein individuals possess not only proficiency in using technology but also the capacity to use it in a manner that ensures safety, responsibility, and ethical conduct.

All-inclusive, the digital resilience framework serves a potent instrument for enabling individuals to effectively and ethically navigate the digital domain. The significance of this approach is rooted in its nature, as it includes various aspects such as digital literacy, online conduct, privacy and security, critical thinking, and emotional well-being. As society progresses in its adoption of digital technologies, the framework discussed herein assumes a role in nurturing knowledgeable and adaptable digital citizenship.

12 Digital Resilience and Regulatory Compliance

Digital resilience, in the context of modern businesses and organizations, is an organization's ability to withstand and recuperate from digital disruptions and threats while ensuring compliance with regulations. This section discusses the fundamental principles of digital resilience and their correlation with regulatory compliance. The capacity to foresee and respond to the dynamic digital infrastructure is a critical element of digital resilience. This is the ongoing surveillance of potential risks, vulnerabilities, and regulatory modifications that could potentially affect an organization's digital activities. By maintaining a state of awareness and taking proactive measures, organizations can diminish the probability of encountering compliance violations and control the potential consequences that arise from noncompliance.

Regulatory compliance refers to the act of conforming to legal requirements, rules, and industry standards that control the utilization and safeguarding of digital data. The legislation about various industries and areas exhibits substantial variation, highlighting the importance for organizations to possess a full comprehension of the unique requirements that pertain to their operations. Failure to comply with regulations and guidelines can lead to significant repercussions, such as potential legal sanctions and detrimental effects on an organization's public image.

The attainment of digital resilience while simultaneously upholding regulatory compliance necessitates the adoption of an all-inclusive strategy. The task at hand includes the deployment of resilient cybersecurity protocols to safeguard digital resources, information, and systems. Also, organizations must develop explicit policies and procedures that follow relevant rules and standards, thereby guaranteeing that staff possess knowledge and receive appropriate training to comply with these guidelines.

Regular audits and assessments are fundamental elements of this method. Organizations must do both internal and external evaluations to evaluate their digital resilience and compliance posture. These evaluations can uncover areas of weakness and opportunities for growth, allowing organizations to rapidly implement remedial measures. Incident response planning is a crucial component; despite the implementation of effective preventive measures, digital incidents may still occur. The presence of a clearly defined and thoroughly tested incident response plan is significant to mitigate the consequences of such incidents and guarantee the organization's compliance to regulatory requirements throughout the action.

Digital resilience and regulatory compliance are closely interconnected in the current digital setting. Organizations should not perceive compliance as an isolated entity but rather as an inherent component of their digital resilience strategy. Organizations may effectively traverse the complexities of digital operations and maintain compliance with the dynamic regulatory framework by adopting a proactive, informed, and adaptive approach. The integration of various measures serves to safeguard the organization from potential legal and reputational threats while also building trust among clients and other interested parties.

12.1 Intersection

Organizations are faced with the challenge of finding an equilibrium between their ability to endure digital incidents and their compliance with regulatory requirements as they traverse the junction of digital resilience and regulatory compliance. This intersection emphasizes the necessity for two seemingly disparate objectives to merge harmoniously. In the present moment, companies must acknowledge that the implementation of digital resilience strategies can effectively contribute to the attainment and sustenance of regulatory compliance. The use of strong cybersecurity measures and proactive risk management methods in digital resilience efforts inherently corresponds to various compliance requirements. By implementing measures to safeguard digital assets and data, enterprises inherently fulfil a core prerequisite of data protection rules. The compliance status of a corporation can be greatly influenced by its ability to efficiently address digital incidents, which is a crucial aspect of digital resilience.

Regulatory bodies commonly examine how corporations address data breaches and other cyber-disasters. The implementation of an effective incident response strategy has the potential to minimize the adverse consequences resulting from a data breach, showcase compliance to regulatory requirements, and diminish the probability of facing penalties imposed by regulatory bodies. On the contrary, compliance to regulations can also serve as a source of guidance and reinforcement for digital resilience initiatives. Regulations often necessitate the implementation of specific cybersecurity practices, data protection safeguards, and reporting requirements. The incorporation of these requirements into the design of digital resilience framework can offer systematic development of an specially resilient digital infrastructure.

The GDPR requires the implementation of data protection measures from the initial design phase and as the default setting. This aligns with the principles of digital resilience, which promote the integration of proactive security measures into the fundamental structure of a company's digital. Businesses enhance their digital resilience by complying with the criteria set forth by the GDPR. The interdependence between digital resilience and regulatory compliance is evident. When enterprises adopt a holistic strategy, they may enhance the resilience and security of their digital infrastructure. The convergence of this confluence presents an encouraging affair to coordinate objectives regardless of streamlining initiatives and eventually enhance a corporation's capacity to flourish all while abiding by the regulatory framework that governs its operations.

12.2 Impact of Non-compliance

Non-compliance with digital regulations and standards can have far-reaching repercussions for businesses. This section explores the complex consequences of non-compliance, emphasizing the diverse aspects that organizations need to take into account. Firstly, it is important to note that failure to comply with regulations and laws might lead to potential legal problems. Regulatory authorities worldwide possess the jurisdiction to levy penalties and punishments upon organizations that do not fulfil their obligations about compliance. The penalties associated with these violations might vary, including both minor fines and significant financial obligations, contingent upon the gravity of the offence and the particular legislation implicated. In instances of significant severity, failure to follow regulations can potentially result in legal proceedings, exacerbating the complexities faced by an organization. In addition to monetary sanctions, failure to comply with regulations can also have detrimental effects on an organization's public image and standing.

Trust is a highly sought after resource for businesses related to digital technology, and individuals who engage with organizations anticipate that their data will be managed responsibly and ethically. Instances of non-compliance, data breaches, or privacy violations have the potential to undermine confidence and damage the reputation of an organization. The process of restoring trust following a compliance failure can be difficult and long. Non-compliance leads to operational problems as an additional outcome. Regulatory bodies possess the authority to mandate that organizations halt specific activities or implement substantial modifications to their operations until they have achieved compliance. The circumstances may result in periods of inactivity, financial loss, and interference with regular operational procedures.

Interruptions of this nature might potentially generate a series of consequences that impact an organization's overall performance and competitive advantage. Non-compliance can also have a detrimental effect on an organization's capacity to compete in the global market. Numerous international rules, such as the GDPR and the California Consumer Privacy Act (CCPA), possess extraterritorial applicability, whereby they extend their jurisdictional scope to include organizations operating outside their territorial boundaries, provided that these organizations handle the personal data of individuals inside said jurisdiction. Failure to comply to these requirements might impede an organization's capacity to broaden its client base and engage in foreign business enterprises. The negative impact on reputation resulting from failure to comply can have far-reaching consequences for the organization's associations with commercial partners and suppliers.

Organizations lacking the ability to exhibit a firm dedication to compliance may encounter challenges in establishing new partnerships or preserving current ones. Organizations that present a compliance risk may encounter reluctance from suppliers and partners to engage in collaborative efforts due to concerns regarding the legal and reputational implications associated with such associations. The ramifications of non-compliance are diverse and can result in significant repercussions for organizations. The scope of consequences includes legal ramifications, harm to reputation, disruptions to operations, and constraints on expansion and cooperation. Hence, organizations must accord high priority to compliance within their digital resilience plan, acknowledging its significance as more than just a legal requirement but as a crucial element for ensuring long-term sustainability.

12.3 Benefits of Regulatory Compliance

The concept of regulatory compliance is commonly perceived as a collection of regulations and responsibilities imposed on organizations. Regardless, it is important to recognize that regulatory compliance also includes several concrete advantages that go beyond basic obedience to legal obligations. This section examines the benefits of regulatory compliance, emphasizing its significance in mitigating risks, nurturing trust, and gaining an edge over competitors.

Regulations are not exclusively a regulatory obligation but rather a significant source of advantageous outcomes. The implementation of risk mitigation strategies not only promotes trust among stakeholders but also confers a competitive edge in the context of a globalized and highly competitive corporate stage. Organizations that use compliance as a strategic advantage position themselves for enduring success and sustainability.

12.3.1 Risk Mitigation

Regulatory compliance confers significant benefits, with risk reduction being a major advantage. Regulations and standards are developed to reduce risks associated with various aspects of an organization's operations. The potential threats included in this context may include but are not limited to data breaches, financial irregularities, environmental consequences, and other associated problems. Compliance standards are developed to build a systematic framework that assists organizations in identifying, assessing, and effectively managing these risks.

One illustration of financial regulations is the Sarbanes-Oxley Act (SOX), which requires companies to implement internal controls and adopt financial reporting practices designed to reduce the likelihood of financial misconduct and fraudulent behaviour. Organizations enhance their ability to detect and prevent financial irregularities, thereby protecting their financial stability and reputation, by adhering to legislative measures such as the Sarbanes-Oxley Act (SOX) and other regulatory requirements.

Similar to this, strict data security procedures are required by data protection laws like the GDPR. Compliance with the GDPR not only ensures the protection of individuals' data but also mitigates the risk of data breaches, which can result in significant financial and reputational harm.

Plus, compliance to environmental legislation and sustainability standards serves to reduce an organization's ecological imprint and mitigate the accompanying risks. By adhering to these norms, firms can effectively reduce the potential for environmental harm, avoid financial fines, and fulfil legal obligations, thereby demonstrating their commitment to ethical corporate conduct.

Regulatory compliance serves as a proactive strategy for risk management. Organizations that spend resources on the implementation of compliance measures demonstrate a higher level of proficiency in identifying potential risks, adopting protective measures to mitigate them, and successfully responding to the occurrence of such threats. This practice not only aims to protect the organization but also develops confidence among stakeholders by showcasing the organization's diligence in mitigating potential threats.

12.3.2 Trust Building

One notable benefit of regulatory compliance is its ability to develop trust and assurance among diverse stakeholders, including consumers, investors, partners, and the wider public. Trust is a highly important intangible asset that possesses the capacity to significantly impact the accomplishments and longevity of an organization. The strict compliance to data privacy legislation ensures that clients can have confidence that their personal information is being handled with great care and following established standards. The development of confidence between customers and an organization over the safeguarding of personal data is anticipated to lead to heightened levels of transactional involvement and the building of long-lasting relationships.

Investors place considerable matters on compliance since it serves as a representation of an organization's commitment to transparency and accountability. Companies that follow financial reporting and auditing standards provide investors with a clearer evaluation of their financial health, reducing uncertainty and perceived risk. Compliance to the regulations plays a role in enhancing an organization's reputation and perception within its specific industry. In the context of the present-day interconnected world, the possession of a reputation that is distinguished by a commitment to compliance with regulations and ethical standards carries considerable importance. Organizations that place a high priority on ethical business practices possess the capacity to attract consumers, partners, and employees who demonstrate a desire to affiliate with such organizations.

Instances of non-compliance or data breaches have the potential to greatly harm the confidence that a business has painstakingly developed over a prolonged period. Instances of non-compliance or security breaches possess the capacity to evoke public indignation, besmirch the organization's reputation, and diminish the trust of stakeholders. The difficult process of restoring confidence after such occurrences necessitates a substantial time investment, highlighting the need for establishing and maintaining a strong commitment to compliance from the outset.

12.3.3 Competitive Advantage

Organizations might gain a competitive advantage by adhering to regulatory compliance requirements. In the context of a progressively complex and interrelated global market, compliance to regulatory standards can distinguish an organization from its rivals through many means. To begin with, following international norms and rules facilitates the global operations of organizations. Numerous pieces of legislation, such as the GDPR, have extraterritorial applicability, whereby they extend their jurisdiction to include organizations that handle the personal data of individuals residing within their purview, irrespective of the geographical location of these organizations.

Compliance to these standards facilitates access to global markets and attracts consumers who prioritize data protection and ethical data management. Also, following regulations and standards might serve as a distinguishing element from the perspective of consumers. There is a growing awareness among ethical customers regarding an organization's dedication to responsible practices, encompassing aspects such as environmental sustainability and data privacy. Organizations that can exhibit compliance to pertinent legislation acquire a competitive advantage by conforming to the values and expectations of their intended audience. Similarly, meeting regulatory requirements has the potential to result in financial benefits over an extended period.

Although there may be an initial financial commitment associated with implementing compliance procedures, the potential cost savings coming from avoiding fines, legal expenses, and reputational harm due to non-compliance can be significant. Compliance promotes operational efficiency and effectiveness, thereby making a further contribution to cost reduction. Compliance can also give rise to innovation. Organizations frequently encounter the necessity to adapt and innovate to effectively address emerging regulatory mandates. This innovation has the potential to facilitate the creation of novel products, services, and procedures, thereby augmenting both regulatory compliance and conferring a competitive edge to the organization within the market.

13 Compliance Challenges

Cybersecurity compliance is a fundamental concept for any organization that handles data or operates in an Internet-exposed environment. The process of managing and transmitting data automatically exposes firms to potential cyber threats. Cybersecurity compliance primarily involves following the norms and regulations that have been established by authorized entities, legislation, or agencies. To achieve compliance, organizations must implement risk-based measures to protect the confidentiality, integrity, and availability (CIA) of their information, regardless of its state, whether it is stored, in transit, or undergoing processing. The task of achieving cybersecurity compliance poses a significant obstacle as a result of the potential convergence and complexity of industry standards and obligations.

Ensuring cybersecurity compliance is key. Companies of all sizes are vulnerable to cyberattacks, highlighting the need for legislation and standards to protect sensitive data. To ensure performance, operational efficiency, and security, organizations must follow regulations. As a result of their perceived vulnerability, SMEs are both susceptible and appealing targets. An additional layer of protection is required for 16 distinct critical infrastructure sectors (CIS), according to the US Cybersecurity and Infrastructure Security Agency (CISA). Reason for designation includes a potential threat to public health, safety, national security, and the economy from breaches in these areas.

Organizational credibility and bottom line can take a serious hit when data breaches trigger a cascade of unfortunate events. Legal proceedings and disputes arising from these violations are on the rise in a wide range of industries. For a cybersecurity programme to be successful, compliance is key. The primary goals of data compliance regulations and cybersecurity measures are the protection of personally identifiable information (PII), commercial data, and health records. A person's date of birth, first and last name, home address, Social Security number, and mother's maiden name are all examples of personally identifiable information.

Information pertaining to money contains private details such as account numbers, expiration dates, CVV, PINs for debit and credit cards, and credit ratings and histories. Information that is safeguarded from unauthorized access includes a patient's medical records, insurance details, appointment details, prescription information, and hospital admission records. Certain private information is subject to various legal frameworks and compliance regulations. These include information about a person's ethnicity, religion, marital status, email address, Internet Protocol address, passwords, usernames, and biometric data such as voice prints, fingerprints, and face recognition.

Organizations face a multitude of tough hurdles when it comes to complying to cybersecurity requirements and legislation. The challenges stem from the complex and dynamic characteristics of the cybersecurity domain. One significant aspect that contributes to these issues is the presence of numerous cybersecurity standards and laws, each imposing its own distinctive set of criteria. As a result, companies often face the challenge of dealing with mandates that may overlap or contradict each other, leading to confusion and increased efforts to ensure compliance. The continuous development of cyber threats, which are becoming more advanced and varied, exacerbates the existing compliance difficulties. Keeping up with developing dangers and adjusting compliance measures accordingly poses a big challenge.

The effectiveness of previous compliance measures may not be sufficient to mitigate emerging dangers in the future. The requirement for compliance necessitates a significant allocation of resources, including financial, technological, and human capital. This places a notable burden on smaller- to medium-sized businesses, which have challenges effectively distributing these resources. The complexity of digital technology infrastructures, which consist of various interrelated elements, amplifies the difficulty of ensuring compliance. The complexity of the situation is further expanded to include cloud services and mobile devices, amplifying the formidable aspect of the undertaking.

The task of effectively overseeing compliance to regulations across a wide range of data formats and ensuring the accuracy and consistency of data poses a significant obstacle, particularly considering the immense quantities of data that are managed by businesses. Numerous organizations depend on external vendors and services to fulfil certain aspects of their operations, posing a challenge in ensuring the compliance of these third parties to cybersecurity standards and the prevention of vulnerability introduction. The dynamic nature of regulations in cybersecurity requires individuals and organizations to remain constantly attentive to keep up with legislative updates and adjust their compliance strategies accordingly. The issue of compliance extends beyond technical considerations and includes the role of human behaviour.

The presence of threats from insiders and human errors has the potential to undermine compliance efforts, introducing complexities in the initiative to uphold a secure environment. The requirement to demonstrate compliance with standards typically necessitates the implementation of thorough documentation and record-keeping practices, which might pose logistical difficulties in terms of managing and preserving these documents systematically. As organizations undergo expansion, the level of complexity associated with regulatory compliance also increases, necessitating the challenging responsibility of guaranteeing that compliance aligns proportionally with the organization's size and scope of operations.

The issue of compliance in cybersecurity presents a complex and long-lasting obstacle, requiring continuous dedication and the allocation of resources. Organizations are required to effectively traverse a dynamic setting characterized by regulatory changes, technological progress, and emerging risks. Simultaneously, they must manage resource constraints and solve the complexities inherent in their operational processes. In light of these significant obstacles, it is crucial to prioritize compliance to regulations to effectively address cybersecurity threats and protect confidential data.

13.1 Complexity of Regulatory Landscape

The complex structure of the regulatory framework in cybersecurity presents a perplexing challenge, characterized by a multitude of conflicting perspectives and a complicated network of established guidelines, requirements, and rules. From our perspective, this discordance has arisen as a substantial obstacle to the achievement of resilient and efficient cybersecurity measures. The undeniable presence of a geopolitical dimension adds to the complexity of this situation. In an age characterized by increasingly intense geopolitical competition, the lack of a cohesive international framework for regulating cybersecurity contributes to a state of disorder. Nations, motivated by their geopolitical interests and perceptions of threats, pursue diverse regulatory trajectories, leading to a fragmented collection of laws and regulations.

Without a universally acknowledged definition of cyberattack or cyberwarfare exacerbates the prevailing lack of consistency, rendering the attainment of regulatory harmonization nearly unfeasible. The absence of transparency in the regulatory framework exacerbates the issue. The complex and subtle characteristics of cyber-risks necessitate a precise and detailed approach in regulatory discourse. Nevertheless, the presence of ambiguity and imprecision commonly seen in regulation texts tends to complicate the efforts of companies aiming to achieve compliance. Consequently, the involvement of legal professionals becomes necessary to interpret the complex legislation, resulting in escalated costs associated with compliance. This outcome is deemed undesirable and should be approached with caution. The existence of numerous regulatory entities, both domestically and globally, gives rise to a complex bureaucratic structure. The presence of numerous entities, each asserting its authority, leads to a state of regulatory confusion.

The lack of a centralized authority to efficiently coordinate and integrate these efforts perpetuates fragmentation, hence posing a huge threat to global cybersecurity. The significance of technical subtleties, which represent an additional dimension of complexity, should not be ignored. The ever-evolving nature of cyber threats is characterized by a notable level of complexity. The challenges faced today are of an unprecedented scale due to the widespread occurrence of advanced persistent threats (APTs), state-sponsored attacks, and the emergence of dangers driven by quantum computing. Regulatory systems, despite their good intentions, frequently struggle to keep up with the swift rate of technological progress. One therefore finds oneself inclined to exhibit a certain level of scepticism towards regulatory initiatives aimed at addressing these difficulties. Numerous policies are formulated with a narrow perspective that emphasizes past risks rather than a proactive consideration of potential future threats.

The lack of foresight, along with the slow and cumbersome nature of the regulatory process, results in regulations that are essentially responsive rather than anticipatory. As a result, firms are compelled to allocate significant resources towards complying to regulations that may potentially lose their relevance in light of evolving dangers. The insufficiency of regulatory frameworks to effectively address the complexities of developing technology is evident. One example to be considered is the emergence of the Internet of Things (IoT), a technological framework that has introduced a period of interconnectedness.

The effects for cybersecurity are significant. However, current regulations are not well prepared to tackle the distinctive issues presented by this paradigm shift. Likewise, the migration of essential components of key infrastructure to cloud-based platforms presents an additional level of complexity. Cloud computing provides enterprises with exceptional scalability and flexibility; nevertheless, it also exposes them to emerging security vulnerabilities. Conventional regulatory frameworks, which are primarily designed for physical environments, encounter challenges in effectively accommodating the intangible characteristics of cloud computing. The magnitude of data generated and processed in the current society is astonishing.

Regulatory frameworks, although admitting the need for data privacy, frequently lack thorough guidance on the management and security of this data. The lack of attention to this matter exposes enterprises to the risk of data breaches, which can have severe implications for both privacy and security. The inherent global characteristics of cyberspace introduce additional complexities. With the digital threats being borderless, regulations that are confined to specific geographical regions are inherently flawed. Cybercriminals leverage this vulnerability to their benefit without being limited by geographical boundaries. The initiative of achieving regulatory harmonization across international borders is a formidable undertaking; however, it must be pursued with resolute commitment. The complexity of the regulatory framework within cybersecurity is a subject of significant concern.

The geopolitical aspect, which is marked by a shortage of worldwide agreement and the existence of conflicting national interests, poses obstacles to international collaboration and coordination. The problematic nature of technical complexities, combined with the failure of rules to keep up with the swift progress of technology, creates a situation in which compliance to regulatory measures is frequently more onerous than advantageous. The insufficiency of legislative frameworks in adapting to evolving technologies, along with the difficulties presented by cloud computing, the volume of data, and worldwide connectivity, intensifies the issue. There is a need to intensify efforts to enhance the efficiency and coherence of regulatory systems, both domestically and globally. It is crucial to adopt a proactive approach that involves anticipating forthcoming dangers and technological advancements. The effective resolution of the complex regulatory challenges in cybersecurity and the enhancement of global security in the face of escalating digital threats on the geopolitical stage necessitates a collaborative and synchronized ambition on a global scale.

13.2 Cross-Border Compliance Issues

The issue of ensuring compliance with cybersecurity and data privacy standards, rules, and regulations across different countries offers a complicated challenge. This multidimensional dilemma provides a significant barrier to the progress of the digital age. This section presents a compelling argument that highlights the burdensome nature of the current regulatory system in a globalized digital society, defined by inconsistencies and shortcomings. The main inadequacy is in the erroneous presumption that the Internet adheres to geopolitical boundaries, but in reality, it transcends them, exerting its impact not only on nation states but also on the global population as a collective entity.

It is essential to highlight the idea that the Internet adheres to the ideals of national sovereignty. The Internet poses a threat to the territorial boundaries upon which traditional governance systems are founded, owing to its inherent qualities. Due to its transcendent nature, the digital world operates outside of national boundaries and laws, showing a disregard for or indifference towards them. Therefore, any attempt to implement localized regulatory measures on a network that exceeds jurisdictional boundaries is inherently incorrect. It is crucial to recognize that the lack of boundaries on the Internet is not a defect but rather a deliberate attribute. The core of its transformative capacity is in its facilitation of worldwide information exchange, the diffusion of ideas, and commercial transactions. Regardless, it is precisely this particular attribute that renders national regulations concerning cybersecurity and data privacy inadequate and ineffective.

The GDPR implemented by the European Union can be considered a pertinent illustration. The GDPR is an admirable initiative to protect the privacy and personal data of individuals who are located in Europe. Still, this inadvertently imposes a substantial and onerous obligation on businesses worldwide. Organizations that operate outside the EU face a complex system of laws, often marked by differing provisions, to avoid the risk of significant monetary sanctions.

The expansion of the jurisdiction of the GDPR beyond national boundaries serves as a notable illustration of the extensive influence of localized legislation on a global level. The disparate and discordant assortment of laws arising from many nation states poses a hindrance to international cooperation in addressing cyber threats. The interconnectivity of the Internet necessitates a global strategy for effectively mitigating cyberattacks. Nevertheless, the presence of a fragmented regulatory framework presents significant challenges when it comes to facilitating the exchange of threat intelligence and impeding international initiatives aimed at apprehending cybercriminals. As a result, the cyber-defence system demonstrates a lack of organization and inefficiency.

Legislation with restricted jurisdictional applicability has global consequences, not only impacting cybersecurity but also exerting a direct influence on economic competitiveness and creative initiatives. The growing frequency of divergent data privacy legislation necessitates that companies invest significant efforts to ensure compliance, often at the expense of their research and development initiatives. Smaller enterprises, in particular, face an unfair distribution of regulatory responsibilities, impeding their growth and their ability to compete in the global market. More importantly, the existence of a fragmented regulatory framework promotes an environment that is favourable for engaging in jurisdictional arbitrage. Corporations, cognizant of the disparities in regulatory frameworks across nations, may strategically position themselves in jurisdictions with more permissive legislation, evading the intended consequences of more strict regulations.

Circumvention affects not just the effectiveness of regulatory measures but also the faith put in the principles that govern the legal system. The absence of geographical constraints in the Internet world multiplies the potential implications of a single security error in data breaches and cyberattacks. A breach that begins in a specific geographic location has the potential to spread its implications beyond national boundaries, influencing individuals and organizations situated at great distances from the original point of origin. The cross-border character of cyber threats emphasizes the need for a coordinated, global strategy for cybersecurity and data protection. As a result, the current regulatory framework is not only insufficient but also harmful to the global community's well-being.

The Internet once hailed as a representation of progress and connection now serves as a testament to the urgent need for international consensus and cooperation in cybersecurity and data safeguarding. The solution to this enigma lies in recognizing the inherently global character of the Internet. The significance of regulatory harmonization and the establishment of a framework that surpasses national boundaries cannot be overemphasized. Efforts concerning cybersecurity and data privacy ought to be driven by an internationally acknowledged standard that is developed through collaborative efforts among nations and draws upon the collective expertise and insights of participating countries.

International organizations, such as the United Nations, should devote an active part in assisting with the standardization process. To successfully address the borderless aspect of the digital world while preserving national sovereignty, a framework must be established via the formation of international accords and conventions. The complications of international compliance barriers in cybersecurity and data privacy is a significant hurdle to the advancement of society's digital capacities. The mistaken belief that the Internet is limited to national borders has resulted in a fragmented regulatory framework that impedes global collaboration, economic competitiveness, and cybersecurity activities. Recognizing the lack of boundaries in the Internet sector and implementing similar regulatory measures on a worldwide scale is critical for the international community. Realizing the full potential of digital advances in terms of security, privacy, and creativity requires collaborative and coordinated efforts.

13.3 Rapidly Changing Regulations

So let's ask, *What is the purpose of cybersecurity?* The question may seem fundamental, but it reflects a major concern confronting corporations on a global scale. This subject holds significant importance due to the persistent prevalence of cybersecurity concerns, despite numerous efforts to strengthen digital systems throughout the past few decades. The complex nature of this phenomenon has necessitated the implementation of strict and rigid regulations that mandate enhanced protective measures and secure infrastructures to ensure the security of data across various geographical areas. Accordingly, the significance of data integrity is growing in importance as firms engage in the exchange of data across diverse IT environments and cloud service providers. Organizations that demand heightened data portability must establish systems that offer increased adaptability, enable higher efficiency, and comply to security and regulatory mandates.

Cybercriminals are always innovating novel methodologies to exploit weaknesses within networks and systems. The topic of rapidly evolving legislation in response to social demands is a highly debated issue, characterized by those who support and oppose this practice. In the subsequent discussion, the contentious character of this matter will be clarified, providing an analysis of the reasons presented by proponents and opponents alike.

We argue that society's requirements exhibit a dynamic nature, undergoing continuous changes in response to technological breakthroughs, economic fluctuations, and cultural revolutions. Hence, it is essential for regulatory frameworks to swiftly adjust to maintain their relevance and efficacy. This particular viewpoint argues that regulations that are rigid and unchanging are fundamentally insufficient to effectively address the complex concerns of the modern age. Similarly, advocates argue that the swift implementation of regulatory modifications empowers governments and institutions to effectively anticipate and tackle emerging concerns.

Through swift implementation of fresh regulations, they can control prospective harm and ensure the protection of the well-being of their populace. The assertion is substantiated by empirical data that presents examples of agile regulatory interventions that have successfully prevented crises and enhanced the welfare of the general public. Nevertheless, critics of this methodology argue that quick alterations to regulations frequently lack thorough consideration and are hastily implemented.

We contend that these modifications have the potential to give rise to confusion, introduce legal uncertainties, and result in unforeseen outcomes. The sudden change in regulatory frameworks has the potential to impose excessive obligations on both firms and citizens, hindering economic progress and curtailing individual liberties. Critics argue that the regularity of regulation modifications might develop an atmosphere of unpredictability, which in turn deters long-term investment and hampers the progress of innovation. Investors and entrepreneurs may exhibit caution when considering involvement in industries that are susceptible to unpredictable regulatory environments, leading to potential missed prospects for economic advancement.

This topic of constant regulatory changes in response to social demands is a subject of debate, characterized by persuasive arguments from many perspectives. Advocates argue that the possession of adaptability and responsiveness is crucial to effectively tackling present-day difficulties. Conversely, critics highlight potential disadvantages, such as legal uncertainties and economic stagnation. The task of achieving an equilibrium between flexibility and reliability in regulatory structures poses a significant obstacle that necessitates ongoing academic study and wise policy formulation.

Data privacy laws refer to legal regulations and statutes that govern the collection, use, storage, and disclosure of personal information by organizations. These laws aim to protect individuals' privacy. The regulations define the requisite measures for safeguarding the privacy of personal, confidential, or classified data. Data breaches refer to incidents where unauthorized individuals get access to sensitive or confidential information stored within a computer system, network, or database. Cybercriminals possess the capability to selectively exploit data sources, systems, or networks, thereby engaging in activities such as data theft, damage, or manipulation.

The concept of data compliance refers to following the legal and regulatory requirements regarding the collection, storage, processing, and sharing of data. The procedure of guaranteeing an organization's compliance to legal statutes, regulatory frameworks, and established benchmarks about its data The complexity of cybersecurity risks Organizations may find themselves overwhelmed by the escalating complexity of cybersecurity threats. Organizations remain vulnerable to data breaches due to inadequate security practices.

Within the dynamics of AI, the advent of expansive language models has incited a flurry of legislative changes. This section explores the interconnected fluctuations in rules concerning AI, massive language models, data privacy, homomorphic encryption, data outsourcing, and zero-knowledge proofs (ZKP). It discusses the complex implications of these emerging paradigms. The emergence of expansive language models, like OpenAI's GPT-3, Google's Bard, and Anthropic's Claude 2, has generated a mixture of enthusiasm and concern. These models, which possess the ability to produce text that resembles human language, have given rise to a wide range of applications, spanning from the comprehension of natural language to the development of imaginative content.

The widespread presence of these entities has resulted in a mass of ethical predicaments, particularly concerning the protection of data privacy. Data privacy, firmly rooted in the fundamental principles of individual freedom, serves as a guardian against the intrusion of personal information. The increasing utilization of expansive language models, which require substantial amounts of data for their training, has heightened this concern. Current regulatory initiatives, exemplified by the GDPR implemented in Europe, establish and safeguard the entitlement to data protection. The extraterritorial scope of the GDPR, which includes companies outside the geographical boundaries of the European Union, exemplifies the global significance of this matter. The practice of data outsourcing exacerbates the complexity of the interaction between extensive language models and the protection of data privacy. Corporations frequently aim to leverage the capabilities of AI while avoiding the responsibility of overseeing extensive data stores.

The practice of data outsourcing, despite its potential for improving operational efficiencies, gives rise to a multitude of vulnerabilities that can be likened to Pandora's box. The concept of data custodianship becomes ambiguous, resulting in a reduction in both accountability and control. As a result, regulatory agencies are faced with the challenge of rethinking concepts related to data ownership and management. To address the risks associated with data exposure, homomorphic encryption has emerged as a prominent means of protection. This cryptographic method facilitates computational operations on encrypted data without the need for decryption. The ability to utilize sensitive material without exposing its raw form is a significant advancement for AI and massive language models. Nevertheless, the current legislations about homomorphic encryption are still in its early stages, requiring the development of frameworks to guarantee its ethical and secure use. The essential requirement for ensuring data security within AI is the utilization of zero-knowledge proofs (ZKP).

These cryptographic protocols enable the verification of a statement without disclosing any underlying data. Within the domains of data privacy and AI, zero-knowledge proofs (ZKP) are considered a significant advancement that brings about a revival in the area of assurance. However, the complex mathematical algorithms and computing requirements of this system require clear regulatory guidelines to ensure that its effectiveness is balanced with its practicality. The confluence of various regulatory streams arises from the transformative capacity of AI. The effectiveness of extensive language models, supported by extensive datasets, enhances their usefulness in several industries, including healthcare, finance, and education. The widespread presence of these technologies necessitates a careful evaluation of their effects on employment. Regulatory bodies are faced with the challenge of addressing the issue of labour market displacement in light of the emergence of AI-driven automation.

The complexities surrounding the governance of AI extend beyond data privacy, including issues of responsibility and accountability as well. The increasing autonomy and decision-making capabilities of AI systems present a complex challenge in determining culpability. The concept of algorithmic transparency resonates with the demand for accountability. The examination of algorithmic decision-making, particularly in fields such as criminal justice, necessitates careful analysis to prevent the continuation of biases and inequalities. The lack of a defined structure in AI requires an adaptable system of regulations. The continuous development of AI algorithms renders static restrictions ineffective. The necessity for regulatory mechanisms to possess flexibility and adaptability is important to align with the continuously increasing frontiers of AI.

The consideration of international harmonization of AI rules is essential for an analysis of the section. The inherent global characteristics of AI require international collaboration to prevent regulatory fragmentation. The necessity of interdisciplinary interactions that incorporate technologists, legislators, ethicists, and legal scholars cannot be overstated in the development of a cohesive regulatory framework.

The current situation AI regulation is characterized by inherent uncertainty and urgency. The pace at which technological advancements are occurring is beyond the development of effective rules, leading to a risky environment. However, the essential requirements of safeguarding data privacy, implementing homomorphic encryption, outsourcing data, utilizing zero-knowledge proofs, and other crucial aspects of governing AI call for prompt and thorough regulatory measures.

The section offers an objective examination of the regulatory modifications prompted by the emergence of AI and large-scale language models. The convergence of several elements such as data privacy, homomorphic encryption, data outsourcing, zero-knowledge proofs, and related aspects highlights the complexities that current regulatory frameworks must address. The necessity to achieve an balance between innovation and ethics highlights the pressing need to establish regulations that include the diverse domains of AI.

13.4 Compliance Costs

The integration of cybersecurity measures and compliance to regulatory frameworks, policies, and legislation are pervasive and essential components of organizational activities. The dynamic nature of cyber threats underlines the need for a systematic strategy to protect sensitive information and maintain the security of information technology systems. Nevertheless, the initiative to achieve compliance frequently presents noteworthy obstacles and involves considerable expenses, giving rise to discussions and disputes within corporations and regulations.

Organizations typically encounter challenges in comprehending the precise needs of cybersecurity rules and regulations due to the use of complex legal terminology. The complex nature of this phenomenon has the potential to result in misunderstandings and mistakes throughout the execution phase, hindering the progress of compliance initiatives. The ever-evolving nature of cyber-risks poses a substantial obstacle to achieving compliance. Regulations are generally characterized by their static nature, in contrast to the dynamic nature of cyber-risks. The misalignment between enterprises and the most recent safety regulations can lead to difficulties keeping pace as compliance attempts frequently fall behind growing risks. Smaller businesses, especially, encounter limitations in resources that impede their capacity to attain compliance.

Achieving compliance may necessitate substantial financial allocations towards technological advancements, human resources, and educational initiatives. The possible consequences of this situation can have a disproportionate impact on small enterprises, thereby impeding both innovation and economic expansion. Numerous firms depend on a diverse range of systems and apps to carry out their operational activities. The task of ensuring compliance with cybersecurity standards across these diverse systems can present significant challenges. The absence of interoperability among diverse technologies may lead to compliance deficiencies. The attainment of compliance frequently requires significant initial capital outlays. Organizations are required to procure and deploy cybersecurity technologies, recruit specialized staff, and organize training programmes. The initial expenses can impose financial constraints and impede the execution of other critical operational initiatives.

Ensuring compliance is a continuous initiative rather than a singular undertaking. To ensure compliance with evolving legislation, organizations must consistently update and modify their cybersecurity measures. This includes continuous expenditures related to software updates, personnel training, and audits. The expenditure associated with securing legal representation to interpret and navigate the many aspects of cybersecurity rules is substantial. Besides, organizations may incur fines and penalties as a consequence of failing to comply with regulations, thereby increasing their legal expenditures. The allocation of resources towards obtaining compliance may have been alternatively directed into initiatives such as research and development or developing market presence. The presence of compliance expenses has the potential to impede growth and innovation.

The correlation between compliance issues and expenses within cybersecurity legislation is a subject of intense debate. There exists a contention that the expenses linked to compliance to regulatory requirements are warranted, given their indispensable role in safeguarding sensitive information and mitigating the risk of cyber-intrusions. From this standpoint, rules function as a protective measure against potentially disastrous violations, which can have extensive ramifications for both entities and individuals. On the other hand, others argue that the costs associated with adhering to regulations frequently surpass the advantages. We contend that the complex nature of restrictions, combined with the swift rate of technological advancements, imposes a burdensome responsibility on businesses, particularly those on a smaller scale. According to some sources, it has been argued that compliance costs have the potential to hinder economic growth and suppress innovation. It has been argued by opponents that complying with regulations does not necessarily ensure the attainment of a secure environment. Even firms that comply completely to regulations can nevertheless become targets of cyberattacks, indicating that meeting regulatory standards does not always guarantee adequate cybersecurity measures.

This prompts inquiries regarding whether the prioritization of compliance hinders the adoption of a more all-including and proactive security strategy. The correlation between issues about compliance and associated expenses within cybersecurity legislation is a subject of debate and complexity. The complexities of the rules, the dynamic nature of the threats, limitations in resources, and obstacles related to interoperability are substantial and tangible. Concurrently, it is critical to recognize the significant impact of compliance expenses, which include early capital outlays, continuous upkeep, legal expenditures, and foregone opportunities.

The controversy stems from the ongoing discourse surrounding the question of whether the advantages associated with compliance to regulations, including the improvement of security measures and safeguarding of data, are more significant than the drawbacks in terms of economic consequences and the ability of organizations to adapt quickly. Achieving an optimal equilibrium between following regulations and ensuring robust security measures continues to be a significant obstacle for politicians, regulatory entities, and other organizations. In light of the dynamic nature of the cybersecurity domain, it is crucial to prioritize the development of inventive and effective strategies for tackling compliance obstacles and financial burdens. This is essential for safeguarding digital resources and maintaining economic prosperity.

14 Regulatory Landscape

The legislative framework that includes cybersecurity and data privacy comprises a complex framework of laws and regulations designed to safeguard sensitive data and uphold the integrity of digital systems. The complex topography of the area requires the implementation of a carefully devised strategy that integrates numerous essential elements. The legislations about cybersecurity and data privacy are undergoing swift and continuous transformations. In a globalized society characterized by heightened interconnectivity, the significance of digital information in various spheres, such as personal and commercial initiatives, has prompted governments and regulatory entities to acknowledge the need to implement frameworks aimed at safeguarding data and mitigating cyber threats. This section centres on the implementation of a regulatory compliance team, regulatory risk evaluation, compliance audits and assessments, and compliance monitoring and reporting as the key components of the regulatory framework concerning cybersecurity and data privacy.

14.1 Building a Regulatory Compliance Team

To adequately mitigate cybersecurity and data privacy concerns, organizations must evaluate regulatory risks. The previously mentioned process involves the identification of potential vulnerabilities and risks that may render sensitive data susceptible to security breaches. Through the evaluation of these risks, organizations can determine the order of importance of their compliance efforts with rules and allocate resources to areas that require the greatest attention. Risk assessment plays a pivotal role in the development of security protocols and tactics for safeguarding data.

To effectively navigate the complex terrain of cybersecurity and data privacy legislation, it is essential for organizations to initially develop a dedicated staff focused on regulatory compliance. The composition of this team should include individuals who have expertise in both the legal and technological dimensions of cybersecurity and data protection. The main duties of this team involve keeping up to date dynamic legislation, evaluating its consequences for the organization, and guaranteeing compliance to all pertinent rules and regulations. Further, the regulatory compliance team must establish a strong collaborative relationship with the IT department, legal counsel, and senior management to devise and execute efficient compliance plans.

14.2 Regulatory Risk Assessment

To effectively address cybersecurity and data privacy issues, businesses must conduct a regulatory risk assessment. The process involves the identification of vulnerabilities and potential threats that can expose sensitive data to security breaches. Through the process of evaluating these risks, organizations can ascertain their relative significance, enabling them to prioritize their compliance initiatives and allocate resources accordingly. It is essential to acknowledge that the conduct of a risk assessment is pivotal in establishing the fundamental basis for the formulation and execution of security protocols and policies to protect data.

The establishment of a regulatory compliance team is a prerequisite for enterprises to effectively handle the complexities of cybersecurity and data privacy requirements. The composition of this team should comprise professionals possessing specialized knowledge in both the legal and technological domains of cybersecurity and data protection. The main duties of this team include the tasks of remaining informed about newly emerging rules, conducting an analysis of the potential consequences these regulations may have on the organization, and guaranteeing compliance to all pertinent legislation. Also, the regulatory compliance team must establish a strong partnership with the IT department, legal counsel, and senior management to develop and implement effective compliance strategies. The adoption of this collaborative strategy guarantees that the organization's cybersecurity and data privacy practices are by the continuously changing regulations.

14.3 Compliance Audits and Assessments

The performance of compliance audits and assessments is of importance in ensuring compliance to regulatory standards. These evaluations involve a systematic analysis of an organization's cybersecurity and data privacy practices to ascertain their compliance to applicable rules. Audits can also assess the efficacy of security measures in safeguarding confidential information. Regularly scheduled assessments are utilized by organizations as a mechanism to identify areas of non-compliance, address flaws, and adjust to evolving regulatory demands. Compliance audits and assessments are important in ensuring organizations following cybersecurity and data protection standards. These processes involve conducting a systematic evaluation of the organization's policies, procedures, and practices to ensure that they are following regulatory requirements.

Regular compliance audits are integral in the identification of potential gaps or shortcomings within an organization's cybersecurity and data protection protocols.

These gaps may occur as a result of alterations in regulations, developments in technology, or changes in the operational activities of the organization. Through the implementation of audits, organizations can take proactive measures to address these issues before they escalate into non-compliance. Engaging auditors who possess the necessary competence in cybersecurity and data privacy standards is essential. These individuals possess the ability to conduct an unbiased evaluation of the organization's efforts to follow regulations and can provide significant insights and recommendations for improvement.

14.4 Compliance Monitoring and Reporting

Consistent monitoring of compliance is important to stay ahead of the ever-increasing legislation about cybersecurity and data protection. Organizations must develop robust processes for monitoring and documenting their compliance with these legal requirements. This involves the consistent monitoring of their internal practices as well as remaining knowledgeable about alterations in regulatory frameworks. The act of reporting on compliance status is important as it demonstrates a commitment to regulatory bodies and stakeholders, building up confidence and developing a sense of accountability. Compliance monitoring and reporting are continuous procedures that guarantee an organization's compliance to cybersecurity and data protection standards. Monitoring involves the ongoing oversight of data management practices, security protocols, and following specified rules and procedures. To ensure efficient oversight of compliance, organizations should use technological solutions such as intrusion detection systems and security information and event management (SIEM) technologies. These instruments have the potential to facilitate the timely detection and response to suspected security breaches.

Regular evaluations and analyses of security logs and incidents are integral components of the monitoring procedure. Meanwhile, it is necessary that organizations promptly notify regulatory bodies, stakeholders, and occasionally the general public of their compliance to regulatory requirements. The notion of transparency is essential to upholding trust and credibility. Compliance reports should provide a precise and all-including depiction of an organization's conformity to regulatory requirements, including any occurrences or violations, as well as the measures implemented to rectify non-compliance. In brief, the regulations of cybersecurity and data protection are complex. To effectively handle the complexities of regulatory compliance, organizations must prioritize the establishment of a carefully structured regulatory compliance team, engage in risk assessments, carry out diligent compliance audits, and uphold attentive monitoring and reporting mechanisms. By adhering to these principles, organizations have the potential to mitigate regulatory risks and protect the confidentiality and integrity of sensitive data in a progressively digitized global environment.

15 Conforming AI to Regulatory Frameworks

One of the primary challenges confronting modern democracy relates to the utilization of information, technology, and automated systems in manners that have the potential to infringe upon the rights of the populace. These methods are often used to restrict opportunities and impede access to crucial resources and services. These concerns have been well reported. Patient care systems used globally have shown deficiencies in terms of safety, efficacy, and impartiality. The use of algorithms in the processes of employment and credit choices has been seen to perpetuate pre-existing inequities or create new prejudices and discriminatory practices. The unregulated gathering of social media data is regularly utilized to jeopardize individuals' prospects, infringe upon their privacy, and extensively investigate their actions, sometimes without their knowledge or explicit agreement. Although these repercussions may be detrimental, they are not always unavoidable.

Automated systems have also yielded substantial advantages, including technological advancements that enhance agricultural productivity, computational models that forecast storm trajectories, and algorithms with the capacity to detect illnesses in individuals. These technologies now occupy a crucial position in significant decision-making processes across diverse sectors as data continues to revolutionize companies on a worldwide scale.

The progression of computational intelligence can induce significant transformations within digital technology. The umbrella term AI includes a diverse set of technologies, ranging from algorithms specifically developed for ML to complex robotic systems. AI has the potential to significantly revolutionize several industries, optimize countless operational procedures, and augment decision-making capacities across a wide range of disciplines. The integration of AI into the fundamental structure of our digital society poses a range of difficulties and possibilities that must be addressed within legal frameworks to achieve success.

In the portion that comes next, an in-depth investigation will be conducted to assess the significance of integrating AI into legal frameworks within the wider context of digital resilience. The need to define a definition for AI emerges due to the wide range of applications and manifestations that are linked with this discipline. Technology has the potential to profoundly transform organizations via the automation of processes, the extraction of valuable insights from extensive datasets, and the facilitation of autonomous ML, all without the need for human participation. The influence of AI is significant and wide-ranging, including several sectors such as healthcare, finance, education, and transportation.

Fortunately, the increasing prevalence of AI technology in our everyday lives presents an array of complex concerns. These include the presence of biases in algorithmic decision-making, concerns about the safeguarding of data privacy and security, as well as ethical quandaries arising from the activities performed by autonomous systems. The development of legal frameworks that can adequately embrace the different qualities of AI is crucial to properly harness its promise while also reducing the threats it presents.

AI systems exhibit a considerable need for data, which has led to concerns over the protection of individual rights and the privacy of data. It is crucial to implement amendments to existing legislative frameworks about data protection and privacy to effectively tackle the distinct difficulties presented by AI. The issues are about the need for transparent procedures in the gathering and utilization of data, the significance of empowering humans to evaluate and challenge automated judgements, and the need for data collection and use to align with principles of justice. The promotion of ethical AI use necessitates the establishment of a harmonic equilibrium between technical advancement and safeguarding individual privacy.

The complexity surrounding the allocation of accountability for mistakes or damages arising from AI judgements grows more complex as these systems acquire greater autonomy. The establishment of unambiguous and definitive regulations inside legal structures is crucial to assist in the evaluation of culpability and responsibility. The objective of this section is to conduct a thorough analysis of the many variables that need to be taken into account when establishing the legal liability of individuals engaged in the creation, implementation, or use of AI systems. The attainment of equilibrium necessitates a nuanced balance between the promotion of innovation in AI and the establishment of accountability for the consequences of AI system deployment.

The correlation between advancements in technology and the legal obligations of individuals is a complex matter. The integration of AI into many sectors gives rise to noteworthy ethical concerns, mostly centred on the possible consequences of biased algorithms perpetuating prejudice and socio-economic injustice. Legal frameworks play a pivotal role in addressing these concerns by imposing requirements for openness and equity in the creation and implementation of AI systems. They provide procedures for continuous monitoring and examination to assure compliance to society's values and standards. Along with that, the incorporation of ethical issues into legal frameworks assumes a necessary aspect in providing guidance for the responsible deployment of AI and mitigating the potential for its inappropriate use.

The fundamental inability of AI to grasp geographical boundaries accentuates the need for global cooperation and the adoption of a standardized governance framework. The alignment of legal frameworks with international standards and agreements is of utmost importance to attain interoperability and consistency in the administration of AI, particularly concerning cybersecurity considerations. It is essential to consider the consequences of AI on worldwide trade, employment, and diplomatic interactions. The establishment of robust global governance frameworks for AI necessitates the urgency of international collaboration.

The challenging task of reconciling the development of AI innovation with the implementation of regulatory measures is a complex initiative that needs carefull attention and resolution. To effectively navigate the rapid advancement of AI capabilities and applications, it is crucial to adopt a versatile approach to implementation. The prioritization of establishing a regulatory framework that encourages innovation while concurrently safeguarding the integrity and security of information systems is of utmost importance. The integration of AI into legal frameworks is crucial to attain digital resilience against the dynamic global regulatory environment.

The increasing prevalence of AI gives rise to both significant advancements and concurrent obstacles. The realization of AI's full potential, the safeguarding of individual rights, the promotion of ethical conduct, and the establishment of accountability all hinge upon the existence of legal frameworks. This part presents an analysis of the legislations related to AI. It offers valuable perspectives on the approaches, difficulties, and international initiatives that seek to effectively incorporate AI into legal frameworks. The above-mentioned acts demonstrate an essential component in the establishment of a resilient, ethically sound, and equitable digital setting for both governments and individuals.

15.1 Safe and Effective Systems

It is essential to ensure the protection of individuals against automated systems that are deemed dangerous or ineffective. Such systems must undergo thorough development procedures that include the participation of many groups, stakeholders, and domain experts. This is necessary to identify and effectively address problems, risks, and possible repercussions. To establish the safety and efficacy of these systems for their intended purpose and beyond, it is essential to conduct pre-deployment testing, perform risk assessments, and maintain ongoing monitoring. If a system is determined to possess inherent risks or lack efficacy, it is essential that its deployment be avoided or its use be discontinued.

Automated systems mustn't be intentionally created or anticipated to pose any risks to the safety of individuals or their communities. Proactive measures must be taken in the design of systems to safeguard users from potential damage arising from unexpected but anticipated uses or effects. It is important to ensure that users are protected from exposure to unsuitable or unrelated data during the whole process of designing, developing, and implementing these systems. Independent assessments and clear documentation must be conducted to validate the safety and efficacy of automated systems. These evaluations should include the measures used to minimize any adverse consequences. It is important to ensure that these findings are disseminated to the public whenever feasible.

15.2 Algorithmic Discrimination and Bias

It must be considered to ensure that algorithms do not engage in discriminatory practices against users and that the design and use of automated systems comply to principles of fairness and equity. Algorithmic discrimination refers to the phenomenon whereby these systems engage in unfair treatment or impose disadvantages on individuals due to their race, colour, ethnicity, gender, religion, age, handicap, or other legally protected categories. Those involved in the creation, development, and operation of automated systems need to adopt a proactive and ongoing approach to mitigate algorithmic prejudice and promote equitable system design.

This includes the implementation of equity evaluations throughout the design phase of the system, the utilization of representative data, the avoidance of proxies for demographic characteristics, guaranteeing accessibility for individuals with disabilities, the execution of pre-deployment and continuous disparity testing, and upholding transparent organizational control. It is advisable to carry out and disclose independent assessments, along with clear and accessible reports that include evaluations of algorithms' influence as well as testing outcomes related to disparities. This practice serves to validate the effectiveness of these safeguards in preventing discriminatory practices.

15.3 Data Privacy

Individuals must be safeguarded against abusive data practices and have authority over how their personal information is used. It is important to include privacy safeguards as a fundamental aspect of the design of automated systems, therefore guaranteeing that data gathering complies to reasonable expectations and alone acquires data essential for specified circumstances. It is essential to get users' consent and uphold their choices about the collection, utilization, accessibility, transfer, and removal of data to the greatest extent feasible. If the existing solution is not viable, it is recommended to develop other measures to protect privacy.

Systems should refrain from including user experience or design elements that hinder user choices or enforce intrusive privacy defaults. The act of seeking consent should be limited to situations where it may be really and comprehensibly provided via concise and easily comprehensible queries. Strict safeguards and regulations must be implemented for data about sensitive fields, while robust supervision mechanisms should be imposed on surveillance technology. Individuals must be provided with access to reports that validate their compliance to their data choices and evaluate the possible effects of monitoring technology on their fundamental rights.

15.4 Notice and Explainability

When automated systems are in use, users must be informed and comprehend the reasons for and how these systems influence their consequences. The individuals involved in the creation, programming, and management of automated systems must provide easily understandable documentation that ascribes the system's operations, defines the parties accountable for its functioning, and offers accessible and timely explanations of the resulting consequences. It is important to inform users of substantial changes in system utilization or functioning. The rationales behind the results should comply to technical validity, possess significance, and be appropriately adjusted to reflect the amount of risk within the given environment. The inclusion of concise summaries of automated systems, written in simple language, together with evaluations of the clarity and quality of the accompanying notifications and explanations, should be included in public reporting.

15.5 Human Alternatives, Consideration, and Fallback

Individuals ought to be provided with the opportunity to choose not to participate in automated systems when it is deemed suitable, and they should also be allowed to explore human alternatives. The decision to opt out should be made by taking into account reasonable expectations and considering accessibility factors. It is essential that users be provided with the opportunity to get prompt human attention and seek resolution via a fallback and escalation mechanism in the event of failures or faults in an automated system or if they choose to challenge or dispute its consequences. Accessibility, equity, effectiveness, and the absence of unnecessary stress for users should be ensured while providing human alternatives and backup mechanisms.

In fields characterized by sensitivity, such as criminal justice, employment, education, and health, it is essential to provide customized procedures that include meaningful monitoring, training, and due regard for human factors when making high-risk judgements. The inclusion of depictions of human governance procedures and evaluations of their promptness, accessibility, results, and efficacy should be included in public reporting.

16 Concluding Remarks

It is important to achieve an equilibrium between the advantageous and detrimental aspects of the many worldwide rules and regulations about security and privacy. Global initiatives are required to guarantee the preservation of security in our digital existence and the protection of our personal information. The enactment of legislation such as the GDPR in the European Union and the Digital Personal Data Protection Act of 2023 in India demonstrates a firm dedication to safeguarding the digital rights of individuals. These rules guide both organizations and governments about the responsible management of data, with a particular emphasis on the principles of openness and accountability.

The primary objective of these regulations is to ensure our safety and security, and they may impose significant regulatory obligations on companies, particularly those operating on a smaller scale. The task of understanding and adhering to data privacy standards may present difficulties due to its complex nature, and the consequences for failing to comply can be severe. The fast rate of technological progress necessitates a constant adaptation of rules and regulations. The ever-changing nature of the digital world might sometimes result in legal loopholes, therefore exposing individuals and their data to potential risks.

Hence, it becomes evident that despite notable advancements in the protection of digital security and privacy, there are challenges that must be surmounted. The process of developing digital resilience is a continuous initiative, requiring a careful equilibrium between safeguarding measures and practical considerations.

Acknowledgments We wish to express our sincere appreciation to Yuvraj Singh, whose invaluable contributions included sectional notes and assistance in the conceptualization of this chapter. His remarks regarding the present state of cybersecurity and privacy legislation in the United States and the legal framework on the Internet in India significantly enlightened us regarding the subject matter that we intended to examine and study. His input has been of immense help and use.

Chapter 4
Nurturing Resilience in Minors

1 Introduction

In the current societal context, handheld devices and edge computing have become a growing presence. Hence, we need to be well aware that there are consequences of technology permeating our lives; while one might be aware of both the advantages and disadvantages, minors are mostly unaware, and the significant time that is usually allotted to them might be a cause of great harm. The constant presence of technical devices in their daily lives has resulted in a growing prevalence of minors utilizing cell phones, tablets, and laptops, leading to a significant allocation of time spent engaging in online activities. Although technology has provided numerous advantages to individuals, it has also presented a range of potential threats, such as cyberbullying, online harassment, exposure to improper content, and cybercrime. The development of digital resilience is necessary to provide minors with the requisite abilities and understanding to effectively defend against the threats present online, thereby mitigating the possible negative consequences that may arise.

Definition 6 (Minor) A minor refers to an individual who has not yet attained the age at which they are recognized as a legal adult, possessing the capacity to autonomously make decisions and engage in activities without the need for parental or legal guardian oversight. The specific age at which this transition to adulthood takes place may differ depending on the applicable legal system, but it typically includes individuals under the age of 18.

S. K. Shandilya et al., *Digital Resilience: Navigating Disruption and Safeguarding Data Privacy*, EAI/Springer Innovations in Communication and Computing,
https://doi.org/10.1007/978-3-031-53290-0_4

The primary objective of this chapter is to analyse the vulnerabilities that are experienced by young individuals during their online interactions. This chapter presents various approaches and strategies that can be implemented to create an atmosphere that promotes the well-being and online resilience of young individuals. These include the cognitive ability to analyse information in a discerning manner, the aptitude to understand and evaluate digital content, the skill to regulate one's emotions, and the proficiency to convey thoughts and ideas with clarity and effectiveness.

The development of critical thinking skills is an essential component of developing digital proficiency. Adolescents bear the significant responsibility of critically evaluating the diverse set of materials they encounter on the Internet. With abundant Internet information, the task of differentiating between credible and dubious sources becomes a substantial obstacle. The minors must determine instances of misinformation or deceit, particularly within environments characterized by an increase of fake news, uncorroborated opinions, and deceptive schemes. The development of critical thinking abilities in young individuals facilitates the development of their capacity to evaluate digital content, hence diminishing their vulnerability to potential threats on the Internet.

Digital literacy is an attribute that is necessary and critical for the development of digital resilience. Minors must acquire awareness of the operational mechanisms of digital technologies, as well as understand the potential threats inherent in the online platforms they engage with. A complete understanding of privacy settings, the possible threats associated with excessive disclosure of personal information, and the innate dangers of engaging with new online entities all contribute towards digital literacy. The addition of these abilities enables young individuals to make well-informed decisions regarding their online conduct, thereby diminishing their susceptibility to potential threats on the Internet.

The capacity to effectively manage and control emotions is a fundamental aspect of developing digital resilience. The digital world offers a multitude of stimuli that have the potential to induce sensations of being attacked and experiencing psychological trauma and pressure. Hence, it's essential for minors to proficiently manage and regulate their emotional reactions when confronted with diverse online circumstances. The ability to accurately identify indicators of stress and anxiety resulting from digital encounters, as well as employing appropriate mechanisms for managing them, emerges as a critical component of emotional regulation.

The ability to effectively manage occurrences of online harassment or cyberbullying, which can result in significant psychological consequences for adolescents, further highlights the need for emotional regulation. The acquisition of emotional control skills enables young individuals to stay safe during their online interactions, hence diminishing their vulnerability to potential online threats.

In the context of digital prowess, the importance of communication abilities is also quite significant. Minors must learn to communicate politely and effectively online. There exists a notable necessity for individuals to effectively articulate their ideas with clarity and assurance while simultaneously demonstrating proficiency in the skill of attentive listening and comprehension of others. The ability to establish personal boundaries and assertively refuse online requests or demands that evoke feelings of discomfort or potential harm is an essential component of these communication abilities. The acquisition of these communication skills enables young individuals to establish and maintain positive online interactions, hence reducing their vulnerability to potential risks and dangers on the Internet.

To promote a culture of digital resilience among young individuals, it is important to adopt an approach that includes the involvement of parents, carers, and educators. These individuals of significant influence are crucial in presenting constructive digital conduct and setting explicit guidelines and anticipations for the utilization of technology. Hence, it is important to consider it a requirement to engage in regular dialogues with minors regarding the benefits and susceptibilities associated with the technology. These discussions should place particular attention on subjects such as digital privacy, security, and accountability. One should be aware that the promotion of digital literacy skills among young individuals can also be effectively supported by interactive platforms that encourage dialogue, offer resources and assistance, and actively encourage seeking support when felt necessary.

2 Digital Environment and Cybercrimes

The perpetuation of cyber-offences against minors, including cyberbullying, grooming, and exposure to inappropriate digital content, can potentially have negative effects on the emotional and psychological well-being of the targeted young individuals. This section aims to provide insight into a range of activities commonly found in cybercrime, including their societal implications and their specific focus on minors.

2.1 Cyberbullying Against Minors

Cyberbullying is a form of harassment that utilizes digital means to frighten, humiliate, or make threats against an individual. Of particular concern is the negative effect it has on individuals under the age of 18, who, given their stage of development, are more susceptible to the emotional and psychological consequences of such mistreatment. The expressions of cyberbullying are many, involving the spread of false rumours, the circulation of embarrassing visual content, the use of abusive language, and even implicit threats of physical violence. The repercussions of cyberbullying on underage individuals are severe and long-lasting. Individuals affected by these circumstances are susceptible to experiencing challenges such as anxiety, despair, and various psychiatric disorders. These factors collectively contribute to negatively impacting their involvement in academic pursuits and societal interactions.

2.2 *Offensive Name-Calling*

Another important aspect of cybercrime targeting minors is aggressive name-calling. This practice involves using derogatory words against minors on Internet. Cyber-violence may inflict psychological and emotional harm on minors, emphasizing the relevance of online abuse. Derogatory name-calling may damage teenage individuals' self-esteem and mental health and create self-doubt and uncertainty. Minors, who are building their self-image, are particularly vulnerable to insulting language. Derogatory words may cause isolation, low self-esteem, and anxiety and despair.

Due to its digital nature, cyberaggression may have a greater impact. Digital name-calling may go beyond narrow social circles and affect a wider audience than physical branding. Cruel words may reach a broad audience and remain online, which can increase emotional distress. This keeps the pain going. Offensive name-calling disrupts schools, which help minors grow. Disparaging words may impair a minor's ability to concentrate, connect with classmates, and engage in school activities. Thus, this might hinder academic progress and social interactions, creating a negative loop.

A thorough plan is needed to stop offensive name-calling. Digital platforms and social media companies must enforce severe content limits. Using algorithms to identify and remove insulting language and reporting and banning offensive individuals may help create safer online environments for younger users. Parents, educators, and carers are equally important. Open discussions on responsible digital behaviour, empathy, and the emotional effects of online interactions may help minors stay safe more effectively. Promoting a culture where minors may report inappropriate name-calling without fear of repercussions helps deter and reject it.

2.3 *Spreading of False Rumour*

Misleading information is part of the cybercrime network targeting minors that demands greater attention. Digital malfeasance involves spreading incorrect and/or inaccurate information about a young individual on several online channels, increasing the risk of emotional injury and social implications. The seriousness of this cyber-transgression emphasizes the need to study its effects on youth welfare and development.

False rumours harm teenage individuals' mental health. Developmentally, adolescents are highly vulnerable to external judgements and opinions. Fabricating or misrepresenting facts may produce severe feelings of shame, humiliation, and loneliness. Rumours spread quickly and widely online, adding to the emotional load as misinformation spreads among friends. Traditional rumours are limited to certain social groups, while the Internet transcends geographical and temporal boundaries. Thus, the quick and extensive transmission of wrong information might perpetuate false rumours, causing enduring psychological discomfort for minors. Lack of restriction on incorrect information dissemination may promote worry and psychological suffering. Inaccurate information may have serious effects on schools. Rumours like these may harm underage individuals' cognitive attention, academic interest, and active participation. Thus, cybercrimes may worsen academic achievement. The spread of false information may also disrupt peer relationships and lead to social isolation, worsening the minor's well-being.

A broad and diversified approach is needed to address misinformation. Digital platforms and social media companies must strictly moderate material to prevent misinformation. Algorithms that identify and highlight fraudulent information and mechanisms for reporting false rumours may help protect digital environments for underage users. Schools are critical in teaching minors digital literacy. Learning to distinguish between real and fraudulent content may help minors surf the Internet more safely.

2.4 *Receiving Explicit Images Without Solicitation*

This sort of digital transgression involves the unlawful sharing of sexually explicit visual content to minors via various online platforms, resulting in significant emotional and psychological effects. The gravity of this cybercrime necessitates understanding its various repercussions on young children's welfare and development. Non-consensual sexual pictures may harm children's mental health. Adolescents are more impressionable and sensitive to unwanted and inappropriate information. Accidental contact with sexual pictures might cause amazement, uneasiness, and a feeling of infringement. The encounter may cause psychological anguish and lower self-esteem and self-worth.

Digital privacy breaches transcend geographical and temporal boundaries, unlike physical privacy breaches. Sexual material may spread widely, worsening youngsters' suffering when these pictures recur or are shared among peers. Receiving unsolicited pornographic photographs disturbs youngsters' education, which is essential to their development. Unexpected exposure to such conditions may cause mental turbulence, making it hard for a youngster to concentrate, study, and socialize. Cybercrimes may harm academic performance and social relationships, increasing their negative impact.

A diversified approach is needed to address unsolicited graphic photo receipts. Online platforms and services must have proactive content moderation to detect and remove explicit material. Advanced algorithms that can recognize and remove explicit photographs, together with rapid reporting and banning, may help safeguard underage Internet spaces. Minors need digital literacy skills from schools too. Promoting open and compassionate discussion between parents, carers, and minors about the emotional effects of unwanted explicit material can improve minors' well-being.

2.5 *Cyberstalking*

Cyberstalking is the unwarranted surveillance, tracking, and intimidation of minor individuals across different online platforms, causing serious emotional and psychological harm. The seriousness of this cybercrime necessitates a thorough understanding of its numerous implications on youth welfare and development. Stalking, especially against minors, may harm their mental health. During identity formation and boundary setting, adolescents are more susceptible to such actions' intrusive and overpowering nature. Cyberstalking may cause anxiety, fear, and vulnerability in minors, eroding their feelings of security and autonomy.

Stalking hurts more due to Internet connectivity and anonymity. Cyberstalking transcends borders, unlike physical abuse. This allows perpetrators to constantly infiltrate a minor's digital presence, making them feel helpless and unable to escape. Being under continual observation may make youngsters feel unsafe and unprotected, which might worsen their emotional distress. Stalking may negatively affect schools, which are essential to minors' development. The unwelcome admission may harm a minor's emotional well-being, affecting their ability to focus, engage in school, and socialize with peers. Cybercrimes may harm academic progress and social ties, increasing their harmful impacts.

A thorough and coordinated effort is needed to combat stalking. Digital platforms and online services must incorporate efficient cyberstalking detection and mitigation mechanisms. Safer digital settings for minors may be achieved by using algorithms that can recognize invasive communication patterns. This may be improved with timely reporting and blocking. Schools and parents have a key role in teaching minors about Internet safety and digital citizenship. Society can safeguard minors by promoting a supportive environment that encourages them to report stalking without fear of retaliation and providing resources to treat the psychological effects of such interactions.

2.6 *Physical Threats*

Cyber-inappropriate behaviour that uses online platforms to threaten minors with physical damage or violence has serious emotional and psychological consequences. This cybercrime's many effects on youth welfare and development must be understood. Internet-transmitted physical threats to minors have serious emotional and mental health consequences. Minors, who are forming their identities and exploring their role in society, are especially vulnerable to these dangers' emotional and psychological effects. Physical threats may cause anxiety, vulnerability, and significant suffering, resulting in a persistent sense of insecurity that affects all aspects of life.

The Internet has changed physical threat communication. Internet platforms may transcend geographical and physical boundaries, unlike older intimidation methods. This allows wrongdoers to issue threats from afar, increasing young victims' uneasiness. Online threats' lasting nature increases their power since individuals may see them as ongoing. Physical threats may disrupt educational surroundings, which help minors grow and develop. Threats may cause psychological anguish, making it hard for minors to focus, learn, and socialize. This may have severe repercussions on academic achievement and social ties.

A comprehensive plan is needed to combat digital physical threats. Online platforms and services must limit threats and violent language. Algorithms that detect and highlight potentially risky information and reporting and banning features for dangerous individuals may help protect digital environments for underage users. In partnership with parents and carers, schools must teach minors appropriate digital conduct and provide them with skills to handle emotional issues.

2.7 Unauthorized Dissemination of Explicit Images

The unauthorized broadcast of sexually explicit visual images by minors via various online platforms is a digital transgression that may have serious emotional and psychological implications. This cybercrime requires a detailed understanding of its complicated effects on minor's welfare and development. Unauthorized transmission of explicit photographs harms minors' mental health. Minors, who are forming their identities, are more sensitive to the emotional and psychological effects of privacy invasions. Explicit photographs shared without consent may cause humiliation, embarrassment, and a deep sense of violation. Privacy breaches may damage minors' self-esteem and well-being, causing psychological harm.

The abundance of obscene photos online makes controlling unauthorized transmission difficult. The digital space transcends geographical and temporal boundaries, unlike traditional privacy intrusions. Minors are uncomfortable because they have no control over the propagation of graphic information. Unauthorized dissemination of sexual photos may disrupt educational settings, which help minors grow and develop. Privacy breaches may cause emotional discomfort in minors, making it harder for them to concentrate, study, and socialize. Academic achievement may be hampered, leading to feelings of isolation and worsening cybercrime effects.

Unauthorized distribution of graphic pictures requires a thorough approach. Digital platforms and Internet services must take full responsibility and use effective algorithms to detect and stop such information. Implementing reporting and removal tools for problematic information is essential to safeguard minors' digital environments. Schools, parents, and carers have an essential role in teaching minors about responsible digital behaviour, consent, and privacy. Advocating for clear and compassionate communication may help young individuals navigate the Internet and report unauthorized graphic picture distribution. All parties must work together to resolve this issue and protect minors.

2.8 Exposure to Inappropriate Content

Inappropriate digital material is a major cyber-risk that harms minors' mental health. Negative material in this category includes graphic sexual actions, unsettling violence, and hatred. These kinds of materials are not suitable for impressionable youth. The prevalence and accessibility of such information online endanger youth health and development. Multiple factors expose minors to inappropriate digital content unintentionally. Profit-driven agendas and ineffective regulations also drive these problems. Digital platforms and websites that utilize deceptive advertising and attractive content to increase user engagement prioritize financial profit above the safety of minors. These methods trick minors into seeing inappropriate stuff. Adolescents can accidentally discover improper content online or on social media. Therefore, preventive precautions are needed to protect minors from accidentally viewing inappropriate digital information.

3 Digital Environment and Minors

Interestingly, psychologists and medical researchers conduct examinations that investigate the effects of minors' engagement with social media on their development across multiple domains, including social, educational, psychological, and neurological dimensions. The field of research in question is continuously developing and increasing in significance, carrying substantial consequences for various parties involved, including young individuals, guardians, educators, legislators, and professionals within the technology sector. The shared obligation to ensure the well-being of minors rests with all of these stakeholders. To fulfil this obligation, it is essential to customize the utilization, features, and authorizations of social media platforms to align with the cognitive and emotional capabilities of minors.

The designs and features that are developed for adults may not be appropriate or suited for children and minors. Hence, parents must exercise active supervision over the social media engagements of minors, actively engage in meaningful conversations, and offer appropriate advice about the encountered information. It is essential to strike a balance between surveillance practices and the recognition of young individuals' privacy requirements. The progressive development of children's autonomy in accessing social media can be attributed to their growth and acquisition of digital literacy abilities.

Understanding the significance of minimizing the exposure of adolescents to social media content that presents unlawful or psychologically harmful activities is key. It is critical to put in place measures to limit minors' exposure to online content that promotes discriminatory attitudes, prejudiced beliefs, expressions of hatred, or instances of cyberbullying, with a particular emphasis on targeting individuals based on their race, ethnicity, gender, sexual orientation, religion, or disability. Maintaining alertness is critical in spotting signs of problematic social media use since it may disrupt minors' daily routines and cause severe psychological suffering in the long term.

The imposition of restrictions on social media usage is necessary to mitigate potential disruptions to crucial elements of minors' overall well-being, including sleep patterns and engagement in physical activities. These factors play a pivotal role in facilitating healthy cognitive development and psychological maturation. Further, it is advisable to promote among adolescents the practice of restricting their engagement with social media platforms mainly to content connected to attractiveness or appearance. It is essential to acknowledge that the impact of social media on adolescents is not inherently favourable or unfavourable. However, the presentation of these effects is contingent on various variables.

Adolescents' online experiences are impacted by their own decisions, such as who they follow and connect with, as well as the underlying aspects incorporated inside social media platforms, some of which are not immediately evident. The generalizability of research on social media's influence on minors is not absolute. Scientific research produces important insights; nevertheless, this information must be combined with a thorough understanding of an individual's unique skills, susceptibilities, and contextual circumstances to develop personalized decisions that meet their needs. Adolescent development is marked by a steady and continuous evolution, beginning with biological and neurological modifications that occur before the commencement of noticeable puberty, which normally occurs around the age of ten. This process continues until significant changes in social circumstances and brain maturity are complete, which usually happens around the age of 25.

Adolescents should use social media in a manner that is congruent with their level of maturity, containing their capacity for self-regulation, cognitive growth, and understanding of potential threats. Due to variations in the rate of maturation among adolescents, it is challenging to pinpoint a definitive age at which they become impervious to the potential risks and opportunities associated with social media. As a result, a continuing study is being conducted to ascertain the optimal age or specific time frame for certain suggestions. In the context of social media usage, it may be inferred that the potential threats are more pronounced during the first stage of adolescence, which is characterized by notable physiological, societal, and psychological changes, compared to the later stages of adolescence and early adulthood.

It is necessary to understand that racism is embedded in social media platforms. Algorithms, which impact users' experiences by determining what information they see, can perpetuate millennia-old racist policies and discriminatory practices. Social media platforms can unintentionally aid the spread of racial hatred, resulting in serious consequences such as physical violence in real-life contexts and threats to individuals' general well-being. The impact of social media on minors is a complicated issue that is dependent on a variety of factors, including personal characteristics, societal situations, and the nature of the platforms themselves.

To promote minors' overall well-being, stakeholders must adjust their social media use to their developmental requirements. Similarly, stakeholders should check the material that minors access regularly and address any threats, such as harmful behaviour and discriminatory practices. It is important to comprehend the progressive process of adolescent development, as well as the significance of continuous research efforts aimed at enhancing recommendations. The responsibility to combat racism on social media platforms has far-reaching consequences for ensuring a secure and inclusive digital space for everyone. The everyday use of technological devices and media has a destructive influence on the overall well-being of children, pre-teens, and minors. This tendency is shown by a greater susceptibility to anxiety, depression, and other mental illnesses. It also makes children more sensitive to potential health problems.

4 Digital Environment and Supervision

The central focus of parenting is an intimate understanding of the child. Guardians exhibit a natural aptitude, which has been perfected through experience and the passage of time, for discerning the challenges of their child's strengths and weaknesses. Understanding the difficulties of their child's perspective is important, especially in the constantly evolving social media. To assist young individuals in safely using digital platforms, guardians are provided with a series of evidence-based guidelines. These guidelines serve as necessary resources for promoting responsible use of social media and protecting the welfare of their offspring. One crucial insight that carers must acknowledge is the distinct vulnerability of adolescent brains to specific characteristics present in social media platforms. Although these platforms present opportunities for significant areas of child development, such as the acquisition of social skills and the nurturing of complex connections, they also pose potential risks. The prevalent use of the button for 'liking' and the utilization of AI to facilitate continuous scrolling can potentially have negative consequences for the maturing adolescent brain.

The process of adolescent brain development commences at approximately ten years of age, initiating a significant period of transformation that continues into early adulthood. In this particular stage, brain regions that are linked to the desire for social recognition experience heightened sensitivity, which might potentially be manipulated by social media sites. Concurrently, the brain regions associated with self-control undergo a process of maturation, achieving complete development alone during the early stages of adulthood. As a result, guardians are required to customize instructions regarding the utilization of social media by their children's development, which may involve imposing limitations on platforms that incorporate metrics such as 'likes' and discouraging excessive screen time. Utilizing the screen time settings commonly found on various devices enables adolescents to establish boundaries, exercise self-discipline, and promote optimal brain development by ensuring a minimum of eight hours of uninterrupted sleep each night.

A diversified strategy is important for the effective administration of social media. The consideration of time limits could be crucial in promoting self-discipline and encouraging balanced use. It is essential to exercise caution when it comes to the use of chat services, particularly in the context of interacting with unfamiliar individuals, and to impose restrictions on exposure to content that is targeted at adults. Enabling social media usage solely within the domestic setting can augment parental supervision for younger minors. Also, it is necessary to maintain vigilant surveillance of social media posts and content, particularly in the early phases of adolescence. It is essential to engage in ongoing dialogues between parents and adolescents concerning the prudent and beneficial utilization of social media platforms. Many adults find the digital platforms foreign and perplexing.

Therefore, it is essential to actively participate in frequent conversations, clarify the functioning of these platforms, and establish a supportive atmosphere that encourages youngsters to openly share their online encounters without concern of criticism. The transmission of influence occurs from parents to children, and this phenomenon remains applicable to social media conduct. Adolescents frequently adopt the behavioural patterns and attitudes exhibited by their parents. It is necessary that adults demonstrate responsible digital behaviour. This involves refraining from utilizing social media during communal family moments or meal gatherings, guaranteeing that discussions about social media align with the intended principles of usage, engaging in conversations with children regarding their individual social media practices, and setting predefined boundaries on personal usage while motivating children to do the same.

Engaging in periodic 'social media holidays' as a family, accompanied by subsequent discussions regarding the difficulties and allure experienced during these digital sabbaticals, serves to strengthen positive conduct. Maintaining a high level of circumspection could be helpful and rewarding for parents; they must be alert to indicators that suggest potentially dangerous interactions of their children with social media platforms. These indicators can be observed as disturbances in individuals' daily routines and obligations, a tendency to prioritize online interactions over face-to-face engagement, difficulties with sleep patterns, persistent usage despite expressed intentions to cease, intense cravings, or even engaging in deceptive behaviour to allocate additional time for online activities. Given these symptoms, carers should engage in a conversation with their children to determine the accuracy of their concerns.

If it becomes apparent that a child is displaying dependency on social media or engaging in detrimental behavioural patterns, the implementation of additional restrictions on technological access is a justifiable and appropriate course of action. In instances where individuals experience psychological distress or have ongoing difficulties regulating their social media usage, seeking the assistance of a mental health expert can potentially facilitate the development of more constructive and beneficial online interactions. The development of social media literacy necessitates a collective effort that involves the collaboration of social media platforms, educational institutions, adolescents, and their carers. Parents and guardians, specifically, have a necessary role in this experience.

The American Psychological Association (APA) supports the implementation of social media literacy curricula that are based on thorough scientific research. The primary objective of these activities is to provide minors with the knowledge and abilities required to effectively use the digital space while also minimizing the risks that may arise. APA is widely recognized as the largest global organization dedicated to the advancement of psychological research and scholarship. APA has been a pivotal figure in establishing standardized methodologies for researching psychological subjects and has contributed to the development of innovative approaches for addressing mental disorders.

The American Psychological Association (APA) utilizes research and education as a means to enhance the global implementation of psychological science, thereby significantly contributing to the comprehension of behaviours linked to the digital space. The nurturing of adolescents as prudent consumers of social media necessitates the implementation of a complete strategy. Carers must possess a keen understanding of the distinct susceptibilities associated with the maturation of adolescent brains. Thus, they should demonstrate a high level of attentiveness, exemplify positive digital conduct, actively participate in constructive conversations, and acquire a sense of proficiency in safely using the social media platforms. By engaging in this practice, individuals enable minors to utilize digital platforms effectively, maximizing the advantages while ensuring their overall welfare.

5 Digital Environment and Guidelines

The current recommendations advanced by medical and psychiatric professionals concerning the integration of technology in child development warrant thorough consideration. The criteria establish explicit parameters that are dependent on the age of the child. It is advised that children who are under 18 months of age should not engage with any type of screen-based media, except for video chatting. Parents of children between the ages of 18 and 24 months should exercise discernment in selecting high-quality digital entertainment and actively engage in co-viewing activities with their young ones. As children transition into the developmental stage spanning from 2 to 5 years of age, it is advisable to restrict their exposure to screens to a maximum of one hour per day, with a specific focus on engaging with educational content of superior quality. When considering children aged 6 and above, there is a requirement for attention to establishing regular limitations about the timing and content of their media consumption.

It is important to recognize the presence of different points of view within the academic community. These opposing viewpoints believe that the current focus on time restrictions fails to adequately account for the important features of the issue at hand as well as the contextual variables at work. As a result, it is stated that parents should concentrate not just on the time of their children's media intake but also on the quality and context in which their children interact with media. It is best to begin with a systematic approach. The massive saturation of technology in our society needs a careful balance of wise limitation and effective empowerment. It is advised that parents instil beneficial technology habits in their children since these behaviours are anticipated to last. One critical idea of guiding is the early development of technical literacy.

Parents are motivated by the desire to instil in their children the idea that digital gadgets are more than just sources of entertainment but rather tools that need intentional and thoughtful usage. Within the context of these talks, it is critical to explain the benefits and downsides of technology use among children while taking into account their cognitive and emotional growth. The educational technique necessitates a consistent approach and should increase in complexity following the child's maturity. The principle of discernment is emphasized as the central theme. While imposing time limitations is often considered wise, it is critical to properly assess the circumstances in which technology is used. The contrast between video chatting and interactive gaming, for example, emphasizes the need for subtle restrictions.

The significance of maintaining a regular sleep habit is emphasized. The usage of electronic gadgets throughout the night has the potential to reduce the overall effectiveness of sleep. As a result, it is advised that mobile phones, tablets, and laptops be used for no more than 30 minutes before sleep. Allowing youngsters to engage with electronic devices in their bedrooms beyond the designated lights-out hour requires careful consideration. The concept of alertness is generally recognized as a necessary principle. When it comes to younger children, parental supervision is rather simple, but as children get older, the practicality of direct monitoring diminishes. As a result, it is critical to engage in open and honest discussions about restricted websites and media. Parents should get acquainted with the digital spaces in which their children engage, allowing for detailed monitoring. A practical complement is the employment of software, specifically intended for content filtering and access control features.

The incorporation of ethical principles inside the Internet domain is a critical component that should not be debated. Behaviours in digital communication typically deviate from those seen in interpersonal, in-person conversations. As a result, parents must emphasize the significance of courteous online interactions. A significant pillar of this philosophy is advocating for the idea of encouraging minors to speak with their parents or guardians about cases of cyberbullying or disturbing Internet encounters. There is a significant desire to prepare youngsters with the necessary cognitive capacities to distinguish between trustworthy and untrustworthy online sources of information. Participating in debates about the principles of authenticity and accuracy in the context of digital platforms is seen as advantageous.

The practice of caution is a corollary, which includes steps such as avoiding downloading stuff from unknown sources, avoiding clicking on suspicious links, and refraining from submitting personal information on apps and websites of malicious or uncertain origins. Similarly, it is advised that children get sufficient guidance on how to avoid communicating with strange individuals via electronic communication platforms and actively encourage them to immediately notify their parents if they encounter such circumstances. The formation and maintenance of authentic interpersonal bonds is a critical and essential function. Some minors who are struggling with social integration may turn to extended Internet use as a substitute for true interpersonal connection. Nonetheless, it is important to emphasize that virtual relationships cannot serve as a suitable substitute for actual physical friendships. Parents should actively encourage the development of social skills and the formation of meaningful interpersonal connections.

In today's world, parents must have a thorough awareness of the many programmes and websites with which their children often engage. The intersection of child development and technology is investigated using a complex approach that takes into account time restrictions as well as a thorough understanding of content and context. As main guardians, parents should take a thorough and knowledgeable approach to instilling long-lasting technology habits and ethical behaviour in their children. The ideas expressed provide the foundation for the suitable and long-term integration of technology in the context of children.

6 Digital Environment and Government's Role

The integration of technology into the daily lives of individuals worldwide, including minors, is a widely recognized and prevalent phenomenon. The acceleration of this tendency has been significantly influenced by the onset of the COVID-19 pandemic, resulting in a substantial transition towards online activities. However, the increase in online participation has brought about simultaneous and increased dangers, notably including cyberbullying, online predation, exposure to dangerous content, and digital addiction. Minors are deemed more susceptible to these threats, hence requiring a heightened focus on safeguarding and guiding them within the digital space.

Given the developmental phase of minors, which is marked by the continuous development of their cognitive and socio-emotional abilities, it becomes apparent that they may lack the necessary experience and understanding needed to traverse the online space safely. Further, their inherent inquisitiveness and inclination towards social engagement make them more susceptible to participating in risky online activities, such as disclosing personal information or talking with unfamiliar individuals.

Governments are obligated to provide resources to initiatives and strategies aimed at developing digital resilience as a result of their fundamental obligation to preserve the well-being of their individuals, especially minors in both the digital and physical worlds. This investment is critical in providing minors with the necessary skills and knowledge to engage with technology successfully and ethically. Governments play an important role in creating digital resilience by encouraging the development of critical discernment skills, allowing the establishment of healthy online connections, and safeguarding the privacy and personal information of children.

The viewpoints and stances of child psychologists and educators coincide with the pressing need to enhance the digital resilience of minors. Child psychologists place significant focus on the ongoing cognitive development of minors, which renders them vulnerable to impulsive and precarious online actions. This vulnerability highlights the importance of providing individuals with the necessary resources to handle the digital spaces responsibly. Educational professionals also agree with this viewpoint, affirming that the incorporation of digital literacy into educational frameworks can enable young individuals to grasp the potential consequences of their online behaviours and engagements.

The significance of the government's involvement in this undertaking is emphasized by its ability to implement regulations that not only mitigate possible threats but also proactively boost digital resilience. The office of the eSafety Commissioner, an agency of the Australian government, has implemented various initiatives, such as the 'eSafety Early Years'[1] programme, aimed at offering educational materials to parents and educators. The objective is to empower young minors with fundamental skills in online safety. Likewise, inside the European Union, the General Data Protection Regulation (GDPR) places significant emphasis on safeguarding the online data and privacy of minors, demonstrating a legal dedication to ensuring their digital welfare.

7 Stance of the World's Governments

Governments throughout the globe have noticed the growing need for digital resilience in preserving the welfare of minors in today's digital environment. The exponential rise of technology and the Internet has resulted in many possibilities for individuals of all ages, including both children and adults. However, it has also exposed children to a slew of risks, including but not limited to cyberbullying, online grooming, and possible exposure to harmful information.

Given the vital need to address these concerns, numerous governments have adopted various ways to increase digital resilience. Several nations have shown

[1] esafety.gov.au.

proactive commitment by integrating digital resilience education into their national educational systems. Some groups, on the other hand, have undertaken special projects and commitments to encourage online safety and a sense of digital citizenship among young individuals.

Nations that have incorporated comprehensive digital resilience education into their formal educational programmes have appropriately acknowledged the importance of equipping minors with the necessary skills and knowledge to navigate the digital domain securely and responsibly. It is worth mentioning that the United Kingdom has implemented compulsory online safety education in both primary and secondary schools. The curriculum provided is extensive and covers a wide range of subjects, such as cyberbullying, online grooming, and the protection of personal privacy. Likewise, the incorporation of digital literacy and safety within the overall educational structure of Australia guarantees that kids receive instruction on the possible risks associated with social media, online gaming, and other digital platforms.

In addition to the formalization of digital resilience education, several governments have made significant investments in various courses and projects aimed at promoting online safety and developing digital citizenship among young individuals. In a similar vein, the US government has implemented the National Cyber Security Alliance,[2] a strategic programme aimed at enhancing awareness and mindfulness regarding cyber-safety among minors, parents, and educators. Also, it should be noted that some governments have not yet incorporated digital resilience education into their educational systems. However, they have implemented various projects and campaigns to promote online safety and promote digital citizenship among young individuals. The 'Digital India'[3]' campaign, initiated by the Indian government, is a comprehensive initiative designed to promote digital literacy and ensure safety among all segments of society, particularly the younger population.

The importance of promoting digital resilience among minors is increasingly seen as a priority by governments across the globe. While the extent of priorities and actions implemented by governments may differ among regions, it is encouraging to observe that numerous countries are actively working towards providing minors with the necessary skills and knowledge required for their safe and responsible use of digital platforms. By persistently devoting resources and efforts to this field, the goal of empowering underage individuals to effectively utilize the benefits of technology while minimizing potential risks is attainable. This section will provide a comprehensive analysis of the many strategies that different nations have adopted to incorporate digital resilience education into their educational frameworks.

[2] staysafeonline.org.

[3] digitalindia.gov.in.

7.1 Australia

The digital resilience framework for minors in Australia is a pioneering effort that prioritizes the promotion of responsible digital citizenship and ethical conduct in the online domain. The framework acknowledges the current reality that minors are currently being raised in a progressively digitalized society, wherein technology and the Internet hold significant sway over their day-to-day experiences. Nevertheless, the expansion of technology and the Internet is accompanied by a corresponding escalation in the potential threats and complexities that accompany their use.

The digital resilience framework for minors has been specifically developed to effectively tackle these difficulties and encourage the development of responsible digital citizenship. The central focus of the framework is the eSmart Schools programme[4,5] which offers extensive resources and training to educational institutions regarding successful approaches to promoting responsible digital citizenship among students. The eSmart Schools programme is an all-inclusive educational initiative that aims to nurture digital citizenship within schools, emphasizing the establishment of a constructive online conduct culture. It also offers educational institutions materials and assistance in formulating comprehensive policies and protocols aimed at tackling online safety concerns, such as cyberbullying and various other threats encountered on the Internet.

The framework acknowledges the importance of empowering children to make responsible decisions in the online domain. In pursuit of this objective, it attempts to facilitate the acquisition of digital literacy competencies among individuals under the age of majority, including the domains of critical analysis, solution-oriented thinking, and judicious decision-making. The framework promotes the development of minors as engaged and accountable digital citizens, capable of utilizing technology and the Internet securely and ethically.

An additional noteworthy element of the digital resilience framework is its emphasis on collaboration and partnership. The framework acknowledges the necessity of a collective initiative including various stakeholders, such as educational institutions, parents, carers, industry representatives, and governmental bodies, to promote responsible digital citizenship. It encourages collaborations among these key actors to further the promotion of digital resilience among minors.

7.2 European Union (EU)

The European Union (EU) has implemented substantial measures aimed at promoting digital resilience and safeguarding the well-being of minors in online

[4] esmart.org.au.

[5] https://www.esmart.org.au/esmart-schools/.

environments. The European Union has established a comprehensive digital resilience framework under its Better Internet for Kids initiative.[6] This framework includes various resources and initiatives to promote responsible online conduct among minors.

One of the fundamental elements of this framework includes the Safer Internet Day[7] campaign, an annual observance occurring in February that aims to enhance consciousness regarding the significance of online safety and promote conscientious digital citizenship. The campaign comprises various set activities, including workshops, seminars, and competitions, and enjoys the backing of a broad spectrum of stakeholders, including industry participants, non-governmental organizations, and governmental entities.

The inclusion of a hotline inside the EU's digital resilience framework serves as a crucial component, enabling the streamlined reporting of illicit online content. The hotline, which is managed by the European Union Agency for Law Enforcement Cooperation (Europol),[8] serves as a medium for individuals on the Internet to report instances of illicit online content. This includes a range of materials, such as child sexual abuse material, terrorist propaganda, and hate speech, among others. The hotline is accessible in all European Union member states and serves as a critical instrument in the fight against cybercrime and the safeguarding of underage individuals.

Alongside the hotline, the European Union (EU) has implemented a network of national Safer Internet Centres which function as repositories of information and provide assistance in matters about online safety. The Safer Internet Centres offer various services, including online safety education, helplines, and awareness-raising campaigns.[9] The centres also collaborate extensively with other stakeholders, including educational institutions, parents, and law enforcement agencies, to promote safer and more responsible use of the Internet by underage individuals. The European Union (EU) has exhibited its commitment to enabling responsible digital citizenship and addressing online crime through its Better Internet for Kids programme, which involves active collaboration with partners across several sectors.

7.3 India

The significance of digital literacy and Internet safety for children in the contemporary technologically driven society has been acknowledged by the Indian government. In addressing this issue, the government has established a digital

[6] betterinternetforkids.eu.

[7] saferinternetday.org.

[8] europol.europa.eu.

[9] digital-strategy.ec.europa.eu/en/policies/safer-internet-centres.

resilience framework that aims to provide young individuals with the necessary skills and knowledge to handle the online space confidently and securely. This framework comprises a variety of activities and recommendations aimed at promoting digital literacy and ensuring online safety for minors, with a special focus on those living in rural regions.

The Digital India initiative[10] is an integral element within the framework of digital resilience. The objective of this initiative is to offer digital literacy instruction to underage individuals living in rural regions, where there may be restricted availability of technology and digital resources. The primary objective of the campaign is to address the issue of the digital divide by offering instructional programmes and necessary resources to empower young individuals residing in rural regions to effectively access and utilize digital technology. By facilitating access to digital resources and offering training opportunities, adolescents residing in remote areas can enhance their connectivity to the global community and effectively utilize the extensive set of online resources at their disposal.

One significant effort within the digital resilience framework is the Cyber Suraksha Kavach initiative, which imparts cybersecurity knowledge and training to students in educational institutions. The primary objective of this effort is to impart knowledge to individuals under the age of majority on the potential threats prevalent in the digital space, including cyberbullying, identity theft, and malware attacks. It also seeks to provide them with the essential competencies required to safeguard their online presence. The government aims to develop a culture of Internet safety and awareness among youngsters through the provision of cybersecurity training.

The digital resilience framework includes principles aimed at facilitating the successful promotion of digital literacy and online safety among students inside educational institutions. These guidelines furnish educational institutions with the essential information and resources required to create efficacious plans for promoting online safety and digital literacy among their student population. By utilizing these principles, educational institutions can establish a secure and favourable atmosphere for students to acquire knowledge regarding digital technology and Internet security.

7.4 Japan

Japan has made notable progress in strengthening the digital resilience of its population, particularly minors, by implementing a framework known as the Basic Act on Cybersecurity.[11] The Act has multiple elements that are designed to enhance cybersecurity awareness and education among minors.

[10] digitalindia.gov.in.

[11] japaneselawtranslation.go.jp/en/laws/view/3677/.

The Basic Act on Cybersecurity framework includes many efforts aimed at promoting awareness and comprehension of cybersecurity among adolescents. An example of such an effort is the Cybersecurity Awareness Month campaign, which aims to provide education and encourage active involvement among children in cybersecurity. The objective of this campaign is to emphasize the significance of protecting personal information and embracing secure online practices. By doing so, it aims to empower young individuals to traverse the digital space with confidence and a heightened sense of awareness.

In current times, there has been a notable increase in the use of digital devices and Internet usage among individuals under the legal age, rendering them susceptible to many cyber threats including but not limited to online harassment, identity fraud, and exposure to harmful materials. To mitigate these concerns, the Basic Act on Cybersecurity framework has implemented safeguards aimed at preserving the privacy of kids and shielding them from potential cyber threats. As an illustration, the Act stipulates that operators of online services are required to implement requisite procedures to safeguard the personal information of children, including the acquisition of parental consent before data collection.

The Act also mandates that educational institutions and schools integrate cybersecurity education into their curriculum, intending to provide minors with the requisite skills and information to safeguard themselves in a digital space. This includes the promotion of secure Internet practices, such as the utilization of robust passwords, refraining from divulging personal information, and exercising caution when encountering dubious emails and websites. The Act promotes the utilization of cybersecurity tools and resources, including antivirus software, firewalls, and content filters, to augment the online safety and security of minors.

Besides, the Basic Act on Cybersecurity framework acknowledges the significance of international collaboration in tackling cybersecurity issues. Japan has demonstrated a proactive involvement in multiple international platforms, including the G7, G20, and the US governments, intending to exchange exemplary approaches and encourage cooperation among governments to enhance global cybersecurity. The Act also has provisions for promoting collaboration with foreign nations and international organizations on topics about cybersecurity.

7.5 The United Kingdom (UK)

The government of the United Kingdom (UK) has acknowledged the significance of safeguarding minors and has implemented substantial measures to improve their safety on the Internet. To attain this objective, the government has implemented an all-inclusive digital resilience framework for minors, which includes diverse educational elements and activities, alongside crucial legislation that seeks to enable age-appropriate design and enforce regulations about online safety.

The integration of digital literacy and online safety into educational curricula is a fundamental element of the digital resilience architecture. The primary objective of the government is to provide young learners with the essential skills and information required to effectively and ethically handle the digital space. Educational institutions are furnished with instructional directives about the proper incorporation of digital literacy and online safety into their academic programmes. This measure guarantees that individuals under the age of majority are provided with a thorough and developmentally suitable education that equips them with the necessary skills and knowledge to handle the complexities of a digital setting.

The UK Safer Internet Centre[12] is a crucial component of the digital resilience architecture for minors, alongside educational programmes. The primary objective of this organization is to offer guidance and access to information about the protection of children and their carers in the domain of online safety. The UK Safer Internet Centre offers a variety of services aimed at assisting young individuals in safely using a digital space. These services include helplines, online safety tips, and resources.

The digital resilience framework includes essential regulations, including the Age-Appropriate Design Code.[13] The Online Safety Bill[14] is of relevance to the Age-Appropriate Design Code that guides organizations regarding the development of online services that are appropriate for individuals under the age of majority, considering their age and developmental requirements. This code advocates for the implementation of design principles that are acceptable for different age groups, intending to ensure the safety and suitability of online services, particularly for minors.

The Online Safety Bill is a significant legislative measure aimed at protecting kids from potential threats on the Internet. The objective of this legislation is to enhance the safety of the digital world for all individuals by implementing measures that deter the dissemination of dangerous content to underage users. Also, it ensures compliance to online safety standards and levies sanctions on organizations that demonstrate non-compliance. The proposed legislation seeks to safeguard individuals under the age of majority from a range of detrimental experiences on the Internet, such as cyberbullying, grooming, and exposure to content that is deemed unsuitable for their age group.

---

[12] saferinternet.org.uk.

[13] ico.org.uk/for-organisations/uk-gdpr-guidance-and-resources/childrens-information/childrens-code-guidance-and-resources/introduction-to-the-childrens-code/.

[14] bills.parliament.UK/bills/3137.

7.6 *The United States of America (USA)*

The Children's Online Privacy Protection Act (COPPA) is the principal legislation in the United States (US) that defines the framework for digital resilience specifically tailored for minors.[15] The Children's Online Privacy Protection Act (COPPA) was officially enacted in 1998 and subsequently changed in 2013. This legislative measure was implemented in response to the increasing concerns around the safeguarding of youngsters' online privacy and security. According to regulations, websites and online services are obligated to acquire specific consent from parents or legal guardians before gathering any personal information from individuals under the age of 13. Following COPPA regulations, website owners are obligated to furnish parents with notification regarding their data collection procedures, as well as allow parents to examine and erase their child's information upon request.

The Children's Online Privacy Protection Act (COPPA) represents a significant measure in safeguarding the personal information of youngsters and advancing their safety in the digital society. Nevertheless, it should be noted that there are other initiatives within the digital resilience framework. The Net Cetera guide,[16] published by the Federal Trade Commission (FTC), represents a noteworthy initiative that seeks to enhance the protection of minors in the online spaces.

The Net Cetera programme includes a diverse set of subjects in the domain of Internet security, including areas such as cyberbullying, social networking, gaming, and mobile devices. This resource provides practical recommendations and guidance for safeguarding personal information, mitigating the risks associated with online scams and fraud, and effectively addressing instances of online harassment. Also, Net Cetera offers educational materials and support for parents and educators in facilitating the development of minors' critical thinking abilities and promoting informed decision-making in the online domain.

Additional components of the digital resilience framework include the National Centre for Missing and Exploited Children (NCMEC),[17] an organization that offers a set of resources and tools intending to mitigate the occurrence of online sexual exploitation of minors. The NCMEC collaborates with law enforcement agencies, commercial industry entities, and the general public to effectively identify and locate children who have gone missing, as well as to actively combat instances of child sexual exploitation. It offers training and materials to law enforcement personnel and other professionals regarding appropriate protocols for addressing instances of exploitation involving individuals under the age of majority.

[15] ftc.gov/legal-library/browse/rules/childrens-online-privacy-protection-rule-coppa.

[16] bulkorder.ftc.gov/publications/net-cetera-chatting-kids-about-being-online.

[17] https://www.missingkids.org/.

8 Stance of Multinational Corporations

In conjunction with non-governmental organizations (NGOs), several multinational businesses (MNCs) are also taking on an active role in addressing the matter of cybercrime that specifically targets minors. This section will address the requisite measures undertaken to enhance the social and digital environment for children, ensuring their well-being and safety.

8.1 Microsoft

Microsoft has implemented a range of strategies to address the issue of cybercrime targeting underage individuals. Microsoft has implemented a significant initiative in combating cybercrime targeting kids by creating a specialized team comprising of professionals in the field of digital crime. These experts engage in collaborative efforts with law enforcement agencies to investigate and prosecute those who engage in Internet-based criminal activities that specifically target minors. The efficacy of this collaborative method in apprehending offenders and dissuading potential participants from partaking in similar illicit behaviours has been well documented.

Microsoft has made significant advancements in the field of AI, exemplified by the development of PhotoDNA.[18] This cutting-edge technology enables the identification and subsequent removal of child exploitation content from Microsoft's platforms. PhotoDNA is an advanced technological solution that utilizes digital signatures to identify and detect photos that contain content related to the sexual abuse of children, commonly referred to as child sexual abuse material (CSAM). The integration of this technology with Microsoft's cloud services and other products, like OneDrive and Outlook, facilitates the automated detection and removal of CSAM information.

Also, Microsoft has established strict policies aimed at preventing the utilization of its services for child exploitation. These regulations serve to restrict users from engaging in the actions of uploading, sharing, or accessing any content that advocates for or enables the exploitation of children. Account cancellation and reporting to law enforcement authorities may occur as consequences for breaching these policies.

In addition to implementing internal safeguards, Microsoft engages in active collaboration with external organizations to enhance awareness and mitigate instances of cybercrime targeting kids. As an example, Microsoft demonstrates its support

[18] microsoft.com/en-us/photodna.

for the National Center for Missing and Exploited Children (NCMEC) through the provision of technological advancements and financial contributions. The company has established collaborations with several entities, like the International Centre for Missing and Exploited Children (ICMEC) and the WePROTECT Global Alliance, to address the issue of online child exploitation and abuse.

8.2 Apple

Apple has taken a range of measures aimed at safeguarding children and minors from online exploitation and abuse, thereby showcasing its dedication to protecting vulnerable populations.

The company has implemented a remarkable set of parental control capabilities as part of its initiatives in this domain. These functionalities enable parents to impose limitations on access to specific applications and content, as well as establish restrictions on screen time. Consequently, this ensures that children and adolescents are shielded from potentially dangerous content and do not excessively engage in online activities. The above-stated controls can be tailored to meet the specific requirements of individuals and may be conveniently accessible and administered via the operating system of the organization.

It has established strict standards that explicitly forbid the utilization of its services for illicit activities, such as the exploitation of minors. The standards provided by the firm specifically specify that any application or service that encourages or enables unlawful conduct, such as the exploitation of individuals under the age of majority, would be eliminated from the App Store. The company places a significant emphasis on adhering to these restrictions and has gained a reputation for removing applications that fail to fulfil its established criteria.

Apple has made significant investments in the field of AI and ML to enhance its ability to identify and eliminate dangerous content with greater efficiency. The organization has successfully devised sophisticated algorithms capable of detecting and highlighting probable occurrences of child exploitation, specifically including visual media including nudity or sexual material involving individuals below the age of majority. Upon identification, the item in question is swiftly eliminated from the platforms operated by Apple, and the appropriate authorities are duly informed.

Also, Apple engages in collaborative efforts with many organizations and stakeholders to enhance awareness and mitigate instances of cybercrime targeting minors. The company actively engages in multiple forums and programmes that are dedicated to advancing digital safety, including the National Centre for Missing and Exploited Children's (NCMEC) Project Safe Childhood. Apple engages in partnerships with law enforcement authorities, wherein it provides its technical assistance and specialized knowledge to aid in the investigation and legal pursuit of those implicated in cybercrimes.

8.3 Facebook

Facebook has implemented many safeguards to combat this issue and safeguard children and adolescents from the risks of online exploitation and abuse. The reporting mechanism adopted by Facebook is regarded as a crucial safeguard, enabling users to report any information that breaches community standards, such as material related to the abuse of children. This mechanism is a substantial factor in facilitating Facebook's ability to rapidly identify and eliminate such content. Facebook has also made significant advancements in the field of AI and ML, enabling it to more effectively identify and eliminate dangerous information, such as materials related to the exploitation of children.

Likewise, Facebook has enacted strict restrictions that explicitly forbid the utilization of its services for illicit activity, such as the exploitation of minors. These regulations function as a preventive measure for individuals who may intend to exploit the platform for malicious intentions. Facebook engages in partnerships with law enforcement authorities to facilitate investigations and apprehend individuals who exploit the social networking platform for illicit activities.

The corporation has adopted strategies to provide education to its users regarding Internet safety, including individuals of all age groups, including children and minors. Facebook has developed a variety of educational resources, including safety instructions, intending to assist parents and children in understanding the potential threats linked with online activities and implementing strategies to minimize such risks. The organization has also allocated resources towards doing research and establishing partnerships to get a deeper understanding of the issue of cybercrime targeting kids, as well as to devise more efficient and impactful remedies.

8.4 Google

Google is the operator of a search engine that is recognized as one of the largest globally. The organization has acknowledged its obligation to mitigate the utilization of its platform for unlawful activities, specifically concerning cybercrimes targeting individuals under the age of majority. Consequently, the organization has adopted multiple strategies to address this problem.

Google has established a thorough reporting system that allows users to report content that is considered unsuitable or unlawful, such as material related to the exploitation of children. The system is reinforced by strict restrictions that explicitly forbid the utilization of Google's services for those activities. Google is demonstrating a proactive posture in protecting children and avoiding Internet abuse by implementing a zero-tolerance policy on the use of its platform for the exploitation of kids.

Also, Google has made significant investments in the field of AI and ML intending to improve the efficiency of its content moderation and detection systems. Through the utilization of these technologies, Google can effectively identify and eliminate detrimental content, such as materials related to the exploitation of children, more efficiently. This measure is of importance in ensuring the protection of underage individuals, given the vast amount of content published by users on Google's site, rendering human monitoring an unattainable ambition.

The corporation has established a collaborative partnership with the National Centre for Missing and Exploited Children (NCMEC) to create a comprehensive framework for the detection and eradication of child exploitation content on its platform. In addition, Google has contributed support for the advancement of technologies aimed at mitigating the risk of online child grooming.

9 Stance of Non-governmental Organizations

The issue of cybercrime targeting kids is an escalating global concern, prompting numerous organizations to dedicate significant efforts towards its prevention and mitigation.

9.1 ICMEC

The International Center for Missing and Exploited Children (ICMEC)[19] is well recognized as a prominent non-governmental organization operating in this particular domain. The primary objective of the organization is to enhance the safety of children worldwide by the elimination of child abduction, sexual exploitation, and trafficking. The organization engages in collaborative efforts with governmental bodies, law enforcement agencies, and non-governmental organizations (NGOs) to formulate policies, enact legislation, and establish training initiatives aimed at safeguarding children against both online and offline threats. The primary areas of concentration for ICMEC's efforts include prevention, policy development, training, and advocacy.

ICMEC operates in the domain of prevention, aiming to enhance consciousness regarding the risks associated with both online and offline child exploitation. Also, the organization works to equip parents, educators, and children with the necessary tools and resources to ensure their safety. The organization engages in collaborations with multinational corporations (MNCs) and other relevant stakeholders to establish and execute optimal strategies for the prevention and mitigation of online abuse and exploitation.

[19] https://www.icmec.org/.

It collaborates with governmental entities in the policy domain to formulate and execute legislative measures aimed at safeguarding children against both virtual and physical threats. The organization offers technical support and training programmes to law enforcement agencies and other relevant parties, aiding them in the successful implementation of certain laws and regulations.

It offers training in the field of child protection, equipping law enforcement authorities, non-governmental organizations (NGOs), and other relevant parties with the necessary tools and resources to effectively detect and address instances of child exploitation occurring both online and offline. The organization engages in the creation and implementation of educational initiatives aimed at equipping parents, educators, and children with the necessary knowledge and skills to ensure their online safety.

In the domain of advocacy, ICMEC works to enhance public consciousness regarding the matter of online and offline child exploitation, as well as to galvanize backing for policies and initiatives aimed at safeguarding children. The organization engages in collaborative efforts with other non-governmental organizations (NGOs) and relevant stakeholders to advocate for the rights of children and facilitate the formulation of efficient methods to address instances of abuse and exploitation, both in online and offline contexts.

The efforts of ICMEC have yielded substantial outcomes in combating cyber-crime targeting underage individuals. The organization has significantly contributed to the formulation of legislation and policies aimed at safeguarding kids from exploitation in both online and offline contexts. In addition to its primary objectives, ICMEC has also facilitated the provision of training and technical support to various entities such as law enforcement agencies, non-governmental organizations (NGOs), and other relevant stakeholders. This assistance has aimed to enhance their capacity to efficiently identify and address instances of child exploitation. The organization has effectively heightened public consciousness on the matter of child exploitation in both online and offline contexts while also successfully rallying assistance for the implementation of policies and initiatives aimed at safeguarding the welfare of children.

9.2 ASACP

The Association of Sites Advocating Child Protection (ASACP)[20] is a notable non-governmental organization (NGO) that has been actively involved in combating this issue since 1996. The organization has been diligently attempting to furnish online enterprises with the essential tools and resources to mitigate the dissemination of child sexual abuse material and safeguard minors from potential harm.

---

[20] http://asacp.org/.

The primary objective of ASACP is to safeguard minors from the adverse consequences of online exploitation. This is achieved by equipping Internet businesses with essential tools and resources that aid in the prevention of the dissemination of information depicting child sexual abuse. The organization has formulated a code of ethics that defines the optimal methodologies for online enterprises to mitigate the dissemination of child sexual exploitation material. The code has provisions for the surveillance and documentation of potentially illicit behaviours, the establishment of mechanisms for verifying the age of users, and the collaboration with law enforcement entities to ascertain the identities and pursue legal action against those engaged in online predatory actions.

ASACP not only offers online businesses the essential tools and resources but also collaborates closely with law enforcement agencies to facilitate the identification and prosecution of those involved in online predatory activities. The organization has a hotline through which members of the public can report any potentially illicit activity about the dissemination of child sexual abuse material. The hotline is operated by a group of knowledgeable professionals who assess and classify the reports before subsequently transmitting them to the appropriate law enforcement agency for further examination.

The global recognition of ASACP's efforts in combatting online child exploitation is evident via the organization's reception of several awards and honours. The organization received the President's Volunteer Service Award in 2019 in recognition of its exceptional efforts in combating online child exploitation.

9.3 WePROTECT Global Alliance

The WePROTECT Global Alliance[21] is a worldwide consortium including non-governmental organizations, governmental bodies, and corporate entities that have pledged their dedication to addressing the issue of online child sexual abuse. The organization emerged as a prominent entity in an attempt to combat cybercrime targeting underage individuals. The composition of its membership includes various entities, such as law enforcement agencies, Internet service providers, non-governmental organizations (NGOs), and technology businesses.

A primary objective of the WePROTECT Global Alliance is to enhance public consciousness regarding the matter of cybercrime targeting individuals under the age of majority. The organization attempts to disseminate knowledge among the general public regarding the risks associated with online exploitation while simultaneously advocating for the adoption of optimal strategies aimed at safeguarding children in the digital domain. It engages in collaborative efforts with many organizations and stakeholders to actively promote the enactment of more robust laws and policies aimed at preventing cybercrime targeting minors.

[21] https://www.weprotect.org/.

The WePROTECT Global Alliance not only attempts to increase awareness but also strives to encourage worldwide collaboration in combating cybercrime targeting minors. The organization facilitates the collaboration of law enforcement agencies and many stakeholders from across the globe, enabling the exchange of information and resources, joint investigations, and the formulation of novel approaches to prevent and combat the online exploitation of children.

An additional key part of the WePROTECT Global Alliance's efforts involves the advancement of technologies aimed at inhibiting the dissemination of child sexual abuse content on the Internet. The organization collaborates with technology companies to facilitate the creation and deployment of solutions designed to identify and eliminate objectionable content while also enabling the identification and monitoring of individuals involved in such activities. It even provides financing for research attempts focused on the development of novel technologies aimed at mitigating online exploitation and safeguarding children in the digital domain.

Since its inception, the WePROTECT Global Alliance has made substantial progress in reaching various important milestones. Subsequently, the programme has experienced substantial expansion, including a multitude of nations, and has yielded a significant number of apprehensions and legal condemnations of perpetrators on a global scale.

10 Limitations of Global Digital Resilience Frameworks

Digital resilience frameworks for minors are specifically crafted to facilitate the acquisition of skills and implementation of methods that enable them to traverse the online spaces securely and responsibly. Although these frameworks possess significant value as resources, they are not without their limits, which include the following.

10.1 One-Size-Fits-All Approach

Digital resilience frameworks sometimes provide broad recommendations that may not adequately consider the varying requirements and perspectives of minors from diverse origins, cultures, and age ranges. In light of the considerable diversity in technological competencies and backgrounds among minors, it may be essential to embrace a customized strategy to ensure the efficacy of digital resilience methods for all youngsters.

It is crucial to acknowledge that the technological experiences of minors are influenced by various aspects, including their cultural and socio-economic contexts, technological accessibility, and unique developmental requirements. Hence, it is essential to take into account the distinct requirements of every underage individual when formulating and executing digital resilience tactics.

To effectively promote digital resilience among minors, it is crucial to establish frameworks that take into account the diverse range of experiences that minors have with technology. By implementing a customized methodology, it is possible to guarantee that digital resilience tactics are suitable, easily accessible, and efficient for all children, irrespective of their variations.

10.2 Lack of Empirical Evidence

Ensuring the evidence-based nature and thorough evaluation of digital resilience frameworks is of importance. Although many frameworks may exhibit a dearth of empirical evidence, it is crucial to acknowledge the necessity of empirical validation in substantiating their efficacy. Hence, researchers and practitioners must give precedence to the assessment of these frameworks to ascertain their foundation in empirical data and their efficacy in promoting digital resilience.

10.3 Rapidly Changing Technology

The dynamic and ever-changing nature of the digital society is marked by a continuous stream of emerging technologies, posing a notable obstacle for frameworks aimed at promoting digital resilience. Consequently, it is essential to provide ongoing assistance and instruction to minors to guarantee their well-being and protection while they use the digital platforms. This calls for a proactive and adaptable strategy in the domain of digital education, which takes into consideration the emergence of potential risks and provides minors with the necessary skills and information to ensure their online safety. To ensure the protection and enhancement of the digital literacy of minors, stakeholders must engage in collaborative efforts and formulate comprehensive plans that effectively adapt to the dynamic nature of the digital environment.

10.4 Limited Parental Involvement

The effectiveness of digital resilience frameworks that exclusively concentrate on the education of kids may be diminished in the absence of parental involvement. It is crucial to acknowledge that the involvement of parents is significant in facilitating the digital resilience of minors. Parents have an essential role in ensuring the safe and responsible navigation of children in the digital world.

Hence, the inclusion of parents as active participants in the schooling process is crucial within digital resilience frameworks. This objective can be accomplished by equipping parents with the requisite knowledge and competencies to facilitate the development of sound digital behaviours among minors, recognizing possible threats, and implementing pre-emptive strategies to alleviate them. By including parental involvement in digital resilience frameworks, it is possible to promote a comprehensive approach towards enhancing digital safety and well-being among minors.

10.5 Insufficient Resources

The deployment of certain digital resilience frameworks may face challenges stemming from insufficient resources, funding, and support. These obstacles have the potential to hinder the effectiveness and scope of these frameworks, hence potentially failing to adequately support all minors who could potentially derive benefits from them. Therefore, it is essential to ensure the provision of suitable resources to facilitate the efficient execution of digital resilience frameworks.

11 Guidelines for Guardians

Current times are characterized by the perpetual evolution of threats where the significance of guardians in facilitating digital resilience among children and adolescents cannot be avoided. The swift progression of technology and its pervasive integration into the lives of minors need the recognition of its significant influence on their growth and maturation. The increasing ubiquity of smartphones, social media, and online platforms has rendered the digital infrastructure an expansive domain for exploration, communication, and education. However, within this seemingly limitless domain, there exist potential dangers and challenges that, if not properly acknowledged and resolved, can jeopardize the welfare of young individuals and adolescents.

It is important that guardians not only oversee but also actively participate in open discussions that promote the development of critical thinking and emotional intelligence. It is crucial to approach these connections with a comprehensive understanding of age-appropriate digital encounters, acknowledging that what is deemed appropriate for adults may not correspond with the developmental milestones of children and adolescents. Guardians need to stay aware of the changing difficulties presented by the digital age. The presence of harmful content, instances of cyberbullying, and exposure to psychologically disturbing material are issues that necessitate the implementation of preventative solutions. By promoting the reduction of exposure to such information and implementing support mechanisms for reporting and addressing negative online experiences, guardians can establish a protective barrier while empowering their charges to effectively traverse these issues.

11.1 Tailoring Social Media Use

An idea that demands significant emphasis is the need for tailored social media experiences, specifically concerning minors and adolescents. There is a growing recognition that digital platforms, which were initially created for adult users, may not consistently correspond with the cognitive and emotional capacities of younger individuals. Given these circumstances, guardians and parents must do a careful and mindful evaluation of the social media platforms to which the younger generation is subjected.

The influence of permissions and privacy settings on a minor's Internet experience is significant. Guardians must engage proactively in the establishment and administration of these permissions to safeguard their children against potential threats such as cyberbullying, exposure to unsuitable content, or interaction with unfamiliar individuals. Through careful configuration of these settings, parents or guardians can establish a digital safety framework for their children, enabling them to navigate the online domain with a heightened sense of security.

11.2 Adult Monitoring and Coaching

Guardians should adopt a proactive approach of constantly monitoring their children's online activity. The monitoring though should not be intrusive but rather serve as a mechanism for safeguarding minors' well-being and directing their conduct on the digital platforms. Effective and transparent communication are a necessity in nurturing healthy relationships between guardians and their children. Guardians must nurture trust with their children, thereby providing a supportive atmosphere that encourages open dialogue about their online encounters and concerns.

Coaching assumes a crucial function in facilitating the navigation of minors inside the digital space. Guardians must allocate sufficient time to impart knowledge to their children regarding the possible threats and advantages associated with the digital platforms. This includes the examination of topics such as privacy, digital etiquette, and conscientious engagement with online content. Through the imparting of these fundamental abilities and principles, guardians enable their children to engage in well-informed decision-making and effectively address the various challenges they may come across in the digital world.

It is also crucial to demonstrate leadership by personal conduct. Guardians must exemplify responsible digital conduct themselves. It is a common occurrence for children to imitate the behaviours exhibited by their guardians. Therefore, exhibiting healthy online behaviours, such as engaging in courteous communication and responsibly managing screen time, can serve as an influential method for instilling these principles in the younger generation.

11.3 Minimizing Exposure to Harmful Content

Guardians must adopt proactive measures to mitigate their children's exposure to inappropriate web content. This requires the implementation of suitable content filters and parental controls on devices and Internet browsers. These tools can prevent or limit the availability of explicit or improper content, thereby establishing a more secure online setting for minors and adolescents.

Guardians must promote responsible consumption of content. Establishing clear guidelines for what is acceptable and unacceptable to view online is also important, ensuring that children are aware of the family's values and boundaries.

It is also essential to maintain open lines of communication. It is the responsibility of the guardians to establish and maintain a continuous dialogue with their children regarding the digital content they come across on the Internet. It is advisable to prompt minors to report any content or experiences that raise concerns while also offering advice on appropriate strategies for managing such circumstances. Guardians can actively contribute to the reduction of exposure to dangerous content by creating a conducive environment that encourages youngsters to openly discuss their online experiences.

11.4 Combatting Cyberbullying and Hate

Guardians must be on the lookout for cyberbullying and hate in their children's Internet encounters. To begin, they should teach their children the value of respectful and empathic online conversation. Encourage children to pause before typing, emphasizing the negative impact that cruel comments can have on others.

Guardians should also regularly monitor their children's online interactions and look for indicators of cyberbullying. If cyberbullying is discovered, it should be addressed as soon as possible and constructively. During these instances, it is important to teach youngsters how to ban or report abusive users and offer them emotional support. Parents should impart to their children the concepts of digital empathy and resilience. Guardians should explain that not all online encounters will be positive but that they may learn to cope with negativity by concentrating on their worth and seeking help when necessary.

Finally, guardians must work with schools and other parents to form a united front against cyberbullying. To effectively tackle this issue, it is necessary to enhance community awareness and facilitate the exchange of resources. By encouraging collaboration, we can ensure that children are provided with a safe and nurturing environment, both on the digital platforms and in their physical surroundings.

11.5 Monitoring Screen Time

Guardians should also remain attentive to indications of excessive or addictive digital conduct. Indicators of potential issues might manifest in several ways, such as the neglect of responsibilities, a deterioration in academic performance, or a retreat from face-to-face social engagements. If these indicators are detected, it is necessary to monitor the screen time. Guardians can promote alternative paths that facilitate the enhancement of physical, mental, and social welfare. Minors should be encouraged to engage in their hobbies and pursue outdoor activities, reading, and interpersonal connections as a means to formulate a well-rounded lifestyle that effectively mitigates the potential negative consequences associated with excessive device usage. Also, it is essential to set a precedent by personal conduct. Guardians must exemplify healthy screen time practices, showcasing that digital devices serve as aids for enrichment rather than incessant sources of diversion. By taking up an equitable stance towards technology, carers can facilitate the development of a wholesome association between minors and electronic devices. Table 4.1 below showcases risks and advises related to screen time and it's monitoring.

11.6 Encouraging Social Media Literacy

Promoting social media literacy is important in nurturing digital resilience among minors. Guardians should commence the process of briefing minors on the fundamental principles of digital footprints and Internet privacy. It is essential to educate them about the enduring challenges associated with the disclosure of personal information online, emphasizing the significance of safeguarding such information.

It is very necessary for guardians to actively assist youngsters in developing their critical evaluation skills when engaging with content on social media platforms. It is also crucial to promote a critical mindset among individuals, urging them to scrutinize the authenticity of sources and discern potential biases. Minors should be educated on the essential skills of fact-checking information and developing a discerning approach towards consuming online content.

Also, it is important to actively participate in dialogues about the challenges of social media on individuals' mental well-being. Guardians must educate their children about the carefully constructed nature of numerous social media profiles and the consequent propensity for comparison and adverse impact on self-esteem. Guardians should promote the facilitation of candid discussions about the affective and stress-inducing aspects inherent in online relationships.

Table 4.1 Risks, guidelines, and effects of screen time

Concerns	Description
Children may be exposed to	• Exposure to violence and risk-taking behaviours
	• Encounter with videos depicting stunts or challenges inspiring unsafe behaviour
	• Presence of sexual content
	• Propagation of negative stereotypes
	• Instances of substance use
	• Exposure to cyberbullies and predators
	• Advertising aimed at children
	• Presence of misleading or inaccurate information
	• Promotion of unhealthy body image
	• Potential desensitization to real-world issues
	• Introduction to harmful ideologies
	• Encouragement of aggressive or antisocial behaviour
	• Possible impact on academic performance
	• Excessive exposure to consumerism and materialism
	• Increased susceptibility to online scams and fraud
Too much screen time may lead to	• Development of sleep problems
	• Attainment of lower grades in school
	• Reduction in reading habits
	• Decrease in time spent with family and friends
	• Insufficient outdoor or physical activity
	• Onset of weight problems
	• Emergence of mood problems
	• Formation of poor self-image and body image issues
	• Development of fear of missing out
	• Reduction in the time allocated for learning alternative relaxation methods
	• Impact on attention span and concentration
	• Potential for addiction or dependency
	• Impaired social skills
	• Disruption of circadian rhythm
	• Decreased participation in extracurricular activities
	• Potential for the development of sedentary lifestyle habits
	• Increased susceptibility to online bullying
	• Potential exacerbation of existing health conditions

(continued)

Table 4.1 (continued)

Concerns	Description
Managing a child's screen time includes	• Restriction of screen use to video chatting with an adult for those below 18 months
	• Limiting screen time to watching educational programming with a caregiver for those between 18 and 24 months
	• Imposing a limit of non-educational screen time for children aged 2–5
	• Encouraging healthy habits and limiting screen activities for those aged 6 and older
	• Mandating the turning off of all screens during family meals and outings
	• Utilization of parental controls to regulate screen content
	• Avoidance of screens as pacifiers, babysitters, or tools to stop tantrums
	• Turning off screens and removing them from bedrooms before bedtime
	• Encouraging outdoor play and physical activities
	• Fostering a balance between screen time and other activities
	• Promoting open communication with the child about their online experiences
	• Encouraging collaborative and educational use of screens
	• Setting specific time limits for different types of screen activities
	• Encouraging the development of digital literacy skills
	• Regularly monitoring and evaluating the content accessed online
Additional tips to consider	• Acquisition of familiarity with programming to ensure age appropriateness
	• Initiation of conversations with the child about the viewed content, emphasizing good behaviour
	• Awareness of advertising and its influence on choices
	• Encouragement of engagement in other activities such as sports, music, art, and hobbies
	• Establishment of a good example with one's own safe and healthy screen habits
	• Impartation of knowledge to children about online privacy and safety
	• Active decision-making regarding the child's readiness for a personal device
	• Promotion of screen usage for creativity and connection with family and friends
	• Consideration of the child or teen's maturity and habits in determining the most suitable plan
	• Encouraging a diversified range of leisure activities
	• Monitoring and discussing online social interactions
	• Encouraging critical thinking about online content
	• Promoting positive online etiquette and behaviour

12 Concluding Remarks

One potential strategy for improving the framework involves including a higher level of personalization. This objective might be accomplished by duly considering the specific and diverse requirements and situations of each learner. It is plausible that students with impairments or from low-income families may necessitate customized and individualized forms of assistance that diverge from those provided to their counterparts. By addressing these particular requirements, the framework can be enhanced in its efficacy, finally resulting in a more favourable outcome for all stakeholders.

Similarly, the digital resilience framework could derive advantages from the implementation of a thorough and all-including training programme designed specifically for educators. This would enable instructors to proficiently impart the curriculum and provide their pupils with the requisite information and competencies to excel in the digital society. By providing teachers with increased authority, the framework can be enhanced and facilitate students' comprehension and proficiency in safely using the digital platforms. However, a crucial approach to enhancing the framework involves its integration with other essential components of the curriculum, such as digital citizenship or media literacy. By implementing this approach, students can receive a comprehensive and comprehensive digital literacy education, enabling them to acquire essential knowledge and skills necessary for making informed and responsible decisions in the online domain.

The regular evaluation and timely revisions of the digital resilience framework are required in maintaining its ongoing pertinence and effectiveness in addressing novel and evolving digital vulnerabilities. It is plausible that due to the ever changing nature of the digital threats, it may be necessary to make regular adjustments to the framework to stay updated on emerging difficulties and ensure the protection of minors' online safety. Establishing legal enforceability for the framework can result in schools being held responsible for their compliance with the specified principles and practices. This would guarantee that these measures are executed with the required level of thoroughness and compliance.

Likewise, the implementation of a more inclusive and cooperative strategy for engaging stakeholders, such as parents, students, and industry professionals, has the potential to result in a digital resilience framework that is more adaptable and customized. By integrating the distinct viewpoints and knowledge of various pertinent stakeholders, the framework can be tailored to accommodate the requirements and considerations of a wider range of individuals. By implementing these innovations, the UK government may enhance the digital resilience framework for minors. This will enable minors to acquire the necessary information and abilities to effectively use the online world securely and responsibly. In contrast, the Government of India has not yet implemented a specific framework for enhancing digital resilience among minors. However, there are various initiatives and programmes in place that aim to promote digital literacy and ensure online safety among the general population. These efforts reflect an increasing acknowledgement of the significance of digital resilience in modern society.

To enhance the effectiveness of several programmes targeting the promotion of digital literacy and online safety among minors in India, a set of essential enhancements could be introduced. First and foremost, these programmes must give precedence to the promotion of responsible digital citizenship. This involves implementing strategies to address the issues of online bullying, hate speech, and the dissemination of false information. This would produce a moral obligation among children and encourage their engagement as proactive contributors in the establishment of a safer and more egalitarian virtual society.

Also, efforts must prioritize the promotion of a heightened level of parental and guardian involvement in the development of digital literacy and the enhancement of online safety for their children. One potential approach to address this issue is through the implementation of seminars or training programmes aimed at providing parents with the necessary knowledge and skills to successfully monitor and support their children's digital activities. By providing parents with such empowerment, children can benefit from receiving thorough and persistent guidance on effectively and responsibly using the digital domain. Also, it is essential to customize efforts to address the distinct requirements and situations of various groups of children, such as those originating from economically disadvantaged households or those with disabilities. This customized strategy would guarantee that every child, regardless of their socio-economic background, is provided with the necessary assistance and tools to excel.

The UK Government's implementation of a digital resilience framework for children is praiseworthy, as it demonstrates their dedication to equipping young individuals with the necessary knowledge and skills to traverse the constantly changing online environment safely and securely. Nevertheless, there are other methods by which the effectiveness of the framework can be enhanced.

It is essential to establish a system of ongoing surveillance and assessment for these activities to ascertain their effectiveness in achieving their intended goals and to assure their continued relevance in light of the ever-changing list of digital risks. Enhanced cooperation with industry stakeholders, such as social media platforms and telecommunications firms, has the potential to expedite the adoption of optimal strategies for ensuring online safety and security. Finally, the government has the potential to initiate public awareness initiatives to enhance knowledge among parents, children, and other pertinent stakeholders regarding the significance of digital literacy and ensure safety while engaging online. These programmes can have a significant impact on promoting a culture centred around digital responsibility and security.

Acknowledgments Gattupalli Sri Viswa Prharsha's numerous contributions aided us in the drafting of this chapter, and we are grateful. The chapter was drafted from his notes and drafts about the issues minors face and the perspectives he offered regarding the challenges faced by children and adolescents in the age of technology. By using these insights, we were able to give a sincere treatment to the subject matter.

Chapter 5
A Study in Attack and Breaches

1 Introduction

In any business environment, information is an extremely valuable asset whose worth has increased significantly in the age of digitalization as organizations expand their international reach. Therefore, it is of foremost importance for enterprises and organizations to prioritize the protection of proprietary information. This includes several aspects such as knowledge archives, infrastructure, client communications, financial transactions, and databases.

Neglecting to follow this practice might perhaps lead to instances of cyberattacks, resulting in the depletion of clientele and the deterioration of their confidence. The level of confidence that individuals and organizations have in their business colleagues to safeguard their sensitive personal data is of central importance. Any deficiency in this regard might lead to not only legal but also financial consequences. To prevent such situations, organizations use experts who are skilled in either preventing or reducing the impact of cyberattacks. The fundamental aspects of security include ensuring the confidentiality, availability, and integrity of systems and data.

The word *security breach* refers to the occurrence of a violation of security and privacy standards that are specifically established to protect the confidentiality, sensitivity, and limited access to information of authorized individuals. The act involves the covert disclosure and distribution of files that are confined inside the breach without the explicit approval of the legitimate owners. These breaches pose a significant threat to several entities, including individuals, prominent organizations, and government agencies. Consequently, it is important to prioritize the protection of oneself and others against the harmful consequences they entail.

Cyberattacks arise as a result of inherent weaknesses in technological systems and user conduct, constituting a fundamental aspect of the issue. The interconnectedness of computers and mobile devices, coupled with the rapid adoption of new

S. K. Shandilya et al., *Digital Resilience: Navigating Disruption and Safeguarding Data Privacy*, EAI/Springer Innovations in Communication and Computing,
https://doi.org/10.1007/978-3-031-53290-0_5

technical breakthroughs, has led to an exponential increase in the attack surface area available for cybercriminals to exploit. The Internet of Things (IoT) and autonomous systems devices demonstrate society's inclination towards convenience, which can have negative security implications. Many of these 'smart home' products display significant vulnerabilities, such as the absence of encryption, making them attractive targets for malicious hackers.

2 Implications and Repercussions

The absence of adequate security testing during the integration of digital goods, services, and tools has resulted in a rise in cyberattacks. Despite the ideal configuration of technology, the security of a website or network might be compromised by the digital behaviour shown by certain users. The inevitability of a cyberattack is contingent upon the absence of adequate security measures implemented at both the individual and organizational levels. Hence, understanding the mechanisms behind security breaches is critical for safeguarding one's well-being as well as that of others. The underlying elements contributing to these breaches are not just traceable to external hackers but may also arise from deliberate attacks or other influencing variables. Therefore, it is crucial to give them careful consideration.

It is essential to acknowledge that security breaches are not just caused by external perpetrators but may also arise from unintentional errors made by personnel or shortcomings inside an organization's infrastructure. The methods by which security breaches transpire are diverse, although there are certain notable recurring incidents.

An example of a security breach might arise accidentally when an individual mistakenly accesses prohibited data while using someone else's computer without proper authority. Although unintentional, this sort of access constitutes a security breach since it involves the unauthorized viewing of information by an individual. In contrast, a malicious actor who has insider access may deliberately use this privilege to illicitly obtain and/or disseminate sensitive data with the explicit objective of inflicting damage on a specific individual or entity.

In the above scenario, the individual with insider status may possess valid authorization to get the information, but they intend to use the data for harmful purposes. The potential occurrence of a security breach resulting from the loss or theft of an unencrypted and unlocked device, which contains confidential information, might significantly expose an organization's vulnerabilities. Ultimately, the security of an individual or network may be compromised by malicious external actors or hackers via several attack vectors, resulting in the unauthorized collection of sensitive information.

The potential consequences of security breaches may arise from inadvertent errors, although the significance of unauthorized access and the theft of personally identifiable information (PII) or business intellectual property for financial gain or malicious purposes should not be ignored. Malicious actors exhibit a method of operation that is characterized by a robust approach, sometimes including a thorough investigation of their targets to identify specific weaknesses.

These vulnerabilities may include employee vulnerability to phishing tactics or the failure to implement necessary upgrades. After acquiring this information, hackers possess the ability to devise strategies aimed at deceiving individuals inside an organization, persuading them to download malicious software, or directly launching attacks on the network. Once unauthorized access is obtained, the malicious actor may extensively explore for important information, a process that can remain undetected for a considerable duration since the typical breach takes more than 5 months to be recognized.

One of the weaknesses often targeted by malicious actors is the use of weak credentials, which is responsible for the majority of security breaches resulting from compromised or feeble login combinations. The reuse of passwords by individuals may render them vulnerable to brute force attacks, which can be used as a means to obtain unauthorized access to a range of sources containing personally identifiable information (PII) or financial data, including but not limited to emails, websites, and bank accounts. Phishing attacks provide a significant security threat due to the potential for malicious individuals to illicitly use stolen personal information to gain unauthorized access to the Internet and banking accounts. It should be noted that malware attacks can bypass conventional security protocols that are generally used to safeguard computer systems. It is worth mentioning that skimming devices, when affixed to petrol pumps or ATMs, can illicitly get sensitive information with each instance of card swiping.

Despite the implementation of protective measures for the system and data, malicious individuals may still abuse third-party providers as a means to infiltrate the network. The adoption of the bring-your-own-device (BYOD) policy introduces another possible avenue for exploitation. Devices that lack adequate protection are susceptible to being compromised by malicious software, enabling unauthorized individuals to get access to the data contained inside the device. This includes sensitive information such as work-related emails, files, and personally identifiable information (PII) about the device owner.

The unavoidable truth is that superficial password changes won't be enough to stop the negative effects of a security breach. The adverse effects of a security breach may result in long-lasting harm to an individual's reputation, financial situation, and several other areas. Security breaches may have severe consequences for commercial enterprises, leading to significant damage to their brand and financial stability. It is worth mentioning that organizations such as Target, LinkedIn, Zomato, Equifax, and Yahoo have had security breaches that have had a detrimental impact on their brand reputation, resulting in their association with the incidents rather than their core business activities.

Governmental organizations face a significant risk of security breaches since the unauthorized disclosure of secret material to foreign groups may pose a threat to national security, infrastructure, military operations, and political engagements. The issue of identity theft is a major concern for individuals, as it presents a formidable obstacle due to the potential exposure of various sensitive information, such as social security numbers and financial information, resulting from security breaches. Once malicious individuals have access to such information, they can engage in several fraudulent acts using the victim's identity, resulting in a series of detrimental outcomes, including damage to their credit, legal complications, and significant obstacles to recovery. While the prevalence of these occurrences is widespread, it is important to note that the detrimental consequences of security breaches extend beyond these specific situations.

This chapter will examine case studies that illustrate the detrimental impacts of cyberattacks while also elaborating on potential remedies to address these complicated problems.

3 Ukraine Powergrid Attack, 2015

The Ukraine Powergrid Attack represents a significant occurrence in which a nation's electrical grid was deliberately targeted and subjected to disruption. In December 2015, a significant cyberattack took place in Ukraine, leading to a pervasive disruption of the electrical power supply. This particular occurrence holds great significance as it effectively demonstrated the susceptibility of essential infrastructure to cyberattacks. The perpetrators used malicious software, commonly known as malware, to illicitly infiltrate the control systems of the power grid, resulting in the disruption of electricity provision to several residential properties. The occurrence has evoked concerns regarding the possible consequences of cyberattacks on critical services and has emphasized the importance of heightened cybersecurity protocols in critical infrastructure systems on a global scale.

3.1 Synopsis

On December 23, 2015, an unidentified entity with advanced knowledge in cyber warfare successfully used its capabilities to impair the functioning of the electrical infrastructure, resulting in a widespread blackout that impacted over 225,000 subscribers in Ukraine [Nas16]. This occurrence, which was unexpected and unforeseen, had significant implications not only for the electrical sector but also raised concerns about the potential use of similar tactics against operators in several sectors.

The cyberattackers, who were recognized as a collective of individuals possessing sophisticated hacking abilities, used a malicious malware known as Black-Energy 3, which exhibited complex characteristics. The virus was used by the attackers to illicitly access the information systems of three energy distribution businesses located in Ukraine. This action led to a notable and temporary disruption in the provision of electricity to end-users. Prykarpattyaoblenergo, a utility company responsible for providing electricity to the Ivano-Frankivsk Oblast, experienced the most significant impact. Approximately 230,000 residents in this region endured power outages lasting between 1 and 6 hours. A total of 30 substations were deactivated, consisting of 7 substations operating at 110 kilovolts and 23 substations operating at 35 kilovolts [Nas16].

Likewise, the destructive cyberattack, including the BlackEnergy 3 virus, had a significant impact on three energy distribution organizations in Ukraine. The companies Chernivtsioblenergo and Kyivoblenergo were also targeted by the attack, albeit with a lesser degree of severity. As per a representative from one of the corporations, the attack was initiated by computer systems possessing IP addresses associated with the Russian Federation. In the year 2019, scholars contended that Ukraine's circumstances were exceptional, characterized by a notably deteriorated infrastructure, a significant prevalence of corruption, the ongoing Russo-Ukrainian War, and a distinct susceptibility to Russian infiltration owing to the long-standing historical ties between the two countries.

The electricity infrastructure in Ukraine was initially established during its affiliation with the former Soviet Union and has subsequently undergone modernization efforts involving the integration of Russian components. However, as of the present year 2022, the power grid remains in an unrepaired state, rendering it susceptible to potential cyberattacks in the future. The Russian attackers possess a comprehensive understanding of the software used within the grid infrastructure, comparable to that of the operators themselves. The incident occurred during a period of increased holiday activity, resulting in a reduced presence of Ukrainian operators on duty. These circumstances facilitated the hackers' execution of their plan.

The cyberattack that took place was a diverse and sophisticated scheme, which included the following sequential stages:

- The initial phase involved a targeted attack on corporate networks, which were compromised through the employment of spear-phishing emails that contained BlackEnergy malware.
- Subsequently, the SCADA (Supervisory Control and Data Acquisition) systems were seized and brought under remote control, enabling malicious actors to switch off various substations, thereby disrupting the electricity supply.
- The perpetrators then proceeded to disable or even destroy various IT infrastructure components, such as uninterruptible power supplies, modems, Remote Terminal Units (RTUs), and commutators.
- To inflict further damage, the KillDisk malware was utilized to destroy critical files that were stored on servers and workstations.
- The scheme's perpetrators also carried out a denial-of-service attack on the call centre to prevent consumers from receiving any updated information regarding the blackout.
- To magnify the consequences of the attack, the perpetrators deliberately turned off the backup power supply at the operations centre of the utility company, which compounded the impact of the pre-existing power outage.

These actions caused an outage of up to 73 MWh (megawatt hours) of electricity, constituting a minuscule fraction of 0.015% of the daily electricity demand in Ukraine [Nas16].

Cybersecurity has transcended the boundaries of being merely the responsibility of IT administrators and network engineers and has now become an indispensable aspect of the safe, efficient, and competitive functioning of businesses. The current age has witnessed a spate of cyberattacks, which are carefully planned and executed to capitalize on the weaknesses of individuals, systems, and procedures, resulting in severe and long-term effects. These attacks are carried out with professionalism, wherein an organization's network infrastructure is exploited in a highly targeted manner.

The times following the annexation of Crimea by Russia in 2014 were characterized by a series of cyberattacks that were targeted at disabling critical infrastructure in Ukraine by groups linked to Russia. One such incident occurred in November 2015, when hackers allegedly associated with Russia successfully infiltrated the Ukrainian power grid, leading to a power outage that lasted for 6 hours and affected approximately 230,000 customers in Western Ukraine. This triumph was followed by the infamous NotPetya ransomware attack of 2017, which was discussed previously, and whose worldwide cost was estimated to be nearly $10 billion [Nas16]. These attacks were branded by the White House as the 'most costly and devastating cyberattacks in history'.

The NotPetya attack was significant as it was the first time that officials from Western countries openly attributed a cyberattack to groups associated with the Russian government. In 2018, the UK's National Cyber Security Centre (NCSC) stated this effect. In the period leading up to Russia's military invasion of Ukraine in 2022, factions with links to Russia continued to launch cyberattacks, albeit on a much smaller scale than the major attacks observed in 2015 and 2017.

In January 2022, Ukrainian government officials disclosed that approximately 288,000 cyberattacks were registered within the first 10 months of 2021, in addition to the 397,000 attacks that were documented in 2020 [Nas16]. Several experts predicted that cyberattacks would play a major role in the conflict that followed the full-scale invasion of Ukraine in February 2022. Some analysts claim that Russia has refrained from launching large-scale cyberattacks against Ukraine due to the use of the same infrastructure by the Russian military.

3.2 *Ramifications*

The cyberattack that took place in 2015 on the Ukraine Power Grid holds significant importance as it marked a pivotal moment in the progression of cyber warfare on a global scale. The electricity grid of Ukraine, which is an essential component of the country's national infrastructure, was subjected to a severe cyberattack resulting in widespread power outages affecting a significant portion of the population. The cyberattack in question not only resulted in immediate repercussions but also had lasting consequences on the worldwide cybersecurity and cyber warfare domain.

This 2015 cyberattack brought to light the significant vulnerability of critical infrastructure to cyberattacks. Prykarpattyaoblenergo and Kyivenergo, two critical power distribution businesses, were subjected to cyberattacks coordinated by state-sponsored hackers. These deliberate actions caused severe disruptions, leading to widespread power outages and ensuing disorder within affected regions. Loosely speaking, the disclosure highlighted the previously underestimated susceptibility, hence illuminating the capacity for cyberattacks to incapacitate critical services.

The consequences of this attack had a broader impact that went beyond the first occurrence of power outages. Industries that crucially depend on continuous access to energy had major financial setbacks, while the considerable expenses incurred in detecting and managing the intrusion highlighted the potential economic upheaval associated with similar cyber disasters. Also, given that the attack transpired amidst a severe winter season, the welfare of the populace emerged as a matter of urgency. The repercussions of power outages were experienced by residential dwellings, commercial establishments, and critical service providers, resulting in adverse effects such as pain and potential health hazards for those impacted.

Also, this attack carried significant ramifications for national security, as Ukraine ascribed it to hacker collectives supported by the Russian government. This leads to an increased level of national security worries and introduces further difficulties to an already stressed state of international relations, hence intensifying geopolitical tensions. This occurrence garnered international scrutiny, highlighting the necessity to strengthen cybersecurity protocols for critical infrastructure on a global scale and instigating dialogues on the pressing need for heightened safeguarding.

After this attack and the growing concerns around it, some nations have implemented or revised rules that enforce distinct cybersecurity requirements for operators of critical infrastructure. The objective is to set explicit directives for the protection of critical services. Governments and organizations have taken measures to allocate resources towards the improvement of cybersecurity, with a particular focus on sectors responsible for managing critical infrastructure. The investments made included enhancements in technology, training of staff, and the formulation of comprehensive strategies to ensure cyber resilience. As a result, this occurrence functioned as a significant catalyst for awareness among both the public and private domains, highlighting the palpable ramifications of cyberattacks on critical infrastructures.

On top of that, the act of ascribing the attack to actors supported by a nation-state has generated concerns regarding the possibility of heightened cyber conflicts among countries, prompting deliberations on establishing standards and protocols for conducting operations in the digital domain. The occurrence also stimulated a degree of cooperation among governments, organizations, and cybersecurity specialists, promoting the dissemination of important information on cyber risks and weaknesses and ultimately enhancing cybersecurity protocols on an international level. Efforts to ensure the security of critical infrastructure and cyberspace persistently draw upon these timeless lessons, adapting to the evolving digital context.

4 WannaCry Attack, 2017

The WannaCry incident, which unfolded in May 2017, is a significant event within cybersecurity. The incident used a type of harmful software known as ransomware, which utilizes encryption methods to limit a victim's ability to access their computer data. Subsequently, the perpetrator demands a financial payment in exchange for supplying a decryption key [US-17].

The incident known as WannaCry involved the extensive dissemination of ransomware, which led to the widespread infection of several computer systems on a global scale. Consequently, this resulted in the disruption of operations for various entities, including hospitals, government institutions, and corporations. The malware effectively exploited a security vulnerability present in Microsoft Windows operating systems, commonly known as EternalBlue. This vulnerability had been earlier recognized by the United States National Security Agency (NSA) and subsequently made public by a group of hackers known as the Shadow Brokers [Log17].

The user's files were subjected to encryption following the penetration of their computers by the WannaCry software, resulting in their inaccessibility and hindering any attempts to retrieve them. Following this, the ransomware presented a message containing a request for compensation in the form of the virtual currency Bitcoin, to facilitate the retrieval of the encrypted data. The comprehensive attack led to significant disruption and financial damages [Her17].

The instance garnered international attention, highlighting the importance of cybersecurity protocols and the importance of timely software updates to protect against such flaws. This occurrence has triggered investigations into the ethical ramifications of offering financial remuneration to those engaged in cybercriminal activities, as well as the obligations of governmental bodies in the development and stockpiling of cyber weapons.

4.1 Synopsis

It was in May in the year 2017 that the WannaCry ransomware, malicious software that is a form of crypto-ransomware, was widely disseminated across computers worldwide that were running the Windows operating system [MC19]. It is worth mentioning that almost 2 months before this catastrophic event, Microsoft had indeed issued a security patch for the EternalBlue exploit that was used by the malicious actors behind the WannaCry ransomware to propagate their malicious agenda. Nevertheless, because numerous Windows users around the world were either reluctant to update their software or operating outdated versions of Windows, the global-scale attack resulted in their susceptibility to this ignominious act of extortion. The extortionists encrypted Windows computers worldwide and subsequently demanded an initial ransom of $300 worth of Bitcoin, which was later augmented to $600 worth. In a matter of just a few hours, the ransomware was able to infect an estimated 230,000 computers across 150 different countries [Rep17].

Although the WannaCry ransomware initially spread rapidly throughout the world, a security researcher by the name of Marcus Hutchins eventually discovered a kill switch that substantially hampered the propagation of the attack. To clarify, crypto-ransomware, the kind of malware used in the WannaCry attack, is a type of ransomware that utilizes encryption techniques to make valuable data inaccessible, or it may block the user's access to their computer. It is important to note that the type of ransomware that blocks the user's access to their computer is known as locker ransomware. The attack known as WannaCry aimed to extort the victims by encrypting their data and threatening to release it only after the payment of a ransom. This type of attack specifically targeted computers that utilized the Microsoft Windows operating system, demanding a ransom in Bitcoin cryptocurrency. The worldwide outbreak of WannaCry, which took place in May 2017, is a ransomware attack that affected a large number of individuals, reflecting its scale and severity [Tea17].

The potent ransomware, known as WannaCry, exhibited a remarkable ability to spread across computer systems that operated on Microsoft Windows. This malicious software could encrypt valuable files, with the demand of a Bitcoin ransom as a prerequisite for their release, which caused a global concern among users. If individuals and organizations had remained vigilant about keeping their software up-to-date and had not persisted in using outdated computer systems, the extensive harm caused by this attack may have been averted. The modus operandi of the WannaCry attack involved unscrupulous cyber perpetrators taking advantage of a vulnerability in the Microsoft Windows operating system using a malicious software tactic that was reportedly developed by the United States National Security Agency under the name of EternalBlue.

The Shadow Brokers, a notorious group of hackers, published this hack before the onset of the WannaCry attack. Almost 2 months before the attack, Microsoft issued a security patch, which acted as a safeguard to protect the computer systems of its users against this exploit. Many users and organizations were remiss in ensuring that they updated their operating systems, thus rendering them susceptible to attack. Those who had neglected to apply the Microsoft Windows update before the ransomware attack were deprived of the opportunity to secure themselves with the security patch, leaving them exposed to the malicious WannaCry ransomware attack due to the vulnerability that was targeted by EternalBlue. In the aftermath of the attack, many individuals assumed that it had originated from a phishing campaign, whereby users were sent spam emails that contained infected links or attachments that tricked them into downloading malware.

EternalBlue served as the critical gateway for WannaCry to proliferate and extend its reach, with DoublePulsar operating as the 'backdoor' implanted in the compromised computers (which in turn facilitated the execution of WannaCry). The attackers demanded an initial ransom of $300 in Bitcoin, later raising it to $600 [MC19, Tea17]. If the affected parties failed to settle the demanded amount within

3 days, the malicious agents behind the WannaCry ransomware warned that they would irrevocably delete all important files. To accept the demands of the malicious actors and submit the ransom fee is unwise, for there exists no guarantee that the data will be reinstated to its original state, and each ransom payment serves to support and perpetuate the criminal enterprise, thereby increasing the likelihood of subsequent attacks.

In the case of the WannaCry attack, it proved wise not to accede to the demands, as reportedly, the code used in the attack was flawed, which meant that the hackers were unable to trace the source of the payment to a specific victim's computer. As such, there is a degree of uncertainty as to whether anyone managed to recover their data. However, according to F-Secure, a company that specializes in security solutions, some individuals may have succeeded in retrieving their information [Pro17]. This serves as a stark reminder of why giving in to the ransom demand is never a viable option when confronted with a ransomware attack.

The impact of the WannaCry attack was significant as it targeted a staggering 230,000 computers globally, and Telefónica, a Spanish mobile company, was among its first victims. It quickly spread to thousands of hospitals and surgeries across the United Kingdom, causing a third of the National Health Service (NHS) hospital trusts to be paralyzed, and ambulances were reportedly rerouted, leaving critically ill patients in a dangerous state. The NHS lost a surprising £92 million as 19,000 appointments were cancelled due to the attack. The WannaCry ransomware attack had a worldwide financial impact, affecting computer systems in 150 countries and causing an estimated loss of $4 billion across the globe [MC19, Her17].

4.2 *Ramifications*

In light of the prevailing circumstances, it is necessary to deliberate upon strategies to safeguard oneself against such attacks. A fundamental measure in safeguarding against ransomware is the consistent updating of software and operating systems. The individuals affected by the WannaCry cyberattack were rendered vulnerable to ransomware due to their failure to maintain regular updates on their Microsoft Windows operating systems. If the individuals had updated their operating systems, they would have acquired the security patch that was previously issued by Microsoft. This patch would have resolved the vulnerability that was utilized by EternalBlue to spread the WannaCry ransomware.

The impact of WannaCry has been significantly reduced primarily as a result of the commendable initiatives of Marcus Hutchins, a renowned British computer security expert. Hutchins skillfully developed a kill switch using sophisticated methods such as reverse engineering and honeypots, effectively impeding the further propagation of WannaCry. A cohort of knowledgeable French scholars has achieved a commendable advancement by explaining a systematic approach to decrypting a portion of the compromised computer systems, without succumbing to the dishonourable practice of acquiescing to the expensive ransom asked by the malicious actors. However, it is important to acknowledge that WannaCry has not been completely eradicated and still presents a possible risk.

5 NotPetya Attack, 2017

The NotPetya incident [Her17, Fay18, Gre18] involved the identification of a malicious computer software program that resulted in extensive disruption and harm to global computer networks. The software, often known as 'NotPetya', displayed attributes consistent with a form of malicious software commonly referred to as ransomware.

The infiltration of computer networks by the NotPetya malware was facilitated through the exploitation of vulnerabilities present in the Windows operating system. Upon gaining access to a computer system, the malicious actor proceeded to encrypt the files stored within, rendering them inaccessible to the authorized user. The individuals responsible for the attack subsequently requested a monetary payment in the form of Bitcoin as compensation for a decryption key that would reinstate the ability to access the encrypted data.

The case in question garnered considerable attention as a result of its widespread ramifications. The impact of NotPetya was observed across a range of entities, including government bodies, global enterprises, and healthcare facilities. The quick dissemination of the malware resulted in major disruptions to critical infrastructure and substantial financial repercussions.

The examination of the NotPetya incident unveiled the complicated characteristics of cyberattacks in the current digital society. The precise motivations driving the attack, the exact identities of the offenders, and their final intentions remained ambiguous, highlighting the difficulties associated with attributing cybercrimes to particular individuals or collectives.

The NotPetya example highlights the significance of implementing cybersecurity measures, such as software patching and personnel training, as a means of safeguarding against ransomware attacks and other forms of cyber threats. This highlights the necessity of global collaboration and the exchange of knowledge to effectively address these challenges.

5.1 Synopsis

Commenced by disseminating destructive ransomware designed to execute the attack, the malicious attack, in question, used a software program referred to as M.E.Doc, which was obligatory for tax filing purposes by the government of Ukraine and, thus, extensively used by most enterprises operating within Ukraine. The perpetrators distributed the ransomware, disguising it as an update to the commonly used M.E.Doc application, thereby infecting the targeted computer systems. In the initial stages, the ransomware remained dormant on the victims' machines, awaiting orders from the attackers to execute the operation.

The ransomware, however, could infiltrate and infect additional systems within networks, thus exponentially magnifying the scope of the attack. The impact of the incursion was not limited to Ukrainian businesses but also extended to a significant number of computers belonging to multinational corporations with a presence in Ukraine. Remarkably, nearly one-fifth of the compromised systems did not belong to Ukraine, and the country that suffered the most significant damage, following Ukraine, was Germany, with approximately 9% of the total compromised systems located in the country [Her17]. It is worth mentioning that the malicious cyberattack inflicted significant losses on the prominent worldwide logistics company, Maersk.

After infecting a sufficient number of computer systems, the attackers ordered the ransomware program to start. Following this, the ransomware encrypted the storage devices of the infected computers and encrypted the computers' restart processes. To regain access to the systems, a ransom payment of about 300 in Bitcoin was demanded. In contrast to prior ransomware assaults, however, resuming system functionality did not become a guarantee upon remitting the ransom, given that the ransomware in question was not specifically engineered to decrypt the encrypted data drives.

Consequently, the victims suffered irrevocable harm in the form of inoperable computer systems and irretrievable data. Due to the singularly unconventional nature of the ransomware, it is estimated that total ransoms paid amount to approximately US$10,000, whereas the total cost of the damage incurred, in terms of lost data, revenue, and other factors, is estimated to be approximately US$10 billion [Fay18]. The malicious attackers strategically timed the execution of the ransomware attack to coincide with Ukraine's Independence Day, which resulted in a significant intensification of the damage caused. The attackers meticulously planned their malicious actions, initiating the attack on a day when most offices were unoccupied, lacking personnel who could potentially hinder the ransomware's surreptitious activities.

The malicious NotPetya attack represents a fusion of older techniques previously used in cyberattacks. The attack, referred to as 'NotPetya', was described as an advanced version of previous methods, incorporating various lateral movement techniques and sharing some code similarities with the Petya ransomware attack of 2016. NotPetya bears an affinity with the open-source application, Mimikatz, created in 2011 to demonstrate that Microsoft was susceptible to cyberattacks. Mimikatz can purloin credentials and heighten privileges.

Besides, NotPetya shares coding with the infamous WannaCrypt ransomware attack that unleashed a global scourge in May 2017. The genesis of the attack is believed to have emerged from M.E.Doc, a widely utilized software in Ukraine due to a Ukrainian government mandate to file tax returns through this software. The attackers initially gained entry to the M.E.Doc update servers, extracted data from them, and subsequently produced a bogus update patch that was disseminated across all M.E.Doc-enabled devices. The attackers then reached a consensus on June 27, 2017, to carry out the assault by disseminating an update to all M.E.Doc users via the M.E.Doc update servers.

At the outset, the perpetrators gained unauthorized access to the M.E.Doc update servers, extracted data from them, and then created a bogus patch update that was distributed to all devices running M.E.Doc software. The attackers initiated the attack on June 27, 2017, by distributing an update to all M.E.Doc end-users via the M.E.Doc update servers. Although the update was marketed to end-users as a regular software update, it failed to include an upgrade for the M.E.Doc software. Instead, it installed the malware software developed by the attackers. Malware infiltrated users' systems subsequent to the implementation of the fraudulent update. By operating inconspicuously in the background, the malware managed to evade user detection while intercepting passwords and acquiring administrative privileges through the use of a credential-dumping application like Mimikatz.

Upon discovering the necessary credentials, the malicious malware proceeded to encrypt every file present across all drivers of the targeted computer system. Upon completion of the encryption process, the malware set a timer for 10 minutes, as a prelude to the seizure of complete control of the system. As the countdown began, a fraudulent message appeared on the computer screen, purporting to notify the user of an error in the C-drive and imploring them not to power off the device.

The computer restarted itself after the timer expired, during which the message remained active. Following the reboot, a ransom message appeared on the screen, declaring that the entirety of the user's files had been encrypted and that a payment of US$300 in Bitcoin was required for their release [Gre18]. Distinguishing itself from typical ransomware attacks, this specific malware was never designed with decryption capabilities. Instead, if granted administrative privileges, it would overwrite all encrypted files, rendering them irrecoverable.

In the absence of elevated privileges necessary for administrative access and the inability to overwrite files, the individuals who paid the ransom were not granted access to the cryptographic keys. The main goal of the attack was to achieve extensive devastation, not to acquire monetary profit by holding users' files hostage. After establishing control over the computer, the malicious software deleted the records pertaining to System, Security, Setup, and Application events in an effort to impede any chances to bypass the intrusion.

The attack's sophistication and devastating impact were attributable to its lateral movement capabilities throughout the network, possessing worm-like capabilities and employing numerous techniques to infect as many computers as possible. As such, merely infecting one computer sufficed to corrupt the entire network. The main goal of the malware during its initial infiltration attempt was to illicitly obtain network credentials or take advantage of ongoing active sessions inside the network. This methodology facilitated a network scan for all devices that were linked to the compromised computer under consideration.

Having completed its scan and identified additional computers, the malicious software used conventional file-sharing methods to dispatch the malicious file to the other systems. Following this, the malware aimed to execute the malicious file remotely on the other computers while the file was in transit. Should it achieve success, this process would be initiated on the other devices that are connected to the network, consequently accelerating its propagation.

The malicious software utilized various techniques to spread itself throughout the network, including taking advantage of legitimate functionalities already present and exploiting vulnerabilities in the server message block (SMB). These vulnerabilities had been used in previous attacks like EternalBlue and EternalRomance, which were widely known. EternalBlue had been used only a month prior in the WannaCrypt attack. In order to access user files and spread like a worm across a network, the NotPetya ransomware combined methods from previous attacks with well-known vulnerabilities. It was deemed a more formidable variant of ransomware on account of its advanced sophistication and detrimental characteristics.

The widespread opinion holds that the NotPetya attack was an element of a hybrid Russian offensive against Ukraine. Relationships between Russia and Ukraine have been tense since 2014 when Russia annexed Crimea. There is evidence to imply that the Russian Federation has provided support to fighters opposing the Ukrainian government in Crimea. Multiple indicators point to Ukraine as the probable main target of this cyberattack. It is worth noting that 80% of the infected computers were situated in Ukraine [Fay18, Gre18].

Also, the M.E.Doc software that was used in this attack was not a typical ransomware, but a tax filing software widely used in Ukraine, indicating that the attack was more geared towards causing widespread disruption and destruction in Ukraine, even though it inadvertently spread to other nations. According to certain experts, the attack targeted Ukrainian enterprises and government organizations, with the attackers underestimating the malware's capability to propagate. The attack was carried out on the eve of Ukraine's Constitution Day, a public holiday when many government and business offices were unoccupied.

The behaviour of the NotPetya ransomware was unusual, as it did not follow the typical modus operandi of ransomware attacks. Malware that displayed a ransom letter promising easy and safe file decryption instead of waiting for a user to interact with it to start the ransom payment procedure overwrote and deleted crucial files and drives on purpose. This behaviour strongly suggests that the malware was not engineered for financial gain, but rather to debilitate the Ukrainian state. The total amount of funds extorted from users through the ransom payments was a mere US$10,000, while the total estimated cost of the damage caused by the malware was over US$10 billion.

The M.E.Doc update system contained a backdoor that was discovered by security experts, which they believe was installed more than 2 months before the NotPetya attack that occurred in June 2017. This indicates a highly organized and skillfully executed attack. Besides, the size of the NotPetya malware, 1.5 gigabytes, suggests that other backdoors may still be undetected [Her17]. The evidence also suggests that the perpetrators of the NotPetya attack have executed similar attacks in the past, and there are speculations that the culprits may be the notorious cyber groups associated with the Russian Federation, including Telebots, BlackEnergy, or Sandworm.

The deduction from the tracks left behind by the NotPetya attack indicates that the same culprits, allegedly backed up by the Russian Federation, were responsible for the December 2016 attack, which had targeted the Ukrainian financial system, as well as for the Petya attack. The United States and the United Kingdom have openly accused Russia of being the architect of NotPetya, with the Russian Main Intelligence Directorate supposedly having masterminded the malware. However, the Russian Federation has refuted these allegations, contending that its systems were affected by the attack.

The effects of the NotPetya attack were far-reaching and extensive. A multitude of enterprises suffered, causing billions of dollars worth of financial damage. It is therefore appropriate to scrutinize the different parties' reactions to the incursion, specifically the Ukrainian government, Maersk, and Microsoft, to identify how they responded in the wake of the attack.

5.2 Ramifications

Before the NotPetya attack, the Petya malware attack had already left a trail of devastation by encrypting the file system and demanding payment in Bitcoin. Remarkably, very little had been done to preclude a similar attack from recurring. The M.E.Doc update servers had already been compromised in the past to propagate a different kind of attack. Yet, the NotPetya attack was allowed to occur on a grand scale despite these warning signs.

The M.E.Doc update servers were compromised by malicious actors on multiple occasions prior to the current intrusion, which Intellect Service has not remedied since 2013. This is in addition to a previous attack of a comparable nature. These incidents represent a substantial deficiency on the organization's behalf. Following the initial compromise of the backdoor exploit, safeguarding and conducting audits of the update servers ought to have been the utmost focus. March 2017 marked the initial availability of the Petya Microsoft upgrade.

Without a doubt, it is a matter of great remorse that the techniques utilized to disseminate the NotPetya malware were already familiar to the cybersecurity community and that the prior instance of a comparable attack, Petya, was still within recent recollection. Therefore, it seems peculiar that more strict measures have not been taken to deal with the incidence of such cyber threats. Our objective in this section is to discuss the course of action that ought to have been pursued prior to, during, and following the attack.

During the attack, the most obvious response was to disconnect computers from the network and to check whether the executable file '`rundll32.exe`' was running in the task manager, as its presence indicated that the computer was infected with the malware, which would proceed to encrypt all data upon restart. Therefore, it is essential to keep computers offline and suspend operations until more details on the attack are available and until steps to contain and combat it can be formulated. It is remarkable to note that power outages provided immunity from the attack to certain sites, as security experts were able to acquire access to already infected but dormant computers and investigate the virus.

As it refers to the salient issue at hand, it is important that companies take proactive measures to inform local authorities and collaborate with them to prevent the NotPetya malware from causing further deleterious consequences. The method of creating a read-only file called 'perfc' and placing it in the Windows directory was an effective way to halt the malicious software's destructive path, as it is the file that the malware initially searches for during execution and, if detected, self-terminates. In the aftermath of the attack, it is important to implement strategic actions that should have been enacted beforehand.

Particularly, measures such as reinforcing the security of the updated servers, updating company computers with the most up-to-date patches, and organizing training sessions to educate employees on cybersecurity best practices should be executed. Employees must be 'cyber aware' and able to identify suspicious emails and various digital threats. Also, businesses should safeguard their data by creating backups to enable partial recovery in the event of a data breach. Companies must use trustworthy security software suites that can methodically scan their file systems and pinpoint and eradicate any malicious files.

6 Marriott Breach, 2018

The Marriott Breach case examined a significant cybersecurity event that occurred in the latter half of 2018, which had a significant impact on the functioning of the Marriott International hotel chain. The security breach led to the unauthorized exposure of personal information about 500 million individuals who had previously stayed at various Marriott properties. The illegal access to the Starwood guest reservation database, a subsidiary of Marriott, has been identified as the cause of the breach. The occurrence of unauthorized access to this database by cybercriminals took place in 2014. However, the breach remained undetected until 2018 when Marriott acquired Starwood.

The compromised data consisted of sensitive information, such as individuals' names, passport numbers, email addresses, and credit card details. Following the identification of the security compromise, Marriott promptly initiated an inquiry and duly informed the relevant authorities. Proactive actions were implemented to mitigate the consequences of the breach, which includes the communication of the occurrence to the individuals impacted and the supply of credit monitoring services to them. The incident drew attention to the increasing concerns surrounding data security in the hospitality industry and led to heightened efforts to strengthen cybersecurity measures inside the company.

Marriott encountered legal and regulatory consequences due to its insufficient safeguarding of guest data, resulting in the imposition of fines by authorities. The Marriott breach case serves as a prime example of the importance of establishing robust cybersecurity protocols to protect sensitive customer information and maintain public trust. This highlights the necessity for companies to consistently remain vigilant in light of the dynamic nature of cyber threats.

6.1 Synopsis

Hoteliers are significantly concerned about the prospect of encountering a data breach, especially when it involves the unauthorized disclosure of sensitive personal data and confidential guest information. Before the emergence of the COVID-19 pandemic, the hospitality industry had already recognized cybersecurity and data protection as significant and formidable risks. The inherent structure of this industry renders it perpetually vulnerable, as evidenced by multiple notable incidents of malware attacks targeting hotels. These attacks have led to the compromising of vast amounts of customers' data, along with substantial financial losses of up to millions of dollars.

An unauthorized attempt to access the guest reservation database of Marriott's Starwood brands was detected by an internal security tool on September 8, 2018. Among the Starwood hotel brands that were considered were W, St. Regis, Westin, and Sheraton. Upon spotting the questionable activities, Marriott undertook an internal investigation, employing forensic procedures whose specific specifics have not been disclosed to the public. In the course of this inquiry, it was found that the Starwood network had experienced a security breach dating back to 2014, a period when it functioned as an autonomous organization. Despite the acquisition of Starwood by Marriott in 2016, the integration of the former Starwood hotels into Marriott's reservation system has not been completed even after over 2 years. Consequently, these properties continued to rely on the IT infrastructure acquired from Starwood. The previously mentioned detail holds significance and justifies the need for subsequent analysis, which will be conducted at a later time.

During their thorough investigation, Marriott determined that the malicious individuals had implemented techniques to encrypt and potentially effectively extract data from the Starwood networks. By the 11th month of that year, the encryption had been successfully decrypted, revealing that the compromised data consisted of information acquired from up to 500 million guest records; however, some of these records may be duplicates or pertain to the same individuals. The exposed data included credit card and passport numbers, as well as other highly sensitive information. After acknowledging the gravity of the situation, Marriott expressed on November 30, 2018, outlining the key aspects of the breach that we have previously discussed. The lodging industry recently faced significant financial consequences, with around $23.8 million in penalties resulting from the data breach that occurred in 2014 [Bre18, San+18]. This financial burden is just one aspect of the challenges faced by Marriott.

The breach resulted in the compromising of credit card details, passport numbers, and birthdates of over 300 million guests that were stored in the brand's global guest reservation database [Bre18]. This incident is considered to be one of the most significant data breaches to date. Although insurance may provide some relief for the financial losses sustained, the damage to the reputation of the Marriott brand is likely to persist for a significant period. This section presents an analysis of the factors contributing to the occurrence of the Marriott data breach and proposes proactive measures that hotels might implement to prevent similar catastrophic incidents. It is important to emphasize that the breach occurred in 2014; however, it remained undetected until 2018 when Marriott was notified by an internal security system about a suspicious attempt to access the internal guest reservation database of the Starwood brands.

The acquisition of Starwood Hotels by Marriott International in 2016 led to the incorporation of 11 additional brands into its existing portfolio of 19 assets [Bre18]. However, this merger also exposed Marriott to a significant security vulnerability. An essential internal security notification, which prompted an investigation, revealed that unauthorized individuals had gained access to the Starwood network in 2014, before its takeover by Marriott. Notwithstanding this cognizance, Starwood refrained from transferring its booking system to Marriott's in 2018, causing the Starwood brands to persist in their dependence on outdated IT infrastructure. This circumstance significantly augmented the magnitude of the data breach. Marriott conducted an internal investigation and determined that the individuals responsible for the incident had encoded and extracted data from the Starwood system. The purloined dataset includes data from around 500 million guest records; however, some of these records may be duplicates [Bre18].

In the data breach disclosure made by Marriott, it was stated that the compromised guest information included a range of personal details such as name, mailing address, date of birth, gender, phone number, email address, passport number, Starwood Preferred Guest (SPG) account details, arrival and departure information, reservation date, and communication preferences. Also, it should be noted that a certain number of guests had their credit card data stolen, although Marriott has chosen not to disclose the exact figure. In late 2018, Marriott made a public announcement regarding the compromise of its reservation system, resulting in the unauthorized access and extraction of a significant number of customer records. These records contained highly sensitive information, such as credit card details and passport numbers, and were obtained by individuals with malicious intent. While Marriott has not yet disclosed the comprehensive timing or technical details of the attack, the available information offers useful insights into the current threat types and provides important lessons for other organizations seeking to protect themselves from cyber threats.

Marriott has chosen not to disclose full details regarding the precise nature of the cyberattack, making it challenging to ascertain the fundamental rationale for the hack. However, the statement made by Arne Sorenson, the CEO of Marriott, during his testimony to the US Senate, offered limited insight into the occurrence. Marriott became aware of the incident due to an atypical database query that was detected by a security instrument. This tool was managed by Accenture, a company that provided IT and information security services for Starwood both before and following the acquisition. The execution of the query was performed by a user possessing administrative credentials. However, upon closer examination, it was discovered that the designated user did not carry out the query, suggesting that someone without proper authorization had gained entry to the account.

After conducting an extensive investigation, it was determined that a Remote Access Trojan (RAT) and MimiKatz, a tool designed to retrieve username/password combinations from system memory, were identified. The attackers used these two tools to get control over the administrator account. The precise source of the RAT (Remote Access Trojan) present on the Starwood server remains undetermined; nonetheless, it is commonly observed that phishing emails are frequently used as a means of distributing such malware. The occurrence of the breach was influenced by not only technological considerations but also basic cultural and business factors.

The primary aspect of the event does not revolve around the cybercriminals' capacity to infiltrate the Starwood network, as it is widely acknowledged by experts that achieving complete protection is essentially unachievable. However, the noteworthy aspect is in 4 years during which the invasion remained undetected. The security culture of Starwood, before its acquisition by Marriott, was deemed inadequate, since its staff regularly had challenges in effectively protecting the reservation system. An alternative unauthorized individual infiltrated the system in the year 2015, and this breach of security remained undetected for 8 months [Bre18].

After the acquisition of Starwood by Marriott, a significant portion of the organization's workforce, including personnel responsible for information technology and security, was dismissed as part of cost reduction strategies. These measures often led to enhanced profitability and synergistic effects, which are the primary motivations driving such mergers. Nevertheless, Marriott was unprepared to transition to its reservation system for the numerous hotels it had recently acquired. Consequently, Starwood's antiquated system, plagued by malware, compromised by hackers, and lacking in consistent maintenance, continued to function for an additional 2 years until the breach was eventually detected.

The data breach encountered by Marriott has the potential to yield severe consequences. A multitude of individuals had their personal information, such as passports and credit card numbers, unlawfully obtained by malicious cybercriminals. Consequently, the victims may face dire outcomes as a result. The security breach at Marriott resulted in a notable concern, particularly about the credit card information. Marriott must be held accountable for yet another instance of security inadequacy, as the encryption keys for the credit card numbers, while being encrypted themselves, were stored on the same server and were also compromised during the cyberattack.

Regarding the compromised passport numbers, most of them were not encrypted, but surprisingly, the Starwood customers did not suffer the full extent of the damage that could have been inflicted by the breach. The reasons for this can be found in the identity of the attackers and their motivations. Marriott's expenditure on the incident amounted to a staggering $28 million, but this only resulted in a small decrease of $3 million in the company's profits by March 2019. However, the company was able to reduce its losses to just $1 million by May of the same year, thanks to its cyber insurance coverage, which covered most of the initial expenses related to the breach [Bre18, San+18]. While cyber insurance is a relatively recent innovation, it has been beneficial for Marriott.

Despite the initial expenses incurred by Marriott following the data breach, the long-term costs resulting from lost revenue due to customers' concerns could amount to billions of dollars. In July 2019, Marriott was dealt a significant blow when the UK's Information Commissioner's Office (ICO) imposed a fine of over $120 million for violating the privacy rights of British citizens under the General Data Protection Regulation (GDPR) [Bre18, San+18]. The ICO specifically noted Marriott's inability to conduct proper due diligence on Starwood's IT infrastructure as the reason for the fine. This fine may just be the beginning, as other jurisdictions may also seek to hold Marriott accountable for their shortcomings.

6.2 Ramifications

The Marriott data breach presents a compelling case study for any individual or organization keen to learn from the mistakes of others. While the precise details surrounding the breach may remain shrouded in mystery, there are several essential lessons that one can glean from the incident. Specifically, it appears that both Starwood and Marriott failed to uphold fundamental security protocols, such as implementing comprehensive defence strategies and maintaining a strict separation between encoded data and the cryptographic keys used to secure it.

Besides, Marriott's failure to operate under the assumption that it had already been compromised is a critical misstep in any cybersecurity strategy. The issues resulting from the merger of Starwood and Marriott further compounded the situation, highlighting the liability of post-merger entities for cybersecurity lapses. The travel industry's cybersecurity measures, while critical, are lagging behind other sectors such as banking, indicating an urgent need for improvement. Lastly, the data breach serves as a reminder that even everyday individuals may become unwitting victims in the larger battle between governments engaged in espionage.

7 RockYou 2021 Breach, 2021

The RockYou 2021 attack case refers to a security incident that occurred on the popular online platform, RockYou. This happened during the year 2021 and resulted in significant repercussions for both the company and its user community. During the incident, malicious forces managed to breach RockYou's security protocols, resulting in the unauthorized acquisition of confidential user data. The incident of this data breach led to the disclosure of personal information owned by users, including usernames, email addresses, and passwords. The attackers were able to successfully carry out the breach by exploiting weaknesses that existed inside RockYou's security infrastructure.

After the occurrence, the compromised user data was used for various illegal purposes, including but not restricted to the fraudulent acquisition of personal identities, unauthorized access to further online accounts, and the continuation of deceptive tactics known as phishing campaigns. This event highlighted the importance of implementing robust cybersecurity protocols within organizations tasked with managing user data, as well as emphasizing the need for users to utilize strong, unique passwords and exercise caution in protecting their online security. The RockYou 2021 attack serves as an illustrative case, shedding attention on the ongoing challenges that enterprises face in safeguarding user data within an increasingly linked digital society.

7.1 Synopsis

The occurrence of data breaches remains a persistent concern in cybersecurity, presenting an ongoing and pervasive danger that evokes strong anxiety among both experts and the general population. The emergence of a substantial collection of passwords, commonly referred to as 'RockYou2021', on the dark web in recent times has raised major concerns owing to its broad scale. The source of the data breach has been ascribed to an anonymous individual on an online forum who distributed a significant compilation file comprising 100 terabytes of textual data. This document includes 8.4 billion records of unlawfully acquired passwords [Wei09, R21]. It is postulated that the procurement of the data was facilitated by a confluence of password-cracking methodologies and prior occurrences of data breaches.

The ramifications of the RockYou2021 data breach are substantial since malicious actors might leverage the compromised credentials to get unauthorized access to sensitive information, perpetrate fraudulent identity theft, and take part in other malicious activities. The RockYou2021 incident, widely recognized as an unprecedented compromise of passwords, poses a significant threat to Internet users worldwide. The collection of compromised passwords, known as the RockYou dataset, is named after the 2009 RockYou data breach, during which 32 million passwords were stored without encryption [R21]. The scale of the revealed credential data, consisting of around 8.4 billion passwords, raises concerns, especially when compared to the estimated worldwide population of Internet users, which is around 5 billion individuals [Wei09].

The individuals presumed to be accountable likely have access to several passwords utilized by a significant number of users. The examination of the data breach has been conducted by security specialists, who have concluded that the compromised credentials are not original and appear to have been accumulated over several years. It is worth mentioning that a considerable proportion of the passwords included in the dossier had been already compromised in prior occurrences of data breaches.

The Combination of Many Breaches, which took place in February 2021, is a significant incident within cybersecurity, representing a significant occurrence of data breaches. The incident involved the exposure of around 3.2 billion records on the Internet. The perpetrator of this offence claimed that all the passwords mentioned had a length ranging from 6 to 20 characters and did not contain any non-ASCII characters or white spaces [Wei09, R21]. Yet, the individual responsible for uploading the data asserted that a total of 82 million credentials had been compromised. Nevertheless, a further investigation carried out by security analysts has revealed that the actual quantity is only one-tenth of the originally stated amount. However, the aggregation of numerous breaches still holds the regrettable distinction of being the most prominent instance of password and credential exposure to date.

Considering the broad spectrum of individuals who may be susceptible to such an attack, organizations should proactively disseminate information about this security risk among their personnel while also enforcing obligatory password modifications for all user accounts.

Likewise, organizations must commence the adoption of multi-factor authentication (MFA) and other essential measures for identity management, if they have not yet undertaken such actions. It is essential for employees and administrators to enact suitable measures to guarantee the formation of strong and secure passwords. Although passwords have been found to have low efficiency as a sole authentication factor, their continued utilization in access management is guaranteed due to their extensive familiarity and broad application. Consequently, your firm needs to furnish users with extensive information about the various alternatives at their disposal. Meanwhile, password checkers can assist employees in assessing the degree to which their passwords conform to the standards established by password crackers and guessing methods.

7.2 *Ramifications*

To safeguard the confidentiality and integrity of personal data, individuals must follow a set of best practices characterized by their robustness and complexity. In the first instance, it is important for individuals to actively engage in the process of thorough inquiry to ascertain whether their data has been compromised. The significance of this matter lies in the fact that the RockYou2021 password list contains data that has been amassed over a prolonged duration. In this particular context, individuals must conduct a comprehensive analysis of the list, regardless of the obsolescence of their data, as the passage of time does not inherently reduce the impact of security breaches. Hence, it is recommended to utilize the services offered by websites that do data breach assessments. Nevertheless, it is important to ensure that these websites possess a solid reputation and are duly validated to mitigate the inadvertent disclosure of sensitive information.

If there is suspicion of a potential compromise of an individual's password, it is crucial to implement actions to modify it across all digital platforms. Implementing this strategy is not just recommended if an individual's password is present in the RockYou2021 password database, but it is also regarded as a best practice in general to mitigate the vulnerabilities connected with potential security breaches. Individuals must use complex passwords that incorporate a combination of letters, numbers, and special characters, as opposed to selecting simplistic options like commonly used keywords or personal names. It is advisable to avoid using identical passwords for multiple accounts since this practice would greatly increase the vulnerability of one's entire digital identity in the event of a security attack.

Multi-factor authentication represents a noteworthy milestone in this field. Through the implementation of supplementary security measures for their online accounts, individuals can significantly enhance the safeguarding of their digital presence and prevent the potential vulnerabilities associated with security breaches. However, despite the deployment of extensive security mechanisms, it remains important for individuals to be vigilant and proactively safeguard themselves against the persistent menace of phishing attempts. It is highly recommended to implement reliable and resilient Internet security measures, as they provide the fundamental basis of a comprehensive security framework that prioritizes the effective protection of personal data.

8 Colonial Pipeline Attack, 2021

The May 2021 Colonial Pipeline cyberattack was not just a highly sophisticated compromise of critical infrastructure. The Colonial Pipeline, responsible for delivering approximately 45% of the region's gasoline, diesel fuel, and jet fuel, plays a critical role in the Eastern United States. With over 5500 miles of pipelines connecting Gulf Coast refineries to distribution centres, major airports, and crucial fuel hubs, this critical infrastructure system ensures the continuous flow of energy supplies [EF23]. These supplies power not only transportation but also businesses, industries, and the daily lives of millions of Americans across the East Coast. The cyberattack on Colonial Pipeline not only became an incident that grabbed national importance but also served as a reminder of the digital vulnerabilities that can potentially disrupt essential services and the broader economy.

8.1 Synopsis

The Colonial Pipeline, which serves as a crucial component of the energy infrastructure in the United States, plays a critical role in guaranteeing the continuous transportation of gasoline throughout the country. The transportation of around 45% of the fuel supply on the East Coast, which includes essential resources such as petrol, diesel, and jet fuel, holds immense importance for the nation's functioning and well-being. The infrastructure comprises an elaborate network of pipes that extends over a distance of 5500 miles [EF23].

This sophisticated system is a major aspect in connecting refineries located along the Gulf Coast to key distribution centres, major airports, and significant fuel hubs. The extensive and complex infrastructure under consideration is not simply a conglomeration of conduits and mechanisms; rather, it functions as the critical circulatory system that supports the economy, facilitates transportation, drives industries, and empowers the daily activities of numerous individuals residing in the densely populated regions of the Eastern seaboard of the United States.

In May 2021, this energy infrastructure was exposed to a malicious cyberattack that caused significant repercussions across the entire nation. The perpetrators, a criminal hacker collective known as 'DarkSide', have a history of bold exploits, especially their tendency to focus on valuable organizations through ransomware attacks. The DarkSide group successfully penetrated the computer networks of the Colonial Pipeline, using previously unnoticed vulnerabilities. The breach executed by the individuals in question exhibited a remarkable level of boldness, granting them backdoor entry to a substantial collection of confidential information, as well as the command mechanisms that regulate the core of this critical infrastructure.

The occurrence of gasoline shortages in several places has led to a state of panic buying and fuel hoarding among the general population. The economic consequences were rapid, resulting in a domino effect that impacted multiple sectors, amplifying weaknesses in supply chains, and generating repercussions throughout the wider economy.

Following the occurrence of this malicious intrusion, the individuals responsible issued a ransom request that caused significant disruption and concern within the executive offices of Colonial Pipeline. An amount of around $4.4 million, denominated in Bitcoin, was requested as ransom in return for a decryption key that purportedly offered the possibility of restoring the company's compromised systems [EF23]. The ransom demand not just includes financial considerations but also encapsulates the ethical and pragmatic dilemmas that organizations face when dealing with cybercriminals who seize control of their digital assets.

8.2 *Ramifications*

Threat actors frequently initiate intrusions by employing social engineering techniques, such as phishing, and utilizing leaked credentials to obtain initial entry. This occurrence highlights the advanced nature of ransomware variants. These malicious programs utilize sophisticated encryption methods, rendering the process of decryption and data retrieval exceedingly difficult, even for highly prepared entities. To effectively address cyber risks, organizations must adopt a proactive approach. This includes the implementation of extensive cybersecurity procedures, which requires providing employee training to augment awareness, engaging in continuous network monitoring to detect anomalies, and conducting frequent vulnerability assessments to find and rectify gaps. The establishment of a clearly defined incident response plan is of importance.

This measure guarantees that organizations can promptly identify, confine, and minimize the consequences of a cyberattack. The implementation of prompt and well-coordinated actions is important to mitigate harm and promote an efficient restoration procedure. The act of determining whether or not to provide a ransom payment is accompanied by a multitude of ethical and practical quandaries. It is important for organizations to thoroughly evaluate the potential dangers and rewards associated with the act of paying a ransom.

Although the utilization of this method may enhance the efficiency of data retrieval, it concurrently possesses the potential to create incentives for those engaged in cybercriminal activities. Also, it does not provide an absolute assurance regarding the secure restoration of encrypted data. The cyberattack on the Colonial Pipeline highlights the necessity of making significant expenditures in cybersecurity techniques and procedures, such as password management and authentication practices, specifically for critical infrastructure.

9 AIIMS Ransomware Attack, 2022

This section will focus on a cyberattack that specifically targeted the crucial server architecture of All India Institute of Medical Sciences (AIIMS), a prestigious healthcare facility in India. The occurrence of the attack resulted in significant disruptions to the delivery of medical treatment, hence inflicting substantial damage upon key patient care services. The Delhi Police commenced a lawful inquiry by lodging a first information report (FIR) with the IFSO special cell. To maintain the provision of important services, AIIMS took the initiative to reassign more staff members.

The occurrence of the breach has a concern on the potential occurrence of data theft, given that the confidential information of around 30–40 million patients may have been compromised. A network sanitization strategy was executed, wherein antivirus solutions were installed on 1200 out of a total of 5000 computers and 20 out of a total of 50 servers. A demand for a ransom amounting to Rs 200 crore was raised, prompting a collective effort to strengthen cybersecurity measures [Onl22, Sta22, Bha23]. AIIMS exhibited a firm commitment to preserving healthcare services through the execution of manual protocols. This occurrence emphasizes the significance of continuous attention and preparedness in the field of cybersecurity to protect healthcare establishments and guarantee the continued provision of essential medical services.

9.1 Synopsis

The growing popularity of ransomware-as-a-service (RaaS) groups has further compounded the challenges faced by international organizations. Criminal groups have emerged as a noteworthy factor, targeting reputable institutions and demanding outrageous ransom amounts. In the year 2022, a considerable number of noteworthy ransomware operations attracted significant media coverage, accompanied by ransom demands of very high magnitude, reaching billions of dollars [Onl22, Sta22, Bha23]. One noteworthy incident pertained to a highly sophisticated attack on the All India Institute of Medical Sciences (AIIMS), a prestigious medical research institution and hospital located in New Delhi, India.

All India Institute of Medical Sciences (AIIMS) experienced a substantial ransomware attack on November 23, 2022, leading to considerable disruption and disarray. The occurrence of the attack resulted in substantial disruption to several components of patient care services, including registration, admission, billing, and discharge procedures. This incident provided a distinct demonstration of the worldwide extent of cybercrime and the challenges faced by law enforcement agencies in apprehending these elusive criminal networks.

The key target of this cyberattack was the 'e-Hospital' application system, which functions as a critical component of AIIMS' digital framework. The program, overseen by the National Informatics Centre (NIC), played a critical role in supervising both inpatient and outpatient operations within the hospital. As a consequence of the attack, the hospital was compelled to shift from computerized to manual operations, leading to the interruption of services for a considerable portion of patients.

The adoption of data backup measures may present a viable strategy for mitigating the consequences of such an attack. However, it is important to note that this matter is significantly more complex. The time dimension of data backups holds significant importance. The act of performing daily backups carries the inherent risk of inadvertently incorporating viruses into the backup data. Cybercriminals often demonstrate a proclivity for postponing the initiation of their encryption mechanisms, occasionally extending the duration over multiple days or even weeks. When the backup procedure is conducted following the compromise of the primary data, the integrity of the backup is also affected.

The team from the National Informatics Centre (NIC) was able to effectively restore the 'e-Hospital' application and database servers at AIIMS within a 5-day timeframe after the cyberattack. The data backup demonstrated significant durability, as it was well isolated from the network and remained unchanged after the event. In the aftermath of the attack, a comprehensive investigation was initiated, which include multiple aspects. The incident prompted the Indian Computer Emergency Response Team (CERT-In), in collaboration with the Delhi Police and its cyber department, as well as the Ministry of Home Affairs (MHA), to commence investigations aimed at identifying the responsible individuals and assessing the extent of the incident's consequences.

The original correspondence from the attackers allegedly lead to a demand for a ransom of up to 2 billion Indian rupees to be paid in the form of Bitcoins [Onl22, Sta22, Bha23]. However, those possessing privileged information regarding the incident who have opted to withhold their identities have contradicted these conjectures, asserting that there is no substantiated proof of any request for ransom being issued. The Intelligence Fusion and Strategic Operations (IFSO) unit of the Delhi Police filed charges of extortion and cyberterrorism, prompting the initiation of a digital forensic examination carried out by the Forensic Science Laboratory (FSL). One of the foremost worries arising from the AIIMS attack was the potential vulnerability of highly sensitive patient data, including electronic health records (EHR). The hospital experienced a substantial disruption to its crucial digital systems as a result of the cyberattack.

This disruption affected several essential tasks, including billing operations, report generation, laboratory testing procedures, emergency department services, and appointment scheduling. This specific occurrence might be regarded as a demonstrative case of a meticulously planned ransomware attack that required a thorough understanding of network components and vulnerabilities within the system. The attackers deliberately focused on the digital infrastructure of AIIMS, aiming to inflict substantial harm and escalate their ransom requests. AIIMS implemented manual data entry techniques as a response to an internal healthcare crisis, intending to maintain uninterrupted patient care amidst the existing digital issues. Simultaneously, the institution actively sought guidance from the Union Ministry of Electronic and Information Technology, showcasing a determined resolve to oppose the requests made by the attackers.

9.2 Ramifications

The ransomware attack on AIIMS in November 2022 exposed the susceptibility of healthcare organizations to cyberattacks. In addition to the financial ramifications and operational disturbances, this occurrence highlights the possible vulnerability of confidential patient information, thereby presenting significant threats to privacy and security.

Given the increasing prevalence of cyber threats, organizations must place a high priority on the development and maintenance of robust cybersecurity resilience measures. This includes comprehensive backup systems that take into account the scheduling of data backups, active network monitoring, and proactive security measures. Also, it is critical to emphasize the necessity of collaboration among law enforcement agencies, government entities, and cybersecurity professionals to effectively identify and ensure the accountability of individuals involved in cybercrime.

The event at AIIMS serves as a warning example for all organizations that are responsible for the protection of critical data and services. In the current age of digitalization, the menace of ransomware assumes significant proportions, necessitating the adoption of proactive measures as the fundamental determinant of an organization's ability to withstand and recover from such cyber threats. As organizations like AIIMS attempt to enhance their cybersecurity measures, the international community must join in addressing the escalating threat posed by cybercrime.

10 Implications of Digital Resilience

Resilience, within its appropriate framework, refers to the capacity of a system or entity to endure, assimilate, recuperate from, or adapt effectively to adverse circumstances or changes in its environment. To assess the resilience of a certain system or entity, it is crucial to first establish the anticipated performance of that system under many conceivable conditions. The assessment of a smart power grid's resilience is contingent upon its ability to mitigate operational failures resulting from both non-malicious and malicious actions, as well as its capability to recover from such failures and reinstate the equilibrium between power supply and demand. The notion of resilience can be further classified into the resilience of individual essential functions inside the system, as each function determines its anticipated level of performance. Consequently, the implementation of resilience inside each fundamental operation enables the entire system to withstand anticipated and unanticipated modifications.

To enhance a system's capacity to endure and adjust to adverse changes, it is crucial to possess a thorough understanding of its three fundamental phases:

(1) Avoidance
(2) Containment
(3) Recovery

The avoidance phase includes the execution of diverse operations aimed at mitigating or eliminating any potential impact on the intended performance of the system. The confinement phase involves implementing procedures that try to restrict the effects to a specific location and prevent their dissemination throughout the entire system. The recovery phase is a sequence of procedures aimed at restoring the system to its expected level of performance.

Essentially, for a system to ensure its ongoing resilience, it is important to use a range of strategies throughout different phases. To safeguard critical infrastructures, particularly healthcare systems, from potential threats such as ransomware attacks, it is important to give precedence to the evaluation of the communication infrastructure, which functions as a central element within the healthcare system. Ensuring the preservation of high availability, data consistency, and system integrity is of utmost importance for this infrastructure since any compromise carries the potential to yield severe consequences.

To increase its capacity to endure and recuperate from attacks, specifically those about ransomware, system engineers must build a well-defined operational framework including all three stages of resilience. Within cybersecurity, the deliberate action of evading or bypassing established limitations, commitments, or duties can result in substantial consequences. The implementation of an integrated and diversified strategy is important to effectively manage these possible dangers and enhance security measures on a wider scale.

This discussion examines various approaches for predicting and preventing cyber invasions, emphasizing the importance of strong permission requirements, intrusion detection, encryption, and containment systems. The anticipation and prevention of cyberattacks are of utmost importance in cybersecurity. One effective strategy that has shown considerable effectiveness involves the examination of network traffic quality and the enforcement of strict authorization restrictions. This specific methodology exhibits considerable efficacy in forecasting instances of ransomware outbreaks.

Organizations increase security measures by consistently implementing upgrades to their software and hardware systems. The incorporation of resilient authorization and authentication methodologies is critical to safeguarding the security of software and hardware systems. The successful execution of this task necessitates the establishment of strict access controls to guarantee that only approved entities are granted entry. The inclusion of both internal and external penetration testing is crucial for the detection of vulnerabilities within the security infrastructure. The use of data encryption is necessary in ensuring the protection of confidential information.

Encryption is a security mechanism that protects data in several states, including its stationary state, during transmission, and when stored on electronic devices. The incorporation of regular key rotation significantly improves the overall security of information systems and embodies a fully automated methodology. The regulation of data exchange on bring-your-own-device (BYOD) systems is a crucial determinant. The establishment of comprehensive systems for the categorization of data on platforms of this nature is of importance and requires immediate attention. Cloud-based solutions provide increased security by implementing limitations on data replication. Besides, the utilization of federated learning (FL) approaches can be implemented to detect ransomware in government healthcare establishments. This novel approach enables collaborative data analysis while ensuring the protection of data privacy by eliminating the need to disclose sensitive information.

The necessity of implementing network segmentation is evident since it involves a variety of technologies like software load balancers, security measures such as firewalls, and role-based authorization protocols. The primary objective is to hinder the mobility of malicious individuals within the network, therefore reducing the probability of lateral infiltration. Logical network partitioning is a technique that leads to the subdivision of a network into distinct sub-functions, facilitating the segregation of segments in the event of a security breach.

The implementation of resilience assists in safeguarding critical infrastructure. System engineers can efficiently meet their cybersecurity responsibilities by integrating security procedures and techniques throughout the preparation, corrective, and maintenance phases. Engaging in deliberate efforts to bypass cybersecurity protections can potentially result in substantial consequences. To efficiently mitigate and minimize these risks, it is crucial to adopt a thorough strategy. This has explored several techniques for forecasting and addressing cyber threats, including rigorous authorization, intrusion detection, encryption, containment, and resilience. Organizations can strengthen their cybersecurity posture and adequately mitigate possible threats through the implementation of subsequent actions, thereby fulfilling their responsibility to protect data and systems.

11 Concluding Remarks

To mitigate the risk of data breaches and the resulting financial losses that can severely impact a company's financial stability, organizations must prioritize cybersecurity in their operations and approach it with great care and attention. This chapter explored the prospects of major enterprises and businesses amidst the prevalence of cyber dangers, as well as the strategies they might utilize to safeguard themselves against the activities of hackers. Over the past few years, several significant instances of cyberattacks have been observed, resulting in increasing financial consequences for organizations. These include the WannaCry ransomware attack in 2017 and the Equifax data breach in 2018. These attacks have highlighted the need to prioritize cybersecurity and allocate resources towards implementing safeguards to protect the data.

The deployment of cybersecurity measures holds significant importance for organizations that manage both their own data and client data. The primary aim of data security is to protect organizational data, including proprietary trade information and sensitive customer data, from potential exploitation by cybercriminals, thereby ensuring the preservation of client privacy. Likewise, compliance to industry and government standards is of significance for firms to maintain lawful operations and safeguard customer privacy.

Cybersecurity also aims to mitigate the potential legal consequences, financial penalties, and reputational harm that might result from security breaches. The ability of large enterprises and businesses to effectively handle the evolving cybersecurity domain will be crucial in determining their future viability in the face of cyberattacks. The progression of cyber risks is occurring at an accelerated rate, necessitating that businesses remain updated on current trends and technology to effectively safeguard themselves.

Companies have a responsibility to enhance the security measures of their automated systems and maintain constant monitoring to swiftly identify any anomalous activities that might potentially signify a cyberattack. Today, it is apparent that the ability of large enterprises and businesses to adapt to the evolving cybersecurity domain will be crucial for their future success. To achieve this objective, firms must allocate substantial resources towards the acquisition and incorporation of sophisticated security measures. These measures include a range of strategies, such as encryption and intrusion detection systems, which are essential for effectively protecting their critical data and networks from malicious individuals or entities. Also, it is important to engage in collaborative efforts with other organizations and governmental entities to exchange relevant information regarding potential dangers and to adopt industry-leading strategies.

These activities are essential in reducing the impact of cyberattacks and ensuring long-term resilience. The increasing implementation of automation and AI in the digital domain offers a significant prospect for organizations to effectively identify and immediately address instances of cyber threats. Nevertheless, it is critical to subject these advancements to security measures to prevent infiltration by cybercriminals. Therefore, it is critical to establish and enforce strong security rules. Companies must give precedence to cybersecurity within their operational framework since the potential consequences are of such magnitude that disregarding or delaying this matter is not a feasible option.

Organizations must adopt a proactive approach to maintain competitiveness and relevance. Organizations must demonstrate their commitment to cybersecurity, including the integration of state-of-the-art technology and methods alongside ongoing evaluation and enhancement of their security stance. Although significant investments in terms of time and money resources are necessary, the investment in cybersecurity is undeniably valuable due to the extremely important role it plays in today's linked society.

Part III
Multidimensional Digital Resilience

Chapter 6
Role of Artificial Intelligence and Machine Learning

1 Introduction

An increasingly prevalent trend among generally accepted computer science textbooks is the inclusion of critiques and discussions on artificial intelligence (AI) and machine learning (ML). Although a direct causal relationship between these concepts and digital resilience may not exist, it is vital to assess the impact that these technologies have on cybersecurity and digital accountability. This is the preferred course of action, particularly in consideration of the increasing significance of this subject in the industrial and academic spheres, which has prompted considerable discussion and consideration.

There is widespread recognition that machine learning (ML) and artificial intelligence (AI) possess the capacity to provide us with tools and techniques that can augment our skills in wide domains. It is essential to consider that these instruments have legitimate societal consequences and effects on human labour dynamics. This chapter does not aim to offer an exhaustive analysis of the topic of ML or its techniques. Rather, it will focus its attention on the ramifications that result from the emergence and utilization of these very valuable methodologies. This discussion aims to delve into cybersecurity, examining its possible benefits for enterprises and the requisite adaptations in digital resilience to efficiently leverage its capacities.

AI includes the ability of machines to replicate human-like capabilities in areas such as perception, cognition, and response. The fundamental assumption of this notion is that human intelligence can be dissected into discrete capacities that can be duplicated through computer programming. AI is a broad term that includes a wide range of concepts and technologies, most notably ML. AI comprises a range of subfields that utilize approaches to replicate specific traits typically attributed to human intelligence. As a demonstration, it is evident that individuals can participate

S. K. Shandilya et al., *Digital Resilience: Navigating Disruption and Safeguarding Data Privacy*, EAI/Springer Innovations in Communication and Computing,
https://doi.org/10.1007/978-3-031-53290-0_6

in oral communication, perceive auditory stimuli, evaluate written texts, and express themselves in writing, thus deriving meaning from various linguistic structures.

Speech recognition and natural language processing replicate these functionalities by converting audio signals into written text and then analysing the text to extract its underlying significance. Another piece of research pertaining to the advancement of intelligent systems that strive to replicate human behaviours is that the discipline of robotics involves the scientific investigation and advancement of machines that possess the ability to execute tasks independently. These technologies simulate the cognitive abilities of humans to navigate their physical environments. Computer vision is a research field that seeks to replicate the cognitive abilities of human beings in perceiving and analysing visual information.

Pattern recognition is a cognitive process characterized by the ability to identify and categorize items based on their similarities and shared characteristics. The use of AI algorithms is prevalent across various fields, demonstrating their wide-ranging capabilities and flexibility in modern society. Simultaneously, a multitude of research activities consistently explore innovative uses for these algorithms. This chapter aims to present a review of the essential concepts and distinctions that exist between AI and ML and that are relevant to our discussion on digital resilience. Thus, this chapter will be concerned with the wide commercial uses of AI and ML while also offering recommendations on how they relate to cybersecurity.

2 History

Let us explore the historical backdrop on this topic. Alan Turing, an English mathematician, held prominence in the early history of AI. Turing's seminal paper, entitled 'Computing Machinery and Intelligence' [Tur50], was published in 1950. This paper presented a methodology proposed by the author for assessing computer intelligence, which subsequently became widely recognized as the Turing test. Turing's proposition sought to evaluate the capacity of a machine to demonstrate conversational abilities that resemble those of an individual.

A mere 5 years after Turing's seminal work, Herbert Simon, Allen Newell, and John Shaw accomplished a notable feat by developing a program called Logic Theorist [McC04]. This program constituted an innovative initiative to simulate human problem-solving capabilities using computational methods.

However, the discipline that is presently acknowledged as AI has not yet been fully defined. The term 'AI' was introduced by John McCarthy, a highly significant individual in the mainstream establishment of AI. He incorporated this specific term into a written proposal that he composed for a summer research conference that focused on the given subject matter. McCarthy's idea postulated the possibility of conducting an investigation which would explore the notion that the complete range of learning and other dimensions of intelligence may potentially be accurately expressed. This, in turn, would facilitate the development of computers with the ability to replicate these characteristics.

McCarthy's proposal outlined a plan aiming to investigate the methods through which machines could acquire language proficiency, develop abstract thinking and conceptualization, demonstrate proficiency in solving complex problems traditionally associated with human intelligence, and potentially enhance their capabilities autonomously. This concept laid the foundation for the subsequent formal development and investigation of the science of AI, influencing its trajectory towards the current understanding of AI.

Moving towards ML, the initial development can be traced to the mid-twentieth century, during which the concept of computers imitating human intellect began to surface. Prominent individuals such as Alan Turing and Marvin Minsky participated in the discourse on AI in the 1950s and 1960s, thereby laying the groundwork for ML. The advancement was hindered by the restricted computational capacity and the scarcity of accessible data. In 1959, Arthur Samuel introduced the term 'machine learning' to characterize his research on software that facilitated computers in acquiring information from data. The late 1950s witnessed significant contributions by Frank Rosenblatt, who invented algorithms such as the perceptron for binary classification. However, it is important to note that these algorithms had certain limits when it came to tackling complex problems.

In the 1980s and 1990s, ML encountered a period of stagnation commonly referred to as the 'winter' of ML, which was characterized by unfulfilled expectations. However, during this particular era, significant algorithms like decision trees and neural networks emerged, establishing the foundation for subsequent advancements. The advent of the Internet and the subsequent profusion of data in the 2000s marked a substantial shift in societal dynamics. Several advanced ML approaches, such as support vector machines, ensemble methods, and deep learning, were created. During the 2010s, the increased computational capabilities and availability of data sets contributed to the widespread adoption and popularity of deep learning techniques, particularly those involving multi-layer neural networks.

The effectiveness of deep learning was demonstrated in the 2012 ImageNet competition, where it exhibited exceptional capabilities in image recognition. ML applications have brought about significant transformations in industries such as healthcare, banking, and autonomous cars. The availability of ML tools has been significantly expanded by the utilization of cloud computing and open-source frameworks, thereby promoting innovation. The incorporation of ML technology into everyday life has been observed in several applications, such as personalized streaming suggestions and voice assistants on smartphones. Current research is dedicated to enhancing the fairness, interpretability, and utility of models.

The origins of artificial neural networks, which serve as the foundation for deep learning, may be traced back to the middle of the twentieth century. Progress in the field was hindered until recently due to challenges related to computational resources and theoretical limitations. The emergence of the underlying concept of deep learning may be traced back to research on brain structure throughout the 1940s. The pioneering work on artificial neurons was conducted by Warren McCulloch and Walter Pitts. The perceptron algorithm saw a surge in popularity throughout the 1960s; however, it subsequently faced a decline in interest due to constraints in its complexity. The resurgence of deep learning in the mid-2000s can be attributed to several factors, including breakthroughs in training methods, the availability of larger data sets, and the development of technology such as Graphics Processing Units (GPUs).

During the period spanning from the late 2000s to the early 2010s, significant advancements were achieved in deep learning. In 2012, convolutional neural networks (CNNs) achieved a level of image categorization performance that exceeded that of human capabilities. The success achieved in numerous domains can be attributed to the capacity of deep neural networks to capture hierarchical data representations. Deep learning has emerged as a critical component in various technological advancements, such as driverless vehicles and medical image processing. Currently, researchers are engaged in the process of refining various methodologies, placing a significant emphasis on the interpretability of models, and integrating ethical considerations into their research.

3 Defining AI and ML

It is widely accepted that the collaborative interaction of computer programs, humans, and machines demonstrates an environment that develops heightened levels of production. The progress of our species has been greatly enhanced through the use of mechanized tools. Throughout history, machines have been key in facilitating various aspects of human existence, ranging from the agricultural revolutionizing wheel to the use of screws in complex construction projects. In recent times, the advent of robot-assisted assembly lines has further contributed to the development of our accustomed way of life. However, despite their significant practicality, humans have expressed concerns about robots, including the possibility of machines acquiring human-like intellect and expressing independence.

3.1 Artificial Intelligence (AI)

The definition of AI involves the development of algorithms that mimic the neural network of the human brain. This enables machines to process and analyse large amounts of data, learn from their activities, and improve future results. AI has different classifications, namely, '*weak/narrow* AI' and '*strong/true* AI', which will be further defined in the subsequent discussion.

What are the defining characteristics of weak AI? Weak AI, generally known as narrow AI, is a specific branch of AI that focuses on generating answers that resemble human reactions by using pre-programmed algorithms. Weak AI tools lack true cognitive capabilities and instead create the illusion of thinking.

Definition 7 (Artificial Intelligence (AI)) AI is the emulation of human-like intelligence processes by computers, including learning from experience, adapting to new information, reasoning to solve problems, and performing tasks similar to human intellect, like understanding natural language, recognizing patterns, decision-making, and creative activities, by deploying algorithms and data-driven methods to enable machines to imitate cognitive functions to boost efficiency, precision, and self-sufficiency across wide domains.

Common examples of poor AI can be observed in the form of voice-activated programs such as Siri, Cortana, Google Assistant, and Alexa. When presented with an inquiry or command, these programs distinguish acoustic signals within the spoken language and subsequently execute a series of predetermined instructions to generate a suitable response. Their understanding of the spoken language or its deeper meaning is limited.

On the contrary, *what distinguishes strong AI?* Strong AI, sometimes known as 'true AI', refers to systems that possess the ability to engage in autonomous cognitive processes. AI systems can do many cognitive tasks such as reasoning, learning, planning, communication, judgement, and exhibiting a level of self-awareness. Thus in nature, AI systems do not merely imitate the human mind; instead, they may exhibit cognitive abilities similar to human minds, albeit in a theoretical sense.

According to scientists, the construction of computers with genuine cognitive skills is deemed feasible by duplicating the structure and operation of the human brain. In deep learning, a subset of AI, researchers utilize neural networks to facilitate the training of computers to achieve enhanced autonomy. However, the attainment of totally autonomous AI, as depicted in works of science fiction, remains a distant prospect. It can be argued that the realization of genuine, strong AI remains more of a philosophical notion than a concrete reality.

3.2 *Machine Learning (ML) and Deep Learning (DL)*

People widely agree that when computer programs, humans, and machines work together, it boosts productivity. Machines have played a big role in our progress as a species, from simple tools like the wheel in agriculture to more complex ones like screws in construction. Nowadays, robots on assembly lines are also shaping how we live. But, even though machines are practical, some worry about them becoming too smart or independent like humans. Figure 6.1 displays a Venn diagram showcasing the relation between the domains of AI, ML, and DL.

Fig. 6.1 AI and its domains

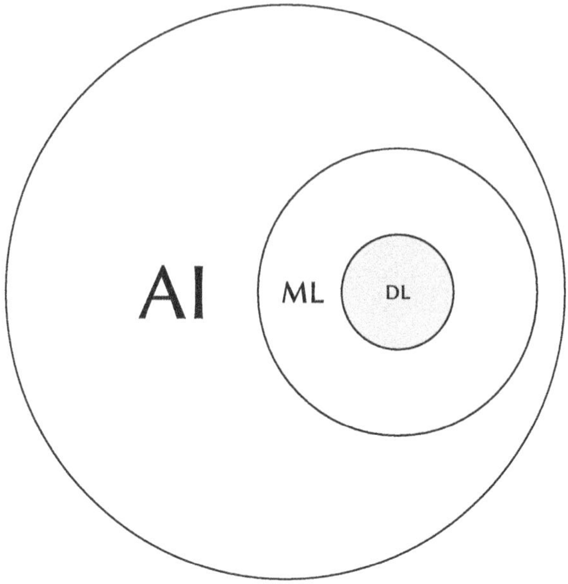

Definition 8 (Machine Learning (ML)) ML is a subset of AI dedicated to creating algorithms and statistical models that enable computer systems to enhance task performance by utilizing data while allowing computers to recognize patterns, learn from instances, and foresee outcomes without direct programming for every situation involving extracting valuable insights from data automatically, progressively boosting system capacities.

The process involves the use of AI methodologies that facilitate computers in acquiring knowledge through experience, hence improving their proficiency in particular assignments. This involves the examination of data through the application of statistical techniques to enhance learning and task performance. In computer science, ML concentrates on the development of computer systems that possess the ability to learn autonomously by processing data, resulting in improved performance in various tasks. Although the terms *machine learning* and *artificial intelligence* are frequently used interchangeably, it is important to note that ML is a fundamental constituent of AI. ML algorithms are frequently classified into two main categories, namely, *supervised* and *unsupervised.*

Let's examine the concept of supervised ML. In supervised ML, an individual assumes the role of guiding an AI program throughout its learning process. The individual provides the algorithm with a set of training data that includes instances and the corresponding expected outcomes for each case. The individual is responsible for determining the factors that will be subjected to analysis and providing input on the accuracy of the computer's predictions. After receiving adequate training or direction, the computer uses the provided training data to make predictions for new data it encounters.

In unsupervised ML, algorithms are provided with training data but do not possess known outcomes for comparison. In contrast, the data is carefully examined to reveal patterns that were previously unrecognized. Unsupervised learning algorithms can cluster comparable data, detect abnormalities within data sets, and establish patterns that connect disparate data points. Semi-supervised ML methods integrate both labelled and unlabelled training data in their training process. The incorporation of a restricted quantity of annotated training data significantly improves the precision of predictions while simultaneously mitigating the laborious and costly process of annotating large volumes of data.

Reinforcement learning, a well-defined methodology, uses a system of rewards and penalties to incentivize favourable actions and discourage unfavourable ones in the process of ML training. In reinforcement learning, the agent, which is the subject of training, engages in the process of perceiving and understanding its surroundings, executing actions, and acquiring knowledge through iterative experimentation. This paradigm refers to the optimal behaviour of intelligent entities within a given environment, intending to maximize the total accumulated reward. Reinforcement learning is considered one of the core paradigms in ML, alongside supervised and unsupervised learning.

Definition 9 (Deep Learning (DL)) Deep learning (DL) is a subset of AI and ML that focuses on the design and development of algorithms and computational models inspired by the structure and function of the human brain's neural networks and involves the construction of artificial neural networks with multiple layers (therefore 'deep' learning) that are capable of automatically learning and extracting hierarchical features from raw data.

Our examination culminates in a review of the notion of deep learning. This signifies a highly sophisticated level of AI investigation, which has enabled the advancement of language models, generative AI, and various other tools in a wide range of industries. Deep learning is a sub-machine learning that aims to replicate the complex workings of the human brain to a high degree of fidelity.

Frequently, a deep neural network, which mimics the interconnected system of human neurons, is commonly utilized in this context. Deep learning contains complex concepts, with its fundamental principles revolving around the repetitive processing of data and the derivation of conclusions that closely resemble human cognitive processes. In instances where an ML algorithm produces an erroneous forecast, human intervention becomes necessary to repair the situation. On the other hand, deep neural networks, or deep learning algorithms, exhibit the capacity to independently evaluate the precision of their predictions. Therefore, it may be argued that deep learning is more suitable for complex problems in comparison to traditional ML models. This is summarized in Table 6.1.

Table 6.1 Comparison of machine learning and deep learning

Machine learning (ML)	Deep learning (DL)
1. Definition	**1. Definition**
Machine learning is a subset of artificial intelligence that focuses on developing algorithms that enable systems to learn patterns and make predictions from data	Deep learning is a specialized subset of machine learning that involves neural networks with multiple layers (deep neural networks) to model and solve complex problems
2. Data representation	**2. Data representation**
ML algorithms rely on feature engineering to represent data in a format suitable for learning	DL algorithms automatically learn hierarchical representations of data, eliminating the need for extensive feature engineering
3. Model complexity	**3. Model complexity**
ML models have moderate complexity and may involve simpler algorithms like decision trees, support vector machines, or regression models	DL models have high complexity, especially in deep neural networks with multiple layers, allowing them to capture complicated patterns in data
4. Training data size	**4. Training data size**
ML models may require a substantial amount of labelled data for effective training and generalization	DL models, with their ability to automatically learn complex features, may require even larger data sets for optimal performance
5. Feature learning	**5. Feature learning**
ML algorithms often require manual feature engineering to extract relevant information from raw data	DL algorithms excel in automatic feature learning, discovering hierarchical features from raw data during training
6. Interpretability	**6. Interpretability**
ML models are generally more interpretable, making it easier to understand the factors influencing predictions	DL models can be less interpretable due to the complex hierarchical features learned, posing challenges in understanding model decisions
7. Computational resources	**7. Computational resources**
ML models may require less computational power, making them suitable for a wider range of devices	DL models, especially large neural networks, demand significant computational resources, often requiring specialized hardware like GPUs or TPUs
8. Transfer learning	**8. Transfer learning**
ML models can benefit from transfer learning, where knowledge gained in one task is applied to another	DL models are well suited for transfer learning, leveraging pre-trained models on large data sets to improve performance on related tasks
9. Use cases	**9. Use cases**
ML is applied across various domains, including image recognition, natural language processing, and recommendation systems	DL excels in complex tasks such as image and speech recognition, natural language processing, and playing strategic games
10. Human involvement	**10. Human involvement**
ML often requires more human involvement in feature engineering and model tuning	DL relies more on automated feature learning, reducing the need for extensive manual intervention

4 Applications

Many people are fascinated by the integration of AI and ML into different areas. The general expectation is that these technologies will bring significant changes to our society, influencing various aspects of our lives. We can already see the impact of AI and ML in different businesses and industries, where they are playing a major role in technological advancements.

Various forms of technology have been utilized in a range of organizations, from small-scale operations supervising garment management to major corporations managing significant assets. These technologies have proven to be beneficial, resulting in favourable outcomes. First and foremost, AI has exhibited its ability to mitigate onerous and monotonous jobs. Significant time savings are achieved by autonomously executing these procedures. Consequently, this enables the reallocation of resources towards initiatives that require concentrated attention, innovative cognition, and intellectual prowess. As a result, the enhancement of productivity contributes to the general growth of organizations.

The purpose of this section is to spell out the significant applications of AI and ML without getting into complex technical complexities. In this context, a preliminary review of application domains will be presented. Although it is not possible to provide an exhaustive overview of all applications of AI and ML due to the rapidly changing nature of this field, the constant stream of new research publications demonstrates the emergence of fresh approaches for implementing AI and ML in various disciplines.

These methodologies are not limited to established disciplines but are expanding into unorthodox domains. Notable illustrations include space technology, applied astronomy, and computational chemistry. The utilization of AI and ML in various transdisciplinary fields showcases the adaptability and variety of these technologies, as they are being used in previously uncharted areas. The emergence of AI and ML has undeniably sparked a paradigm shift, leading to significant progress across various sectors. The ability of these technologies to optimize processes, improve efficiency, and enable innovative applications remains influential in shaping our society, as advancements in these fields continue to reveal new possibilities for implementation.

4.1 Agriculture

The agricultural sector, which is critical to guaranteeing enough nutrition for the world's population, has reached a tipping point where it meets technological improvements. In this arena, the potential for integrating AI and ML is significant. Agriculture necessitates the management of several sophisticated processes at the same time over enormous agricultural lands that may span hundreds of acres. Crop production is affected by a variety of interconnected elements, such as weather patterns, sunlight availability, animal activity, fertilizer and pesticide usage, planting procedures, and irrigation timelines.

This is an important issue in ML. Farmers, cooperatives, and agricultural enterprises are increasingly using data-driven approaches in the current setting of technology developments. The recognition that good data analysis is critical to achieving financial success within agricultural cycles is the driving factor behind this transformation. AI and ML approaches are increasingly being used to supplement manufacturing and increase output quality. At first, surveillance systems with AI and ML capabilities were used to methodically evaluate real-time video feeds from agricultural areas. These devices quickly recognized instances of human or animal trespassing and gave timely alerts.

By using AI that can easily adapt to varied sizes of agricultural activities, this technology enhances crop resilience against unforeseen harm and unauthorized entry. By analysing current data, the use of modern technology such as sensors and drones improves the accuracy of agricultural output projections. This enables agricultural specialists to derive unique insights from unmatched statistics, allowing tailored crop yield recommendations. Supervised machine learning algorithms are critical to the yield mapping process because they efficiently find patterns in large and complex data sets. These algorithms provide farmers with a wealth of information about predicted crop yields before they begin planting by merging 3D mapping with soil data from sensors.

To improve pest control strategies, global corporations and agricultural organizations use drone data and in-ground sensors. The utilization of infrared data gathered by drones in combination with sensor data linked to plant health enables AI systems to efficiently foresee and accurately pinpoint cases of pest infestations. The use of AI- and ML-enabled smart tractors, robots, and robotics is required to solve the agricultural manpower crisis, especially in distant agricultural regions.

These technologies have the dual effect of reducing manpower shortages while also increasing the effectiveness of field operations. The use of AI-powered track-and-trace systems in supply chains provides significant advantages in terms of improving operational efficiency. These technologies significantly reduce inventory losses while increasing visibility and control throughout agricultural supply networks. The use of AI increases the use of environmentally friendly pesticides by accurately determining the areas affected, resulting in lower costs and increased agricultural productivity. The application of AI in price prediction, especially with a focus on yield rates, plays a critical role in easing the design of strategic crop pricing plans.

Practice ensures that farmers and cooperatives may engage in mutually beneficial discussions that end in profitability. AI is a key element in a variety of agricultural sectors, including the detection of irrigation leaks, the improvement of irrigation systems, and the evaluation of the impact of recurrent watering on crop output. The use of AI in these situations improves the overall efficiency of agricultural practices. AI is used to effectively supervise livestock well-being, assess their actions, track their daily routines, and check their food intake. The rise of AI in the agriculture industry opens up new opportunities for improving cow well-being and production. Crop and soil condition monitoring is an essential component of agricultural management.

Soil composition has a substantial influence on crop health and output, and monitoring crop development phases may lead to increased production efficiency. Previously, human observation was used to assess soil quality and crop health, which was prone to mistakes and time limits. A more successful strategy involves the use of drones, also known as unmanned aerial vehicles (UAVs), to collect aerial picture data. Following that, AI-powered computer vision models are used to thoroughly examine the data to conduct crop and soil monitoring. The use of visual sensing AI enables the processing of agricultural data, allowing crop health assessment, accurate production forecasting, and early detection of crop malnutrition. This is beyond the scope of human action.

Using AI insights helps farmers to handle particular difficulties in a timely and effective way. Several applications have proved the efficiency of computer vision in crop health management. One use of AI is in the assistance of labour-intensive activities, such as the monitoring of the development stage of wheat heads. To speed up the developing process, the researchers used a two-step detection approach. This approach included the use of photographs of wheat at various stages of heading that were collected under various lighting conditions. The model outperformed human observation, eliminating the need for farmers to undertake regular field trips. Similarly, the application of AI simplifies the evaluation of tomato ripening on a broad scale.

The value of computer vision in assessing soil quality is obvious. This includes assessing soil texture and organic matter content using image data obtained with a portable microscope. The approach presented herein achieves a level of accuracy comparable to that seen in laboratory analysis while avoiding the time-consuming and labour-intensive features often associated with older procedures. The combination of AI with computer vision technology demonstrates outstanding performance in the detection and diagnosis of plant diseases and pests. Using deep learning-based image recognition, models may autonomously detect diseases and pests, automating the identification process.

The agricultural business has seen significant development as a consequence of advancements in computer vision, which have reduced the need for labour-intensive occupations and, in many instances, have exceeded human talents in monitoring and diagnosing issues. AI, in combination with ML approaches, plays a critical role in the identification and quantification of pests, especially elusive flying insects that pose difficulties in human collection and evaluation. The researchers used sticky traps to capture insects in the air while using real-time imaging technologies to identify and roughly estimate their presence.

While the emphasis has mostly been on botanical organisms, it is important to recognize that agriculture includes a broader variety of activities than the growing of specific crops such as wheat, rice, tomatoes, and apples. Animals play an important part in the agricultural business and so need close monitoring. Cattle, poultry, and pigs are exposed to monitoring using computer vision and technology with AI capabilities. The use of drone-based technologies allows for the use of overhead cameras and computer vision algorithms to assist remote monitoring of cattle health and behaviour. The approach used effectively decreases the need for continual physical presence while improving the ability to recognize and fix difficulties in real time.

As a result, we can clearly state that the intersection of AI and agriculture creates a compelling partnership with the potential to dramatically transform this critical sector. Given the multiple challenges posed by changing climates, environmental concerns, and rising global food demand, the agricultural industry has the potential to undergo considerable change by using AI technologies. The previously mentioned technology has the potential to improve operational efficiency, enhance sustainability, and enable more smart resource allocation within the sector. The use of this technology necessitates a reorganization of the agricultural environment in which farmers assume the roles of AI developers. This transformation will result in an era of improved agricultural productivity, food safety, and environmental awareness.

4.2 Autonomous Systems

The use of AI facilitates the creation of self-governing technologies by researchers and professionals in science and engineering. These technologies can operate autonomously, adapting to ever-changing surroundings. The incorporation of complex systems that combine AI with robotics has experienced notable progress in recent years, notably within the framework of the Fourth Industrial Revolution (IR4.0).

The integration is achieved through the use of the Internet of Things (IoT), which enables the interaction of several domains spanning AI and robotics. AI algorithms are integral in augmenting networked systems, which are frequently found in several domains, including surveillance and autonomous robotics. These algorithms are proficient at analysing data to simplify the decision-making process by providing valuable insights. In our current globalized society, there is an increasing need for monitoring of agricultural practices and human activities.

The attainment of this objective can be accomplished by employing a distributed surveillance system that integrates AI-powered robotic drones. The drones effectively collect data, examine it, and choose appropriate courses of action. The main subject of inquiry in this work refers to the development of innovative architectural designs that are suitable for interconnected autonomous, surveillance, and robotic systems. The effective execution of these ideas requires access to abundant data collected from a variety of sensors, including lidar and cameras. The main aim is to provide insight into the most recent developments in agricultural surveillance, pedestrian identification, and targeted monitoring. The current objectives involve a set of devices, spanning from individual cameras to extensive networks of interconnected cameras.

These devices are enhanced by AI to attain elevated levels of accuracy and precision. The solicited contributions include original research publications, reviews, and methodologies. Possible research topics include various domains, including image and video processing for surveillance, recent advancements in the integration of AI with image and video surveillance to promote security, innovations in crop monitoring techniques, and methodologies for observing plants on a large scale.

The preliminary objective is to examine the use of AI in autonomous navigation and decision-making for drones, namely, unmanned aerial vehicles (UAVs) and unmanned ground vehicles (UGVs). We investigate the application of AI technology in several sectors, such as security surveillance, agricultural monitoring, and the advancement of autonomous drones. The notion of autonomous surveillance systems contains the amalgamation of unmanned aerial vehicles (UAVs) with video surveillance technology to detect and oversee human subjects while simultaneously utilizing AI to facilitate self-governing and distributed drone surveillance operations.

The application of autonomous computer systems in transportation is widespread; however, it is primarily focused on assistive technology rather than full autonomy in consumer automobiles. The classification system for vehicle autonomy levels comprises a spectrum ranging from zero to five levels of automation. While self-driving cars have received significant attention, it is crucial to acknowledge that AI can be utilized in other areas, including aviation, cargo transportation, and emergency services.

The application of advanced autonomous technology can bring about substantial advancements in medicine, science, and space exploration. The technical track includes a range of fields, including ML, AI, computer vision, embedded systems, sensor systems, control systems, and robotics. Autonomous systems have been developed to enhance the quality, safety, and security of human existence through their wide applications. Autonomous cars possess the capacity to address environmental contaminants, enhance accessibility for those with restricted mobility, lower transportation costs, reinforce safety protocols, and diminish occurrences of criminal behaviour. Humanoid robots possess the ability to provide aid to individuals in need, perform dangerous tasks, and undertake activities that go beyond human skills. The process of decision-making in these applications is enhanced through the utilization of complex data acquired from various sensors, such as radar, lidar, charge-coupled device (CCD) cameras, and micro-electromechanical systems (MEMS) devices.

Autonomous AI (AAI) refers to a highly advanced subset of AI that focuses on the self-governing capabilities of machines. These systems leverage sophisticated algorithms and use ML approaches to analyse data, acquire information, and make autonomous judgements. These technologies find use in various domains, including autonomous transportation, healthcare systems, and industrial production. Artificial general intelligence (AGI) differentiates itself from conventional AI in its capacity to function independently, without reliance on human interaction or input.

Reflex agents, classified as a subset of AI, demonstrate reactive behaviour in response to stimuli through the application of predetermined rules. These agents demonstrate a high level of suitability for the execution of activities that are straightforward and not complex. AI surpasses these methods by operating independently and augmenting decision-making capabilities through the application of ML techniques. Autonomous systems engineering involves the systematic development of AI systems that possess self-governance capabilities, achieved through collaborative efforts across multiple disciplines.

Autonomous intelligence, as shown by self-driving cars, drones, and robotic automation, functions independently by employing sophisticated algorithms to analyse data, make decisions, and enhance productivity. In contrast, assisted intelligence functions in collaboration with human agents, augmenting efficiency, as evidenced by the utilization of voice assistants and data analytics tools. AI has a wide-ranging impact on several sectors, including healthcare, manufacturing, fraud detection, and investing. These elements are required in improving efficiency, decision-making processes, safety measures, and cost savings.

AI possesses the capacity to optimize several procedures; yet, its integration also involves various challenges including safety, security, ethics, job displacement, and the complexities of human-machine interaction. The significance of proficient design and shared comprehension in promoting cohesive collaboration cannot be overemphasized. The promise of AI in the future includes enhanced capabilities, heightened safety protocols, and increased innovation across various industries. Effective handling of issues involves the proactive reduction of safety concerns, the thoughtful examination of ethical dilemmas, and the promotion of smooth integration between humans and technology.

4.3 Education

AI has infiltrated the fabric of modern daily life, transforming sectors across the board. Its influence is most visible in sports, construction, and education, where it has resulted in a paradigm change, transforming both operational and educational techniques. This section looks into the innovative opportunities AI has brought to education, specifically through the integration of educational technology (EdTech) that uses AI to deliver personalized learning experiences, identify knowledge gaps, and provide precise feedback through digital learning solutions.

The use of AI in education has resulted in a revolutionary learning paradigm that challenges traditional methodologies. The introduction of AI-powered educational software has substantially impacted conventional learning techniques, allowing digital courses, online reference materials, and virtual classrooms to be accessed via mobile devices. As more EdTech companies adopt AI technology, a thorough assessment of its uses and advantages in education becomes critical. To begin this investigation, a quick examination of prospective opportunities in the AI-driven educational technology industry is required.

AI is pushing education into unknown territory, questioning traditional pedagogical practices, and charting the course of educational technology breakthroughs. AI systems, defined by their complex algorithms, analyse large data sets to provide specialized and customizable learning experiences. This includes personalized learning routes, immediate feedback, and exposure to immersive technologies such as augmented and virtual reality, all of which have practical applications in educational settings.

Conversational AI in education, as shown by chatbots and virtual tutors, improves self-directed learning by offering timely advice. These AI-powered chatbots use natural language processing and ML methods to catalyse a fundamental change in the mechanics of student learning, providing quick and targeted help throughout the educational experience. A graphical depiction shows organizations' significant expenditures in AI applications such as educational app development, robots, virtual assistance, natural language processing, computer vision, and ML.

Given the educational benefits provided by AI technology, a thorough examination of the several ways AI is transforming education is required. The discussion that follows defines a few unique uses of AI and its accompanying educational advantages, which are backed by much research. AI plays a critical role in education by simplifying numerous tasks and improving the overall learning experience. The analysis that follows describes ten critical cases in which technical advancements provided by AI have had a substantial influence on the learning process and educational practices.

- **Personalized Learning:** Previously constrained by the limitations of traditional institutions, the educational approach of tailoring training and learning experiences to individual needs and interests has been greatly expanded by the application of AI in online education. AI ensures that instructional software is customized to each learner's specific requirements, utilizing auxiliary technologies such as ML to accommodate individual learning preferences.
- **Task Automation:** The utilization of technology to automate repetitive tasks, commonly known as task automation, has gained traction in educational institutions and online learning environments. AI systems not only oversee and perform value-added tasks but also tailor teaching methods to meet the unique needs of individual students, showcasing the wide applications of AI in education.
- **Intelligent Content Production:** AI and ML technologies contribute to the creation of distinctive educational materials, enhancing the teaching and learning process. AI-assisted intelligent content creation manifests in information visualization, simulating realistic visual study environments, and optimizing laboratory research. The resulting digital courses, requiring less storage space, cater to the needs of remote learners.
- **Adaptive Accessibility:** Adaptive accessibility, defined as the ability of a system or environment to accommodate individuals with wide requirements and constraints, is a key feature of AI in education. AI's role in ensuring unrestricted access to information and promoting diversity in educational settings

is evidenced by multilingual support and features addressing visual or hearing impairments.

- **Identifying Flaws in the Educational Environment:** It is noteworthy to acknowledge AI's significant contribution to addressing and mitigating environmental impact through remote learning. Despite concerns that AI may replace human instructors, the symbiotic relationship between AI and education enhances both traditional face-to-face and online teaching methods, presenting a balanced perspective.
- **Resolving Skill Deficiencies:** The introduction of student upskilling programmes in software and application development, facilitated by AI and ML, provides cost-effective avenues for individuals and organizations to enhance their skills. The impact extends beyond the student population, developing a sense of motivation and collective commitment to innovation among the existing workforce.
- **Skill Acquisition Methods:** AI in education influences learning and development by automating the process of skill acquisition and incorporating cognitive processes similar to human learning. Data analysis is utilized to offer tailored feedback, crucial for generating effective learning experiences in both educational and professional contexts.
- **Data-Driven Feedback System:** AI in education utilizes data analysis to provide feedback based on current facts, enhancing student satisfaction, reducing bias in the learning process, and identifying areas where skills may be lacking. The personalized feedback system operates twenty-four-seven, offering unique feedback to each student and employee.
- **Conversational AI:** The application of conversational AI, exemplified by AI-powered chatbots, enhances user engagement in personalized learning. Chatbots excel at addressing enrollment queries, facilitating access to learning materials, and providing ongoing support throughout the day, catering to the preferences of global learners who opt for online education and corporate training.
- **Secure and Decentralized Learning Systems:** Educational platforms prioritizing data security while decentralizing power across multiple nodes or organizations constitutes a technological advancement in education. Despite lingering concerns related to data security, inconsistent data access, and outdated certification methods, the potential of decentralized AI-based solutions holds promise for positive change in education.
- **AI in Assessment:** The utilization of AI in assessment has the potential to disrupt traditional review processes by automating evaluation procedures. AI software solutions seamlessly integrated into examination and interview processes use algorithms to actively monitor individuals, identifying and alerting supervisors to any problematic behaviour. This proactive approach addresses user concerns, enabling EdTech organizations and corporations to effectively leverage AI technology.

The many uses of AI in education highlight the essential role that technology plays in changing traditional paradigms. The evidence-based investigation of the applications defined AI's transformative influence, offering insights into its potential to optimize learning experiences, improve efficiency, and increase diversity in education. As the educational environment changes, the wise use of AI advertises to be an incentive for innovation and progress, ushering in a new age in pedagogical practises and learning techniques.

4.4 Cybersecurity

Attention has been drawn to AI and ML because of their potential to change cybersecurity. It is important to exercise caution when considering this assumption. While it is expected that there will be noteworthy gains, it is crucial to acknowledge that progress may occur at a moderate pace. When considering these improvements for a future characterized by full autonomy, it is evident that they signify significant advancements that surpass previous capabilities. To have an understanding of the influence of AI and ML on cybersecurity, it is important to take into account the prevailing issues that currently exist within this discipline.

AI can address certain characteristics that have traditionally been considered standard. System setup errors made by humans are significant vulnerabilities in cybersecurity. Getting the right system configuration might be difficult, especially with the help of a skilled IT workforce. As technological advancements progress, the complexity of security measures increases. Adaptive tools can facilitate the identification and resolution of issues that arise throughout the evolution of network systems, including the incorporation of novel technologies such as cloud computing into pre-existing systems. The concept of human efficiency also presents certain obstacles.

The process of establishing and maintaining settings across several endpoint machines can be a labour-intensive task that requires a significant investment of time and effort. In the same way, the ability to effectively address dynamic threats necessitates adaptability, a quality that can be facilitated through the use of AI and ML technologies. An additional concern is the phenomenon of alert fatigue resulting from an overabundance of security messages. AI has the potential to assist in the management of warnings and provide recommendations for actions, thereby reducing the impact of decision fatigue. The prompt emphasizes the criticality of swift threat response within cybersecurity.

The speed of malicious attacks is increasing as a result of automation. ML can efficiently examine attack data, resulting in the generation of succinct reports and suggestions. The capacity of AI to differentiate trends in historical data can be essential in the detection of emerging threats and the proactive anticipation of potential attacks. IT and cybersecurity teams are currently facing significant challenges due to staffing shortages. The utilization of AI-driven tools has the potential to mitigate employment requirements through the augmentation of human initiatives. ML is essential in facilitating processes such as the identification of data patterns, thereby relieving individuals from engaging in repetitive and mundane activities.

The need for adaptability cannot be overstated, particularly in situations where human teams may possess little knowledge or skill in specialist domains. AI can be customized to specific requirements by utilizing appropriately trained algorithms. The attainment of genuine AI, characterized by its ability to engage in abstract reasoning, continues to be an elusive objective. ML constitutes a fundamental element of AI in cybersecurity. The process involves analysing patterns within the data to make predictions about future events. Although not comparable to human intellect, it demonstrates exceptional proficiency in pattern recognition and the presentation of data in a coherent manner.

ML comprises a range of techniques, including data categorization, grouping, action recommendations, possibility synthesis, and predictive forecasting. ML has found various applications in cybersecurity. Notable examples include the classification of data privacy, the analysis of user behaviour and system performance profiles, the identification and banning of behavioural bots, and the utilization of predictive forecasting techniques. Notwithstanding the potential, there are persistent restrictions. The use of data in ML gives rise to problems regarding privacy.

The task of assembling proficient teams specializing in AI and ML continues to be a significant obstacle. The abilities of decision-making, creativity, and critical thinking continue to be human capabilities that are not yet capable of being recreated by AI. To effectively adapt to the evolving outlook of cybersecurity, it is important to allocate resources towards the acquisition of technology that is oriented towards future advancements.

The integration of AI into human teams can serve as a valuable complement, enhancing the overall capabilities and efficiency of cybersecurity operations. It is essential to remain well informed and up-to-date on the prevailing data privacy legislation, as compliance with these regulations is crucial for maintaining the security and integrity of sensitive information. AI possesses the capacity to enhance security protocols while also affording cybercriminals autonomous entry into systems.

The subsequent discussion examines the favourable consequences of AI on cybersecurity within various fields. Organizations specializing in vulnerability management face the challenge of effectively managing and prioritizing the multitude of vulnerabilities that emerge regularly. Current vulnerability management methods primarily adopt a reactive approach, responding to security breaches only after hackers have successfully exploited the identified vulnerabilities. The use of AI and ML has the potential to enhance risk management through the enhancement of vulnerability databases.

Similarly, the use of AI-powered tools, such as user and event behaviour analytics (UEBA), enables the examination of user activity on servers and endpoints. This analysis allows for the identification of anomalies that may indicate potential hidden attacks. This enhances security measures before the formal identification and resolution of vulnerabilities. The practice of threat hunting involves proactively searching for and identifying potential security threats within a network or system. Traditional security tools use signatures or attack indications to identify and mitigate potential attacks, which is particularly effective in dealing with familiar and established threats. These methods continue to demonstrate limited efficacy in countering yet-to-be-identified risks. AI has the potential to significantly enhance the effectiveness of traditional methods, resulting in a success rate increase of up to 95 per cent.

However, it is important to note that there may be instances where AI produces falsely positive outcomes. The optimal approach involves integrating AI with conventional techniques, resulting in detection rates that approach near-perfection and reduce the occurrence of false alarms. AI enhances the effectiveness of threat hunting through the integration of behaviour analysis. Profiling apps within an organization's network by using endpoint data analysis exemplifies a prominent case. The topic of discussion is network security. The focus of traditional network security revolves around the development of security policies and understanding the network environment. Two aspects require careful consideration:

- The formulation of policies that distinguish between what is considered valid and invalid.
- Understanding the application and workload environment can help detect fraudulent network connections.

The difficulties associated with formulating and maintaining policies for several networks are readily evident. AI provides network security enhancement through the utilization of pattern recognition in network traffic, the provision of guidance on security policies, and the implementation of functional workload grouping. Data centres are facilities that house computer systems and other components, such as telecommunications and storage systems.

These centres are designed to store AI, which can oversee and enhance essential operations within data centres, including the management of power consumption, internal temperature regulation, bandwidth utilization, and cooling mechanisms. The integration of AI into data centre architecture leads to improvements in both security and efficiency through the provision of valuable insights. The increasing prevalence of AI and ML in cybersecurity establishes a framework for a growing competition between attackers and defenders. Both factions recognize the potential of AI and ML to enhance their capabilities. Adversaries utilize AI and ML techniques to identify weaknesses and execute complex attacks, while defenders leverage the same technologies to proactively anticipate and counteract such attacks.

AI and ML have been a significant area of interest and research in recent years. AI and ML are rapidly becoming essential components of attackers' resources, facilitating automation and coordination inside conventional cyberattack frameworks. These activities comprise many stages of a cyberattack, including reconnaissance, target identification, weaponization, payload deployment, exploitation, installation, command and control, and data exfiltration. Similarly, the use of AI and ML enables the execution of customized attacks through the dissemination of persuasive phishing emails, thereby evading the alert of even the most cautious users. The use of AI by malicious actors to develop malware that imitates harmless software, together with the implementation of evasive polymorphic malware, highlights the new methods used by attackers to exploit AI technology.

The topic of concern refers to fraudulent activities and aggressive strategies. AI and ML technologies assist individuals engaging in fraudulent activities by generating fabricated transaction data, such as counterfeit credit card transactions. The deceptive behaviours in question possess the ability to closely resemble genuine transactions, resulting in monetary deficits. However, defenders have available options to address this issue by utilizing AI and ML techniques to identify and mitigate intrusions. The ability of AI to identify complex patterns that are not easily perceptible to human observers is of importance.

The system is capable of detecting anomalies or suspicious behaviours within applications, processes, requests, responses, and network traffic, thereby providing alerts to mitigate ongoing attacks. The topic of discussion refers to the evolution of security measures. As AI and ML become increasingly integrated into businesses, there is a growing need to establish defensive measures against adversarial strategies that specifically target these technologies. The phenomenon of adversarial ML presents a significant risk, as it involves the deliberate manipulation of training data or the introduction of dangerous inputs by attackers.

Defenders utilize AI and ML techniques to enhance the features of antivirus software, firewalls, and security orchestration, automation, and response (SOAR) platforms. These adaptive methods facilitate the detection of threats in real time, automate incident response, and provide improved capabilities for cybersecurity. The concept of inevitable progress refers to the notion of continuous advancement that is unstoppable and inevitable. The irreversible nature of the rising involvement of AI and ML in cybersecurity is evident. Both perpetrators and defenders leverage these tools, leading to widespread adaptation. The cybersecurity domain is currently seeing a significant shift as these technological advancements are critical in shaping the trajectory of protecting digital environments.

4.5 Energy

The recent increase in economic advancement and the growing preference for convenience among individuals has resulted in a corresponding rise in the need for energy. In consideration of the limited availability of energy resources and the importance of conserving them for future generations, there is a significant benefit to the optimization of energy systems. However, the attainment of this objective is not straightforward owing to the complex characteristics of integrated energy systems. Computer-aided decision-making has the potential to significantly impact the identification of the optimal balance between energy supply and demand.

It is worth noting that the energy sector has the potential to see substantial benefits through the use of AI and ML techniques. The application of ML has resulted in significant progress in various sectors, including the energy industry. The utilization of ML in the energy industry includes a wide range of applications, including but not limited to intelligent grid technology, predictive maintenance for equipment faults, and forecasting the generation of wind and solar electricity. The use of AI and ML is increasingly being adopted to improve customer engagement.

By utilizing these technologies, energy businesses can provide customized information to consumers that is specifically matched to their individual needs and preferences. This involves utilizing data analysis techniques to comprehend the patterns of individual energy consumption and providing advice on modifying behaviours to decrease energy usage. Microgrids are self-sufficient energy networks that may operate independently from traditional grids. These systems utilize AI and ML techniques to effectively manage energy distribution and optimize usage. Microgrids are increasingly being recognized for their capacity to enhance energy security during emergencies and enable the seamless integration of renewable energy sources into energy networks, beyond the capabilities of conventional setups.

The occurrence of theft and fraudulent practices within the power industry results in significant financial consequences for the energy sector, amounting to billions of dollars on a global scale. AI and ML techniques are utilized to autonomously detect anomalies associated with power theft and energy fraud, enabling energy firms to promptly address and resolve these concerns. Not only does this practice ensure the protection of their assets, but it also effectively mitigates energy waste, leading to financial savings. The instantaneous delivery requirement presents distinct issues in energy trading. AI and ML present viable answers by accurately forecasting energy consumption and providing up-to-date pricing information to dealers.

This enables energy merchants to make well-informed judgements on the procurement and sale of energy. The advent of blockchain technology has facilitated the optimization of power purchase agreements, enhancing the efficiency and security of transactions. This technology facilitates the reduction of transaction time and expenses, thereby contributing to a more economically and securely structured platform for energy exchange. The growing energy storage business offers prospects for improved energy management. The integration of intelligent energy storage devices within power grids enhances the efficiency of energy management.

Besides, these systems facilitate the establishment of virtual power plants, which effectively guarantee the provision of energy even in situations of limited supply, reducing the need for constructing new power plants. Predictive analytics enables the anticipation of changes in energy demand, hence aiding the planning of future infrastructure. Energy businesses utilize these analytics methodologies to forecast equipment failures, mitigate unforeseen outages, and maximize asset replacement strategies. The use of AI and ML techniques is augmenting productivity within the energy industry. ML algorithms are essential in aiding oil and gas firms to optimize well placement and enhance production. As a result, there is an improvement in energy efficiency and a more easily controllable energy system.

AI enhances the operation of energy grids by effectively managing the distribution of energy among various sources, storage facilities, and consumers. The implementation of this approach results in a decrease in energy waste and an increase in customer engagement concerning energy consumption. The adoption of renewable energy sources, such as wind and solar power, is on the rise. However, the intermittent nature of these sources requires the implementation of real-time management strategies.

AI plays an integral role in predicting the availability of renewable energy and effectively regulating energy systems in response. Robotic systems are also included in energy systems to perform maintenance and monitoring activities, enhancing operational efficiency and reducing expenses. Cybersecurity precautions are important to safeguard energy grids against potential intrusions. AI and ML are critical in enhancing the security of power grids by utilizing data analysis techniques to anticipate potential cyberthreats and effectively respond to them to minimize associated risks.

Smart grids are advanced systems that incorporate sensors, data analytics, and energy administration technologies to collect data from a variety of devices to improve energy efficiency. These grids provide continuous monitoring of energy usage and the implementation of automated reactions, thereby reducing energy consumption during periods of high demand. The convergence of AI and ML within the energy sector presents significant opportunities for disruptive advancements. These technologies facilitate effective energy management, optimize manufacturing processes, and promote consumer involvement, ultimately contributing to the development of a more environmentally friendly and technologically sophisticated energy ecosystem.

4.6 Entertainment

The depiction of AI in a variety of science fiction films has played a significant role in shaping its importance within cultural cinema. The prospect of using AI for the creation of films presents a captivating idea. AI has demonstrated its aptitude in music generation by composing complex compositions. This artistic initiative serves as a prime example of the expanding function of AI within art. The integration of AI into the entertainment industry has the potential to revolutionize the way audiences engage with various forms of media, including films, television shows, videos, sports events, and games. According to industry analysts, there is a prevailing prediction that AI has the potential to spearhead the imminent industrial revolution.

It has been observed that AI can improve efficiency through the automation of repetitive processes, thereby providing a competitive advantage. The use of AI in task automation enables producers to allocate additional time towards enhancing their abilities and producing engaging content. Also, AI has the potential to empower production organizations to make informed choices in marketing and advertising through the analysis of critical data. Therefore, the media and entertainment industries face the need to enhance both the volume and calibre of their content to appeal to a broader clientele and generate increased value. AI is essential in the media industry by facilitating task acceleration, personalization of experiences, and improvement of marketing initiatives.

Media enterprises utilize AI technology to optimize their operations in several sectors, leading to reduced operational expenses, enhanced productivity, and amplified financial gains. To address this difficulty, entertainment companies are leveraging sophisticated technology, such as AI. The following are several instances of AI applications within the entertainment sector that are actively transforming the industry; customized contents for prominent platforms in music and video streaming, such as Netflix and Spotify, have achieved considerable success by effectively providing customized content to a wide range of individuals with varying preferences. These platforms utilize AI algorithms and ML models to analyse the behaviour and characteristics of individual users, thereby customizing content recommendations based on their references. Netflix utilizes AI to augment video suggestions by tailoring them to individual user preferences.

The process of creating a film, sometimes referred to as movie production, involves a series of interconnected activities that include pre-production, production, and editing. The film production process involves a multitude of responsibilities, such as scriptwriting, location scouting, shot outlining, visual storyboarding, budgeting, filming, scheduling, and editing. AI possesses the capability to automate several tasks, including but not limited to screenplay analysis, visual storyboarding, timetable development, and resource management. AI tools can automatically synchronize video clips, and with further advancements in technology, they may someday possess the ability to make material independently.

The generation of accurate and multilingual subtitles is a must for media firms seeking to distribute their content globally. The manual creation of subtitles is a laborious process that is susceptible to inefficiencies and inaccuracies. AI systems streamline the process of creating subtitles by examining speech patterns and producing accurate subtitles in several languages. This feature assists filmmakers in guaranteeing precise synchronization between audio and subtitles. The process of assigning metadata tags to effectively manage substantial quantities of content and improve its discoverability poses a significant problem. Video analysis tools powered by AI meticulously examine the content of films and distinguish various objects, images, or locations included in them. These tools then assign appropriate tags to the identified elements, thereby enhancing the search functionality of the movies. Search optimization, often known as search engine optimization (SEO), is a process that aims to improve the visibility and ranking of a website. The considerable amount of digital content available on the Internet frequently impedes the effective retrieval of information.

AI is key to improving search results by facilitating visual search capabilities and optimizing content classification. Similarly, AI can automate the process of movie selection and categorization, resulting in a substantial enhancement in efficiency. The merging of virtual reality (VR) and augmented reality (AR) technologies inside the entertainment industry enhances user involvement. AR and VR apps offer immersive experiences that enhance user engagement in domains such as gaming and sports. AR and VR possess the capacity to significantly transform the way we perceive visual content, perhaps facilitating the future development of holographic viewing for sports and movies. The utilization of automated content development is indicative of the impact of AI on both the creation of content and the engagement of audiences on various media platforms.

The progress of computational intelligence has facilitated the development of many AI tools that use methods such as NLP and text analysis to efficiently generate audio, written, and visual material. By using these capabilities, media organizations may enhance the speed of content production while reducing the need for physical labour. While AI-produced content enables efficient mass manufacturing, it lacks the consistent quality exhibited by content developed by human beings. Hence, a recommended strategy involves integrating AI-driven content generation with human authors. The attainment of a customized user experience posed difficulties before the integration of AI. The latest social media platforms such as Facebook, Instagram, and Twitter utilize ML and deep learning networks to study individual behaviours and traits.

This facilitates media organizations' acquisition of an in-depth understanding of consumer preferences and interests, leading to enhanced precision in content recommendations. Consequently, this phenomenon leads to heightened user involvement, broadens the user demographic, and augments website traffic. In a similar vein, electronic commerce systems utilize AI algorithms to provide consumer-centric recommendations, thereby enhancing sales and delivering tailored customer experiences. AI also assumes a role in the implementation of measures aimed at limiting access to content that is not suitable for certain age groups, hence strengthening protective measures for young viewers. The identification of copyright infringement is an additional domain in which AI demonstrates its indispensable value. The proliferation of social media platforms has resulted in instances of copyright infringement, presenting significant obstacles for the media and music industries.

AI technology is integral in the identification and prevention of copyright infringements. For example, YouTube, a famous platform for streaming videos, utilizes Google's ML capabilities to effectively handle copyright concerns. The use of AI is being applied to online music platforms such as Spotify, YouTube Music, Pandora, and Apple Music, enhancing the level of service and user satisfaction. Streaming platforms such as Netflix, Hulu, Amazon Prime Video, and Disney Plus utilize AI to improve the quality of streaming and provide tailored content suggestions. In the current age of extensive information distribution, AI has emerged as a promising approach to mitigate the negative effects of disinformation and fake news.

Still, despite the anticipated effectiveness, AI in this particular domain has demonstrated a degree of uncertainty. The use of predictive analytics has emerged as a method by which AI optimizes marketing efforts. Through the examination of historical data and the projection of patterns in viral media content, the use of AI-powered predictive analysis enables marketers to generate superior online content, differentiate target audiences, and develop online ads that possess greater potential for conversion. AI's role extends beyond advertising, as it also plays a significant part in increasing search optimization.

The use of this technique not only facilitates the creation of specific and tailored content but also plays a role in enhancing search engine optimization (SEO) strategies, thereby maximizing visibility and user interaction. AI exerts a substantial impact across wide fields, including chatbots, customer service, and virtual assistants. Although there may be early uncertainties about the use of AI in entertainment, its unquestionable significance becomes apparent. AI has been utilized in various capacities within the entertainment industry.

AI, as an illustration, has influenced film production, resulting in the production of renowned science fiction blockbusters like *Terminator* and Marvel's *Avengers: Age of Ultron*. However, the originality lies in the integration of AI into filmmaking itself. In the year 2016, IBM Watson utilized AI technology to produce the inaugural movie trailer for the film *Morgan*, which was produced by Fox. This development represents a significant achievement, indicating the possibility for AI to participate in various aspects of scriptwriting, scheduling, editing, and potentially even the filmmaking process. Similarly, the impact of AI extends to the advancement of computer-generated imagery (CGI), a technology that finds extensive application within the film industry.

The presence of AI in gaming is readily apparent through the provision of gaming experiences that are characterized by their responsiveness and adaptability. Non-player characters (NPCs) serve as a prime illustration of this phenomenon. AI enables non-player characters (NPCs) to anticipate and respond to the actions of players, facilitating the development of an immersive gaming experience. The integration of AI with virtual and augmented reality (VR and AR) technologies contributes to the advancement of gaming by providing more immersive and entertaining experiences. The effect of AI extends to music production as well. AI systems can produce novel compositions, while voice recognition technology facilitates intelligent vocal synthesis. Another characteristic that musicians use is that artificial composers contribute to the process of creating music.

The emergence of copyright concerns over AI-generated music has become apparent, while the majority of AI-powered voice-over tools are experiencing a notable increase. The process of automatically generating subtitles has also been revolutionized by the introduction of AI. By utilizing ML and natural language processing techniques, it is possible for machines to automatically generate video subtitles. YouTube has incorporated AI technology to offer both human and auto-generated captions, benefiting both content creators and consumers in a variety of ways.

The potential of AI extends beyond pessimistic forecasts of human displacement. On the contrary, AI functions as a valuable collaborator, augmenting a wide range of jobs. The potential of human-AI collaboration in the media and entertainment sectors is vast and boundless. The assured engagement of AI in several fields is a direct consequence of technological advancements. The usage of AI in the media and entertainment industries, namely, in the dissemination of visual content, is experiencing significant growth. AI is integral in enhancing user experiences by facilitating content production and personalization. The service of AI in targeted online advertising facilitates the achievement of maximum audience reach.

Similarly, AI is critical in supervising the distribution of online material and facilitating the organization of media to enhance the efficiency of user searches. The integration of AR and VR technologies in film production has gained significant attention in recent years. The convergence of AR and VR technologies, in conjunction with AI, presents opportunities for engaging and immersive experiences within entertainment. Within film production, which includes a multitude of complex procedures, the use of AI and automation serves to streamline many operations, including screenplay categorization, the visual storyboarding process, and timetable planning.

The integration of AI technology facilitates the efficient streamlining of text, visual, and voice searches, resulting in relevant outcomes. The influence of AI on the ranking of search engines necessitates marketers to strategically organize their content to effectively target certain users. The advent of AI has significantly transformed the entertainment industry, introducing enhanced intelligence, interactive experiences, and novel ways to express creativity. As AI assumes a prominent role, organizations must develop strong strategies to effectively leverage its capabilities.

4.7 Finance

The integration of AI has led to significant changes in various sectors, including the financial industry. The advancement of technology has resulted in the efficient completion of routine activities through automated procedures, leading to a significant change in the way work is coordinated inside various organizations. The financial sector is not immune to this pervasive tendency, as technology-driven solutions have become the foundation for several operating procedures. The effectiveness of doing manual evaluations on large amounts of data and the risks connected with erroneous information have decreased due to time limitations and a need for expedited results.

AI is regarded as a catalyst for facilitating smooth and uninterrupted customer interactions, eliminating the necessity for repetitive tasks, minimizing the occurrence of human errors and unreliable results, optimizing expenditure, automating various processes, identifying fraudulent activities, and offering personalized recommendations. The implementation of AI in the banking sector has resulted in numerous significant benefits. The integration of AI into finance is evident in various disciplines.

The usage of AI-powered algorithms enables the application of advanced computational methods to credit evaluation processes, resulting in the automation of credit applicant assessment and enhancing overall efficiency. This technology effectively captures data embedded inside reports and facilitates lending operations, thus reducing the need for extensive human intervention. Both retail and commercial lending organizations benefit from the efficient examination of documents and improved accuracy in credit evaluation. Robo-advisors powered by AI have taken on the responsibility of delivering personalized financial advice, offering investing enthusiasts a guide for managing their portfolios. These financial advisors skillfully utilize spending patterns and savings goals to provide financial advice and adjust investment portfolios by using a mixture of assets, such as stocks or bonds.

The utilization of AI enhances the efficiency of debt collection operations, improves the management of invoices using automated technology, and expedites the process of account reconciliation in commercial banking. The benefits of these improvements are evident in the acceleration of processes, a reduction in errors, and a noticeable improvement in the loan recovery rate. The insurance industry is heavily influenced by AI, which plays a significant role in determining the pricing mechanisms that form the foundation of insurance premiums. These pricing mechanisms rely on the assessment of clients' risk profiles. The application of calculus results in cost reductions and enhances customer satisfaction.

Similarly, AI facilitates the management of assertions through rapid examination, adjustment, inquiry, and analysis, resulting in increased effectiveness and the detection of deceptive tactics. AI supports audits and regulatory compliance by helping in detecting fraudulent activities and strengthening the security of the system. It contributes to the commitment to regulatory compliance and the improvement of operating costs. The strength of AI resides in its ability to efficiently analyse vast numbers of documents, ensuring compliance with established regulations.

AI-enabled tools are key in the implementation of 'Know Your Customer' (KYC) procedures by effectively identifying abnormal patterns and areas of potential risk without requiring human participation. AI also contributes its knowledge to customer inquiries, analysing possibilities for increasing sales and promoting related products, as well as predicting trends in customer churn. In trading, AI has taken on the responsibility of providing valuable insights into the complex fabric of market movements and conducting prospective analysis. This entrusts financial institutions with the necessary tools to make informed and prudent decisions. The recommendations provided by AI are customized to accommodate the risk preferences demonstrated by individuals, enhancing their investing strategies.

AI has brought about a significant transformation in the finance industry, leading to a period of renewed growth and development. This has been marked by the implementation of efficient operational processes, strengthened decision-making structures, and enhanced customer satisfaction. These advancements have been made possible through the utilization of automation and the extraction of valuable insights from data. The seismic effects of AI's presence are felt throughout several sectors, including lending, investments, operations, insurance, compliance, customer service, and trading.

The integration of AI is now widely accepted as an essential element in contemporary finance, serving as a fundamental building block for future advancements in this esteemed discipline. *Have you ever used a chatbot to guide you through the complexities of opening a savings account? Has a financial institution ever implemented a procedure to obtain your verification in instances of anomalous credit card activity?*

AI is currently seeing rapid expansion, exerting its revolutionary influence on various businesses and sectors. Finance and banking firmly hold a prominent position in this procession, effectively utilizing the capabilities of this groundbreaking technology. Undoubtedly, AI can be utilized to thoroughly assess the eligibility of an individual seeking a loan. Financial institutions and software systems utilize ML algorithms to efficiently and accurately evaluate loan eligibility, providing customized solutions promptly. The value of AI lies in its capacity for objectivity, which enables it to make prompt and accurate decisions.

The persistent challenge of risk mitigation applies to a wide range of industries, including the banking sector. ML algorithms are capable of identifying discernible patterns, accurately identifying areas of danger, optimizing resource allocation, and enhancing predictive capabilities through a thorough analysis of data. AI is at the forefront of technological advancements in fraud detection. These systems are specifically built to analyse spending habits to identify any irregularities or deviations from normal patterns. When such anomalies are detected, the systems promptly provide detailed alerts to bring attention to any fraudulent activities. The application of AI enables the efficient and precise assessment of potential customers based on a variety of parameters, including data collected from smartphones, without the influence of human bias.

Portfolios and financial data are subjected to expedited analysis through the utilization of algorithms infused with AI, facilitating prompt access to trends and information. Within the trade, AI extensively analyses vast data sets, thereby accelerating various procedures and providing valuable insights. Virtual assistants and conversational agents enhance customer service encounters by facilitating tasks such as balance inquiries, payment schedules, and personalized advice in a seamless manner. Cybersecurity also greatly benefits from the utilization of AI, since it possesses the ability to swiftly identify deviations and rapidly notify organizations, thereby mitigating instances of security breaches caused by human error.

The use of AI in analysing expenditure patterns significantly enhances predictive models used to assess the risk associated with loans. The significance of this phenomenon is apparent in areas where traditional methods of evaluating creditworthiness are insufficient but digital networks are widespread. The efficiency and productivity of client engagement are consistently enhanced by the constant availability of virtual assistants and conversational agents. The implementation of automated processes for repetitive operations serves as a symbol of AI's capabilities, resulting in reduced errors and a staff that is freed from mundane responsibilities.

The adaptability of AI is demonstrated by its ability to adjust its capabilities to accommodate activities of different lengths, making it a versatile tool that can be applied to both short-term and long-term jobs. The use of AI in underwriting judgements, particularly when it favours marginalized communities, is regarded as a commendable practice that leads to the attainment of sound judgement. The integration of AI yields various benefits, including economic gains, improved consumer experiences, and enhanced decision-making capabilities. Eventually, the integration of AI gives rise to a new approach that offers cost-effective solutions, enhances customer experiences, and leads financial institutions into a domain marked by informed decision-making based on data-driven analysis.

4.8 Healthcare

AI is extensively utilized across various medical domains, including the management of chronic conditions such as cancer and the interpretation of radiographic images. The main purpose of this effort is to optimize the process of developing influential innovations that enhance the standard of patient care. To ascertain productive resolutions for these circumstances, AI offers numerous advantages when compared to traditional approaches in analytics and clinical decision-making. The use of AI algorithms enhances the precision of systems by utilizing trained data and facilitating the understanding of treatment variations, care processes, diagnostics, and patient outcomes.

The fundamental stages of healthcare contain the management and coordination of medical information and documentation, including patient records and medical histories. Digital automation is a widely used methodology that involves the utilization of robotic systems to collect, store, and retrieve data, thereby enhancing the accessibility of standardized information. Robots exhibit remarkable aptitude for executing tasks that require repetitive actions.

Robotic systems are critical in facilitating a wide range of operations, including the expedited interpretation of medical images such as X-rays and computerized tomography (CT) scans as well as efficient data entry processes. These systems have exhibited notable effectiveness, especially in cardiology and radiology. It is projected that in the forthcoming years, the healthcare sector will progressively depend on robotic systems to execute fundamental tasks, thereby enabling healthcare practitioners to allocate their attention to complex situations. The procedure of formulating a treatment regimen involves the use of multiple data sources, including patient records and scientific research, to provide personalized recommendations.

Precision medicine is a multidisciplinary domain that encompasses the integration of genetics and genomics to detect anomalies and establish correlations between genetic variations and diseases. AI is of significant importance in expediting the timely identification of medical ailments, including cancer, through scrutinizing bodily scans and integrating genetic data. Wearable health monitors, such as Fitbit devices, can monitor and record various health-related data, thereby providing valuable assistance to healthcare professionals and specialists. AI assesses healthcare systems to identify deficiencies in treatment and operational inefficiencies. By using this procedure, AI can enhance healthcare facilities by reducing preventable hospitalizations and enhancing overall efficiency.

The advent of technology has precipitated substantial transformations in pharmaceutical research. The application of this technology includes a wide range of disciplines. However, the primary objective is to find and validate pharmacological targets. The primary stage of pharmaceutical development is investigating the functional significance of prospective molecular targets, such as genes or proteins, in pathological conditions. The effectiveness of pharmacological medications is determined by this factor. The methods used include the examination of functional genomics, structural genomics, proteomics, in vitro experiments utilizing cell-based models, and in vivo research conducted on animals. AI utilizes data analytic techniques to scrutinize information obtained from an extensive pharmacological knowledge library. This database includes several aspects, such as drug candidates, gene expressions, protein interactions, and clinical data.

The main objective is to forecast the effectiveness of treatment for these specific targets. Different methodologies, including deep autoencoders, relief algorithms, and binary classification techniques, are used to prioritize prospective pharmacological targets. AI utilizes computer vision and ML techniques to analyse and comprehend the three-dimensional configuration of protein structures. This is achieved by leveraging two-dimensional microscopy data. This process requires the examination of compounds and the refinement of potential leads, employing methodologies like high-throughput screening, virtual screening, and combinatorial chemistry.

AI-driven virtual screening uses extensive compound databases sourced from publicly available repositories. Various ML algorithms, such as naive Bayesian classifiers, k-nearest neighbours, support vector machines, random forests, and neural networks, have demonstrated their effectiveness in identifying lead compounds. Preclinical research is conducted to evaluate the safety and effectiveness of novel medicinal compounds. AI is necessary in the analysis of molecular mechanisms by employing unsupervised clustering-based machine learning techniques to understand RNA sequencing data.

Pharmacokinetic and pharmacodynamic modelling, another developing field, is utilized to forecast the reaction of drugs, utilizing in vitro and preclinical data. The application of ML techniques enhances the accuracy of combination therapy forecasts. The Deeptox algorithm [Kla+16] demonstrates a high level of accuracy in predicting the toxicity of substances. In silico methodologies use transcriptome data to forecast pharmacological qualities, thereby augmenting the usefulness of clinical trials through disease identification, molecular effect prediction, and facilitation of patient adherence.

AI facilitates the implementation of risk-based monitoring strategies and the identification of suitable patients for clinical trials, thereby facilitating the discovery of biomarkers associated with various illnesses. Some of the challenges and potential advancements in this field include the conversion of laboratory findings into formats that are compatible with existing systems, the assurance of data integrity, and the implementation of robust security procedures. The patenting process and ensuring data protection have complexities. The evaluation of novel compounds in conjunction with established agents is a formidable task; the use of algorithms can facilitate the acceleration of drug discovery. The potential for transformative impacts is positioned to fundamentally disrupt the field of drug discovery and the pharmaceutical business.

The application of ML in disease detection holds significant relevance. The process of traditional disease diagnosis necessitates a significant depth of medical expertise that is often obtained through years of study and practice. Diagnostic procedures may require much time and effort. In specific domains, there exists a deficiency in specialized knowledge, which poses obstacles to the healthcare sector and leads to delays in the diagnostic process.

ML has shown notable advancements, particularly with the advent of resilient deep learning algorithms. These improvements have resulted in significant accomplishments in the automation of disease identification as well as the improvement of affordability and accessibility in diagnostic methods. Algorithms can identify patterns that bear a resemblance to the cognitive processes used by medical practitioners. Algorithms require a substantial number of precise instances in digital format to comprehend.

The utilization of digital diagnostic data provides significant advantages, as illustrated by the detection of lung cancer or strokes through CT scans, the evaluation of heart risks via electrocardiograms, and the identification of diabetic retinopathy in eye photographs. The power of algorithms to analyse large volumes of data has made them strong competitors for diagnostic specialists. The consistent provision of timely and cost-effective outcomes has the potential to enhance the global accessibility of quality radiological diagnostics. ML offers a promising solution to the cost constraints encountered in pharmaceutical development. This novel approach has the potential to revolutionize analytical processes by enhancing efficiency, expediting project deadlines, and minimizing costs. ML exhibits substantial usefulness at every stage.

The crucial phases include the identification of appropriate targets for prospective therapeutic interventions. ML systematically analyses a wide range of data sources intending to achieve desired objectives, and hence the beginning of various projects. Algorithms are critical in the discovery of pharmaceutical candidates, facilitating the rapid assessment of molecular fitness and structural features. This leads to a reduction in the time required for candidate identification. The rapid advancement of science and technology enhances the likelihood of significant advancements. The clear utility of ML is observed in clinical trials. The optimization of participant selection serves to enhance the efficiency of research. The recognition of inconclusive findings catalyses timely adjustments and the refinement of research methodologies. The utilization of ML techniques has brought about a resurgence in biomarker research for diagnostic purposes, enabling the automated identification of biomarkers with enhanced efficiency and precision.

ML, although demonstrating remarkable effectiveness in pharmaceutical research, serves as a valuable adjunct to healthcare experts rather than a complete substitute for their expertise. This technology facilitates the analysis of complex data, optimizes decision-making processes, and augments capabilities. In addition to its application in drug development, ML has been utilized in personalized healthcare to customize therapies based on statistical studies. The ability of ML to analyse patterns facilitates accurate predictions of patient reactions, thereby assisting in the personalization of treatment. Through the process of comparing instances, professionals in the field can enhance therapeutic approaches, which in turn have the potential to increase overall well-being.

The CRISPR-Cas9 system [Ran+13], which is considered revolutionary in gene editing, has the potential to significantly advance scientific research. However, despite its promising capabilities, there are still obstacles that need to be addressed to mitigate any unexpected repercussions that may arise from its use. The utilization of ML techniques enables the prediction of interactions between guide RNA and target DNA, thereby expediting precise DNA alteration and enhancing the precision of gene editing procedures. The convergence of predictive modelling, pharmaceutical research, and healthcare presents a multitude of potential opportunities. AI and ML algorithms are key in facilitating discovery, inquiry, and intervention, enhancing effectiveness and accuracy, and holding significant potential for advancements in medicine.

4.9 Manufacturing

The incorporation of AI and ML has become an essential component of the manufacturing industry, operating in close conjunction with the Industrial Internet of Things (IIoT) and closely aligned with the ideas of Industry 4.0. The use of AI in the manufacturing sector includes a wide range of applications, among which ML has emerged as a prominent subset. Within process manufacturing, an industry characterized by intense competition and complex systems, AI and ML have emerged as strategic instruments that promote innovation and enhance profitability.

ML is of utmost importance in wide fields, including predictive analytics, robotics, predictive maintenance, and automated operations, among others. The utilization of this technology in these sectors leads to improved effectiveness, economic benefits, and greater security within industrial enterprises. The importance of AI and ML in process manufacturing is substantial. These technologies enable the integration of many data sources, perform analysis, and then provide important insights and predictions that significantly enhance the decision-making process.

ML functions by performing analysis on large data sets to detect complex patterns and develop trends. The previously stated findings are later used to develop models that can create predictions regarding future occurrences. The application of ML in manufacturing facilities enables the precise prediction of market demand and supply fluctuations, the determination of optimal maintenance strategies, and the timely detection of early signs of anomalies. The integration of AI and ML has significant implications for industrial companies.

The integration described above presents a range of advantages, such as increased operational efficiencies, better understanding of market trends, compliance with regulatory requirements, reinforced safety protocols, improved product quality, streamlined production processes, enhanced visibility within the supply chain, early detection of faults, thorough analysis of root causes, and extended equipment lifespan. To enhance the efficiency of AI and ML implementation in the manufacturing sector, it is crucial to emphasize some fundamental techniques. The strategies involve proficient data management, precise objective establishment, substantial utilization of AI and ML technologies, evaluative appraisal of competencies and capabilities, and the establishment of a corporate culture that emphasizes data-driven decision-making.

The integration of AI and ML has the potential to significantly improve industrial productivity. This purpose can be achieved through several tactics, such as improving production quality, predicting market fluctuations, mitigating unforeseen circumstances, and strengthening safety protocols. The iterative process occurs continually, resulting in the persistent development of value and the gradual enhancement of revenue generation over an extended period. The significance of AI and ML in manufacturing, namely, in process manufacturing, is undeniably significant.

The integration of the system, in conjunction with the Industrial Internet of Things (IIoT) and within Industry 4.0, presents numerous benefits. The advantages include improved operational efficiency, increased market intelligence, compliance with regulatory requirements, enhanced safety protocols, improved production quality, optimized manufacturing processes, enhanced visibility within the supply chain, early detection of faults, systematic analysis of underlying causes, and extended equipment lifespan. The potential benefits of integrating AI and ML into industrial enterprises are significant and long-lasting. This objective can be accomplished by employing intelligent implementation approaches and embracing data-centric practices. The present integration of AI and ML is revolutionizing and enhancing the modern manufacturing industry.

4.10 Weather Forecasting

Weather forecasting has been revolutionized by AI, which has led to improved prediction methodologies and dissemination strategies. Since the 1970s, the accuracy of weather models used by broadcasters has been steadily improved by using AI technologies. These models, which rely on complex algorithms run on high-powered computers, are given an extra boost by ML. The influence of AI may be seen in many areas, such as in customized smartphone apps that provide up-to-the-minute weather reports. Content triggers powered by AI in these apps perform actions automatically in response to changes in the weather. When temperatures exceed certain limits, for instance, users are alerted for their protection. This individualized strategy improves readiness for weather-related occurrences by ensuring timely and relevant information transmission.

Through the examination of raw data in real time, AI is also helping to spot potential lightning strikes, hurricanes, and floods. Users can then take the necessary precautions thanks to timely automated warnings that warn them of oncoming severe weather. AI enhances weather broadcasts by adding lifestyle material, such as health, wellness, and scientific information, to complement meteorological reports. The use of AI in traffic monitoring also improves the reliability and timeliness of reports, letting viewers make better plans for their trips. Broadcasters can provide more accurate reports on traffic conditions by using AI-derived insights. Using AI to improve weather predictions has several benefits. Because AI algorithms are so good at quickly processing large data sets, they can provide reliable forecasts in a short amount of time.

Meteorologists can better serve the public, businesses, and governments by analysing real-time data. This speed is particularly useful in situations where it is necessary to make quick decisions, such as when the weather is rapidly changing, as in the case of severe storms or impending hurricanes. Better and more reliable weather predictions can be made with the help of AI. Also, ML is integral in improving the accuracy of weather forecasts. Insights are gleaned from both historical and real-time data by AI algorithms, with ML algorithms being particularly adept at this. They pick up on tiny patterns and connections that more traditional methods of analysis can miss. The cutting-edge weather models can accurately represent complex weather phenomena that are challenging to recreate using conventional methods.

As a result, thanks to these developments, forecasts have improved, allowing individuals and organizations to make better decisions in light of current data. AI-based weather forecasting also has the benefit of increasing operational efficiency. By streamlining data processing and other operational tasks with the help of AI, meteorologists are free to focus on doing what they do best:

- Evaluating weather data
- Analysing weather data
- Communicating weather data

This change allows for more in-depth research into the consequences of weather patterns, resulting in new insights and actionable suggestions for a wide range of stakeholders.

Gains in productivity thanks to AI-driven processes also translate to more efficient use of resources and lower overall costs for weather forecasting operations. ML algorithms analyse large amounts of atmospheric data to find useful insights, reveal previously unseen trends, and adjust meteorological models for improved performance, making them the backbone of AI weather forecasting. A great deal of this information comes from sources like the National Weather Service. While similar, different weather models can depict a wide range of possible outcomes on a global scale due to their differences. The butterfly effect, however, makes it difficult to achieve 100 per cent accuracy in weather forecasting. In 1972, Edward Lorenz asked, 'Does the flap of a butterfly's wings in Brazil set off a tornado in Texas?' This phenomenon illustrates the difficulty of trying to quantify the impact of small events on the atmosphere.

There will always be some degree of error in weather forecasting due to factors like the complexity and randomness of the environment and our inability to pinpoint the location of every air particle. Even with these obstacles, though, accurate weather predictions may be possible with the help of AI. Through the use of complex weather models able to analyse large amounts of data, AI enables meteorologists to create much more accurate predictions than was previously possible. Multiple elements, such as data quality and quantity, AI model complexity, and the nature of the weather phenomenon being projected, contribute to the reliability of AI-powered weather forecasts.

In line with the general tendency in weather forecasting, AI predictions made for shorter periods (a few days) tend to be more accurate than those made for longer periods (weeks or months). AI is limited in its ability to predict extreme weather events because it is trained to look for patterns in historical data. The forecasts improved threefold in comparison to other weather services once they incorporated AI and human knowledge. However, several persistent obstacles remain in the way of using AI for weather forecasting. Due to the complexity of weather systems, pinpoint accuracy is impossible to achieve. Interconnected factors such as atmospheric conditions, geographical features, and chaotic behaviours make accurate forecasting of weather patterns extremely challenging.

Also, some weather stations or forecasting companies may not be prepared to exploit AI-driven insights, which could limit the full potential of AI in weather forecasting. For AI to be effectively integrated, the necessary infrastructure, resources, and training must be made available to collect, analyse, and interpret massive amounts of weather data. Investing in AI is critical for weather forecasting organizations because it equips meteorologists and forecasters with the knowledge and tools they need to fully utilize AI.

While AI is essential for improving the quality and timeliness of weather forecasts, human expertise is still required for translating and disseminating that data. Expert human meteorologists can take numerical meteorological data and turn it into useful insights and actionable consequences. A meteorologist can provide context for the effects of a snowstorm on morning traffic and commuters if, for instance, an AI system forecasts high snowfall due to an impending blizzard. Integrating AI-improved forecast accuracy with distinctive human stories has the potential to greatly improve individuals' daily experiences.

The usage of AI in weather forecasting dates back to at least the 1970s. Supercomputers running complex machine learning algorithms may one day improve the reliability of weather predictions. The world needs a combination of AI and human knowledge that can provide the most accurate forecasts possible across a wide range of geographic areas. To do this, AI takes into account data from over one hundred different prediction models throughout the world, adjusting variables based on location, time, weather type, and the accuracy of recent forecasts. AI is useful for improving meteorological efficiency, but it is not yet ready to take on the role of human meteorologists. Expertise from humans is still needed to glean practical insights from data that is relevant to viewers and provides local context. The meteorologist's continued importance is promoted by their storytelling and presentation skills and the fact that they serve as a distinctive identity for various commercial media outlets.

5 Advancing Cybersecurity Through AI

Present-day security professionals face an abundance of obstacles that hinder their ability to protect data, administer user access, and expeditiously detect and mitigate threats. Such issues arise due to various variables, including the presence of highly skilled adversaries, the continuous expansion of potential targets for attacks, the overwhelming volume of data to be managed, and the growing complexity of infrastructure systems. The utilization of AI technologies in this particular context has several benefits, one of which is the capability to ensure the security of data within hybrid cloud systems. AI systems provide the capability to identify shadow data, detect anomalies in data access, and notify security staff regarding potentially risky behaviours shown by those who are interacting with the data. This functionality enables the prompt identification and resolution of problems, thereby conserving significant time.

Before the incorporation of AI, the field of cybersecurity predominantly depended on traditional approaches, with a particular emphasis on signature-based detection systems. These systems operate by comparing incoming network traffic with a database of known threats or malicious code signatures. Upon the occurrence of a match, the system promptly triggered an alarm and implemented actions aimed at impeding or isolating the identified threat. Although skilled at mitigating established risks, this methodology proved inadequate for addressing emerging and undiscovered risks.

Perpetrators can effectively evade detection systems that rely on signature-based methods by modifying code or developing novel iterations of malware that have not yet been documented. These systems tended to produce a significant number of false positives, whereby lawful network activity could be mistakenly flagged as malicious if it shared similarities with a recognized security issue. As a result, security analysts have allocated significant efforts to the examination of false positives, placing a burden on organizational resources.

During the previous era of cybersecurity, manual examination played a crucial role. Security analysts are actively involved in the practical assessment of security alerts and records, to identify trends or indicators that may suggest the occurrence of security breaches. These procedures are characterized by a significant investment of time and a strong dependence on the analyst's expertise in detecting potential dangers. Rule-based systems were also utilized, involving the creation of rules or policies that defined permissible behaviour within the network. The violation of these regulations resulted in the activation of an alert. While these systems demonstrated usefulness in specific situations, their limited flexibility and adaptation to evolving threats were frequently observed. In general, the field of pre-AI cybersecurity predominantly used a responsive approach, relying on manual examination, detection systems based on signatures, and mechanisms based on rules.

Regrettably, this particular approach demonstrated inadequate effectiveness when confronted with unknown dangers, frequently resulting in an excessive number of false-positive outcomes and placing a burden on available resources. AI-infused cybersecurity diverges from traditional techniques across various dimensions. Historically, the field of cybersecurity has predominantly relied on signature-based technologies that were specifically developed to address established threats. As a result, previously unknown and unexplored dangers remained unnoticed. On the other hand, AI solutions utilize ML algorithms to identify and mitigate both familiar and unfamiliar risks in real time. The algorithms undergo training using a significant amount of data, which includes historical threat data as well as real-time network and endpoint information.

This training facilitates the ability of AI-based systems to detect complex patterns that are not perceptible to human observers, enabling real-time identification and response to threats without the need for human involvement. For example, ML algorithms can analyse patterns in network traffic to identify abnormal activity that may indicate a cyberattack. This can then prompt security staff to be alerted or even trigger automated actions to minimize the potential risk. One more difference between AI-driven solutions and conventional approaches is their inclination towards ongoing learning and development. In response to emerging dangers, ML algorithms have the potential to undergo retraining using up-to-date data to enhance their ability to detect and mitigate these threats.

As a result, AI systems continue to adapt to the dynamic threat environment, offering increasingly advanced cybersecurity measures as time progresses. The integration of AI into the field of cybersecurity represents a significant and revolutionary change in the strategies and methodologies used by organizations. AI solutions provide enhanced protection against both familiar and unfamiliar risks by utilizing ML algorithms to accurately identify and rapidly mitigate potential threats. This results in enhanced safeguarding of confidential information and essential infrastructure. The primary focus of AI cybersecurity revolves around the timely identification and mitigation of potential threats. AI systems provide the capability to analyse vast data sets, thereby revealing discernible patterns that serve as indicators of potential cyber dangers.

These algorithms greatly improve the precision and ranking of detected threats. This is particularly relevant in light of the increasing occurrence of cyberattacks, which can be attributed to the widespread adoption of Internet of Things (IoT) technology. Despite the considerable advancements made in cybersecurity, there are still existing vulnerabilities that highlight the important role of AI in enhancing the security of data and devices. The integration of AI with cybersecurity is seen in various applications. An example that can be cited is network threat analysis. As the reliance of organizations on digital services continues to grow, the necessity for changes to internal networks becomes increasingly critical. The extensive scale of these networks necessitates a significant allocation of security resources to protect the data they contain. The amalgamation of AI with cybersecurity allows for a surveillance system that monitors both incoming and outgoing network traffic. This integration greatly assists in the identification and detection of cybercriminal activities and potential threats.

Malware detection, which is regarded as a critical component, confers substantial benefits when AI is implemented. The prospective application of AI in cybersecurity is the proactive detection and identification of malware before its ability to carry out detrimental actions. AI exhibits the capacity to differentiate among various classifications of malicious software, a critical aptitude given the extensive set of behaviours that are included under this class of threats. AI has already shown its usefulness in supplementing malware analysis. The presence of several dangers requires the implementation of efficient management strategies, an area in which AI solutions demonstrate exceptional performance. Through the automation of jobs and the improvement of prioritization, AI enables analysts to allocate more time towards addressing high-risk risks and formulating successful strategies.

The progression of AI is paralleled by the progress of cyberattacks, which are characterized by increased complexity and frequency. Hackers are currently engaged in the development of AI tools to identify network vulnerabilities. They are actively creating sophisticated malware that possesses intelligent capabilities to tailor attacks based on user data. AI exhibits the potential for enhancing cybersecurity measures, although it also presents the possibility of being exploited for malicious purposes. Therefore, organizations must prioritize the protection of their data, particularly when they heavily depend on legacy systems that are challenging to replace or update.

6 Advantages and Benefits

The field of AI is continuously expanding the capabilities of machine-enabled functionalities. This cutting-edge technology facilitates the ability of machines to operate autonomously, leading to the efficient completion of repetitive tasks. AI is important in the development of a cutting-edge work environment that promotes efficient cooperation between organizational systems and humans. Hence, it may be argued that the advent of rising technology does not render human resources obsolete but rather enhances and reinforces their contributions. Unquestionably, AI allows enterprises to allocate resources towards more advanced and complex undertakings. In this section, we will attempt to explore the potential advantages of the subject matter, enabling an understanding of AI and its associated benefits.

6.1 AI and Availability

AI has evolved as a disruptive technology with revolutionary effects in wide fields, introducing an array of capabilities and efficiencies. The continuous availability of AI is a notable benefit that carries significant effects across various industries, including healthcare, finance, manufacturing, and customer service. The perpetual availability of AI systems can be attributed to their characteristics, as they function ceaselessly without requiring breaks or periods of rest. The continuous accessibility of AI offers a range of advantages that lead to increased efficiency, greater decision-making capabilities, and streamlined operational procedures. Within healthcare, the perpetual accessibility of AI serves as a fundamental source of assistance.

AI-powered solutions can surpass temporal limitations in medical diagnostics and patient monitoring by continuously analysing data and generating valuable insights. For example, an AI system that monitors a patient's vital signs can immediately identify abnormalities and notify medical workers, facilitating appropriate intervention. The ability to maintain constant alertness without succumbing to human fatigue provides a significant benefit since it has the potential to avert situations that could endanger lives. Also, telemedicine platforms that utilize AI-driven chatbots offer patients continuous access to medical information and support, addressing health-related inquiries promptly and at any time. Financial institutions also leverage the uninterrupted functionality of AI for their benefit.

The dynamics of the stock market, currency exchange rates, and investment opportunities undergo continuous evolution. AI systems can efficiently analyse extensive data sets, exhibiting exceptional processing rates while simultaneously detecting and discerning trends, patterns, and anomalies. The continuous analysis provided around the clock enables traders to make well-informed judgements that surpass the limitations of human capabilities, ultimately resulting in the optimized administration of investment portfolios. Also, fraud detection systems powered by AI function with persistence to identify atypical transactions, thereby mitigating financial losses through timely identification and alerting of suspicious activity, irrespective of the temporal context.

AI-driven automation in the manufacturing industry greatly benefits from its perpetual availability. Factories and assembly lines that are outfitted with AI-powered robotics ensure uninterrupted work, obviating the necessity for scheduled breaks or shift changes. The continuous functioning of this process optimizes productivity, minimizes periods of inactivity, and guarantees a uniform level of product excellence. Also, the use of AI plays a role in enhancing predictive maintenance, which is considered a fundamental aspect of optimizing industrial productivity. Machines that are equipped with sensors facilitate the transmission of real-time data to AI systems. These AI systems then analyse the data patterns to make predictions regarding maintenance needs. This process serves to minimize the occurrence of unplanned outages and optimize the scheduling of maintenance activities.

The availability of AI-driven solutions has led to a paradigm shift in customer service, enabling round-the-clock support. Chatbots and virtual assistants are capable of interacting with clients at all hours, addressing queries, furnishing information, and even enabling transactions. The ongoing engagement between customers and service providers enhances customer satisfaction by minimizing any delays in response and facilitating a smooth provision of support. The use of AI-powered language translation tools effectively addresses the issue of language barriers in global markets. These technologies provide real-time communication in many languages, thereby promoting the development of worldwide commercial partnerships that surpass the limitations imposed by different time zones.

The reputation of twenty-four-seven AI availability is also seen in data analysis. Business enterprises are faced with an overwhelming amount of extensive data sets that necessitate analysis. The tireless computational skills of AI enable uninterrupted data analysis, facilitating the discovery of insights and trends that may otherwise remain concealed amid the immense expanse of information. The fast capture of time-sensitive possibilities and the early identification of potential threats are crucial. As a result, the process of strategic decision-making becomes more knowledgeable and adaptable, thereby enhancing competitive advantage in dynamic market environments. It is important to recognize the constraints and obstacles linked to the perpetual accessibility of AI.

The issue of ethical considerations arises concerning the possibility of excessive dependence on AI systems, which could lead to the marginalization of human involvement and decision-making. Privacy concerns arise when AI systems engage in the continuous processing of personal data, hence demanding the implementation of strict security measures to protect sensitive information. Similarly, the continuous functioning of AI systems necessitates the presence of resilient technological infrastructure and optimized energy utilization, thereby raising concerns about sustainability and the potential environmental consequences. The continuous accessibility of AI is a significant benefit with far-reaching consequences in multiple industries.

The sectors of healthcare, banking, manufacturing, customer service, and data analysis all derive advantages from the consistent operational capabilities of AI systems. This advantage facilitates prompt medical interventions, well-informed financial decisions, optimized industrial processes, greater customer experiences, and improved data analysis. To fully use the continuous accessibility of AI, it is important to confront and overcome the ethical, privacy, technical, and sustainability obstacles that arise. The ongoing advancement of AI holds the potential to significantly transform various industries and redefine the potential for achieving smooth, efficient, and successful operations.

6.2 AI and Scalability

AI represents a notable technical progression with its use across several sectors, including healthcare, banking, transportation, and entertainment. In all the above examples, scalability is a crucial characteristic that significantly enhances the significance of AI due to its capacity to successfully manage expanding workloads. Scalability refers to the ability of a system to effectively and efficiently adjust and operate at an optimal level in response to increasing demands of growing complexity. In AI, the concept of scalability is important in guaranteeing the extensive and enduring implementation of AI solutions.

AI systems are engineered to replicate human cognitive capabilities, including learning, reasoning, and problem-solving, through the analysis and manipulation of extensive data sets. The advantage of AI's scalability is evident in its capacity to effectively handle increasingly large data sets while maintaining optimal performance. The integration of AI applications into various aspects of human life is leading to a significant increase in the generation and consumption of data. This expansion is expected to follow an exponential trend. The utilization of scalable AI frameworks is crucial to effectively manage the influx of data while minimizing delays and inefficiencies. The arena in which the scalability benefit of AI is prominently demonstrated is data analysis.

In several domains, such as finance and marketing, organizations handle vast amounts of data regularly, which includes information on client behaviour, market trends, and financial transactions. The utilization of scalable AI technologies, such as parallel processing approaches, allows organizations to efficiently derive significant insights from extensive data sets. With the continuous accumulation of data, the scalability of AI guarantees the maintenance of accurate and timely analysis, thereby augmenting decision-making processes. The concept of scalability is fundamental to the advancement of AI-driven systems that depend on real-time processing. Autonomous vehicles, for example, warrant consideration.

These vehicles produce a wide range of real-time data, including sensor inputs, GPS coordinates, and object identification data. To guarantee the safety and operational efficiency of these vehicles, AI systems must process this data with the shortest delay. Scalability facilitates the smooth incorporation of additional processing resources in response to the increasing complexity of real-time AI tasks. This capability empowers autonomous vehicles to efficiently and securely navigate complex settings.

The role of AI in predictive analytics is further enhanced by its scalability, in addition to its capability for real-time processing. To make accurate predictions, predictive models utilized in healthcare settings for tasks such as illness diagnosis or patient outcome forecasting necessitate substantial quantities of training data. With the increasing number of patients and the growing volume of medical data, the utilization of scalable AI algorithms becomes crucial in enabling the effective training and deployment of predictive models.

The scalability of AI technology allows medical personnel to effectively utilize it for making well-informed decisions regarding patient care, particularly in urgent situations where time is a crucial factor. Similarly, the ability of AI to scale plays a major role in driving progress in the fields of NLP and language translation. The increasing prevalence of the Internet and the process of globalization have led to a significant rise in the need for instantaneous language translation and comprehension. Transformer-based architectures, which are scalable AI models, facilitate efficient and precise language translation, thereby facilitating seamless cross-lingual communication. The expansion of user-generated content across many languages has been accompanied by the implementation of scalable AI systems.

These AI systems are key in reducing language barriers and facilitating enhanced global communication. The concept of scalability also includes the economic considerations associated with the implementation of AI. The computational demands of complex AI algorithms may pose challenges for conventional computer resources, leading to elevated operational expenses. In contrast, scalable AI frameworks facilitate the allocation of computational jobs among a cluster of machines, enhancing the efficiency of resource utilization and diminishing operational costs.

The cost-effective scalability of AI solutions incentivizes a greater number of companies to embrace these technologies, thereby stimulating innovation and forging competitiveness across a wide range of industries. It is important to recognize that the attainment of scalability in AI is accompanied by several obstacles. The process of developing algorithms that can effectively allocate jobs across numerous processors or nodes necessitates deliberation. To fully leverage the scalability of AI systems, it is important to address key factors such as load balancing, data synchronization, and communication overheads.

Also, the increasing need for computational infrastructure to achieve scalability gives rise to concerns regarding energy consumption and the sustainability of the environment. The scalability of AI is a notable benefit, as it can effectively handle increasing workloads and larger amounts of data without compromising its performance. This characteristic positions AI as a highly influential tool in various fields. Scalable AI frameworks facilitate the effective resolution of complex difficulties for industries and researchers, including a wide range of applications such as data analysis, real-time processing, predictive analytics, and language translation. The growing reliance on data and interconnectedness around the world highlights the significance of AI scalability in developing innovation, optimizing decision-making processes, and augmenting our general well-being.

6.3 AI and Accuracy

In contrast to human beings, AI systems do not experience fatigue or succumb to distractions. AI systems possess the capability to efficiently handle a significantly larger volume of information and constantly follow predetermined algorithms for data analysis and decision-making. These attributes significantly enhance their propensity to yield precise outcomes with a high degree of reliability. The progress made in AI has made substantial contributions to the advancement of a wide range of technologies, resulting in transformations across sectors and everyday existence. One of the noteworthy benefits that have arisen from these breakthroughs is the enhanced precision and diminished error rate of AI systems. This has extensive consequences throughout several sectors, including healthcare, finance, manufacturing, and entertainment.

This subsection aims to explore the complexities surrounding the impact of AI's improved accuracy and mistake-reduction techniques on modern society, emphasizing its role in tackling complex problems. AI systems, which consist of ML algorithms and neural networks, have been developed to replicate human cognitive processes. This allows them to acquire knowledge from data and execute tasks with ever-higher accuracy. The capacity to acquire knowledge and adjust based on facts is fundamental to the enhanced precision of AI. In contrast to conventional rule-based systems, AI systems possess the capability to effectively analyse extensive volumes of data and identify complex patterns that may elude human analysts. In medical diagnostics, AI-enabled technologies have exhibited exceptional precision in detecting initial indications of illness through the analysis of medical imaging data.

This has resulted in prompt interventions and enhanced patient outcomes. The ability to analyse large data sets not only accelerates the process of decision-making but also enhances accuracy by reducing the likelihood of human errors caused by fatigue and oversight. AI systems have demonstrated their effectiveness in minimizing errors by employing iterative learning and ongoing refinement processes. The iterative process of AI training facilitates the acquisition of knowledge by models through the recognition and rectification of errors, leading to a steady reduction in inaccuracies.

In language translation, AI-powered platforms have progressed from producing slightly distorted translations to delivering coherent and contextually precise renderings. The evolutionary process serves as evidence of the error mitigation capabilities of AI, which progressively enhances its algorithms through the assimilation of human feedback. In sectors that heavily rely on accuracy, such as the manufacturing industry, quality control systems powered by AI have significantly reduced the occurrence of errors by identifying abnormalities that human inspectors may overlook.

This not only enhances the quality of the product but also reduces the operational expenses associated with defective production. Due to their ability to process complex calculations with remarkable speed and dependability, AI systems exhibit increased precision and a reduced margin of error. AI functions continuously without interruptions or instances of diminished focus, thereby mitigating errors that can arise from human tiredness or a lack of concentration. This phenomenon is observable in financial trading, as AI systems swiftly execute complex trades, thereby mitigating the occurrence of errors stemming from delayed responses. AI-powered autonomous vehicles demonstrate enhanced precision in manoeuvring through traffic and evading accidents, surpassing human drivers in their capacity to analyse extensive volumes of up-to-the-minute information.

These applications highlight the consistent accuracy of AI and its capacity to surpass human talents in situations prone to errors. However, the journey towards attaining increased precision and minimizing errors in AI systems is not without obstacles. The accuracy of AI is contingent upon the calibre and variety of the data from which it acquires knowledge. Instances in which facial recognition algorithms exhibited racial bias due to imbalanced training data serve as illustrations of how biased or inadequate data can result in erroneous conclusions.

The maintenance of data integrity and variety is integral to improving the accuracy of AI systems across various populations and circumstances. Besides, the opaqueness inherent in many AI models, namely, deep neural networks, can impede the comprehension of decision-making processes. The presence of opacity in a system might pose challenges to diagnosing errors and hinder the process of identifying vulnerabilities. The advancements in AI have brought about significant benefits in terms of enhanced accuracy and decreased error rates. The revolutionary impact of AI systems lies in their ability to acquire knowledge from data, iteratively improve their algorithms, and perform complex computations with consistency.

This has led to significant transformations in sectors and disciplines that prioritize precision and reliability. The utilization of AI has yielded improved results, increased operational effectiveness, and reduced instances of errors across various industries, ranging from healthcare to manufacturing. It is important to adopt a cautious approach when it comes to the development and implementation of AI, placing significant emphasis on factors such as data quality, bias mitigation, and the promotion of openness. As AI progresses, the ability to leverage its capabilities to improve precision and reduce mistakes will play a major role in defining a future that is more efficient, dependable, and technologically advanced.

6.4 AI and Big Data

The proficient management of large data sets, popularly known as 'big data', has arisen as a significant issue and potential avenue for advancement. Concurrently, the use of AI methodologies inside this particular framework confers a significant benefit. This subsection will explore the importance of effectively managing large volumes of data and how the integration of AI enhances this capability. The term 'big data' refers to data sets that possess significant volume, complexity, and diversity, exceeding the processing capabilities of conventional database systems.

A wide range of sources contribute to the generation of these databases, which include social media interactions, sensor networks, and scientific investigations. The issue is not solely attributed to the extensive quantity but also to the wide range of data formats, including text, photos, videos, and other forms, which are commonly referred to as unstructured data. The management of big data sets includes several duties such as data collection, storage, processing, and analysis, all of which necessitate the utilization of scalable and efficient methodologies. AI is a field of study and research that focuses on the development of intelligent machines capable of doing tasks that typically require human intelligence.

AI, commonly referred to as the field dedicated to the development of intelligent machines, covers a wide set of techniques that facilitate the replication of human-like cognitive abilities in computers. These abilities include but are not limited to learning, reasoning, and problem-solving. In big data, AI offers a supplementary set of tools to effectively manage and extract valuable insights from these vast databases. One of the key benefits that AI offers in large data sets is its capacity to identify and analyse patterns, correlations, and insights that may not be immediately evident using conventional approaches. ML, which is a subfield of AI, assumes a crucial role in the extraction of significant insights from large-scale data sets.

Algorithms such as decision trees, support vector machines, and neural networks can acquire knowledge from extensive data sets to provide predictions, classifications, and choices. In medical diagnostics, the utilization of ML models that have been trained on patient data can significantly contribute to the early diagnosis of diseases. This work would be considerably more challenging to do without the computational capabilities provided by AI. Also, AI approaches enable the automation of data processing pipelines.

In conventional data analysis, a significant portion of the procedure involves manual manipulation and interpretation. AI can automate operations such as data cleansing, feature selection, and outlier detection. This automation expedites the data analysis process and mitigates the potential for human error. Another crucial aspect in which AI demonstrates exceptional performance is in NLP. NLP facilitates the computational comprehension, interpretation, and generation of human language by computers. Proficiency in handling textual data is a priority when working with unstructured big data, as it comprises a substantial proportion of such data.

Sentiment analysis, a well-recognized application in natural language processing (NLP), is capable of assessing the prevailing public sentiment towards certain products, services, or policies through the analysis of social media postings and reviews. The ability to analyse vast amounts of data provides businesses and policymakers with useful insights that can inform their decision-making processes.

However, it is crucial to recognize that, although AI has significant benefits, its incorporation into large-scale data poses certain obstacles. The primary importance lies in guaranteeing the quality and dependability of insights generated by AI. The presence of biases within the training data has the potential to be perpetuated, resulting in outcomes that are skewed.

Also, the complexity of AI models requires significant computational resources, hence requiring the implementation of effective parallel processing and distributed computing approaches, especially when dealing with extensive data sets. To fully harness the advantages of AI in the management of large-scale data sets, it is essential to adopt an interdisciplinary approach. The convergence of statistics and domain knowledge is necessary to effectively address the various aspects of this integration. Computer scientists are required to develop algorithms that possess the ability to successfully scale and demonstrate robustness in the presence of noise.

Statisticians are important in the validation of the integrity of insights given by AI, ensuring protection against overfitting or misinterpretation. Also, it is significant to note that domain specialists possess the requisite contextual knowledge that is essential in ensuring that the obtained insights are in line with real-world situations and can be effectively put into action. The efficient handling of large volumes of data is significantly enhanced with the integration of AI. Innovative solutions are required to address the issues presented by large data sets, their wide nature, and their complex characteristics.

AI possesses a distinct edge over traditional methods because of its adeptness in pattern recognition, automation, and natural language understanding. This advantage is not attainable through the use of traditional methods alone. The utilization of AI, facilitated by ML and natural language processing techniques, allows for the extraction of significant insights, which in turn brings advantages to several domains such as healthcare and business. However, it is necessary to adopt a collaborative methodology that incorporates computational, statistical, and domain-specific knowledge to ensure the responsible and efficient integration of AI with big data. The continuous advancement of technology is expected to have a transformative impact on many sectors, scientific research, and the way decisions are made.

6.5 AI and Automation

The AI transformative factor can revolutionize sophisticated problem-solving and significantly affect mundane and repetitive jobs. The automation of these seemingly mundane processes, which frequently require substantial human labour and time, can be effectively achieved through the application of AI techniques. This presents a huge advantage that not only improves productivity but also enables experts to concentrate on more innovative and valuable pursuits. AI comprises a range of technologies that facilitate the emulation of human intelligence by computers, enabling them to do tasks that were conventionally reliant on human intervention.

The automation of repetitive jobs is a particularly intriguing aspect of AI, as these tasks are distinguished by their predictable and monotonous nature. The completion of these duties is crucial; nonetheless, they can be significantly time-consuming, prone to errors, and demanding of resources when performed manually. The integration of AI in the automation of various processes brings forth enhanced efficiency, heightened precision, and increased reliability. Data entry and processing display the inherent value of AI in automating routine processes. The process of manually entering and handling substantial amounts of data is not only laborious but also prone to errors.

The utilization of AI-driven automation enables the flawless execution of data extraction, transformation, and loading processes, reducing the potential for human errors and significantly expediting the overall procedure. Consequently, this enables professionals to effectively distribute their cognitive resources towards more complex and analytical aspects of their professional activities. Also, the advantages of AI apply to tasks that necessitate the ability to recognize patterns and classify information. In several domains, such as image processing, AI algorithms can undergo training processes that enable them to identify complex patterns and characteristics within images.

This ability facilitates the accomplishment of tasks like quality control in manufacturing or the analysis of medical images. By assigning these responsibilities to AI systems, professionals can shift their focus towards the development of sophisticated algorithms or the creation of creative apps, thereby promoting intellectual advancement and technological development. Cybersecurity presents a convincing demonstration of the capabilities of AI in automating operations that involve repetitive actions. The task of recognizing and addressing security concerns necessitates ongoing attentiveness and a prompt reaction.

AI-powered security systems possess the capability to independently oversee network activity, identify deviations from normal patterns, and perhaps anticipate forthcoming security risks by leveraging past data. The implementation of real-time threat assessment not only lightens the burden on cybersecurity professionals but also enhances the resilience of digital infrastructures against developing vulnerabilities. Customer service has experienced a significant transformation as a result of the automation possibilities offered by AI. AI-powered chatbots and virtual assistants can interact with clients by responding to frequently asked questions, fixing issues, and even facilitating transactions. The implementation of automated customer interaction not only improves the overall user experience but also relieves human agents from engaging in repetitive encounters.

This allows them to dedicate their attention to addressing complicated and emotionally complex client concerns that require human intervention. However, the incorporation of AI to automate routine processes is not without its obstacles. The advent of automated systems that assume jobs that were previously executed by individuals gives rise to ethical concerns. The possible displacement of human labour gives rise to inquiries regarding the stability of employment and the socio-economic consequences it entails. Professionals are faced with the critical task of achieving an equilibrium between the advantages of automation and the ethical obligation to safeguard employment prospects for the workforce.

Within education, tools powered by AI have exhibited their capacity to optimize administrative duties, enabling educators to focus on enhancing their instructional approaches and delivering individualized support to students. Automating the grading of assignments through the use of AI algorithms presents a viable solution to relieve educators from the arduous responsibility of assessing repetitive evaluations. This technological advancement not only streamlines the grading process but also grants educators additional opportunities to actively interact with students in more substantial educational initiatives.

In software development, AI has demonstrated its efficacy in automating the procedures of testing and debugging. The process of software testing, characterized by its iterative and time-consuming nature, can be accelerated by the utilization of AI-driven testing frameworks. These frameworks use AI to systematically investigate different code pathways and detect potential software defects. The acceleration of the software development lifecycle is accompanied by an improvement in the quality of the final output, resulting in the production of more robust and reliable software systems.

The incorporation of AI in the automation of monotonous and repetitive processes offers a significant benefit that extends to wide fields. The implementation of AI automation enables professionals to utilize their cognitive abilities for creative problem-solving, innovation, and value-added activities, enhancing efficiency, accuracy, and scalability. It is important to adopt a prudent and equitable stance towards AI-powered automation, taking into account ethical concerns and the possible displacement of human labour. As the development of AI technologies progresses, it is increasingly crucial for professionals to effectively utilize their potential, establishing a mutually beneficial association between human expertise and machine capabilities.

6.6 AI and Hazard Mitigation

One major advantage of AI resides in its capacity to reduce risks in domains that pose substantial dangers to human engagement. By utilizing AI-enabled technologies, it becomes feasible to mitigate the potential hazards connected with tasks that pose significant risks to human beings. This section explores the domain of AI's advantage in risk mitigation, emphasizing instances when AI applications are essential in the management of dangers and the preservation of human safety. AI systems present a feasible solution in situations characterized by hazardous surroundings where human interaction has the potential to result in significant repercussions. These environments include various situations, including mining activities conducted in hazardous conditions, exploration of the deep oceans characterized by numerous uncertainties, and interventions carried out after catastrophic natural disasters.

The use of artificially intelligent robots serves as a testament to how technological progress can enhance safety standards. An exemplification of the advantageous utilization of AI in risk mitigation can be observed in the coal mining sector. The practice of conventional coal mining is accompanied by a multitude of risks, including the possibility of structural collapse, the presence of hazardous gases, and various dangerous circumstances that can endanger human life. The use of AI-powered robots that are equipped with cutting-edge sensors and imaging technologies enables the execution of mining operations without posing risks to human safety. These robotic systems exhibit the capability to manoeuvre within restricted areas, evaluate the stability of underground passages, and identify hazardous levels of gas. This practice not only guarantees the effective retrieval of resources but also serves as a protective measure for miners, mitigating the risk of life-threatening circumstances.

An additional noteworthy use of AI might be observed in marine life exploration. The oceanic depths include a multitude of enigmatic phenomena, yet the initiative to investigate these unexplored regions involves significant hazards. Underwater vehicles integrated with AI can submerge into the depths of the ocean, gathering valuable information on marine ecosystems and geological formations. The previous vehicles were outfitted with advanced sensors that facilitate the collection of real-time data while also possessing the ability to navigate around barriers and adjust to changing circumstances. The use of AI in this particular context functions as a mechanism to enhance our comprehension of the complexities of the oceans while mitigating the risks connected with human divers encountering severe pressures and unexpected underwater currents.

The potential of AI to mitigate risks becomes particularly apparent when considering natural disasters. During occurrences such as earthquakes, hurricanes, or wildfires, rescue activities pose significant risks for human responders. AI technologies have a significant impact on improving the effectiveness and security of these operations. Unmanned aerial vehicles (UAVs) integrated with AI algorithms can rapidly assess areas affected by disasters, offering critical information to support rescue teams in formulating efficient strategies. The use of AI-driven prediction models can anticipate patterns of disasters, empowering authorities to proactively implement actions to mitigate the potential harm inflicted upon individuals. Although AI shows potential for mitigating risks, it is important to acknowledge that there are limitations and ethical problems associated with its use.

AI systems, although proficient in executing predetermined tasks, may exhibit a deficiency in adaptability and discernment compared to their human counterparts. This has the potential to give rise to unanticipated difficulties in situations characterized by dynamism and unpredictability. The dependence on AI systems may unintentionally diminish the focus on developing human proficiency in managing dangerous circumstances. Ethical considerations hold significant importance as well; as AI becomes more incorporated into areas that include a higher likelihood of risks, concerns around accountability and the allocation of decision-making duties come to the forefront. The question of accountability arises in instances where an AI-powered system fails to execute a critical judgement.

Achieving a balance between human supervision and automated intervention is crucial to optimizing the advantages of AI in risk mitigation while simultaneously upholding ethical principles. The utilization of AI offers a significant benefit in the reduction of risks across wide fields. The prioritization of human safety in dangerous and possibly life-threatening conditions can be achieved through the deployment of AI-powered devices. The relevance of this advantage is highlighted by the cases of AI-driven robotic intervention in coal mining, exploration of the deep oceans, and disaster rescue operations. Still, it is critical to identify the constraints and ethical implications linked to the integration of AI. The ongoing advancement of AI holds significant promise for transforming the field of risk management. This technology offers an option as it enables individuals to handle hazardous settings with more confidence.

7 Challenges and Considerations

AI has rapidly emerged as a significant influence, fundamentally transforming multiple aspects of human existence. The significant impact of technology on modern businesses and daily life has resulted in a wide range of favourable opportunities as well as obstacles. This section will attempt to examine the challenges that AI must address as well as the promising opportunities it offers in professional and personal domains.

7.1 AI and Data

The issue of data quality and availability has become a major problem within the rapidly evolving domain of AI. To fully harness the capabilities of AI systems, it is important to ensure the convenient availability of substantial volumes of superior data for training. Regardless, enterprises continue to encounter substantial obstacles in navigating the complexities surrounding data quality and accessibility. The presence of imperfect data has extensive implications, resulting in a decrease in the general efficiency and productivity of AI systems. Enhancing the quality of data and implementing administrative practices are crucial for identifying effective solutions to this issue.

To assure the adequacy of data for issue-solving purposes, businesses must undertake a process that includes not only the acquisition of a substantial amount of data but also the verification of its accuracy, completeness, and relevance within the specific problem domain. The achievement of this objective necessitates the methodical implementation of data cleansing techniques, normalization procedures, and data set enrichment. To overcome the limitations imposed by data centres, it is critical to extend internal data repositories with relevant external sources. Within AI, the issue of data quality is a significant challenge for organizations. The effectiveness of AI algorithms is positively correlated with the precision of the training data they are provided with.

Accordingly, the presence of inadequate data quality can initiate a series of adverse effects, including reduced operational efficiency and imprecise predictive analyses. Anomalies, errors, and inconsistencies are prevalent in real-world data owing to its complexity. Hence, ensuring data quality is not merely a suggested procedure but rather an integral requirement for the development of resilient AI systems. Data cleansing techniques are a key component of data quality management, involving the elimination of errors, duplications, and inaccuracies within a given data set. This strategy might be likened to the process of purifying raw materials before their utilization in the production of a finished product.

Enterprises can systematically reduce the size of their data sets by using techniques such as outlier detection, duplicate elimination, and mistake correction. Likewise, the enhancement in data quality not only contributes to the development of more reliable AI systems but also enhances their ability to extrapolate patterns from the available data. In the initiative to enhance the quality of data, the process of data normalization is also important. The data obtained from a wide range of sources and locations possesses heterogeneity. Essentially, the process of normalization ensures that data representations are rendered consistent and can be readily compared.

Through this transition, algorithms exhibit enhanced capability in identifying underlying patterns by mitigating the influence of uneven scales and units. The process of normalization involves the integration of wide information into a cohesive structure, enhancing the flexibility of AI models in terms of their capacity for learning and making predictions. Although internal data is crucial in the training of AI, it is important to acknowledge its limitations and consider a broader perspective.

An adequate understanding of the problem domain can be attained through the integration of supplementary data sources within the process of data enrichment. The augmentation of the existing data corpus can be achieved by including additional data sources, such as public databases, industry reports, and expert annotations. This improvement facilitates the ability of AI systems to address the constraints associated with insufficient data volume, thereby expanding their knowledge and capabilities as they acquire new information. It is critical to meticulously select and incorporate the additional data in a manner that aligns with the problem's context and avoids introducing any extraneous information. The presence of substantial quantities of data is an essential need for the development of proficient AI systems.

The quantity and coverage of internal data repositories in numerous companies are insufficient. In these instances, intelligently acquiring data from other sources emerges as a compelling alternative. The enhancement of data availability can be significantly augmented by using open data efforts, establishing collaborations, and engaging external data producers. It is important to exercise diligent oversight to uphold the integrity of data and comply with the prescribed limitations on data usage during the process of integrating external data. The correlation between data quality and availability is significant, as both aspects are integral in influencing the performance of AI systems.

To fully harness the promise of AI, organizations must confront these challenges directly through substantial investments in enhancing data quality and implementing intelligent data procurement strategies. The optimal conditions for effective learning and accurate prediction are created when data is thoroughly cleansed, standardized, and enhanced in a coordinated manner. By integrating more data sources, the constraint of restricted data availability has been overcome. Organizations that embrace these principles will have the capacity to develop AI systems that distinguish themselves as pioneers within the AI sector.

7.2 AI and System Integration

Integration of AI solutions into existing business operations represents a formidable challenge in a technological setting that is constantly evolving. Enterprises' effective integration of AI technologies into their current infrastructure, such as customer relationship management (CRM) or enterprise resource planning (ERP) systems, presents a significant challenge in this project. The successful implementation of this integration process necessitates careful and thorough planning as well as effective collaboration between AI specialists and the IT department or third-party integration experts.

This collaboration is crucial to guaranteeing the seamless operation of these systems without any disruptions. Organizations undertaking the process of integrating AI solutions are faced with the complex challenge of aligning these innovative AI systems with their existing technological infrastructure. The successful completion of this initiative requires the application of strategic foresight, as an unorganized approach may lead to operational inefficiencies, inconsistencies in data, and potential system breakdowns. One notable example involves the incorporation of AI into CRM or ERP systems, which play a fundamental role in facilitating organizational operations.

The primary objective is to provide a smooth flow of information between the AI solution and the foundational systems, with the aim of minimizing disruptions and maximizing the potential benefits that AI may offer. To effectively address this difficulty, organizations must build a collaborative framework that involves a close partnership between their internal IT department or external integration specialists and the team responsible for AI implementation. The integration of this synergy is crucial to facilitate a seamless alignment between the AI system and the pre-existing technological infrastructure. To excel in their field, AI specialists must have a grasp of the complexities of CRM or ERP systems.

This asks for a deep understanding of the fundamental structure, data schemas, and communication protocols that underlie these systems. This understanding facilitates the development of a systematically organized integration strategy that effectively addresses potential conflicts and obstacles. The integration process gives rise to security as a primary concern. Ensuring the secure movement of data between AI systems and pre-existing systems is important to protecting sensitive information. Businesses must conduct thorough evaluations of the security mechanisms included in both the AI solution and the pre-existing systems. To mitigate unauthorized access and data breaches, it is important to diligently implement encryption, secure application programming interfaces (APIs), and authentication procedures. The collaboration between experts in AI and the cybersecurity team is crucial in strengthening defence systems and, hence, enhancing the organization's data integrity and regulatory compliance.

The implementation of APIs or webhooks is a crucial approach to streamlining the transfer of data across different systems. APIs act as middlemen that make it easy for data to flow smoothly. They make it possible for AI solutions and CRM or ERP systems to talk to each other while keeping each part's unique qualities. Thorough attention to detail is essential in the design and implementation of APIs, as it needs compliance to standardized protocols and data formats to guarantee compatibility. By adopting APIs, organizations have the opportunity to develop an efficient and seamless exchange of data, which serves as the foundation for the integration of AI capabilities with their current systems.

The acquisition of a holistic AI ecosystem necessitates a thorough comprehension of data compatibility. AI systems frequently depend on a variety of data formats, structures, and semantics that may differ from those that are in CRM or ERP systems. This disparity has the potential to hinder the process of integration. To overcome this difficulty, it is necessary to implement a data mapping approach. This involves establishing associations between the data used by the AI system and the data kept within pre-existing systems.

Data mapping is important in facilitating the seamless translation and transformation of data, thereby ensuring a comprehension and processing of information between the AI system and the pre-existing systems. A scientific perspective on integration necessitates a thorough assessment of the holistic performance of the system as a whole. Conducting testing is essential to detecting bottlenecks, inconsistencies, and potential problems that may occur as a result of the integration process. Rigorous testing involves the creation of scenarios that replicate real-world interactions, allowing for the identification and resolution of faults before the integrated system is deployed.

Ongoing monitoring and refinement following integration are equally crucial to maintaining continuous operational excellence. The incorporation of AI systems into pre-existing technology frameworks poses is a complex task that requires careful design, teamwork, and technological expertise. The successful integration of AI solutions with CRM or ERP systems relies on mutually beneficial collaboration between AI specialists and IT departments or integration experts. The integration initiative relies on security as a fundamental component, which is accomplished by implementing strong protocols, mapping data effectively, and strategically using APIs. The outcome of effectively integrating AI into current systems results in an enhanced organizational environment. This environment is characterized by the interaction of technology, enabling organizations to fully use the capabilities of AI while building upon the existing foundation established by their pre-existing systems.

7.3 AI and Professional Competence

AI is extraordinarily complex and multidimensional, requiring a rare collection of abilities and in-depth knowledge. In the current business environment, organizations are faced with the significant task of identifying and hiring skilled employees who possess a deep understanding of AI. This expertise is crucial for effectively creating and implementing AI-based solutions. The complexity of AI arises from its wide range of subdomains, which include ML, natural language processing, computer vision, and robotics. Proficiency in mathematical foundations, algorithmic subtleties, and programming languages is essential in each discipline, necessitating specialized talents that are not often possessed by all experts. As a result, businesses frequently face significant challenges when attempting to discover individuals who possess this distinctive combination of skills.

The limited availability of AI knowledge poses a complex challenge for businesses, impeding their capacity to fully utilize AI's promise for innovation and gain a competitive edge. The limited availability of resources is a significant obstacle to the advancement of AI-driven applications, as it hampers the creation of precise and effective algorithms, models, and systems that serve as the foundation for these applications. Also, the shortage of resources has an impact on the implementation phase. During this phase, the process of transforming theoretical concepts into practical applications requires a high level of coding proficiency and specialized knowledge in the relevant field. The current lack of experience in this field necessitates the implementation of an approach that incorporates multiple strategies to mitigate the difficulties presented by the complex terrain of AI.

To begin with, firms must adopt a proactive approach towards investing in training and development efforts for their current personnel. This method recognizes the existing institutional knowledge held by present employees, enabling them to engage in a process of learning that leverages their pre-existing skills and abilities. By developing an environment that promotes ongoing learning, organizations have the potential to progressively narrow the disparity between their existing knowledge and the growing requirements within AI.

Likewise, the adoption of a strategic approach to establishing collaborations with scientific institutions is deemed to be a wise course of action. By engaging in partnerships with prestigious universities and research centres that have established themselves as leaders in AI, individuals and organizations can gain valuable access to the latest advancements in research and valuable insights.

These collaborations may materialize through the establishment of collaborative research initiatives and the organization of workshops and seminars, with the primary objective of disseminating the most recent breakthroughs to the employees of the company. These initiatives not only function as forums for the diffusion of information but also facilitate a mutually beneficial partnership between academics and industry. Also to develop scientific collaborations, enterprises should contemplate the involvement of third-party vendors who possess expertise in offering AI training and certification programmes.

Loosely speaking, the providers provide carefully devised educational materials that address many aspects of AI, including fundamental principles as well as sophisticated methodologies. The act of enrolling personnel in these programmes serves the dual purpose of developing specialized expertise and guaranteeing that the training is administered by experienced professionals with practical knowledge of AI. An additional strategic option to address the scarcity of expertise is to outsource the development and deployment of AI to suppliers with specialized knowledge in this field.

These suppliers possess a team of experienced AI professionals who are capable of taking on the task of developing customized AI solutions that are specifically designed to meet the unique needs of the business. This method effectively reduces the burden on organizations to individually acquire scarce AI skills while still allowing them to leverage the capabilities of AI technologies. The limited availability of specialized knowledge and skills in AI presents a notable obstacle for enterprises seeking to harness the revolutionary capabilities of this technology. The complex and diversified nature of AI, which includes various subfields and complex algorithms, requires individuals with specialized talents and extensive expertise that are not universally accessible.

To overcome this challenge, businesses must implement a thorough approach that includes various strategies. These strategies include allocating resources towards training and development programmes for current employees, establishing partnerships with scientific institutions, collaborating with third-party vendors who provide specialized training services, and contemplating the outsourcing of AI projects to proficient vendors. By adopting these tactics, organizations can proactively tackle the scarcity of knowledge and position themselves to fully exploit the range of possibilities driven by AI for innovation and growth.

7.4 AI and Creativity

In recent years, there has been significant progress in AI, leading to the development of many applications such as self-driving cars and virtual assistants that respond to human verbal commands along with generative AI. Although AI holds numerous potential benefits, there is concern regarding its potential to impede human creativity and innovation.

Concerns have been expressed over the potential of technology to exert a dampening influence on human creativity. Upon closer examination, it becomes evident that this assertion is inherently paradoxical. In contrast, computers have the potential to facilitate creative thought rather than hinder it. AI can acquire the skill of identifying human emotions, including frustration. However, it is important to note that machines cannot empathize and experience emotions themselves. Human beings possess the remarkable ability to experience emotions, granting them a significant edge over emotionless AI systems in various domains, such as the professional environment.

Although AI has made significant advancements in various domains, it still falls short of the deep emotional and cognitive capabilities exhibited by humans. The generated output exhibits the capacity to emulate the stylistic elements of renowned musicians or poets, although it generally lacks the emotional resonance commonly found in human-authored compositions. The limitations of AI become evident in its capacity to grasp aesthetic and emotional aspects since it is constrained by its inability to comprehend meaning beyond established patterns. At present, AI cannot independently understand artistic expressions such as music and literature.

However, AI is integral in facilitating the generation of novel concepts and improving the process of learning by actively involving users. For AI to make progress, it must operate in collaboration with humans. The integration of humans and AI can significantly enhance the potential for human creativity and invention, a proposition that holds immense significance. The advancement of AI is being driven by collaborative efforts among humans, wherein ML is critical in enhancing comprehension of creativity and enabling individuals to effectively harness their creative capabilities.

As researchers and computer scientists, it is pressing to address the fundamental inquiries. It is widely acknowledged that large language models like GPT, Claude, and Bard possess the ability to replicate human responses, emotions, and languages. However, an appropriate question arises as to whether these models are capable of generating genuinely novel content. For example, *can an individual with a large language model produce truly innovative music rather than merely replicating and combining existing material derived from training data?* Similarly, *can they compose an original poem?*

Another area of doubt worth exploring relates to the distinguishing factors between great musicians such as Ludwig van Beethoven and Antonio Vivaldi in comparison to a music synthesizer. Similarly, it is of interest to determine the dissimilarities between celebrated literary figures like Rabindranath Tagore, William Shakespeare, and Lord Alfred Tennyson and a sophisticated large language model. Similarly, another area of examination of the distinctions between the artworks of Pablo Picasso, Vincent van Gogh, and Leonardo da Vinci and those generated by AI offers a fascinating subject for investigation. These queries will not only facilitate our sense of the nature of human creativity but also shed light on the constraints of AI.

AI includes various labour-saving applications, including automatic idea development and data processing. It is a dangerous proposition to assume that AI will not eventually supplant the arts. AI, at its nascent stage, currently exhibits limitations in surpassing the cognitive capacities of human beings. The absence of originality and innovation originating from human cognition is evident in this context. AI is limited in its ability to exhibit empathy, intuition, and an emotional understanding, hence rendering it suitable for augmenting human creativity rather than supplanting it.

When it relates to the creation of music, AI serves as a compelling substitute for human musicians. To devise effective strategies to overcome limitations, it is important to possess an understanding of the constraints associated with AI. The integration of human-annotated data sets with deep learning algorithms has the potential to greatly augment the inventive capacities of AI. These activities include artistic practices such as drawing and composing music. There exists a divergence of expert perspectives regarding the potential for AI to attain an understanding of beauty and creativity. There exists a belief among many individuals that AI has the potential to enhance human imagination, although this objective remains quite distant. The impact of AI across several industries is gaining recognition as our understanding of AI advances.

The use of AI in voice assistants, minimally invasive surgical procedures, and interactive companions exemplifies the enhancement it brings to the human experience. The comprehension of creativity poses a challenge for AI, while it has been observed to enhance human creativity. There is a consensus among specialists in AI that engagement in creative activities should not be seen as a precursor to the emergence of autonomous creative intelligence when making predictions of AI. The objective is to ascertain the extent to which AI can drive innovation.

ChatGPT, an AI language processing system, serves as a notable example. Recent technological advancements have significantly transformed online communication and collaboration. ChatGPT mimics human interaction by using advanced machine learning methods and processing natural language inputs. The influence of ChatGPT on creativity is under consideration. The level of originality exhibited can either facilitate or hinder the development of creativity and uniqueness. The powerful

AI technology utilized by ChatGPT can enhance the efficiency of examination processes and facilitate brainstorming sessions.

The disproportionate dependence on this technology could potentially stunt the progress of innovation. The rapid responses of chatbots have the potential to restrict creative cognition and emotional vulnerability in the face of difficulties. The spectrum of potential results highlights the ongoing discourse surrounding the impact of AI, specifically language models such as ChatGPT, on creative and critical thinking. There are multiple perspectives regarding the impact of AI on creativity. According to a particular theory, the provision of tools and resources by AI has the potential to enhance creativity.

An alternative viewpoint expresses concern about the potential decrease in creativity and independent judgement resulting from excessive dependence on AI-generated content. Achieving a balanced equilibrium between the use of AI and the nurturing of human capabilities is of major importance. With time, it becomes important for human agency to exert influence over the function of AI in facilitating innovation and developing critical thinking rather than restricting these processes. The progression of AI faces an assembly of obstacles, including issues related to misinformation, privacy concerns, economic transformations, bigotry, prejudices, and the concept of singularity. AI development is faced with numerous challenges as a result of the associated threats.

Thus, it can be seen that using AI and the tools that go with it, like ChatGPT, Bard, DALLE, and other similar technologies, has the potential to make an average individual much better at doing many things. It is crucial to acknowledge that these tools cannot empower an ordinary individual to mimic or imitate the artistic genius of Leonardo da Vinci, the literary expertise of William Shakespeare, the poetic nature of Lord Alfred Tennyson, or the compositional brilliance of Wolfgang Amadeus Mozart. Genuine creativity goes beyond mere replication, embracing the generation of original and imaginative creations that have not previously been experienced by the human race.

It may be empirically seen that the utilization of generative AI tools doesn't end up resulting in the transformation of an individual with average creative abilities into a model of creativity. Instead, they support the concept that individuals who possess innate creative brilliance are going to persist as such. Prominent figures in multiple fields are unlikely to experience significant creative progress through the utilization of generative AI or other techniques, given that their inventive powers and intellectual prowess significantly exceed the capabilities of trained models. Therefore, it is readily apparent that although AI can enhance the capabilities of average individuals, it is unable to compete with the innovative thinking of visionaries.

7.5 AI and Resources

The incorporation of AI solutions in today's technological environment has attracted significant interest due to its potential to transform multiple industries. Although the prospects of improved efficiency, advanced decision-making, and cost reduction are appealing, it is important to recognize that the deployment of AI solutions carries financial considerations. This section aims to explore the complex nature of expenses related to the creation and implementation of AI solutions, with a specific emphasis on the challenges encountered by small- and medium-sized enterprises (SMBs). The assertion that the pursuit of developing and deploying AI solutions is inherently characterized by a significant requirement for financial resources is irrefutable. This financial obligation arises from multiple aspects.

Initially, the procurement of state-of-the-art hardware and software components specifically designed to facilitate AI operations involves a significant financial investment. To meet the computational requirements of AI algorithms, it is important to acquire and sustain high-performance GPUs and other specialized gear. Simultaneously, the acquisition of advanced software frameworks and tools specifically developed to streamline the development and implementation of AI models exacerbates the financial demands. In addition to the necessary hardware and software requirements, personnel expenses represent a substantial factor that contributes to the entire cost.

The implementation of AI solutions necessitates a workforce that possesses a skill set in various domains, including data science, ML, and software engineering. These experts frequently receive higher salaries as a result of the specialized nature of their expertise. The process of recruiting and keeping such employees requires not only providing financial compensation but also incurring continuous expenses for training and enhancing skills, given the dynamic nature of AI technology. To effectively address the financial difficulties at hand, it is advisable for organizations, particularly SMBs, to adopt a prudent strategy by assessing the return on investment (ROI) provided by AI solutions.

It is essential to carefully consider the expected advantages concerning the associated expenses. The benefits that are often highlighted include increased productivity, enhanced decision-making facilitated by data-driven insights, and the possibility of cost savings through automation. It is important to thoroughly evaluate the possible benefits concerning the financial investment necessary for the development, implementation, and upkeep of AI solutions. Enterprises can consider the use of cloud-based AI solutions as a means to address the cost challenges associated with using AI technology.

Cloud computing, demonstrated by prominent providers such as Amazon Web Services (AWS), Microsoft Azure, and Google Cloud, provides an opportunity to avail AI capabilities under a flexible payment model based on usage. This eliminates the requirement for significant initial investments in hardware and software infrastructure. SMBs can greatly benefit from this strategy, as it provides equal access to AI capabilities without requiring excessive financial investment. SMBs face a significant financial challenge when it comes to implementing AI.

Before implementing the integration of AI, the entities must carefully consider their budgetary constraints. SMBs should conduct thorough cost-benefit evaluations, carefully considering the possible competitive advantages that AI can provide in comparison to the financial costs it may incur. It is crucial to engage in prudent financial planning that includes the evaluation of long-term cost forecasts, assessments of potential return on investment, and the assessment of associated risks. Another approach to mitigating the financial burden of AI implementation is by developing collaborative ecosystems.

SMBs have the option to explore partnerships or consortiums as a strategic approach to combining their resources and expertise. Through the practice of distributing financial responsibility, organizations have the opportunity to collaboratively allocate resources towards the implementation of AI solutions, thereby reducing individual financial strain. Collaborations of this nature facilitate the exchange of knowledge, the sharing of risks, and the improvement of affordability, potentially augmenting the viability of integrating AI for SMBs.

AI solutions have the undeniable capacity to bring about significant transformations, with the potential to restructure many industries, enhance operational effectiveness, and expand decision-making capabilities. However, it is important to acknowledge the significant financial implications associated with the development and integration of AI solutions, especially for SMBs. By doing a thorough analysis of ROI, carefully assessing the feasibility of cloud-based solutions, and actively participating in collaborative initiatives, businesses may effectively navigate AI expenses. The combination of prudent financial planning and a dedicated focus on leveraging the capabilities of AI has the potential to result in wise investment decisions that generate significant long-term advantages. The integration of AI into various domains is heavily reliant on the strategic alignment of technological aspirations and financial prudence as the digital world undergoes constant evolution.

7.6 AI and Ethics

A primary concern is the delicate equilibrium between privacy and security, which has significant societal and cultural consequences. Also, ethical and legal considerations accompany the development and application of AI systems, necessitating the implementation of careful safeguards to address potential threats. This section explores the previously stated concerns, providing an in-depth analysis of their complexities and proposing effective approaches for firms to traverse them. AI has provided businesses with an opportunity to leverage extensive data sets for predictive analysis and informed decision-making. The intersection of data-driven technology and the protection of individual privacy has drawn significant concerns.

Organizations are faced with the delicate task of balancing the use of personal information to improve their services while upholding the privacy rights of individuals. This becomes particularly significant in light of the growing strictness of data privacy legislation, as illustrated by the General Data Protection Regulation (GDPR) implemented in Europe. To effectively tackle this challenge, enterprises must embrace the concepts of privacy-by-design, thereby integrating privacy considerations into all phases of AI system development. Organizations may effectively protect sensitive information and simultaneously leverage the advantages of AI by employing strategies such as data anonymization, encryption, and access controls. Further, it is important to note that the connection between privacy and security includes more than just the safeguarding of data.

The increasing integration of AI systems into critical infrastructures and sectors such as healthcare and finance has the effect of greatly magnifying the possible repercussions of security breaches. The establishment of strong security measures that undergo testing and frequent updates is necessary to safeguard AI systems from cyberthreats and maintain their resilience. Simultaneously, it is important to assess the cultural consequences of security measures. Keeping a balance between strong security measures and a user-friendly interface is the most important thing you can do to prevent bad things from happening like users not wanting to use AI-based solutions or being hesitant to do so. At a more societal level, the implementation of AI systems has the potential to exert a substantial influence on social norms and cultural dynamics.

The implementation of AI-based automation has the potential to cause significant changes in employment dynamics, necessitating a thorough examination of the potential consequences of job displacement. To address these issues, businesses should prioritize activities aimed at reskilling and upskilling their employees. By using this strategy, it will be possible to create a workforce that can work cooperatively with AI systems rather than becoming obsolete because of them. It is also important to ensure that the development of AI technology considers cultural sensitivity, thereby mitigating the potential for biases that may reinforce prevailing socioeconomic disparities. To address this issue, it is important to assemble heterogeneous groups of developers and utilize extensive data repositories to prevent the unintended reinforcement of biases by AI systems.

The implementation of AI systems necessitates the consideration of ethical and legal factors. The opacity inherent in numerous AI algorithms gives rise to concerns regarding bias and equity, particularly when the outcomes produced by AI systems have significant effects for individuals. To effectively tackle this issue, it is important to stress transparency and explainability in AI research. Organizations are required to develop comprehensible and analysable models that facilitate the detection and correction of biased patterns. It is important to establish ethical rules for the creation and utilization of AI to define explicit parameters for permissible conduct.

The occurrence of considerable instances involving data misuse and breaches provides substantiation for the assertion that data privacy is intertwined with both ethical considerations and legal frameworks. Businesses must ensure that their data gathering and usage practices are in strict compliance with the applicable laws and regulations. The ethical management of data necessitates the acquisition of informed consent, the anonymization of data whenever feasible, and the implementation of solid security protocols.

Not only does this practice ensure the protection of individual rights, but it also develops a sense of trust between businesses and their clientele. The deployment of AI systems also gives rise to a variety of legal problems. To avoid disputes and conflicts, it is essential to establish clear definitions for intellectual property rights, especially when AI-generated content is involved. It is important to establish clear guidelines about the responsibility for decisions or acts created by AI, particularly in situations where the well-being of individuals or crucial operations are at risk.

Policymakers and legal researchers must engage in collaborative efforts with developers of AI to construct a framework that effectively and appropriately assigns responsibility. To effectively manage the complexities associated with these difficulties, firms must develop a corporate culture that prioritizes both responsibility and accountability. It is important to regularly undertake audits of AI systems to proactively detect and swiftly address any potential ethical or legal problems that may arise. Also, engaging in partnerships with regulatory organizations and industry counterparts can facilitate the development of universally accepted norms within the sector, thereby promoting the adoption of ethical practices in AI.

The AI setting places significant emphasis on ethical and legal considerations. The presence of biases within AI algorithms has the potential to perpetuate existing societal imbalances and erode trust in these systems. Hence, organizations must place importance on the principles of fairness and diversity when engaging in data collection and algorithm development. The use of transparent and explainable AI models not only develops accountability but also enhances the process of decision-making and risk assessment. Within law and ethics, the dynamic characteristics of AI technologies necessitate flexible regulatory frameworks. Close collaboration between policymakers, legal experts, and AI developers is critical to guarantee that legislation remains aligned with the rapid progress of technology.

The evolution of intellectual property laws, liability frameworks, and data protection legislation is necessary to effectively tackle the distinctive issues presented by AI-generated material and judgements. Businesses must have a proactive approach to effectively addressing ethical and legal challenges. The establishment of explicit rules for the development and utilization of AI lays the groundwork for the implementation of responsible practices. It is critical to regularly conduct audits on AI systems to discover and address possible faults, thereby promoting a culture of ongoing enhancement. The successful integration of AI technologies in the business sector requires a thorough comprehension of the complex relationship among privacy, security, societal dynamics, ethics, and legal considerations.

Organizations need to carefully strike a delicate equilibrium between the advantages of AI and the privacy rights of individuals by implementing privacy-by-design principles and adequate security mechanisms. The proactive reskilling of individuals, the promotion of diversity in AI development, and the thoughtful examination of potential effects on social norms are important because of the societal and cultural consequences involved. The deployment of AI necessitates careful attention to ethical and legal issues, which are fundamental throughout the entire process. These considerations include the principles of transparency, explainability, and conformity to set rules. In AI integration, businesses must maintain a state of constant alertness to effectively confront and overcome the problems associated with this process. By doing so, businesses may fully use the revolutionary capabilities of AI technology, all the while ensuring the protection of individual rights and upholding social values.

8 Explainable AI

There is currently a significant surge in interest surrounding AI in the modern society. The majority of businesses are either in the process of incorporating AI, currently utilizing its capabilities, or repositioning previous rule-based systems as technologies enhanced by AI. As organizations increasingly integrate AI and advanced analytics into their operational processes and automate decision-making, there is a growing need for clarity in understanding the underlying logic behind these models' judgements. *How can we effectively maintain transparency while leveraging the operational advantages provided by AI?* The emergence of explainable AI (XAI) presents itself as a suitable solution in this particular context.

Explainable AI refers to the approaches and strategies utilized in the implementation of AI technology that enables human professionals to gain an understanding of the results produced by the solution. The concept being discussed presents a clear contrast to the idea of the 'black box' in ML, wherein the individuals responsible for designing these AI systems are unable to explain the rationale behind particular actions. XAI can be understood as a reflection of the societal expectation for a justification. In this section, we will further explore the earlier discussed topic and examine the advantages and disadvantages associated with it. The global proliferation of AI has necessitated the growing importance of understanding the inner workings of AI algorithms and systems, as well as the processes by which they arrive at specific outputs. The following subjects will be addressed, and a subsequent discussion will be conducted on each of them. Figure 6.2 displays a Venn diagram showcasing the interrelation of XAI.

AI is characterized by its ability to receive inputs and generate outputs through a series of computational processes. When exploring the concept of explainable AI, we are essentially examining the impact of input variables on the final result. Explainable AI, also known as interpretable AI or explainable ML, refers to AI systems that can be effectively supervised and understood by human operators or the methodologies used to attain this level of comprehensibility. The primary emphasis is frequently on clarifying the underlying reasoning behind decisions or predictions obtained from AI, making them understandable and transparent.

XAI addresses the tendency to view ML as an opaque system when the creators of the AI are unable to explain a specific decision. The fundamental aspect of XAI revolves around enhancing the proficiency of users of AI-driven systems by providing them with a deeper comprehension of the reasoning processes used by these systems. The possibility exists that XAI could serve as a representation of

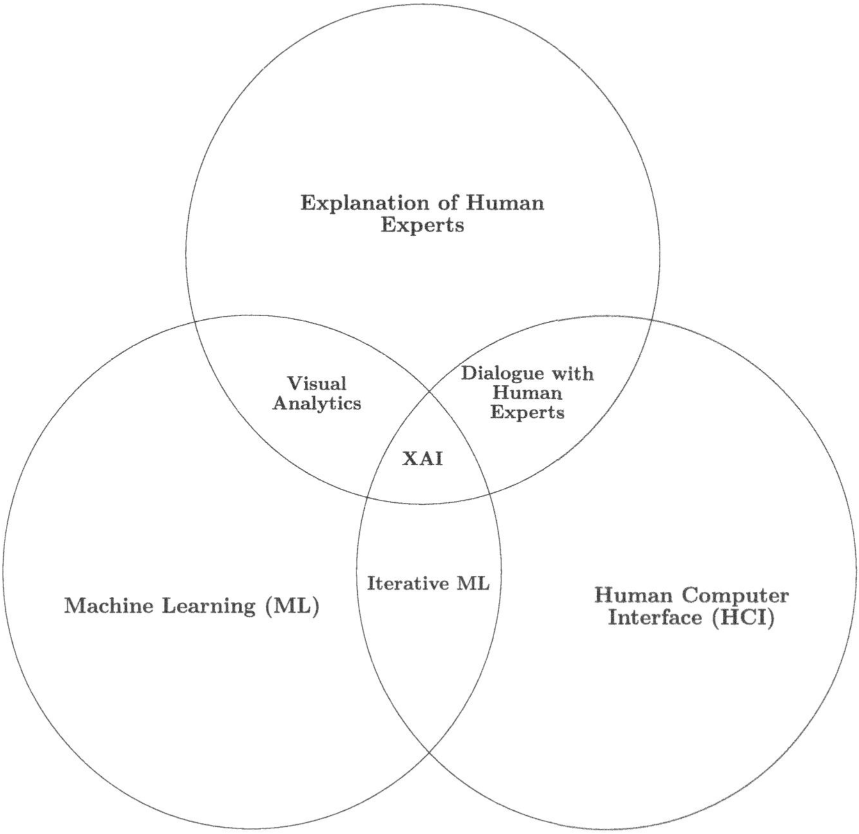

Fig. 6.2 Interrelation of XAI

the societal entitlement to receive explanations. Even in scenarios devoid of legal mandates or regulatory duties, XAI can augment the user experience while using a product or service by fostering user confidence in the AI's capacity to render dependable conclusions.

The objectives of XAI comprise providing an analysis of past, current, and future behaviours while revealing the underlying facts on which these actions are based. This process not only confirms established information but also questions it and generates new hypotheses. The ML algorithms that form the foundation of AI can be classified into three principal categories:

(1) White Box
(2) Grey Box
(3) Black Box

White box models provide interpretable results that can be easily understood by experts in the field. On the other hand, black box models are difficult to explain,

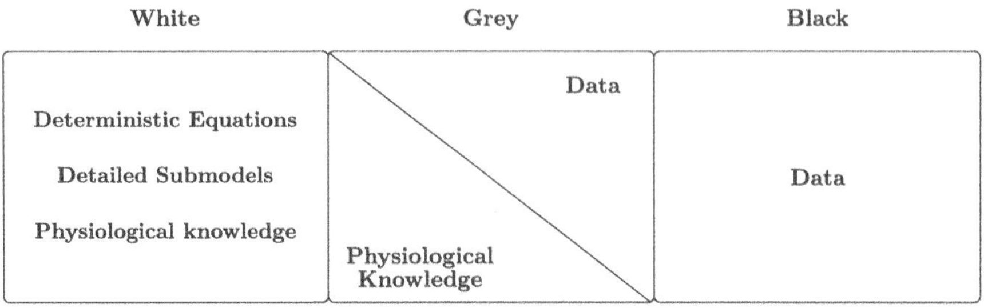

Fig. 6.3 White box versus black box

and even domain experts may not understand the reasoning behind them. Grey box models fall somewhere in between the two, combining some of the advantages of both white box and black box models. One can get a better understanding of these concepts and their interrelations using Fig. 6.3. Explainable AI algorithms are guided by three fundamental concepts:

(1) Transparency
(2) Interpretability
(3) Explainability

To be transparent, designers must be able to explain and justify the processes they used to get model parameters from training data and generate labels from testing data. What we mean by "interpretability" is the ability to explain the decision-making process and the ML model in a way that humans can grasp. It is possible to view explainability as a collection of traits within comprehensibility that have contributed to a certain decision-making process, even though there is no universally accepted definition of the term. Adhering to these principles allows algorithms to become more capable of authenticating, tracking, improving, and coming up with new solutions.

In some cases, white box ML algorithms, which by nature possess interpretability, can yield high-accuracy results. This attribute holds great importance in fields like medicine, defence, economics, and legislation, where the understanding of judgements and the development of trust in algorithms are of significant value. Scientific researchers argue that symbolic regression holds significant potential, especially within supervised ML, as a promising direction for future advancements. In this context, algorithms are used to systematically analyse mathematical expressions to choose the model that best fits the given data sets.

AI systems are specifically designed to optimize actions based on mathematically defined objectives established by the system's designers. The objectives may involve instructions such as 'optimize the precision' in evaluating the favourable nature of data within the experimental data set. Although AI systems can learn important general principles from the test set, there is also a risk that they may internalize inappropriate rules, limiting their ability to be applied outside the training set. XAI provides human evaluators with the chance to thoroughly examine these rules, allowing them to evaluate the system's ability to generalize to data from the real world outside the test set.

The concept of explainability holds great importance in AI, as it is responsible for making judgements that can have significant impacts on enterprises, both positive and negative. Similar to the practice of designating decision-makers inside an organizational structure, it is important to understand the decision-making process of AI. Many organizations desire to harness the potential of AI, yet they frequently have concerns about granting AI the authority to make significant decisions, primarily due to a prevailing lack of trust.

The concept of explainability functions as a remedy, providing valuable insight into the operational methods used by models while making decisions. The primary difficulties lie not in the development of the most effective model but rather in obtaining the agreement of stakeholders that the model surpasses human decision-making. This involves using suitable inputs to derive effective solutions. Oftentimes, individuals in positions of authority require an understanding of the decisions at hand.

XAI aims to provide clear explanations of its objectives, reasoning, and decision-making process in a manner that is understandable to anyone without specialized knowledge in the field. XAI enhances human understanding of the fundamental rationale behind AI and ML algorithms, developing increased confidence and reliance on these technologies. The concept of pertinence is identified as significant within fairness, accountability, and transparency (FAT) in ML, especially as AI becomes increasingly incorporated into various domains.

XAI is essential and integral to developing trust among enterprises that aim to leverage AI technology. It achieves this by providing transparency into the behaviour of AI models, enabling the identification of possible problems. ML models are commonly categorized as either white boxes or black boxes, with the former type providing more transparency and interpretable results. XAI provides information regarding the decision-making process of an AI system, including its capabilities, boundaries, requirements, alternatives, and the necessary level of trust.

The importance of XAI is emphasized by its contribution to the promotion of algorithmic accountability and the improvement of decision-making processes in domains such as health informatics and mortgage lending. This ensures the achievement of fair and unbiased outcomes. The implications of addressing bias mitigation and trust-building procedures become increasingly significant as AI continues to expand its presence.

8.1 Application

Representatives of explainable AI (XAI) can be observed within several areas, including the healthcare industry. In the present scenario, comprehensible AI systems assume a key function in aiding physicians in patient diagnosis and formulating trust between healthcare practitioners and AI technology. Clinicians' capacity to understand the reasoning and approach utilized by AI in formulating diagnostic conclusions is important in developing trust and confidence. XAI is utilized in finance to evaluate and render decisions on financial solicitations, including loan and mortgage applications, by either providing approval or refusal. It facilitates the detection of instances of financial fraud.

Within the military domain, the integration of AI systems into military operations requires the construction of a distinct and well-defined means of communication between military personnel and equipment with AI capabilities. The significance of this matter cannot be overstated, its essential function in guaranteeing the security and dependability of those systems. XAI is used in autonomous vehicles to provide clarity on the decision-making process connected to driving, with a specific focus on decisions that impact safety. The promotion of passenger security is enhanced through the provision of knowledge regarding the fundamental ideas and reasoning that guide the driving choices made by the vehicle. This level of understanding allows passengers to evaluate the efficiency of the vehicle's performance under different circumstances.

The significance of XAI is of importance in sectors that have the potential to directly impact individuals' lives. Within healthcare, AI exhibits the potential to discern and categorize patient fractures through the examination and analysis of X-ray and magnetic resonance imaging (MRI) pictures. However, healthcare personnel may demonstrate hesitancy in adopting AI technology if they have doubts about the system's dependability or if they have little understanding of its diagnostic methodologies. A system possessing the capacity for explanation provides healthcare personnel with the potential to conduct evaluations of diagnoses and utilize that information in their prognostic evaluations.

The Defence Advanced Research Projects Agency (DARPA) is now making significant advancements in XAI within its third-wave AI systems. As to the findings of the DARPA, the process outlined involves the application of AI for categorization and comprehension. The NIST has outlined five distinct forms of explanation:

(1) Informing algorithm subjects
(2) Developing societal trust
(3) Fulfilling regulatory obligations
(4) Promoting system enhancement
(5) Serving the interests of algorithm owners

The process of engaging in communication with individuals impacted by algorithmic outcomes involves the clear articulation of decisions, such as the approval or rejection of loan requests. Concurrently, the establishment of social trust is achieved through providing justifications that authenticate the underlying principles and methodologies. This practice increases individuals' self-confidence by providing them with crucial information about the algorithm, including its intended goal, procedural origins, data sources, inherent benefits, and associated limitations.

Also, the obligation to comply to regulatory frameworks is particularly significant in organizations that are subject to oversight, particularly in cases where AI assumes a central role. The importance of adhering to current regulations cannot be overstated when it comes to safeguarding the integrity and accountability of AI applications. An illustrative instance of this phenomenon can be observed in the utilization of AI algorithms within autonomous vehicles. These algorithms are mandated to effectively exhibit their compliance to well-defined traffic regulations and societal conventions.

The process of assisting the improvement of a system is an additional component of efforts directed towards offering explanations. To fully understand why AI systems sometimes give wrong results, it is important for technical professionals who are interested in how algorithms work to learn a lot about where these problems come from and why they happen. The acquisition of this information is of importance in the iterative enhancement of AI systems, leading to the reinforcement of their general effectiveness and reliability.

Individuals or businesses that possess algorithms, that are commonly used in diverse sectors, stand to gain substantial advantages from the explanation of outcomes generated by AI. One potential strategy for organizations offering streaming services is to promote user engagement and improve customer retention through the provision of explanations that define the rationale behind content recommendations. The significance of this initiative becomes evident when individuals, commonly known as end users, attain a thorough understanding of the conveyed information through these explanations.

Recognition of the wide range of user profiles is important while trying to understand the explanations provided by AI. Various user groups may necessitate tailored explanations that are specifically crafted to cater to their distinct cognitive capacities and information needs. An illustration can be observed in the divergent requirements imposed on the operator of an autonomous vehicle and an AI developer. Individuals who prefer a concise explanation may find satisfaction in the assertion that the AI erroneously classified the plastic bag as a rock and responded accordingly. Regardless, those with more extensive comprehension desire a more deep examination of the underlying causes of this misclassification.

It is important to differentiate between the validity of explanations and the precision of outputs to effectively deliver explanations. The capacity of an AI system to proficiently express the fundamental rationale behind its results sets it apart from the sheer accuracy of those results. This case arises when a loan approval algorithm justifies its decisions by taking the applicant's income and financial commitments into account, although the final decision is primarily based on the applicant's postal code. In these instances, the explanation supplied lacks specificity, despite the potential accuracy of the conclusion.

AI systems may face restrictions in terms of knowledge in two unique ways. At the outset, it is conceivable that the information provided by the user may exceed the system's domain expertise. As an illustration, a classification system that is specifically designed for categorizing bird species ought to outline that an image portraying an apple does not belong to the categorization of avian creatures.

Also, when seeing a visually ambiguous image, the system must acknowledge its limitations in precisely identifying the bird or demonstrate a much-reduced level of confidence in its identification.

8.2 Advantages

Explainable AI (XAI) offers numerous advantages across various domains. The earlier-mentioned benefits pertain to the act of making well-informed decisions, mitigating risks, enhancing confidence, enhancing user acceptance, reinforcing governance, expediting system improvements, and developing the global progress and practicality of AI. The ability of XAI to promote transparency and instil confidence is mostly derived from its underlying mechanism of operation.

The process of moving from unprocessed data to practical insights involves a critical component of building trust. The imposition of an expectation for individuals to place complete trust in a self-governing process from its inception is sometimes seen as excessive. Hence, it is advisable to provide users with the opportunity to advance through successive stages of understanding steadily. The design of the user interface (UI) enables users to navigate through different stages of the process, beginning with the initial inputs, thereby facilitating a progressive unveiling of information. This sequential query creates a unique degree of transparency and develops trust. This technique accommodates a wide range of users, including both seasoned individuals characterized by a propensity for critical inquiry and novices driven by a strong inclination towards knowledge acquisition. By addressing the requirements of both novice and expert users, this approach develops a feeling of confidence while simultaneously enhancing the user's efficiency and comprehension.

The interactive approach described above facilitates a productive feedback loop that improves the algorithms of AI and ML, thereby ensuring the ongoing iterative improvement of the system. The notion of XAI is experiencing a surge in prominence as enterprises increasingly recognize the necessity of comprehending the decision-making mechanisms used by opaque AI models, i.e. 'black box' models. The growing interest in the concept of XAI offers several notable benefits and enables more effective decision-making by leveraging greater capabilities. One notable advantage of XAI lies in its capacity to offer lucid and understandable explanations regarding the many factors that impact the expected outcomes.

The explanations offered by XAI supply individuals with readily comprehensible and transparent reasoning for the decisions rendered by algorithms of AI. The level of clarity offered allows users to fully grasp the specific aspects that impact projections, therefore aiding in the formulation of educated strategies to modify anticipated results. Based on the knowledge gained, organizations possess the capacity to enhance their offerings to mitigate customer attrition. Likewise, there exists the possibility of attaining efficient model optimization.

The utilization of XAI permits the expeditious enhancement of AI model optimization by offering a framework that facilitates continuous monitoring and evaluation. The degree of visibility provided allows for the acquisition of useful insights into the model's performance, the key elements that influence its outputs, and the amount of accuracy demonstrated by the model. Conversely, black box models demonstrate a lack of transparency, hence presenting challenges in distinguishing the root causes of discrepancies observed between predicted and actual results. The application of XAI explanations enhances the establishment of trust and the reduction of bias by offering a deeper understanding of the patterns detected in the data by AI models.

The provision of transparency allows the MLOps team, which is tasked with the management of ML processes, to proficiently identify errors, assess potential biases, and verify the integrity of the data. This effect amplifies the level of trust in AI systems and promotes a proactive approach to addressing biases. The achievement of heightened AI adoption is attainable. XAI is integral in promoting the widespread acceptance and utilization of AI systems. It achieves this by developing understanding and instilling trust among many parties involved, including internal stakeholders, customers, and partners.

As comprehension and confidence in AI models increase, their use in predictive, prescriptive, and augmented analytics systems becomes increasingly prevalent. Ensuring regulatory compliance is of utmost significance in upholding the ever-evolving regulations, and the incorporation of explainable AI plays a key role in facilitating this undertaking. The ability to critically analyse the underlying reasoning behind decisions made by AI systems enables compliance with legal obligations, such as the California Consumer Privacy Act (CCPA), which mandates transparency in the outcomes produced by AI systems. Explainable AI offers a range of advantages that have a positive impact on several aspects, such as decision-making processes, model optimization, trust reinforcement, bias reduction, adoption rate increase, and legal compliance assurance. The amalgamation of these characteristics presents AI with an elevated degree of efficacy, clarity, and societal use.

8.3 Disadvantages

Obtaining user approval for AI-powered solutions and systems is the priority. In circumstances where individuals possess the capacity to exert influence on processes of decision-making, obtaining user approval becomes especially significant. However, the development of effective approaches to explain the functioning of these systems poses considerable challenges. One of the challenges that necessitates attention is the difficulty of striking a balance between the simplicity of algorithmic transparency and the effectiveness of complex, non-transparent models. Attaining this state of balance requires a thorough examination of privacy and the safeguarding of sensitive data, concurrently with the advancement of transparency. An additional issue emerges during the task of discerning pertinent information for users while considering their varying levels of expertise. The task of delving deeper into a subject and presenting concise yet meaningful explanations might present substantial difficulties.

A significant fraction of explanation methodologies primarily prioritize explaining the internal mechanisms of AI decision-making, often disregarding the wider real-world context within which these decisions occur. As a result, this methodology may lead to explanations that exhibit a deficiency in practicality or applicability. This raises the question: *What are the issues that explainable AI seeks to address?* There are various AI and ML models that exhibit a bizarre quality, leading to outcomes that are challenging to explain. The ability to provide a clear and thorough explanation of the rationale behind chosen courses of action and the outcomes achieved is essential for developing trust, advancement, and adoption of AI technologies. Through the usage of data, models, and processes, operators and users can attain a comprehension and mastery of these systems, thereby enabling optimization through lucid and well-substantiated reasoning.

The role of explainability is of foremost significance in facilitating the identification and remediation of faults, biases, and risks through simplifying their communication. It is crucial to recognize that XAI does include significant limitations. In contrast to alternative methods of attaining transparency, this particular technique possesses the capacity to oversimplify and misinterpret complex systems, hence initiating discussions regarding the advancement of AI systems that integrate more interpretable or causally accurate models. XAI systems often exhibit diminished performance when compared to opaque models. The training methodology used for AI systems that express their thinking is notably more complex when contrasted with those applying opaque models.

The transparency in an XAI system can potentially compromise the confidentiality of secret material when it is given to such a system. Also, despite the primary objective of XAI being increasing trust in AI, there may still be individuals who maintain a sceptical attitude towards the system, even when presented with understandable arguments for its decision-making processes. The objective of establishing trustworthiness involves clarifying the attributes and rationale that underpin the consequences of AI, to formulate human trust in the AI model. Transferability refers to the act of explicating an AI model in a manner that amplifies its comprehension and flexibility across wide problem sets or domains.

The concept of informativeness refers to the dissemination of thorough clarifications to users regarding the operational mechanisms of an AI model, thereby reducing the likelihood of misunderstandings or misinterpretations. This term is complexly connected to the ideas of human agency and autonomy, as it relates to enabling individuals to make decisions that are informed by thorough and precise knowledge. The development of confidence in the implementation of an AI model is supported by incorporating qualities such as robustness, stability, and explainability within the model, thereby increasing human confidence. Privacy awareness is key in the advancement and execution of AI and XAI methodologies, as it ensures the safeguarding of sensitive information. Data anonymization is a widely utilized technique for achieving this objective.

Actionability refers to the capacity of XAI to provide users with recommendations for modifying their behaviours to attain wide consequences. These recommendations are followed by a complete explanation of the rationale underlying the proposed outcomes. So, the provision of customized explanations aims to cater to users with varying levels of knowledge, facilitating their understanding of the behaviours and predictions of AI systems through explanations tailored to their specific roles, objectives, and preferences.

9 Responsible AI

Responsible AI (RAI) refers to the ethical advancement, evolution, and utilization of AI technology for beneficent objectives, aiming to empower individuals and businesses while ensuring equitable outcomes for customers and society at large. This approach allows enterprises to develop trust and improve the implementation of AI with a sense of confidence. Responsible AI facilitates the advancement, evolution, and deployment of ethically sound AI systems and solutions.

Ethical AI operates following its predetermined aims, following ethical principles, and promoting human accountability and comprehension. Institutions possess the capability to expand or alter their ethical benchmarks on AI; yet, these qualifications must include specific foundational attributes. The previously mentioned requirements include attributes such as robustness, impartiality, transparency, accountability, resilience, confidentiality, and sustainability.

The benefits of responsible AI include the ability to set fundamental objectives and build a regulatory structure. This makes it easier to build AI systems that help AI and its related industries grow and thrive. The integration of accountability mechanisms within AI systems serves to eliminate unintended biases, promoting impartiality and inclusivity in algorithms and the underlying data. The build-up of confidence among employees and customers necessitates ensuring the transparency of AI. To achieve this goal, it is important to make progress in the development of explainable AI, which is characterized by its dedication to ensuring transparency throughout its operations and capabilities.

The implementation of staff empowerment strategies within organizations facilitates the ability of workers to voice concerns or raise issues related to AI systems, thereby effectively overseeing the technology while also promoting innovation. The preservation of data privacy and security is achieved by executing a plan that prioritizes privacy and security, thereby guaranteeing that personal or sensitive data is never used in an unethical fashion. The establishment of an ethical framework for AI effectively addresses the management of potential risks and social advantages, therefore ensuring the protection of the interests of stakeholders, personnel, and the wider community. Thus the usual question is how does responsible AI differ from explainable AI? A summary is provided in Table 6.2 to show a few basic differences.

The establishment of trust and confidence in the management of principles and governance necessitates the precise expression and efficient dissemination of a goal

Table 6.2 Comparison of XAI and RAI

Explainable AI (XAI)	Responsible AI (RAI)
1. Focus XAI focuses on making AI models and their decisions interpretable and understandable by humans	**1. Focus** RAI focuses on ensuring ethical, fair, and unbiased AI practices, considering the broader societal impact of AI systems
2. Interpretability XAI emphasizes providing insights into how AI models reach specific decisions, enhancing transparency and interpretability	**2. Ethical considerations** RAI considers ethical considerations in the development and deployment of AI, addressing issues related to bias, fairness, and societal impact
3. Human understanding XAI aims to enhance human understanding of AI decisions, enabling users to trust and verify the outcomes	**3. Accountability** RAI emphasizes accountability and responsibility in AI development, deployment, and use, ensuring compliance with ethical standards
4. Model complexity XAI deals with explaining complex AI models, simplifying them for user comprehension without necessarily addressing broader societal issues	**4. Fairness and inclusivity** RAI addresses fairness and inclusivity concerns, working to mitigate biases and ensure AI benefits all segments of the population
5. User empowerment XAI empowers users to understand and trust AI systems, fostering user acceptance and collaboration with AI technologies	**5. Legal and regulatory compliance** RAI focuses on complying with legal and regulatory frameworks, ensuring AI systems comply with established guidelines and standards

and principles that prioritize the conscientious execution of AI. Simultaneously, the organization implements a transparent governance system to augment trust in AI technology. The risk mitigation strategy involves enhancing compliance with existing regulations and adopting a proactive approach towards anticipated future obligations. Policies are formulated to efficiently mitigate risk and are operationalized through the utilization of a risk management framework, which includes standard reporting and monitoring protocols.

The primary objective of the advancement of enabling technologies and procedures is to seamlessly integrate concepts such as fairness, explainability, robustness, traceability, and privacy into AI systems and platforms. A strong commitment from leadership is essential in the culture component, as it emphasizes the prioritization of responsible AI as a crucial requirement inside the organization. This commitment is reinforced by training, ensuring that all personnel possess a comprehension of ethical principles in AI and the standards necessary for attaining success. The proactive detection of bias in AI before its widespread application is of utmost importance.

Algorithmic assessment is a systematic approach that assists in the discovery and reduction of potential risks and unexpected consequences that may arise from the use of AI systems within an organization. This technique enables the establishment of trust and the development of supportive frameworks for decision-making driven by AI. The process of choosing use cases is essential for focusing the assessment on those that have the highest levels of risk and impact. The algorithmic assessment includes four key stages:

(1) Firstly, the establishment of fairness objectives for the system, considering the varied end users
(2) Secondly, the evaluation and identification of disparities in potential outcomes and sources of bias among different users or groups
(3) Thirdly, the mitigation of unintended consequences through recommended corrective strategies
(4) Finally, the implementation of controlling and monitoring mechanisms to address future disparities as the AI system evolves

9.1 Principles

The concept of responsible AI includes the moral creation and implementation of AI systems, taking into account ethical and legal dimensions. The principal aim of responsible AI is to utilize AI systems in a manner that ensures security, dependability, and compliance with ethical principles. The practice of responsible AI involves the promotion of transparency and the proactive addressing of difficulties, such as the issue of AI bias. Supporters of responsible AI aim to build a universally recognized set of guidelines that enable organizations around the world to prioritize human-centredness, interpretability, and comprehensibility in their AI initiatives.

The incorporation of responsible AI practices facilitates the attainment of fair and just outcomes, reliability, and openness. It is important to acknowledge that the current construction of dependable AI standards relies on the judgement of data scientists and software engineers responsible for designing and implementing AI models within an organization. As a result, the strategies implemented to address discrimination and promote transparency differ across companies. Similarly, the procedure for implementation can also vary. For example, the chief analytics officer or specialized AI teams may be entrusted with the task of developing, executing, and supervising a company's responsible AI framework. It is recommended that the organization thoroughly document its framework on its official website, providing an outline of its strategy towards ensuring accountability and promoting the use of non-discriminatory AI applications.

The sphere of AI governance is witnessing the emergence of responsible AI, which holds significant importance. The concept of 'responsibility' involves ethical considerations and the democratization of AI. Oftentimes, the utilization of data sets for training ML models results in the introduction of biases inside AI systems. Biases may emerge due to the presence of inadequate or flawed data as well as the predispositions of individuals participating in the training of ML models.

AI programs that exhibit bias possess the capacity to detrimentally impact individuals, resulting in unjust denials of financial loan applications or potentially erroneous medical diagnoses. The growing ubiquity of software programs integrating AI functionalities highlights the importance of AI standards that surpass abstraction, as initially proposed by science fiction author Isaac Asimov in his 'Three Laws of Robotics [Asi88]'. The adoption of responsible AI practices serves to limit the presence of AI bias, facilitate the establishment of transparent AI systems, and develop user trust in such systems.

The implementation of responsible AI is guided by a range of principles that may vary among different organizations. For example, Microsoft and Google each follow their own unique sets of principles. The National Institute of Standards and Technology (NIST) has developed version 1.0 of an AI Risk Management Framework that exhibits certain similarities with the principles adopted by Microsoft and Google. The seven principles put forward by the NIST contain several components, which are as follows:

(1) **Accountability and Transparency:** The promotion of transparency in AI systems contributes to the establishment of confidence and facilitates the effective resolution of problems associated with the outputs of AI models. The implementation of transparency measures also strengthens the level of accountability among developers of AI.
(2) **Explainability and Interpretability:** The concepts of explainability and interpretability pertain to the capacity to describe and understand the activities of AI, thereby providing insights into the functionality and credibility of an AI system. The objective of explainable AI is to furnish individuals with a comprehension of the underlying reasoning behind the outputs generated by AI systems.
(3) **Fairness with Managed Bias:** The concept of fairness is closely associated with the management of prejudice in AI systems since it seeks to address issues related to bias and discrimination. The underlying objective of this principle is to attain parity and fairness, but the complexities arise from the divergent values prevalent inside different organizations and cultures.
(4) **Privacy Enhancement:** Privacy protections serve to protect the autonomy, identity, and dignity of users. Responsible AI systems prioritize and follow fundamental principles such as anonymity, secrecy, and user control.
(5) **Security and Resilience:** The development of responsible AI systems prioritizes security and resilience to effectively mitigate possible threats, such as adversarial attacks. The individuals in question should possess the requisite capacity to proactively avert, alleviate, and recuperate from various forms of hostile action.
(6) **Validity and Reliability:** The concepts of validity and reliability are crucial in responsible AI systems, as they pertain to the ability of these systems to continue their performance even under unforeseen circumstances. By exhibiting consistent functionality without any instances of failure, responsible AI systems demonstrate their validity and reliability.

(7) **Safety:** Safety is a crucial aspect of responsible AI, as it relates to the prevention of harm to human life, property, and the environment.

The ongoing scrutiny of ethical AI systems is necessary to ascertain an organization's commitment to the fair and dependable implementation of AI. This highlights the importance of an organization adopting a maturity model as a framework to guide the advancement and implementation of AI technologies. The establishment of responsible AI requires strict compliance to development criteria that align with the concepts of responsible AI design. Due to the heterogeneity of these principles across different firms, it is crucial to undertake a review. Compliance to standards defined at the organizational level is crucial for the development of AI.

Ensuring the design of AI models has objectives that prioritize the development of models characterized by safety, reliability, and compliance to ethical standards is of importance. Organizations possess the capacity to align themselves with specific objectives and principles in the process of building responsible AI, whereas the available options are dependent upon their guiding principles. The concept of explainable data relates to the approach by which information is conveyed in a manner that is readily comprehensible to individuals.

Thorough documentation is required to facilitate the understanding and resolution of deficiencies in design and decision-making processes. The promotion of a varied work environment is a key undertaking as it develops inclusivity and facilitates productive discussion, thereby decreasing bias. It is recommended to use interpretable features, which refer to data attributes that can be easily comprehended by individuals. The process of development is distinguished by a methodical approach that highlights guaranteeing transparency and inclusivity during the phase of model design. The key subject being examined relates to the notion of explainable AI.

It is advisable to give preference to the use of white box or explainable AI systems that provide a thorough understanding of the decision-making process. The achievement of effective integration of responsible AI necessitates the establishment of systematic and replicable governance protocols, including the subsequent components: to achieve desirable results and assure the reliability and effectiveness of ML models, it is important to comply with established best practices within the ML domain.

Responsible AI is always evolving, necessitating organizations to be aware of the knowledge acquired via its application. The process involves understanding and implementing fair practices, integrating ethical considerations in technology, and considering other relevant factors.

9.2 *Application*

Responsible AI is currently gaining recognition as it explores the ethical and societal consequences of AI systems. It places focus on the importance of creating and implementing AI technology in a manner that aligns with human values of fairness, transparency, and responsibility. In the given setting, the integration of exhaustive perspectives into responsible AI systems demonstrates promising prospects for varied applications in several domains. One sphere in which AI showcases its ability to assert strong perspectives is in content moderation and online discourse.

The exponential growth of hate speech, the dissemination of disinformation, and the circulation of objectionable content present a substantial obstacle for social media platforms in their initiatives to develop an environment that is favourable to productive and meaningful discussions. AI systems that possess firm viewpoints have the potential to enhance the efficiency of identifying and highlighting such content. By imparting instruction to these systems to recognize not only explicit transgressions but also subtle expressions of destructive material, online platforms can develop an atmosphere that is distinguished by more regard and inclusiveness.

In healthcare, diagnostic systems that utilize AI can benefit from viewpoints by providing more reliable and accurate diagnoses. These tools can examine medical data, such as photos and patient histories, and provide well-supported and justified findings on potential diagnoses. When adequately trained on a wide data set that truly reflects the population, these systems possess the capacity to support healthcare professionals in making more informed decisions, ultimately leading to improved patient care. The transportation sector is an industry in which the implementation of responsible AI with robust opinions can have a significant impact.

Autonomous vehicles, characterized by their capacity to navigate complex and dynamic environments, are obligated to prioritize the safety of their occupants. By integrating several perspectives, these vehicles demonstrate the ability to make critical decisions in ambiguous situations, such as ensuring the safety of individuals in unexpected conditions. This methodology guarantees the alignment of AI decision-making with human moral intuitions, enhancing the overall safety of autonomous cars. The integration of AI systems, typified by robust viewpoints, holds the potential to yield substantial benefits for education.

Personalized learning platforms can use these technologies to provide tailored feedback to specific students. By examining and understanding the unique learning patterns of students, along with their specific strengths and limits, AI has the potential to offer detailed and valuable assessments of their academic performance. Consequently, this develops an enhanced and effective educational encounter for the learner. The utilization of this technology possesses the capacity to augment the proficiencies of educators by enabling a more all-including methodology in addressing the specific requirements of learners.

AI-driven algorithms often are essential in influencing investment choices within the financial industry. The incorporation of strong viewpoints into these algorithms possesses the capacity to enhance transparency and reduce risks. An AI system can provide precise justifications for its recommendations about specific investments, thereby enhancing investors' understanding of the underlying rationale behind such assessments. The potential impact of implementing this strategy lies in its ability to address the prevalent issue of opacity in algorithmic trading, thereby developing more trust and confidence among market participants. The integration of AI into criminal justice systems mandates a central emphasis on ethical considerations.

The existence of robust perspectives within AI holds the capacity to offer significant support to judges and legal professionals by furnishing information regarding the likelihood of an individual reoffending. However, it is crucial to ensure that these systems do not perpetuate any biases that could be present in the previous data. Continuous monitoring and intervention are essential in the deployment of responsible AI practices to prevent the occurrence of biased outcomes and promote fairness in the decision-making process. The integration of robust viewpoints into responsible AI systems presents significant opportunities across various industries.

The enhancement of technology's positive influence on society can be attained by harnessing the potential of AI endowed with robust perspectives. This can effectively facilitate more constructive online discourse, enhance medical diagnostic capabilities, improve transportation safety measures, and facilitate tailored educational experiences. However, it is important to adopt a prudent methodology while handling these applications, giving significance to ethical considerations, equity, and openness during the stages of creating and executing such systems. The implementation of responsible AI possesses the capacity to exert significant influence in several areas, thereby contributing to the advancement of a more informed, inclusive, and morally sound future.

9.3 Advantages

The concept of responsible AI has garnered significant attention as a basic strategy for tackling the ethical and societal effects associated with AI systems. The concept includes a variety of principles, such as transparency, justice, and accountability, to guarantee that AI technologies are created and used in manners that provide advantages for society at large. The benefits of responsible AI are significant, as it can address and reduce negative consequences, promote creativity and progress, and maintain ethical principles within the dynamic domain of AI. One of the key benefits associated with responsible AI lies in its ability to effectively address and minimize potential adverse consequences for society.

The integration of AI systems across several sectors gives rise to concerns about bias, infringement of privacy, and unforeseen effects. Responsible AI aims to actively tackle these concerns by implementing mechanisms that guarantee fair treatment of humans across various demographic groupings. Responsible AI is essential in promoting fairness in various domains such as criminal justice, lending, and healthcare by effectively detecting and addressing biases present in AI systems. The promotion of openness and accountability in AI systems is a key aspect of responsible AI. In an epoch characterized by the opaqueness of complex AI algorithms, it becomes important to comprehend the process by which judgements are rendered.

The promotion of responsible AI involves the advancement of AI models that are capable of providing explanations, allowing users to gain an understanding of the underlying logic behind the suggestions or conclusions provided by AI systems. The provision of transparency in AI systems serves two important purposes:

(1) Establishing confidence among users and enabling them to identify
(2) Rectify any potential flaws or biases that may arise

As a result, the overall dependability and effectiveness of AI technologies have significantly improved. The advantages of responsible AI include ethical principles. In contrast to the prevailing belief that ethical issues impede creativity, it may be argued that the integration of responsible AI serves as a driving force for developing innovation.

Responsible AI develops the exploration of novel pathways by AI developers through the imposition of ethical principles and frameworks, which serve to establish boundaries that mitigate the risk of potential misuse or harm. The presence of a dual nature of innovation, constrained by ethical considerations, serves to ensure that advancements in technology are in harmony with societal norms and are held responsible for their possible impacts. Also, the incorporation of responsible AI practices are necessary in ensuring the sustainability of AI systems. The reliability and consistency of AI systems are essential as they become more integrated into essential systems such as autonomous vehicles and medical diagnostics.

The concept of responsible AI places significant importance on the implementation of rigorous testing, constant monitoring, and auditing. These practices serve to minimize the probability of catastrophic failures and develop a sense of trust in the dependability of AI-driven solutions. The prevailing consensus among researchers and experts regarding the merits of responsible AI is overwhelmingly positive and by its fundamental principles. There is a general argument asserting that the advantages of prioritizing ethical practices in AI far surpass any potential disadvantages. The proponents urge the establishment of an ecosystem that integrates scientific breakthroughs with social responsibility to mitigate the possible misuse of AI and its negative consequences.

The promotion of collaboration among AI developers, ethicists, and politicians is regarded as an active strategy for moulding the societal consequences of AI, thereby giving rise to the concept of responsible AI. Some individuals raise concerns with the pragmatic use of responsible AI, highlighting difficulties in establishing universally applicable ethical principles and adapting them to various circumstances. Amidst these deliberations, there is a prevailing agreement that responsible AI offers an essential structure for steering the progress and implementation of AI. The primary focus should be placed on the continuous process of refining and adjusting ethical standards as our comprehension of the consequences of AI becomes more profound.

The use of responsible AI offers various benefits that span both ethical and practical considerations. One notable benefit relates to the mitigation of bias in AI models. The appropriate use of AI requires the assurance that AI models, algorithms, and the underlying data utilized in their development are devoid of biases and uphold a thorough representation. The implementation of a thorough technique in this particular situation yields improved results and effectively addresses the challenges associated with data and model drift. This practice serves to address potential harm to users resulting from biased outcomes of AI models, thereby upholding principles of fairness and equality from ethical and legal perspectives. Also, the use of responsible AI significantly amplifies the degrees of transparency and comprehensibility inside AI models.

Organizations promote trust and assurance among their stakeholders and clientele through the implementation of acceptable AI practices. The increased degree of transparency enables users to comprehend the decision-making mechanisms utilized by AI systems, thereby promoting a perception of reliability. This practice enhances the process of democratizing AI by enabling its accessibility to both corporate entities and individual users, promoting a more equitable technological setting.

The conscientious integration of AI also presents opportunities of great importance. Both developers and users can articulate their uncertainties and worries about AI technologies. The employment of a dialogue-oriented methodology offers a dual objective of proficiently addressing prospective concerns and promoting the advancement and execution of morally upright AI solutions.

By engaging in active involvement within this iterative process, ethical AI establishes the foundation for a technologically advanced future that follows moral principles. One crucial element of ethical AI involves the commitment to prioritizing the protection of privacy and the assurance of data security. The foundational concept emphasizes the significance of not using personal or sensitive information for actions that are considered unethical, irresponsible, or illegal. The concept of responsible AI places significant importance on the protection of privacy and security, effectively reducing the potential for data misuse and upholding the confidentiality of individuals' personal information.

This strategy develops an environment that is marked by trust and accountability. Responsible AI is essential in mitigating potential risks and ensuring safety. The application of responsible AI offers advantages to numerous stakeholders, employees, and society as a whole by creating well-defined ethical and legal boundaries for AI systems. The establishment of well-defined limits mitigates the probability of AI systems veering into hazardous regions.

The adoption of a prudent risk management strategy ensures the concurrency of technology and the welfare of society. There exists a multitude of advantages associated with the implementation of responsible AI, which includes both ethical principles and practical considerations. The collaborative effort to mitigate bias, boost transparency, empower stakeholders, protect privacy, and manage risks together amplifies the advantages of responsible AI. As society grapples with the complex process of incorporating AI, the adoption of responsible AI principles assumes utmost importance in shaping a future that is characterized by technological advancement and ethical integrity.

9.4 Disadvantages

The proper development and deployment of AI have become a prominent concern in recent years due to its integration into multiple areas of human existence. The concept of responsible AI, which frequently includes ethical considerations, openness, and fairness, is widely recognized for its capacity to address and minimize the possible adverse effects of AI systems. This particular methodology is not devoid of its downsides and obstacles, necessitating a thorough analysis to effectively leverage the advantages of AI while mitigating its limitations. A notable limitation of responsible AI is its capacity to impede innovation and hinder progress.

The implementation of stricter ethical norms and laws is crucial to mitigate the potential harm caused by AI systems. However, it is important to acknowledge that these measures may unintentionally impede the progress of creative solutions. For example, the implementation of strict standards for data privacy and security could potentially limit the availability of useful data sets, hindering the training and progression of AI algorithms. The situation at hand necessitates an equilibrium between ethical considerations and the importance of developing technological progress. Also, the implementation of responsible AI has the potential to further amplify the problem of bias inside AI systems.

The process of recognizing and reducing prejudice, although intended to eliminate bias and promote fairness, can be complex and demanding. The effort to address bias in AI systems may result in excessive correction, which could potentially compromise the accuracy and effectiveness of those systems. These concerns are particularly evident in fields characterized by strongly entrenched historical prejudices, such as criminal justice and hiring procedures. The pursuit of equity in AI necessitates a nuanced strategy to mitigate inadvertent repercussions. Another issue to consider connects to the possible increase in costs that may arise from the implementation of responsible AI.

Compliance to ethical principles, the thorough conduct of audits, and the maintenance of openness can impose a greater financial burden on businesses, particularly those of smaller scale that possess fewer resources. The financial burden associated with AI development may discourage certain businesses from participating in it entirely or lead to the use of suboptimal strategies to comply with regulatory requirements, thereby undermining the desired advantages of responsible practices. Also, the widespread implementation of AI on a worldwide scale presents challenges in establishing a generally applicable framework for ensuring ethical AI practices.

The need to adapt ethical rules to various circumstances arises from the presence of distinct cultural norms, legal systems, and societal values. Yet, the act of customizing AI systems may result in fragmentation and inconsistent execution, compromising the cohesive development of AI technology. Achieving an equilibrium between overarching concepts that apply universally and adjustments that are specific to particular contexts is crucial, though complex. Wide perspectives exist among researchers and specialists about the issue of responsible AI. Some proponents argue that the potential benefits of responsible AI exceed its drawbacks. They contend that responsible AI can mitigate severe outcomes, such as the malicious exploitation of AI or the continuation of detrimental biases.

The proponents urge a cautious and deliberate strategy, placing significant emphasis on developing collaboration among AI developers, ethicists, and legislators. It is widely considered that the systematic resolution of the difficulties presented by responsible AI can be achieved through interdisciplinary conversation. On the other hand, there exists a group of individuals who voice concerns over the potential negative consequences of adopting an excessively cautious stance towards responsible AI, as they believe it may slow technical advancement and impair society's development. The argument put forth is that an abundance of regulations could potentially impede the independence of AI engineers and restrict the pursuit of inventive applications.

These individuals espouse a paradigm that promotes a greater degree of adaptability, enabling the exploration of new ideas and approaches while also maintaining fundamental ethical principle. The authors contend that achieving an equilibrium between regulatory measures and developing innovation is important to fully harness the capabilities of AI. The drawbacks associated with responsible AI highlight the need for delicate equilibrium in managing the complex terrain of AI advancement. Responsible practices have a particular role in mitigating harm and promoting equity; yet, it is important to acknowledge that they can unintentionally impede innovation, perpetuate bias, and impose financial constraints.

There exists a wide range of perspectives about this problem, with certain individuals placing significant emphasis on the necessity of careful regulation and collaborative efforts, while others argue in favour of adopting a more flexible approach to encourage and facilitate innovation. The ongoing influence of AI on our society necessitates urgent attention and effort to confront these obstacles. This initiative is crucial to effectively utilize the advantages of AI while mitigating its potential negative consequences.

10 Digital Resilience with AI

Frequently, complex ecosystems consist of a heterogeneous assortment of discrete elements. These problems frequently involve socio-technical domains. The rise of collaboration and sophistication can be attributed to the interaction between interconnected components and their ability to withstand disturbances, which frequently expose them to cascading effects. The conventional analysis of complex systems can shed light on a variety of aspects, such as organizational hierarchies, socio-technical systems, infrastructure frameworks, and supply networks. Modern methodologies take into account the impact of certain functional elements within a given locality, such as data pipelines, human decision-makers, memory buffers, and energy suppliers.

Also, they take the topological characteristics of the entire ecosystem into account. In this context, AI is used to detect vulnerabilities or flaws in networked systems, which are often challenging to assess using conventional learning methods. Within this theoretical framework, for example, organizations seek to leverage AI to assist chief financial officers (CFOs) in augmenting cash flow while simultaneously protecting and preserving current financial resources. The incorporation of AI facilitates the implementation of a data-centric, adaptable methodology that exploits exhaustive data from a variety of crucial organizational operations and business sectors.

AI is necessary for mitigating the potential negative effects of major disruptions or catastrophes, such as the COVID-19 pandemic. This is achieved by enhancing the capability to monitor and control global cash visibility, optimizing resource allocation, and facilitating client-vendor relationships. AI models have emerged in the healthcare industry as an instrument to assist organizations in comprehending the effects of significant disruptions to their supply chains. Organizations use AI-based scenario modelling techniques to develop strategies for effectively managing logistics within the healthcare system and consumer product interactions. The topic of resilience in AI and data science has garnered considerable interest and is now the subject of scientific and professional research.

The adoption of AI and data science to improve human decision-making is largely attributable to their rapid development. However, the efficacy of AI implementations depends on procedures and processes that may not have consistently demonstrated a substantial degree of resilience, as will be explained in the following section. Numerous vulnerabilities could compromise the integrity of the data pipeline that supplies data to AI engines. Conventional cybersecurity issues, such as elevated access, injection attacks, interference, and DDoS attacks, as well as emerging attack techniques, such as infected data and malicious manipulation, have the potential to result in negative outcomes for future AI systems.

In situations that require safety, such as autonomous driving, and time-sensitive applications, such as financial trading and remote operations, these compromises may result in significant consequences that degrade the application's general robustness. The emergence of novel vulnerabilities resulting from AI algorithms, especially deep algorithms, is naturally linked to the problem of data integrity. These vulnerabilities may involve malicious intent or the presence of tainted data. Deep algorithms, in contrast to conventional algorithms such as probabilistic, stochastic, physics-based, and feature-driven models, function by analysing data in higher dimensions, often concealing their flaws, whereas the flaws of the latter are evident and simple to comprehend. Unintended consequences may arise from the use of experimentation by clever attackers or end users.

The primary focus of this particular framework is the level of trust that human end users have in AI systems. Significant scepticism has the potential to undermine the overall robustness of the ecosystem as a whole. There are two distinct types of trust:

(1) Tangible trust, which is based on the rational justification of actions or decisions
(2) Emotional trust, which involves having faith in an algorithm without engaging in dialogue

In the belief that planning will effectively mitigate the majority of challenges, organizations frequently strive to strengthen their programs to enhance resilience against incidents. In the event of adversity, the use of strategies based on well-established best practices is capable of persuasively demonstrating the responsible nature of their initiatives to both regulatory bodies and the general public.

Significant progress has been made in enhancing resilience; however, deficiencies in the current system are still observed on occasion. Typical business operations are adept at addressing minor disruptions, whereas major incidents are typically addressed by implementing the organization's continuity strategy. The concept of resilience is predicated on the execution of deliberate strategies, methodical procedures, and strict compliance to established norms, with a particular emphasis on the recovery and recouping of resources during times of adversity. The ability of AI to find and assess system vulnerabilities and flaws make it prevalent in this scenario. However, the presence of complex and difficult circumstances has compelled businesses to demonstrate greater adaptability to effectively deal with unexpected or difficult situations.

Numerous corporate leaders acknowledge that relying solely on a reactive strategy is insufficient for effectively addressing unanticipated disruptions and anticipated uncertainties. Beyond asset recovery protocols, the development of agility, adaptability, learning, and regeneration is essential for organizational resilience. This measure ensures preparedness for complex and difficult situations. Due to its complexity, volatility, and unpredictability, the operational environment of this era presents numerous obstacles to the adaptation process. Recent disasters have highlighted the need for businesses to be able to rapidly adapt to new threats, disruptions, and disturbances, such as pandemics and climate change.

The implementation of AI-driven scenario modelling enables businesses to actively partake in innovation when confronted with resilience challenges, potentially developing a proactive approach to obstacles that were previously insurmountable. The promotion of resilience through the deliberate incorporation of AI has become a necessity for corporate boards, governmental bodies, and society as a whole, despite the difficulties associated with establishing an economic justification for doing so. Digital resilience and AI interact to contribute to the transformative character of the modern society. Digital resilience is the capacity of individuals, communities, and systems to endure and recuperate from digitally related disruptions.

The observed phenomena may be compared to the resilience exhibited by an ecosystem, which is characterized by its ability to adapt and endure in the face of multiple obstacles. Let us now explore AI. AI is an expression of human creativity and innovation, including the development of computational systems that simulate cognitive functions such as learning, reasoning, and problem-solving, among others. These innovations demonstrate the peak of technological progress, with the potential to radically transform multiple industries and improve the condition of humanity. The relationship between digital resilience and AI is complex and of great significance. An increase in our digital resilience overlaps with the societal adoption of AI-driven technologies.

AI provides the necessary capabilities for predicting, mitigating, and addressing disruptions. In cybersecurity, AI can rapidly identify and analyse patterns of adversarial behaviour, thereby facilitating preemptive actions to mitigate cyberthreats and strengthen our digital security protocols. However, this connection presents numerous obstacles. The increasing reliance on AI has the potential to interfere with optimal decision-making and diminish human capabilities. It is widely acknowledged that AI has the potential to increase productivity. Regardless, it is indispensable to recognize that AI may inadvertently enable a complacent attitude towards managing unexpected digital disruptions.

Maintaining digital resiliency requires achieving the optimal balance between automated AI responses and human participation. The development and implementation of AI necessitate ethical considerations. The incorporation of biases within AI algorithms may result in the unintended amplification of digital vulnerabilities and the ensuing compromise of resistance. The importance of ensuring justice, transparency, and accountability in AI systems in our digital environment cannot be overstated.

The ongoing discussion surrounding digital resilience and AI is dynamic. Due to the increasing prevalence of AI integration in our digital world, we must recognize and fulfil a dual responsibility. To strengthen our digital resilience, it is essential to optimize the utilization of AI's capabilities. In other words, it is essential that we diligently confront the ethical and practical dilemmas associated with AI. For the establishment of a long-lasting digital future, it is essential to develop a symbiotic connection between technological proficiency and human intuition.

11 Concluding Remarks

In a rapidly evolving digital environment, where technology is deeply integrated into all aspects of our daily existence, it is important to emphasize the importance of including resilience. AI and ML have rapidly emerged as important mechanisms in our defence against cyberattacks. This phenomenon is attributed to the escalating complexity of cyberthreats, along with the deep interconnectedness of our current digital world. The primary objective of this chapter was to undertake an examination of the current and prospective roles that AI and ML will assume in enhancing our digital security measures. This analysis examined the primary domains in which these technologies are balanced to have an impact, explaining their capacity to enhance digital resilience and elevate cybersecurity to advanced levels.

AI and ML have emerged as major contenders poised to spearhead transformative transformations in the domain of threat detection and mitigation. The current proficiency of the system primarily resides in identifying existing vulnerabilities.

However, its full potential will be realized when it becomes capable of proactively identifying zero-day vulnerabilities and disclosing novel attack pathways. This ambitious undertaking necessitates the creation of complex deep learning models and the acquisition of expertise in advanced anomaly detection techniques. The successful completion of a task of such complexity necessitates the concerted efforts of highly skilled individuals, wherein professionals specializing in the domains of data science and cybersecurity collaborate to advance the boundaries of knowledge. Intelligent systems are anticipated to emerge through the process of seamlessly integrating the pooled wisdom and skills of various entities. Not only can these systems identify but they can also predict cyberthreats before they manifest harmful intent.

The seamless integration of AI and ML into security operations presents a vision of a future characterized by increased automation and autonomous threat response. In the present scenario, security orchestration and automation systems, which have lately been equipped with AI capabilities, are approaching the point of becoming highly skilled performers, efficiently coordinating responses without requiring human intervention. In the current age characterized by the growing prevalence of digital technologies, it is necessary for enterprises to adopt AI-based security solutions and guarantee their seamless integration with existing infrastructures. The proposed merger is expected to enhance threat detection capabilities and provide a new phase of streamlined incident response. This new approach aims to mitigate the vulnerability of human error and exhibit typical agility in promptly addressing security risks.

The increasing integration of AI models in the field of cybersecurity necessitates their strict compliance with the principles of transparency and explainability. Professionals within the security industry must possess a comprehension of the fundamental mechanisms behind AI-generated assessments, as this understanding serves as the foundation for establishing trust. The path ahead necessitates a persistent pursuit of novel methodologies and technology, aiming to guarantee that explanations produced by AI are lucid and comprehensible. The promotion of openness not only develops trust but also enables human experts to critically evaluate the insights generated by AI. This ensures that the collaboration between humans and machines operates cohesively, thus enhancing cybersecurity.

The domain of user training and awareness initiatives is poised for a notable transformation that will be facilitated by the capabilities of AI. We are currently at the threshold of a significant shift in our understanding and approach, wherein there is a growing recognition of the need for ongoing user education that effectively adapts to the always-evolving environment of threats. When employing a progressive approach that incorporates real-time feedback systems, users are equipped with the resilience required to withstand sophisticated cyberattacks. In this particular context, training modules powered by AI emerge as the vanguards of simulation. The modules in question serve to replicate a range of potential threat situations and provide users with a set of cyber hygiene best practices.

The potential offered by blockchain technology and decentralization in enhancing data integrity and identity verification is undeniably groundbreaking. The incorporation of tamper-proof records and safe verification procedures in these technologies establishes a robust defence mechanism against digital malfeasance due to the synergistic effect resulting from their integration. However, as we progress, we will encounter other obstacles including scalability, energy efficiency, and the complex network of regulations that now exist. Hence, we must concentrate our collective initiatives towards the enhancement of user-friendly interfaces and the establishment of standardized protocols for the wide range of applications facilitated by blockchain technology. This will enable the complete realization of the promise of these technologies within the domain of cybersecurity.

In the context of cybersecurity, it is important to devote sufficient attention to the ethical implementation of AI. The system upholds the preservation of user information confidentiality and algorithm integrity through the use of impartiality measures. The future needs the implementation of strict industry standards and regulatory frameworks to effectively monitor and ensure the ethical utilization of AI. To align with these rigorous criteria, corporations and other entities must willingly subject themselves to the rigorous process of ethical AI assessments. In the domain of cybersecurity, the utilization of AI is anticipated to yield positive outcomes due to ongoing initiatives in this domain.

The maintenance of a robust cybersecurity foundation necessitates the commitment of two key elements:

(1) Industry-wide collaboration
(2) Exchange of threat intelligence

For the continued advancement of cooperation, it is important to enhance its structure and efficiency. This may be achieved by establishing standardized procedures for the sharing of crucial threat data, which will help define the bounds of collaboration. Public-private partnerships have become significant collaborations in the context described, as they bring together resources, knowledge, and intelligence to establish a strong defence against cyberattacks.

As we approach the imminent rise of quantum computing, we must pay attention to the urgent call for preparedness. The development and expeditious deployment of encryption algorithms that possess resistance against quantum computing are of importance. The advancement of research in the field of post-quantum cryptography is important, and critical systems need to initiate the transition towards adopting encryption methods that can withstand the computational power of quantum computers. As an ongoing initiative, our primary objective is to exceed the capacities of quantum computing to uphold the integrity of our digital environment.

The course of cybersecurity and digital resilience is linked to the gradual integration and advancement of technology such as AI and ML. These two immense institutions, poised to revolutionize our security measures, are preparing to enhance our defences against escalating threats. To fully reach their potential, organizations, governments, and the cybersecurity professional community must make dedicated investments in extensive research and development, ethical frameworks, collaborative initiatives, and proactive security measures. AI and ML are poised to assume key roles in safeguarding the digital infrastructure, thereby enhancing its safety and security. These technologies will be important in upholding constant care in the face of escalating threats. In the context of our globally interconnected society, it becomes necessary for us to collectively address the dynamic threats with confidence and adaptability, thereby facilitating a more secure trajectory for the digital age.

Chapter 7
Thriving in the Quantum Era

1 Introduction

In this chapter, we will discuss the fundamental principles of quantum computing while scrutinizing the effect it has on digital resilience. The complexity of the information provided in this chapter will be lucid, to facilitate understanding for readers of all backgrounds. Those with an established background in mathematics can decide to dive further into portions that include more complicated concepts in mathematics.

Researchers are interested in how quantum computing, a cutting-edge technology, may affect digital resilience and the digital world. This chapter examines the effects of quantum computing on digital resilience, including its pros and cons. Digital resilience suffers as firms and governments try to protect their information security infrastructures. Existing encryption techniques are unsafe; hence, quantum-resistant cryptographic algorithms must be created. Quantum computing threats to digital communications security and integrity must be mitigated by this study.

Digital resilience in the larger context of quantum computing refers to the ability of digital systems to withstand and recover from security breaches, malicious attacks, or threats that occur within the quantum computational environment. This requires the capability to guarantee the ongoing operation and integrity of digital processes despite potential challenges posed by quantum technologies. The execution of complex calculations by quantum computing, which is exponentially faster than classical computers, poses significant challenges and opportunities for enhancing digital resilience. When examining the impact of quantum computing on digital resilience, it is essential to consider several key factors. This chapter will begin with an examination of the emergence of quantum algorithms, followed by an analysis of the quantum properties that the system must possess and the quantum primitives that have been developed and implemented.

S. K. Shandilya et al., *Digital Resilience: Navigating Disruption and Safeguarding Data Privacy*, EAI/Springer Innovations in Communication and Computing,
https://doi.org/10.1007/978-3-031-53290-0_7

Quantum computing also opens new optimization avenues. Grover's quadratic speedup while searching unsorted databases affects cybersecurity and data retrieval. Quantum computing can optimize supply chain management, complicated data processing, and other digital activities, improving digital resilience. These improvements must be handled carefully. Error correction and decoherence plague quantum computers, which are still developing. The search for fault-tolerant quantum computing continues, and large-scale, error-resistant quantum computers are still far off. Creating a quantum-resilient digital infrastructure demands significant funding and planning. Strategy and knowledge of the shifting threat environment are needed to design quantum-safe communication protocols and quantum-resistant cryptography systems.

Besides cybersecurity, quantum computing has far-reaching digital effects. Quantum computing may change ML and optimization. QML algorithms that use quantum parallelism solve complex problems quicker, which will affect data-driven sectors. As companies adapt to new environments, legislators must balance innovation and security. A secure and robust digital environment in the quantum age requires proactive quantum-safe standards and guidelines.

To review the structure of the chapter, the introductory section attempts to explore the basic ideas that differentiate quantum mechanics from classical physics. The focal area of research under consideration relates to post-quantum-resistant cryptography, also known as quantum-resistant cryptography. The principal objective of research in this domain is to develop cryptographic frameworks that exhibit resilience against prospective assaults from both conventional and quantum computers. These frameworks have been meticulously designed to ensure their robust security measures, while also facilitating smooth integration with established communication protocols and networks.

The severe problem arises from the probable introduction of large-scale quantum computers, which present significant challenges. Historically, there has been uncertainty surrounding the creation of highly effective quantum computing devices. However, a consensus has emerged among scientists who see this effort as a substantial technical undertaking. Certain scientists have made predictions suggesting that computationally powerful quantum computers could become available in the next two decades, resulting in the obsolescence of most existing public-key encryption techniques.

Given the extensive duration required for the development of modern public-key cryptography infrastructure, we should begin preparations to enhance our information security systems in anticipation of the forthcoming age of quantum computing. The need for preparedness is crucial, irrespective of our capacity to properly forecast the advent of the quantum age. Cryptographic methods predicated on the presumption of integrity by at least one individual are much more advantageous compared to those that depend on the reliability of the majority. The former proposition depends on a less restrictive assumption, which seems more attainable or credible in comparison to the latter.

In the traditional context, cryptography is based on the assumption that the mathematical problems it relies on are often difficult to solve on average. This implies that meeting a randomly chosen problem that poses a significant challenge is extremely probable. An instance of a difficult problem in mathematics is the factorization of an integer, which may be represented as the product of two prime factors, indicated as $n = p \times q$. This kind of difficulty is sometimes referred to as 'average-case hard', which sets it apart from 'worst-case hard' problems, which exhibit high complexity in at least one occurrence. So far, the standard of average-case hardness imposes a far higher level of rigour and demands in comparison to worst-case hardness.

Some problems may exhibit a small number of very complicated examples, whereas the majority of occurrences are readily solved. In the present setting, a problem is considered worst-case hard, but it may not attain average-case hardness as a result of the limited availability of highly demanding situations. Nevertheless, it is necessary to note that every problem that exhibits average-case hardness also necessarily exhibits worst-case hardness. The fundamental contention is that designating a task as worst-case hard involves a less strict and thus more rational assumption in contrast to designating it as average-case hard.

Significantly, in the area of lattice-based problems, which are used a lot in cryptography, showing the worst-case complexity of these problems is enough to prove the security of systems, so there's no need to show their average-case complexity. There is a possibility that specific complex situations may arise that pose substantial difficulties, whereas the bulk of situations are comparatively easier to handle. While these complex instances may be categorized as 'worst-case hard', it does not imply that they are inherently 'average-case hard', as the probability of meeting them randomly is limited. However, it is noteworthy that every problem classified as average-case hard inevitably exhibits worst-case hardness. Hence, positing worst-case hardness as an assumption is comparatively more permissive and hence more rational than positing average-case hardness.

The discipline of lattice-based cryptography emerged with the seminal contributions of Miklós Ajtai in 1996 [Ajt96], representing a significant turning point in its development. Following this idea, modern technological networks and platforms need to include safeguards that enhance the resilience of current data against prospective post-quantum threats. Scientists in academia and industry demonstrate a strong commitment to dedicating resources to conduct research and development activities. They engage in a comprehensive evaluation of the efficacy and suitability of existing lattice-based algorithms.

In the last 10 years, researchers in cryptography and mathematics have set out on the arduous undertaking of translating complicated mathematical problems into concrete cryptographic structures. At present, there is an extensive spectrum of mathematical problem classifications that are being examined, each originating from different mathematical disciplines. To begin this part, we are going to start by explaining the particular category of problems that are fundamental to the field. The terminology in question has its origins in the cryptography systems used in this particular sector. These systems heavily depend on security validations that are linked to the resolution of complex mathematical problems rooted in the subject.

1.1 Quantum Superposition

This section delves into the concept of quantum superposition, which denotes the simultaneous existence of particles in multiple distinct states. This phenomenon holds significant implications within information processing. To comprehend this concept, consider an analogy involving the simultaneous touching of two separate points on the surface of a pond. In such a scenario, waves emerge from each point, eventually converging to form a composite wave pattern representing a state of superposition. Analogously, within quantum mechanics, particles such as electrons and photons exhibit wave-like properties that allow their combination into what is termed a 'superposed' state.

While ripples on a pond are generated through the movement of water, quantum waves are purely mathematical abstractions. These waves are expressed as mathematical equations that define the probabilities of a particle being in a specific state or possessing particular properties. These equations provide insights into the probability of attributes such as an electron's velocity or position. When an electron exists in a state of superposition, its states can be envisioned as distinct potential outcomes, each associated with a specific probability of observation. For instance, an electron might simultaneously inhabit a superposition of two velocities or occupy two separate spatial locations.

Understanding superposition has the potential to drive advancements in quantum technologies, such as quantum computing. Comprehending the concept of quantum superposition can be challenging. Traditional explanations have turned to analogies, such as a coin displaying both heads and tails simultaneously, or the well-known Schrödinger's cat thought experiment. In the latter, physicist Erwin Schrödinger proposed a scenario where a cat within a sealed box could be considered both alive and deceased concurrently until the box is opened, at which point observation randomly determines the cat's state.

Superposition can be likened to an equation with multiple valid solutions. Analogous to how the equation $x^2 = 4$ has two solutions ($x = 2$ or $x = -2$), superposed wave functions entail more complex solutions, which can be approached with a similar analytical mindset. This leads to the question: *How can scientists experimentally confirm the existence of superposition?* Numerous experiments have undeniably validated superposition.

One illustrative experiment involves the use of light filters—devices that selectively block light based on polarization. Examples of such filters can be found in items like polarized sunglasses and camera lenses. The light we encounter originates from various sources and comprises a combination of diverse waves with differing polarizations, characterized by peaks and troughs oriented in distinct directions. Consequently, the light exists in a superposed state including these various polarizations.

As light waves interact with their environment, their properties undergo modifications. For instance, light reflecting off the surface of a lake or snow-covered terrain often acquires horizontal polarization. When this polarized light encounters a filter that only allows vertically polarized light to pass through, the reflection is impeded, reducing glare. Superposition occurs when multiple filters are manipulated in different configurations to extract additional light properties.

As light traverses a horizontal filter, its probability of passing through a second horizontal filter remains certain. As the orientation of the second filter shifts toward the vertical axis, the probability of both filters permitting light passage gradually diminishes. At a diagonal orientation (45 degrees), half of the light prevails, dwindling to zero when the filter aligns vertically. Superposition is evidenced by the introduction of a diagonal filter between horizontal and vertical filters, enabling some light to pass through the system unimpeded. In this manner, the diagonal filter serves to "reset" the light's superposed state by enhancing the probability of vertical polarization.

1.2 Quantum Entanglement

This section will clarify the premise of quantum entanglement, which refers to the phenomenon where particles become associated in such a way that the state of one particle has an impact on the state of another particle, irrespective of the physical distance between them. Quantum entanglement is a fascinating and enigmatic phenomenon within quantum mechanics. Though two or more particles establish a discernible correlation, their quantum states stay entangled, even though there is a significant separation in space. This phenomenon gives rise to a condition of entangled quantum behaviour, in which changes to the state of one particle immediately reveal information about the other particles, regardless of the distance between them. Any alteration made to one particle is accompanied by a matching resonance in the remaining particles.

The idea of entanglement is considered fundamental in quantum mechanics and has the potential to be of great importance in the development of future quantum technologies. Similar to other phenomena seen in quantum mechanics, entanglement mostly occurs at the microscopic level of atoms and subatomic particles. Entanglement is a phenomenon that occurs in particles like photons and electrons; despite being far apart, they remain connected. The identification of this emergent phenomenon was seen by researchers throughout the first decades of the twentieth century as they attempted to comprehend the complexities inherent in quantum mechanics. The investigation by the researchers showed that it was crucial to introduce a theoretical concept known as a quantum state to explain the complex workings of very small systems.

In quantum mechanics, events possess inherent uncertainty. An exemplary illustration is an inherent challenge of accurately determining the exact location of an electron inside an atom; one can only hypothesize on the probability associated with its potential presence in various places. The concept of a quantum state involves the representation of probabilities associated with several characteristics of a particle, including but not limited to its location and momentum. Therefore, the quantum state of an electron contains the whole list of possible positions it might potentially occupy, along with the corresponding probability assigned to each different site.

Quantum states possess a fascinating characteristic known as entanglement, whereby the measurement of one state influences the other. In 1935, Albert Einstein, Boris Podolsky, and Nathan Rosen [Ein+35] conducted groundbreaking research that revealed that particles involved in strong correlations lose their individual properties and adopt a collective state. The term 'quantum entanglement' is now used to describe the condition of amalgamation being discussed. The experiment conducted shed light on the observation that measurements performed on one particle have an immediate impact on the other, even though the two particles are somewhat far from each other. The notion in question was given the word 'entanglement' by Erwin Schrödinger, a renowned scientist in quantum mechanics.

Schrödinger saw it as a fundamental aspect of quantum mechanics and emphasized its deviation from the operational laws that control our daily world. According to Einstein, Podolsky, and Rosen, the phenomenon of entanglement manifests itself instantly. The acquisition of knowledge from a single quantum state then leads to the acquisition of knowledge about the states of all entangled particles. The transmission of information seems to occur across significant distances, perhaps extending over intergalactic distances, and at speeds exceeding the conventional speed of light.

The expression 'spooky action at a distance' was used by Einstein to succinctly describe this phenomenon, which he saw as indicative of an unresolved element within the quantum theory framework. However, several experimental studies have repeatedly shown that entangled particles exhibit reciprocal impact regardless of their spatial distance, providing more evidence for the validity of quantum mechanics. The unsolved nature of this phenomenon, sometimes referred to as 'spooky action', continues to perplex researchers. Nevertheless, the fundamental idea of causality is still valid despite the immediate reciprocal impact that entanglement over physical distances demonstrates.

An external observer situated at a distance from the entangled system cannot distinguish if a nearby observer has influenced the entanglement of the system, and conversely, to definitively demonstrate the assumption, a communication interaction must occur at the velocity of light. The speed of light still restricts the transmission of such information, despite the ability to determine the state of a distant particle. Several distinct approaches have been developed for entangling particles. One method involves exposing particles to reduced temperatures, causing their quantum states to overlap due to the inherent uncertainty in their locations, resulting in their indistinguishability.

An alternative approach leverages inherent mechanisms occurring at the subatomic level, such as nuclear disintegration, to autonomously produce entangled particles. By using pairs of entangled particles, a transmitter and receiver may create a secure link. These particles function as the fundamental basis for producing private cryptographic keys that are exclusively available to the sender and recipient, thereby facilitating the encryption of communications. Both parties involved quickly detected any attempt to sabotage the entanglement and intercept the communication. The phenomenon of entanglement has significant practical value in quantum computing.

Quantum computers can collectively tackle complex problems through particle entanglement. For instance, a quantum computer with just 10 qubits can represent the same amount of information as 210 traditional bits. Entanglement is used in quantum teleportation, a process that may be analogized to traditional communication but is specifically designed for the quantum world. During this procedure, a sender undertakes the task of preparing a particle to transmit information, namely by manipulating its quantum state. The state of this entity is then combined with the quantum state of another entangled particle, resulting in a corresponding change in the far-away entangled particle.

The receiver proceeds to quantify the modifications occurring in the quantum state of the distant particle. Afterwards, using traditional communication at the velocity of light, the sender conveys information about the modifications to the state of the distant particle. This process enables the duplication of the quantum state at a different spatial position. While the complexity of this approach may seem daunting for transmitting individual bits of information, it offers a reliable method of communication. Any attempt to intercept the signal leads to disturbances in the entanglement, which become apparent when comparing the regular signal with the changes in the state of the entangled particle.

1.3 No-Cloning Theorem

Within the field of physics, there is a fundamental theory known as the no-cloning theorem. This concept posits that the creation of an identical and replicated form of a random and unfamiliar quantum state is unachievable. The concept described carries substantial consequences, especially in the field of quantum computing. The theorem in question originates from the no-go theorem formulated by James Park in 1970. This theorem demonstrates the inherent impossibility of achieving a direct and flawless measurement method without inducing disturbances. Wootters, Zurek, and Dieks [WZ82, Par70, Die82] independently arrived at the previous conclusion in 1982.

Significantly, it is crucial to note that these theorems do not preclude the possibility of intermingling the state of one system with another, as the concept of cloning specifically refers to the creation of a divisible state with identical components. An example of interlinking two qubits without violating the no-cloning theorem is achieved through the use of the controlled-NOT gate and the Walsh–Hadamard gate. This phenomenon arises due to the absence of a distinct and well-defined state that can be attributed to an individual component inside an entangled state. The no-cloning theorem, as commonly understood, exclusively applies to pure states. When confronted with situations that involve mixed states, the no-broadcast theorem, which has a wider scope, becomes relevant.

The No-Cloning Theorem is a fundamental principle in quantum mechanics that restricts the ability to create an exact copy of an arbitrary quantum state. This theorem assists in understanding the unique and non-classical behaviour of quantum systems. To explain this theorem mathematically and formally, we will start by introducing the necessary concepts and then present the theorem itself.

In quantum mechanics, a quantum state is described by a mathematical object called a state vector, typically denoted as $|\psi\rangle$. This state vector exists in a complex vector space, known as a Hilbert space, which captures all the possible states the system can be in.

The No-Cloning Theorem addresses the question of whether it is possible to create an operation that duplicates an arbitrary quantum state. In more precise terms, it examines whether there exists an operation that can take two distinct quantum states $|\psi\rangle$ and $|\phi\rangle$ and produce two copies of one of them, say $|\psi\rangle|\psi\rangle$. The no-cloning theorem can be stated as follows:

Theorem 1 (No-Cloning Theorem) *For any two distinct quantum states $|\psi\rangle$ and $|\phi\rangle$ in a given Hilbert space, there does not exist a quantum operation U that can perfectly clone these states:*

$$U(|\psi\rangle \otimes |0\rangle) = |\psi\rangle \otimes |\psi\rangle$$

$$U(|\phi\rangle \otimes |0\rangle) = |\phi\rangle \otimes |\phi\rangle$$

Where $|0\rangle$ represents a blank state, and $\otimes$ signifies the tensor product, or, in simple words; it is impossible to create a universal quantum cloning machine that perfectly duplicates an arbitrary quantum state.

In more understandable language, the theorem informs us that it is impossible to replicate an arbitrary quantum state with complete precision. In classical physics, the act of duplicating information is a very uncomplicated procedure. However, within the world of quantum mechanics, the presence of quantum entanglement and the uncertainty principle render this task unattainable.

The previously stated theorem holds a significant influence in quantum cryptography and quantum computing. The usage of this mechanism effectively functions as a deterrent against specific forms of unauthorized interception inside quantum communication protocols. Also, the fact that quantum states can't be copied is a key part of quantum computing algorithms because it makes it possible to manipulate information safely and efficiently by using the principles of quantum mechanics.

The No-Cloning Theorem contains vast consequences for diverse elements of quantum computing, including cryptography, error correction, algorithms, and communication protocols. Scientists and engineers working in the field of quantum computing might enhance the security and efficiency of quantum technology by comprehending and accepting the constraints imposed by this theorem.

- The use of quantum encryption protocols, notably quantum key distribution (QKD), heavily depends on the no-cloning property. This principle plays a crucial role in ensuring high-level security in communication channels. If someone tries to intercept and duplicate the quantum states being transmitted, the act of cloning leads to observable errors, alerting the parties involved about potential tampering.
- Quantum error correction addresses errors in quantum computing systems caused by factors like noise and decoherence. These correction codes are tailored to protect quantum information from the detrimental effects of errors. The no-cloning theorem shapes the design of these codes since they exploit cloning limitations to identify and rectify flaws.
- Quantum cryptography includes various protocols such as quantum digital signatures and quantum coin flipping. These protocols utilize the principles of the no-cloning theorem to establish secure cryptographic procedures that are resistant to tampering. The integrity of communication is ensured by these protocols, which stem from the foundational idea that quantum information cannot be accurately duplicated.
- Quantum teleportation involves transmitting the quantum state of a system from one place to another using entanglement and classical communication. The concepts of quantum teleportation are rooted in the no-cloning theorem, which forbids the direct replication of an unknown quantum state.
- Quantum algorithms, including Shor's algorithm for large number factorization and Grover's algorithm for unsorted database search, exploit quantum properties to achieve greater computational efficiency compared to classical algorithms. The no-cloning theorem, which limits the duplication of intermediate quantum states during computation, shapes the formulation and execution of these algorithms.
- QML refers to using quantum computers for processing and analysing data through specific algorithms. The design of these QML algorithms is influenced by the no-cloning theorem, which sets boundaries for the manipulation of quantum information. These algorithms must be carefully crafted to avoid actions that violate the constraints imposed by the no-cloning theorem.
- Quantum simulation involves using computers to replicate and study complex quantum systems that are challenging to simulate using classical computing methods. The no-cloning theorem affects the execution of quantum simulations, as it restricts the replication of the quantum states being simulated.

The No Cloning Theorem has consequences for quantum computing. The field of quantum computing capitalizes on the distinctive characteristics of quantum states, such as superposition and entanglement, to execute specific computational tasks more efficiently than classical computers. The limitations imposed by the theorem about the cloning of quantum states have numerous applications and consequences within the field of quantum computing, which we will examine using a straightforward academic approach.

2 Quantum Computing

Due to the convergence of crucial technologies like high-performance computing, AI, and quantum computing, the state of the computational world as it is today will soon undergo a significant transformation. Quantum computing is a notable departure from conventional paradigms, presenting the potential to address complicated problems that go beyond the computational constraints of classical systems. The technology can reveal longstanding scientific questions and surmount constraints to augment industrial efficiency.

Definition 10 (Quantum Computing) Quantum computing is a complex computational framework that leverages the principles of quantum mechanics to perform information processing and manipulation.

Quantum computing does not fundamentally provide a new computational paradigm that exceeds the theoretical limitations of a Turing machine. The component of quantum computing is its ability to significantly accelerate the resolution of numerous problems compared to conventional methodologies, owing to its inherent complexity. The unique characteristic of quantum computing in computer science is its ability to use concepts drawn from quantum mechanics to address complex problems that present challenges for conventional computers. Quantum computers have a significant influence on computational efficiency, especially in scenarios necessitating expeditious data processing.

The main focus of their research is on addressing difficulties that go beyond conventional computer paradigms. One advantageous attribute of quantum computing originates from the use of concepts derived from quantum mechanics, leading to the creation of robust and strong systems. The computational results demonstrate a significant level of resilience and protection. The use of quantum computing showcases remarkable efficacy in domains marked by complex numerical calculations and sophisticated data manipulation methodologies. The use of this technology includes a wide range of disciplines, including but not limited to the domains of medicine, healthcare, manufacturing, computational security, and finance.

The fundamental principle behind quantum computing is the use of interconnected quantum bits (qubits) contained inside quantum registers. The manipulation of quantum bits (qubits) is accomplished by using quantum circuits. The computational process consists of two primary steps, namely preparation and observation. During the preparatory stage, the formation of pertinent states generally occurs via the input of data. Subsequently, the observation phase involves the derivation of outcomes or resolutions from the quantum system. The concept of a quantum algorithm is an essential feature of quantum computing. The procedure is deliberately designed to optimize the arrangement of computing phases to increase the probability of attaining the desired results.

As noted, quantum computers are characterized by their use of quantum bits (qubits) rather than conventional electrical components, thereby manifesting a fundamental differentiation from classical computers. Quantum computers outperform binary systems by using qubits, which abide by the principles of quantum mechanics. The notions outlined above demonstrate significant differences when compared to classical physics, especially when studying events that occur at the atomic and subatomic levels.

The observed rise in the number of quantum computing activities conducted globally indicates substantial advancements within the sector. Yet, it is anticipated that a significant duration would be necessary for their contributions to have a transformational influence on technology. To date, there has been a lack of substantial evidence indicating that quantum computing experts have achieved superior performance compared to conventional computers.

The emergence of complexity may be attributed to the elaborate processes inherent in the fabrication of hardware components. Quantum computers use qubits as the foundational elements for encoding data, using diverse technologies such as superconducting rings, optical traps, and photons. Certain techniques need strict cooling measures, but others may function at typical ambient temperatures. The underlying principle of this phenomenon is the incorporation of quantum bits (qubits) into probabilistic combinations, referred to as superpositions, of distinct digital states.

Quantum algorithms are specifically designed to manipulate and govern the phenomenon of superposition transition. This transition is subject to the laws of quantum mechanics, which encourage interaction and improve the computational power of qubits. The attainment of accurate calculations requires a significant number of qubits, a noteworthy goal that has yet to be achieved.

The behaviour of qubits and their interactions may be influenced by many disturbances, including temperature fluctuations and electromagnetic wave interference. These disturbances can lead to the degradation of data, often known as decoherence. To achieve a high level of accuracy, it may be necessary to assign a significant number of qubits only to implement error correction algorithms.

2.1 Development of Quantum Computing

Richard P. Feynman and Yuri Manin first demonstrated the theoretical foundations of quantum computing in the 1980s [Fey82, NC12, Man80]. This strategy aims to address the computing difficulties brought about by quantum systems that conventional computers are unable to adequately describe. The conceptual basis for this theory originates from Feynman's acknowledgement of the constrained computational capabilities of classical computers in reproducing complex quantum mechanical processes.

In 1981, Feynman, while associated with MIT, made a notable observation that highlighted the substantial obstacles faced by classical computers when attempting to accurately simulate quantum processes. Feynman introduced an initial foundation for a quantum computing paradigm to tackle this challenge and facilitate the modelling of dynamics in quantum systems. The adoption of this paradigm shift has shown the potential for a substantial increase in computing efficiency when compared to conventional methods.

The evolutionary progression of quantum computing occurred gradually. Before the introduction of the foundational Shor algorithm [Sho94] by Peter Shor in 1994, the existing technique had facilitated the remarkable capability of quantum computers to efficiently factorize very large numbers, surpassing the constraints imposed by conventional algorithms through exponential efficiency. The proposal that the Shor algorithm could effectively and efficiently solve massive factorization problems, hence representing an imminent danger to modern cryptographic systems, brought about significant interest and anticipation within quantum computing and cryptography research communities.

Further advancements have further heightened the level of interest in quantum computing. It is influential to mention that in 1996, Lov Grover presented a quantum database search approach [Gro96] that provides quadratic acceleration for applications that mainly depend on random or brute-force searches. In 1998, a significant achievement was attained with the successful development of a working 2-qubit quantum computer through technological efforts. This achievement facilitated the implementation of critical quantum algorithms, such as Grover's algorithm, which stands out as a notable example. This occurrence represented the beginning of a competitive struggle to establish leadership in quantum computers, distinguished by an abundance of rapidly expanding applications.

The beginning of progress was seen in 2011 when D-Wave Systems[1] presented the first quantum computer that was available for commercial use. In 2017, a notable milestone was reached with the introduction of the first commercially viable quantum computer by IBM. This event signified a significant change in the quantum computing field. In the year 2019, IBM accomplished a significant achievement by introducing the 'IBM Q System One,[2]' which is considered the foremost commercially functioning quantum computer globally. The accomplishment at hand was made possible by a collaborative effort among many firms, which together engaged in the design and fabrication of this innovative quantum computing wonder.

The introduction of the IBM Q System One signalled the beginning of a period characterized by unparalleled promises in quantum computation. This revolutionary invention has provided alternatives for an extensive number of transformational applications and advancements that can significantly alter several aspects of everyday life for humans. In November 2022, a significant advancement occurred with the unveiling of 'Osprey [CN22]', IBM's newest and most significant computational quantum system. The above statement highlights the continuous advancement of quantum computing in pushing the bounds of processing power.

[1] www.dwavesys.com/learn/quantum-computing/.

[2] research.ibm.com/interactive/system-one/.

2.2 Classical versus Quantum Computing

Comparing classical computing with quantum computing reveals a significant difference. Classical computing depends on the principles of Boolean algebra and operates by using logic gates. Within this particular environment, the processing of data is constrained to binary states, precisely denoted by the values of 0 or 1, which are represented mathematically by individual bits. These discrete units live only in a certain state at a specific instant.

The conventional computer architecture, characterized by the use of a collection of transistors and capacitors in the millions, is subject to limited speed constraints while transitioning between states. In contrast, quantum computing uses a dual-mode logic gate that incorporates both an XOR gate and a QO1 mode (the ability to change 0 into a superposition of 0 and 1). The provided configuration enables the conversion of the binary value 0 into a quantum state known as a superposition, which covers both states 0 and 1 simultaneously. Data processing involves the use of subatomic particles, such as electrons or photons. Each of these particles has an electric charge or polarization, indicating a binary value of either 0 or 1.

The particles, often known as qubits, are the basic constituents of quantum computing. The distinctive characteristics and actions shown by quantum systems serve as the foundation upon which the possibility of achieving quantum supremacy in quantum computing is established. Modern cryptography systems have been supported by four key elements. Mathematical problems typically provide security through the implementation of secret keys, which renders them relatively straightforward to figure out with the key but difficult to decrypt without it. Algorithms function as defensive measures against attacks, which sometimes require tedious trial-and-error efforts, but cryptography analysts might discover alternative approaches.

Also, it is essential to combine secure algorithmic selection with secure execution to ensure the secrecy of the secret key. Finally, the human element becomes of great importance, emphasizing the need to strengthen defensive measures against deceptive social engineering strategies. The implementation of extended AES and ECC encryption keys has effectively preserved the integrity of cryptographic systems. Nevertheless, the exponential increase in processing power, particularly in quantum computing, calls for a comprehensive reevaluation.

Quantum computers differ from traditional digital computers since they use quantum bits (qubits) rather than ordinary binary digits for computational purposes. These machines possess the capacity for advantageous applications, such as the simulation of complex biological systems. Nevertheless, they provide a concurrent challenge to existing cryptographic protocols. Currently, the operation of quantum computers is complex and involves significant computational and financial expenses. The capacity of qubits in these machines is still restricted, with the most significant quantum computers supporting about 200 qubits.

However, there has been a gradual increase in the number of qubits. Disturbing forecasts suggest that the compromise of RSA-768 encryption may be achieved with a very small number of stable qubits, estimated to be about 4000. It is anticipated that a more significant number of stable qubits will be necessary to break more resilient iterations of RSA encryption. As a consequence, our dependence on current cybersecurity measures is temporary, given their gradual inadequacy in response to the advancing capabilities of quantum computing.

2.3 Challenges in Quantum Computing

With the publication of the ground-breaking algorithm by Peter W. Shor in 1994, it was demonstrated that quantum computers are capable of subverting a considerable number of conventional asymmetric encryption methods. Regardless, a prominent obstacle in quantum computing revolves around the need to build reliable qubits that have an essential function in maintaining the accuracy and dependability of computational operations. For one to accomplish successful decryption of modern encryption methods, someone must have a minimum threshold of 4000 stable qubits.

The recent development of quantum computing presents a significant obstacle to the existing security framework, presumably leading to severe consequences. The escalating progress and extensive use of quantum computing present substantial risks to the security of data and robust infrastructure, considering the global dependence on encryption. With the emergence of functional quantum computers, there will be a significant increase in processing speeds, resulting in exponential acceleration. As a result, the mathematical complexities of algorithms will be much reduced in comparison to earlier situations. However, the progress made in quantum-resistant cryptography is a crucial milestone in cybersecurity, with the potential to greatly enhance data security in the next few years.

Quantum computing has significant challenges that include error correction, output retrieval, and interference problems. Interference is a substantial obstacle in quantum calculations, potentially resulting in the failure or inability of such computations. The occurrence of decoherence, which involves the entanglement of a quantum system with its surrounding environment, presents a notably complex problem. To mitigate the potential threats of interference, it is necessary to implement strict safeguards that effectively counteract external effects throughout the computing process. Significant advancements have been achieved in the use of resilient magnetic fields for safeguarding qubits. However, the persistent challenge of interference continues to impede the development of quantum computing.

One apparent constraint of quantum computing can be observed in the need for error correction to achieve accurate computational processes. The disparity between qubits and digital bits makes typical error correction approaches unsuitable. A single error made while computing could jeopardize the integrity of an entire operation. Despite notable progress in error correction techniques, particularly in the use of a 9-qubit system with one qubit specifically allocated for computing, further research and development initiatives are required to achieve further advancements.

The process of extracting output data from quantum computers poses a significant obstacle owing to their inherent vulnerability to data corruption. One such approach is using database search techniques that are dependent on the probability curve waveform of the quantum computer. This particular approach enables the quantum state to effectively transition into an accurate solution throughout the process of measurement. Nevertheless, there are larger concerns that arise in quantum computing, namely with the guarantee of security and the incorporation of cryptographic mechanisms that can withstand quantum attacks. The prolonged retention of quantum information has presented significant historical obstacles; however, recent progress has improved the practicality of quantum computing.

Succinctly, it can be expressed that conventional computers would require an extensive span of millions of years to successfully factorize a number consisting of 300 digits. Conversely, Shor's approach serves as a marked demonstration of its effectiveness in actually compromising modern cryptosystems. The implementation of this technique has significantly reduced the time needed to compromise cryptosystems from a span of years to only a few hours. This accomplishment has generated increased curiosity in quantum computing and its wide-ranging applications.

One such approach is the use of symmetric encryption for data and asymmetric encryption for keys. The use of lattice-based encryption presents a favourable strategy for addressing the challenge of quantum-resistant cryptography. In this particular technique, lattices are used in the creation of cryptographic algorithms. The use of post-quantum algorithms, such as lattice-based encryption, facilitates the enhancement of the security of the first key exchange process, which is then followed by the implementation of an asymmetric algorithm to guarantee the security of the primary message.

The recent development of quantum-resistant encryption has presented an achievable solution for addressing the security concerns that arise from the rapid improvements in quantum computing. The encryption method described below utilizes fundamental laws of quantum mechanics, including the Heisenberg uncertainty principle and photon polarization, to protect communications against illegal interception. Quantum-resistant cryptographic keys have been produced by quantum operations, establishing a degree of inherent security and robustness against tampering or interception.

3 Qubits and Quantum Gates

The research into the subject of quantum computing incorporates the analysis of fundamental characteristics at the atomic and subatomic levels, thereby providing an in-depth exploration of the complexities inherent in quantum mechanics. The effectiveness of quantum computing arises from its capacity to concurrently encode both binary states of 1 and 0 via the use of qubits, a feat made feasible by the fundamental laws of quantum mechanics. Quantum computing leverages the superposition states of qubits, in contrast to the binary bits used in conventional computing. Qubits, which are derived from elementary particles such as electrons, transmit information using properties such as spin.

Superposition and entanglement are fundamental principles that underpin quantum computing. Entanglement makes it possible for the quantum states of qubits to be linked instantly, while superposition makes it possible to explore a wide range of possible combinations within qubits. To facilitate the functioning of quantum computers, a certain arrangement is required to accommodate qubits, signal transmission, and classical control.

To optimize their performance, qubits are maintained at temperatures in the milli-kelvin range by the use of cooling methods such as dilution freezers. The phenomenon of entanglement, whereby qubits become interconnected and may instantaneously affect each other regardless of distance, is a fundamental characteristic of quantum computing. The potential of this entanglement is significant for drug research and cryptography.

The substantial processing capability of quantum computing arises from the exchange between superposition and entanglement phenomena. The quantity of qubits accessible for computational purposes is seeing a fast expansion. The benefits of quantum computing are remarkable. The computational speed of financial simulations and models has revolutionized the field, surpassing that of ordinary computers. The system demonstrates exceptional proficiency in decryption, factorization, and complex simulations, thereby contributing to the advancement of drug research. The use of quantum computing significantly improves the optimization of algorithms in AI and ML, marking the advent of a transformative age.

Qubits are fundamental units in quantum information. In contrast to conventional bits, which are limited to being in either a state of 0 or 1, qubits leverage the phenomenon of quantum superposition to occupy a coherent combination of both states simultaneously. The distinctive characteristic described serves as the fundamental basis for quantum gates, which bear similarities to conventional logic gates yet possess significant differences as a result of quantum mechanics.

In classical computing, a bit can represent either a binary 0 or a binary 1, analogous to a switch that may be in either an off or on state. On the other hand, a qubit utilizes the principles of superposition, allowing it to simultaneously embody both the 0 and 1 states. The emergence of this phenomenon may be attributed to the inherent quantum properties of particles, which enable qubits to be present in a superposition of probabilistic states. From a mathematical perspective, this may be represented as:

$$\alpha|0\rangle + \beta|1\rangle$$

Here, α and β are complex probability amplitudes, and $|0\rangle$ and $|1\rangle$ denote the basis states. The absolute squares of the probability amplitudes determine the probability of measuring the qubit in either state upon observation.

3.1 Single-Qubit Gates

Abstract mathematical notions give conception to a fundamental element of quantum information known as a qubit. A qubit, also known as a quantum bit, is characterized by its ability to exist in a distinct superposition of states, setting it apart from conventional bits. This section offers a description of the fundamental constituents of quantum operations, often denoted as gates. The primary emphasis of this discussion is on single-qubit gates, namely the Pauli-X, Pauli-Y, and Pauli-Z gates. These gates exhibit the capacity to change the states of qubits, resulting in various transformations.

3.1.1 Pauli-X Gate

The Pauli-X gate, also known as the bit-flip gate, functions as a one-qubit unitary operator. This gate represents a permutation of quantum states, precisely exchanging the probabilities associated with the qubit being in its basis states. This exchange of probabilities takes place while maintaining quantum coherence.

The Pauli-X gate is represented by a matrix with a structure of

$$\begin{pmatrix} 0 & 1 \\ 1 & 0 \end{pmatrix}$$

Thereby showcasing its straightforward nature. This transformation can be likened to a classical NOT gate, but its impact extends into the quantum domain. For instance, when a qubit is initially in the $|0\rangle$ state, applying the Pauli-X gate results in its conversion to the $|1\rangle$ state, effectively enacting a quantum-level negation.

Quantum gates are the analogues of classical logic gates in quantum computing. However, they wield a distinctive potency, operating on qubits in a manner beyond the confines of classical logic. These gates manipulate qubits by transforming their quantum states, a process analogous to rotating or flipping the quantum state vector.

The Quantum NOT gate, also known as the Pauli-X gate, serves as an elemental quantum gate that simulates classical bit-flipping. It acts on a single qubit, exchanging the amplitudes of $|0\rangle$ and $|1\rangle$. Its operation can be succinctly described as:

$$X|0\rangle = |1\rangle, \quad X|1\rangle = |0\rangle$$

The effect of this gate is similar to a classical NOT operation, but the quantum version includes both 0 and 1 states simultaneously.

Another crucial quantum gate is the Hadamard gate, denoted as H, which endows a qubit with the ability to occupy a superposition of $|0\rangle$ and $|1\rangle$ states. Its operation can be represented as:

$$H|0\rangle = \frac{1}{\sqrt{2}}(|0\rangle + |1\rangle), \quad H|1\rangle = \frac{1}{\sqrt{2}}(|0\rangle - |1\rangle)$$

The Hadamard gate facilitates the exploration of different computational paths simultaneously, a phenomenon with immense potential in quantum algorithms such as Grover's and Shor's algorithms.

3.1.2 Pauli-Y Gate

Turning our attention to the Pauli-Y gate, similar to its Pauli-X counterpart, this gate exerts influence on qubit states via a unitary transformation. In contrast to the Pauli-X gate, the Pauli-Y gate not only swaps probabilities but also introduces a phase factor. This phase factor leads to the emergence of distinct quantum phenomena. The Pauli-Y gate introduces a rotation of π radians about the y-axis on the Bloch sphere, a geometric representation of qubit states. Specifically, this gate can be denoted by the matrix

$$\begin{pmatrix} 0 & -i \\ i & 0 \end{pmatrix}$$

highlighting its departure from the Pauli-X gate. When applied to the $|0\rangle$ state, the Pauli-Y gate induces a transition to the $|1\rangle$ state while introducing a phase shift of π radians, thus showcasing its unique nature.

3.1.3 Pauli-Z Gate

Also, our exploration includes the Pauli-Z gate, a quantum operator that imparts a phase shift of π radians around the z-axis of the Bloch sphere. Notably, the Pauli-Z gate maintains the probability distribution of qubit states while adjusting their phases. The corresponding matrix representation,

$$\begin{pmatrix} 1 & 0 \\ 0 & -1 \end{pmatrix}$$

This captures its operation, reflecting its simplicity similar to the Pauli-X gate. When applied to the $|0\rangle$ state, the Pauli-Z gate remains unchanged. However, for the $|1\rangle$ state, it induces an inversion while upholding the probabilistic characteristics intact.

3.2 Multi-Qubit Gates

The CNOT gate, which is representative of two-qubit gates, establishes a connection between a control qubit and a target qubit. This connection gives rise to entanglement, a perplexing quantum phenomenon where qubits become intertwined in correlated states even when they are separated by distance. In practice, the CNOT gate influences the transformation of the target qubit based on the state of the control qubit.

If the control qubit is in the state $|1\rangle$, the target qubit undergoes a Pauli-X gate operation, resulting in a quantum-level flip. Conversely, when the control qubit is in the state $|0\rangle$, no transformation occurs. The CNOT gate can be mathematically represented using a matrix, which provides insights into its role in computational processes involving entangled qubits.

The CNOT gate is crucial in achieving quantum computational supremacy, as it forms the basis for constructing quantum circuits capable of solving computational tasks that surpass classical computing capabilities. These tasks include error correction, number factorization, and simulations, all of which are essential in quantum algorithms. These efforts significantly contribute to the progress of the cryptography and optimization fields. The consequence of the CNOT gate extends beyond classical simulation, as it enables the creation of fundamental quantum gates for quantum Turing machines, which are the quantum counterparts of classical Turing machines.

The possibility of quantum computing lies not only in the discrete operation of individual gates but also in their coherent interactions. By sequentially combining multiple gates, complex operations on qubits can be achieved. These interactions follow the principles of unitarity, requiring quantum gates to be reversible to maintain information integrity throughout the computational process.

Quantum circuits visually represent interactions between quantum gates. These circuits use horizontal lines to indicate qubits and boxes to depict quantum gates. The sequential arrangement of gates visually represents the transfer of quantum information. The concept of abstraction offers a tangible framework for developing and understanding quantum algorithms.

3.3 Quantum Algorithms

This section attempts to provide a fundamental understanding of the fundamental quantum algorithms [Cle+98], specifically Shor's algorithm for factorization and Grover's algorithm for search problems. A quantum algorithm is an algorithm specifically built to function within the limitations of a viable quantum computing model. The most often used model for this purpose is the quantum circuit model of computation.

In contrast, a classical algorithm, also known as a non-quantum algorithm, refers to a finite series of instructions or a systematic approach to problem-solving. Each step or instruction inside this algorithm is designed to be performed on a classical computer. Similarly, a quantum algorithm includes a methodical process in which each step may be implemented on a quantum computer. While it is true that classical algorithms can also be executed on quantum computers, the term 'quantum algorithm' typically refers to algorithms that possess inherent quantum characteristics or rely on fundamental aspects of quantum computation, such as quantum superposition or quantum entanglement.

The existing architecture for data security is based on the assumption that factoring numbers with a thousand or more digits is a very challenging process. The notion remained prevalent until the year 1995, at which point Peter Shor proposed a concept that effectively used the principles of quantum mechanics. Shor's proposition suggested that the use of quantum mechanical processes may potentially facilitate the factorization of integers within a polynomial time frame.

This is a significant deviation from the exponential complexity often associated with conventional techniques. The runtime of traditional methods, such as the general number field sieve (GNFS) [Pol93], experiences exponential growth as the number of digits in the integer being factored increases. On the other hand, Shor's method, which is a quantum algorithm designed for integer factorization, has polynomial-time complexity. From a practical standpoint, it can be seen that the duration needed to factorize an integer exhibits a polynomial growth pattern concerning the number of digits possessed by the integer.

Conceived by an Indian-American computer scientist, Grover's Algorithm [Gro96] is of significance in quantum computing following in the footsteps of Shor's Algorithm [Sho94]. This algorithm excels at expediting unstructured search problems, achieving quadratic speedup. Likewise, it serves as a potent tool or subroutine to impart quadratic runtime enhancements across a spectrum of other algorithms.

Consider a scenario where a collection of n items must be sifted through to identify a solitary unique item. Classical computation necessitates, on average, examination of $\frac{n}{2}$ items to locate the unique entity, with worst-case instances requiring scrutiny of all n items. Quantum computing, however, leverages Grover's amplitude amplification to substantially curtail the number of steps to approximately $\sqrt{n}$, a quadratic acceleration when compared to classical methods.

The Deutsch–Jozsa algorithm [DJ92] is a quantum computational construct crafted to tackle the 'Deutsch-Jozsa problem'. This challenge involves discerning whether a given Boolean function is "constant" (yielding identical outputs for all conceivable inputs) or "balanced" (generating differing outputs for at least one input pair), all while minimizing the number of queries. For n-bit inputs, classical algorithms mandate a minimum of two function calls and a maximum of $2^{n-1} + 1$ to definitively ascertain the nature of the function. In contrast, a quantum computer adeptly resolves this with a solitary call to the function $f(x)$.

Much like the Deutsch–Jozsa problem, the Bernstein–Vazirani algorithm [BV97] tackles a specific query within a black box framework. This puzzle involves deducing s in the equation $f(x) = s \cdot x$, where s represents an undisclosed string and '$\cdot$' denotes the bitwise product (i.e., AND) operation. Conventional methods require multiple invocations of the function $f(x)$, followed by iterative extraction of s's bits. This necessitates n calls to the function for a complete recovery of the string. Quantum computation, however, solves this puzzle definitively after just one invocation of the function $f(x)$.

Simon's algorithm [Sim97] marked a significant achievement in quantum computing by demonstrating an exponential acceleration compared to the most efficient classical counterpart when addressing a specific problem. This challenge revolves around a function $f(x)$ that yields either unique outputs for each input (one-to-one) or pairs of distinct inputs producing a common output (two-to-one).

There is an unknown string s that has an effect on any input strings x and makes $f(x)$ the same as $f(x \oplus s)$. If s has a non-zero value, $f(x)$ turns into a two-to-one mapping, otherwise, it becomes a one-to-one mapping. The objective is to find out whether f is a one-to-one or a two-to-one mapping by discovering the hidden string s. Although conventional methods can determine the value of s for the function f with certainty, it requires up to $2^{n/2}$ function calls. Quantum computing significantly reduces this requirement to a little over n calls, providing a 100 per cent accurate solution.

Quantum Phase Estimation (QPE) [Kit95] is a fundamental subroutine in various quantum algorithms. This algorithm's core function involves approximating the phase associated with an eigenvalue of a unitary operator. In simpler terms, given a unitary operator U and an eigenvector $|\psi\rangle$ such that $U|\psi\rangle = e^{2\pi i\theta}|\psi\rangle$, where θ denotes the unknown phase angle, QPE strives to deduce θ. This algorithm finds application across domains like quantum simulation, quantum chemistry, and optimization, firmly establishing its utility and significance.

4 Quantum Cryptography

This section explains quantum cryptography techniques, which are firmly rooted in the principles of quantum mechanics and serve to ensure robust mechanisms for encryption and decryption. The rise of advanced technologies, particularly electronic communication, has firmly established itself as a central pillar of the modern society. Within this context, the principles of confidentiality, integrity, authenticity, and non-repudiation in data transmission and storage have elevated cryptography to a foundational discipline within information technology. Cryptographic systems are broadly classified into symmetric and asymmetric categories.

Quantum computing theory has undergone exhaustive investigation and represents a significant challenge to present-day asymmetric cryptography. Also, specific quantum algorithms can exert an impact on symmetric cryptography, albeit with the potential for enhancement through the utilization of larger key spaces. As society embarks on an age marked by the continuous evolution of quantum computing, formidable security challenges will inevitably arise. Quantum computers, once operational, will wield exponential computational speeds, effectively diminishing the complexity of algorithms that were once deemed formidable. In this context, the emergence of quantum-resistant cryptography marks a significant stride forward in cybersecurity, with the potential to ensure data security over extended timeframes.

Definition 11 (Quantum Cryptography) Quantum cryptography is a distinct area of study within quantum computing that utilizes fundamental principles derived from quantum mechanics to facilitate the establishment of secure communication channels between multiple parties.

This class of cryptography harnesses the principles of quantum mechanics and quantum algorithms to encode and transmit data with an exceptionally high level of security, resistant to attacks. Quantum-resistant cryptography, a nascent technology poised for substantial traction in the years ahead, holds the potential to revolutionize secure communication by anchoring its security measures in the fundamental laws of physics, diverging from the mathematical algorithms or computing technologies commonly used in security.

The potentialities offered by quantum-resistant cryptography are manifold, enabling the accomplishment of cryptographic tasks that were previously deemed impossible through traditional (non-quantum) means of communication. This juncture in technology positions humanity at the precipice of a paradigm shift, driven by the convergence of advanced computing and cryptography. This impending transformation includes the establishment of a robust and sophisticated system capable of withstanding the disruptive potential of quantum computing while optimizing the capabilities of high-performance computing. Innovative algorithms have been introduced to dismantle prevailing asymmetric cryptographic schemes, which rely on the complexity of prime number factorization and the discrete logarithm problem. Even elliptic curve cryptography, which is currently regarded as secure and efficient, succumbs to the vulnerabilities posed by quantum computers.

Consequently, the demand for cryptographic algorithms resilient to quantum computation is becoming increasingly essential. The application of digital signatures and critical Internet protocols, including HTTPS (TLS), integral to secure online activities such as browsing, banking, and shopping, faces significant repercussions due to advancements in quantum computing. This arises from the inherent threat that quantum computing poses to symmetric cryptography systems like the Advanced Encryption Standard (AES) [DR99], which are commonly combined with asymmetric cryptography (e.g., RSA) in HTTPS implementations.

The majority of data are encrypted using asymmetric encryption, which serves as the standard of modern security protocols. This security paradigm hinges on the premise that classical computers cannot efficiently factor equations involving large prime numbers. However, the emergence of quantum computers, which possess the efficiency to factor in large prime numbers, raises concerns about the viability of asymmetric cryptography. This threatens existing cryptographic methods against adversarial attacks. The appeal of asymmetric encryption lies in its capacity to encrypt data using a recipient's public key, with only the intended recipient in possession of the corresponding private key. This framework with two distinct keys is founded on the belief that certain mathematical computations are easier to perform than to reverse.

Confronting the challenge of quantum-resistant cryptography necessitates exploring key lengthening strategies, with agreement converging around the use of 2048-bit keys as a minimum for quantum defence. However, the increase in key size increases both the complexity and cost of cryptographic operations. In the foreseeable future, quantum computers are projected to compromise various existing public key infrastructures (PKIs), such as RSA, whose security relies on the computational infeasibility of integer factorization. For instance, a quantum computer equipped with 4098 qubits can break a 2048-bit RSA key using Shor's algorithm [GE21]. Notably, the largest number successfully factored using Shor's algorithm and a quantum computer is 56153 [DB14], which is the product of 233 and 241.

Quantum computers have ushered in a paradigm shift in computational methodology. In contrast to digital processors that rely on binary digits for computations, quantum computers harness the principles of quantum mechanics through the utilization of quantum bits or qubits. While these revolutionary machines hold the promise of substantial benefits, such as simulating complicated biological systems for advanced medical therapies, they also render previously arduous mathematical problems trivial. Regrettably, some of these problems constitute the foundation of Modern cryptographic standards.

At present, quantum computers pose challenges in terms of operation and prohibitive costs, with the most substantial machines featuring a mere 200 qubits. However, the rapid expansion of qubit numbers raises significant concerns. Disturbing projections suggest that a relatively moderate count of 4000 stable qubits could potentially undermine encryption schemes such as RSA-768, and the situation is even more dire for larger RSA versions in practical usage. This progression highlights the inevitability of the obsolescence of cybersecurity measures that have served as our defence for years. Recognizing the imminent threat of quantum computers to cryptography, the National Institute of Standards and Technology (NIST) initiated a competition in 2016 to formulate new, quantum-resistant algorithms.

The recent introduction of the first four algorithms in 2022 represents a substantial milestone in the ongoing initiative to guard against the impending quantum computing threat. NIST's challenge includes the identification of algorithms that satisfy the dual criteria of quantum resilience and suitability for Modern Internet applications. Amidst the robust computational power and memory resources of conventional computers and smartphones, the increase in compact Internet of Things (IoT) devices—characterized by limited computing capabilities, kilobytes of memory, and energy-efficient operation—adds a layer of complexity to the equation.

For key exchange, NXP's CRYSTALS-Kyber [Bos+18], celebrated for its modest key size and rapid processing speed, emerged as a selection. This algorithm operates based on the learning-with-errors (LWE) problem over module lattices. In digital signatures, CRYSTALS-Dilithium [Duc+17] and FALCON [Pre+20], both lattice-based cryptographic algorithms, were introduced, alongside SPHINCS+ [Ber+19], a hash-based signature scheme, to ensure mathematical diversity. While these algorithms have been designated, several factors warrant attention before their practical deployment. The nature of key exchange and the architecture of security certificates stand as key considerations and hardware support assumes equal significance. While benchmark assessments for embedded systems are underway, dedicated instructions, hardware acceleration IP, and security chips tailored for quantum-resilient hardware are expected to materialize in the coming years. Table 7.1 shows a concise summary of a few well-known post-quantum cryptographic methods.

Table 7.1 Post-quantum cryptographic methods

Category	Method	Definition
Hash-based	XMSS	Let H be a hash function; S_K and P_K represent the secret and public keys, respectively. The signature σ is derived as $\sigma = (I, R, \{R_i, S_i\}_{i=1}^{h})$ where I is the index, R is the root, and R_i, S_i are the authentication path components
Lattice-based	NTRUEncrypt	The NTRUEncrypt scheme operates in a lattice defined by a polynomial ring $R = \mathbb{Z}[x]/(x^N - 1)$. Encryption involves a public key h and a message m, with ciphertext c computed as $c = \text{Round}(hr + m, q)$ where r is a random polynomial and q is the modulus
Code-based	McEliece	Let G be the generator matrix of a linear code, S be a random invertible matrix, and P be a permutation matrix. The public key is $K_{\text{pub}} = (G, S)$ and private key is $K_{\text{priv}} = (G^{-1}P, S^{-1})$. Encryption involves multiplying the message vector by G and adding a random error
Multivariate polynomial-based	Rainbow	Define multivariate quadratic polynomials F_i and G over a finite field. The public key is a set of F_i polynomials, and the private key consists of the corresponding G polynomials. Signatures involve solving a system of multivariate quadratic equations derived from F_i polynomials
Hash-based	SPHINCS+	Utilizes stateless hash-based signatures; the public key consists of a set of hash functions, and the private key is generated by selecting random seeds for these hash functions. Signatures are produced using the hash functions and the private key

4.1 Lattice Cryptography

The present focus revolves around the potential capacity of a quantum computer to compromise established cryptographic norms. Consequently, the development of novel cryptographic standards that can maintain their security even in the presence of quantum computers has become a necessary initiative. Lattices stand as essential within the mathematical study; their understanding and thorough examination have been documented over centuries.

Mathematicians such as Johann Gaub, despite precluding the foresight of quantum computers, significantly contributed to their significance. These inquiries last as insights, deepening our understanding of lattice problems. Lattice problems offer adaptability, notably in cryptography, progressively displaying their prospect. In the current cryptographic scenario, lattice-based problems command the 'post-quantum' cryptographic literature, crafting systems invulnerable to robust quantum computers. The NIST's competition highlights Lattice's dominance among contenders.

In contrast to surviving traditional methods, lattice-based cryptography (LBC) stands out as a thoroughly explored field in post-quantum cryptography. This distinction arises from the fact that this cryptographic paradigm is founded on mathematical problems associated with lattices. In this context, a lattice can be visualized as a grid resembling graph paper, formed by points situated at the intersections of infinitely extending straight lines. The ever-extended mathematical nature of a lattice constitutes a defining attribute.

Commencing from an initial collection of points i.e., 'vectors', numerical values are combined in integer multiples, leading to the establishment of the lattice structure. The complexity becomes apparent when attempting to identify lattice points situated near zero or other designated points. While performing this task is straightforward in two dimensions, it transforms into a hard problem in higher dimensions, such as 400 dimensions. Within this context, one group of points could function as the private key, while another set of more distant points could serve as the public key.

Deriving the private key from the public key involves an exhaustive inquiry of probable combinations, a time-intensive and unfeasible effort, even with the involvement of quantum computers. It is hypothesized that quantum computers would still necessitate an impractical amount of time to solve complex lattice based problems. Accordingly, lattice-based approaches continue to be the only feasible candidates in the assessment conducted by the NIST for offering fundamental encryption primitives, key encapsulation, and digital signature schemes.

Prominent instances of lattice-based algorithms surround CRYSTALS-KYBER, an algorithm for public key encryption and key establishment, as well as CRYSTALS-Dilithium, designed as a digital signature algorithm. By furnishing standardized post-quantum cryptographic (PQC) algorithms, researchers can devise cryptographic systems that guarantee security against both quantum and classical computers, while upholding compatibility with existing communication protocols and networks.

Loosely speaking, lattices, precise geometric structures, can be conceptualized as organized arrays comprising points extending infinitely. These discrete entities, referred to as 'vectors', are defined by numerical coefficients denoting their spatial coordinates. A lattice materializes as an aggregation of uniformly distributed vectors, culminating in a coherent arrangement. Lattices attain significance as generators of mathematical problems, thereby forming the foundation for robust cryptographic frameworks.

Given that lattices are infinite structures, they pose problems for computers constrained by finite memory. The basis of a lattice, which consists of a compact set of vectors spanning the entire lattice, can address this limitation. The 'origin', represented by a vector with all coordinates set to zero holds significance within a lattice. Vectors situated far from the origin are designated as 'long', whereas those close are termed 'short'. Geometrically, a lattice comprises uniformly spaced points in a vector space. In cryptography, effectively managing these expansive entities within limited memory presents a challenge. Basis vectors offer a succinct approach, acting as a condensed set capable of generating any lattice point.

Consider a positive integer n. Algebraically, a lattice $\mathcal{L}$ can be defined as a finite rank-free module over the integers ($\mathbb{Z}$), having a rank of n and accompanied by a positive definite quadratic form called q. While geometrically, a lattice is a discrete subgroup of $\mathbb{R}^n$ or set $\mathcal{L}(b_1, b_2, b_3, \ldots b_n)$ covering all possible linear combinations $\sum x_i b_i$, where $x_i \in \mathbb{Z}$ and b_i is a linearly independent vector over $\mathbb{R}$. Here, $(b_1, b_2, b_3, \ldots b_n)$ denotes the basis vectors. Importantly, lattices sharing the same basis are equivalent, preventing two lattices from having identical bases. In lattice-based cryptography, a Finite Abelian group operating modulo the lattice $\mathcal{L}$ is used. Besides, lattices sharing the same basis are considered equivalent, and the concept of identical bases is excluded. Although lattices themselves are infinite, lattice-based cryptography operates within a finite Abelian group functioning modulo lattice $\mathcal{L}$.

More formally, a lattice is defined as a discrete subgroup of Euclidean space with certain integral linear combinations closed under addition and subtraction. Consider a real Euclidean space $\mathbb{R}^n$. A lattice $\mathcal{L}$ in $\mathbb{R}^n$ is defined as a discrete subgroup of $\mathbb{R}^n$ with the following properties:

(1) **Discreteness:** There exists a positive real number λ such that, for any vector $\mathbf{v} \in \mathcal{L}$, the Euclidean norm of $\mathbf{v}$ is greater than or equal to λ: $||\mathbf{v}|| \geq \lambda$ for all $\mathbf{v} \in \mathcal{L}$.

(2) **Closure under Linear Combinations:** For any vectors $\mathbf{v}, \mathbf{w} \in \mathcal{L}$ and any integers $m, n \in \mathbb{Z}$, the linear combination $m\mathbf{v} + n\mathbf{w}$ is also an element of $\mathcal{L}$.

Symbolically, we express a lattice $\mathcal{L}$ in $\mathbb{R}^n$ as:

$$\mathcal{L} = \left\{ m\mathbf{v} + n\mathbf{w} \mid \mathbf{v}, \mathbf{w} \in \mathbb{R}^n, m, n \in \mathbb{Z} \right\}$$

where $\mathbb{Z}$ represents the set of integers. For context, the expression $m\mathbf{v} + n\mathbf{w}$ denotes the integral linear combination of vectors $\mathbf{v}$ and $\mathbf{w}$, where the coefficients are given by m and n; additionally, λ represents a positive real number ensuring the discreteness of the lattice, such that the Euclidean norm of any vector in the lattice is greater than or equal to λ. Interestingly, this definition covers the rudimentary properties of lattices, describing their discrete nature and closure under integral linear combinations.

Several common lattice problems are exploited in cryptographic primitives, including the Shortest-Vector Problem (SVP), the Approximate-Shortest Vector Problem (α-SVP), the Shortest Independent Vector Problem (SIVP), the Closest-Vector Problem (CVP), the Approximate Closest Vector Problem (CVP), the Bounded Distance Decoding (BDD), and the Shortest Integer Solution Problem (SIS). The Shortest Vector Problem (SVP) stands as a fundamental problem in lattice-based cryptography. Given a lattice $\mathcal{L}$ in $\mathbb{R}^n$, the SVP seeks to identify the shortest non-zero vector $\mathbf{v} \in \mathcal{L}$, where vector length is defined as the Euclidean norm, while in nature, the problem concerns finding a vector $\mathbf{v} \in \mathcal{L}$ that has the minimum Euclidean norm among all non-zero vectors in $\mathcal{L}$.

Formally, SVP can be defined as identifying a non-zero vector $\mathbf{v} \in \mathcal{L}$ such that $\|\mathbf{v}\|_2 \leq \|w\|_2$ for all non-zero vectors $w \in \mathcal{L}$. The SVP poses computational difficulty; no known algorithm can efficiently solve it within polynomial time. The challenge arises from the exponential growth of the search space for potential vectors within the lattice, as the lattice's dimensionality increases. The SVP and its variations are central to several cryptographic schemes founded on lattices, including the Learning with Errors (LWE) problem, key in constructing post-quantum cryptographic protocols.

Equally noteworthy are challenges like the Shortest Independent Vector Problem (SIVP), Closest-Vector Problem (CVP), Approximate Closest Vector Problem (CVP), Bounded Distance Decoding (BDD), and Shortest Integer Solution problem (SIS), each presenting distinct hurdles. These lattice-based computational problems serve as foundational pillars for lattice-based cryptography and pose substantial barriers to adversarial entities. They hold a critical role in constructing a robust encryption framework capable of withstanding quantum computing attacks, thereby being indispensable for ensuring the security of forthcoming cryptographic systems.

The notable problem in lattice-based cryptography is the SVP, which revolves around generating a point with small coordinates using a basis comprising long vectors. While this may seem relatively plain, the task becomes exceedingly complex when dealing with lattices of high dimensions. For instance, when addressing lattices with 10,000 coordinates for cryptographic applications, the intent of identifying a combination of basis vectors that yield points with small coordinates becomes arduous work. The complexity of this challenge is such that even with the computational power of a quantum computer, which surpasses classical capabilities, efficiently translating the SVP remains beyond grasp.

Numerous algorithmic techniques for solving lattice problems include methodologies like Gram–Schmidt orthogonalization [PT91], the Lenstra–Lenstra–Lovász (LLL) lattice basis reduction algorithm [Len+82]—recognized as the most prominent polynomial-time algorithm—BKZ [KZ77], and Babai's Round-Off Algorithm. In lattice-based cryptography, more expansive lattices are used, leading to more cost-effective operations. This cryptographic approach capitalizes on high-dimensional geometric structures to obscure information, thereby crafting challenges that are perceived as insoluble without the requisite decryption key, even for universally fault-tolerant quantum computers. To establish the practicability of such cryptographic systems, a prototype must be constructed and subjected to rigorous testing under diverse adverse conditions. The viability of this cryptosystem hinges squarely on the computational intractability of the underlying problem.

Assuming its position as a candidate for post-quantum cryptography, the problem must demonstrate a level of complexity sufficient to overcome quantum computational frameworks. Various modern cryptographic schemes utilize concepts such as the SVP, CVP, and BDD to establish the hardness inherent in these problems, leveraging the concept of Lattice Equality as a one-way trapdoor function. Prominent cryptosystems include Piekert's Ring-Learning with Errors (Ring LWE) Key Exchange [Lyu+13], the Micciancio Cryptosystem [Mic01], NTRUEncrypt [Hof+09], GGH [Gol+97], and other CVP-based cryptographic schemes.

4.2 Code-Based Cryptography

Code-based cryptography is a cryptographic methodology that rests upon the properties of error-correcting codes to ensure the confidentiality of encrypted information. This technique revolves around exploiting the complex challenges tied to the decoding of linear codes, with the ultimate objective of securing data through encryption and decryption processes. Its significance is particularly pronounced in post-quantum cryptography, where traditional cryptographic techniques are susceptible to attacks from quantum computers. Post-quantum cryptography includes cryptographic systems designed to withstand attacks from quantum computers while functioning on classical computers.

In this context, code-based cryptography emerges as a strong candidate. Originally conceived by Robert J. McEliece [McE78], code-based public-key cryptography stands apart from RSA and ECC methods due to its resilience against quantum computer attacks. Both classical and quantum computer attacks against code-based cryptography involve exponential time complexities. McEliece's public-key cryptosystem offers fast encryption, but its adoption has been limited due to the relatively smaller key sizes compared to RSA and ECC alternatives. However, this limitation doesn't undermine the feasibility of code-based cryptography; its competitiveness is framed in a pre-quantum setting.

At the core of code-based cryptography lies the application of linear error-correcting codes, mathematical constructs designed to identify and rectify errors in transmitted data. These codes are encapsulated within matrices that transform the original message into a coded representation. To encrypt a message, a randomized error vector is introduced into the coded representation, resulting in a perturbed coded form—the ciphertext. This encrypted message exhibits apparent randomness and is disconnected from the original message.

The decryption process in code-based cryptography involves extracting the original coded form by removing the appended error vector. This is achieved through a decoding algorithm that seeks to identify the nearest valid coded representation to the perturbed version. If the error vector is sufficiently small, the decoder can rectify the errors and recover the initial coded representation, which corresponds to the decrypted message.

The security of code-based cryptography hinges on the inherent complexity of decoding linear codes, even when adversaries have access to substantial amounts of ciphertext. The security level is influenced by the code's dimensions and the complexity of the decoding algorithm. To break the encryption, an attacker must decode the perturbed coded representation to retrieve the error vector. This computational task can be rendered infeasible through careful parameter selection.

The central objective of code-based cryptography is to minimize the size of encryption and decryption keys. This is pursued by exploring alternatives to McEliece's classical Goppa codes [Ber73]. The aim is to uncover novel designs, optimal parameters, and configurations at various security levels. Code-based cryptography serves as a robust post-quantum cryptographic framework, facilitating the creation of essential cryptographic components like encryption, signatures, zero-knowledge proofs, and hashing. Implementations of these components are characterized by efficiency, and their security is well established.

This discourse focuses on the fundamental code-based cryptographic primitives, centres around the McEliece and Niederreiter public key encryption [Nie86] systems, the CFS digital signature scheme [Cou+01], and the Stern zero-knowledge identification protocol [Ste94], along with different variations. This exploration is made possible by revisiting foundational concepts in algebraic coding theory, elaborating on linear codes, binary Goppa codes, decoding methodologies, and crucial properties.

McEliece cryptosystem introduced an asymmetric encryption scheme based on the elaborate complexity of decoding generic linear codes. A notable aspect of the McEliece system is the quadratic complexity evident in both encryption and decryption procedures concerning the block length. This operational efficiency positions it favourably in comparison to the RSA system. Also, the challenge of polynomial-time quantum decoding of general linear block codes remains unresolved, with its NP-hard nature established. However, a limitation of the McEliece system is the relatively large size of the public key, exceeding that of RSA and elliptic curve counterparts. In response, alternative strategies within code-based cryptography have been proposed, including the incorporation of Generalized Reed–Solomon (GRS) codes. These codes, combined with monomial transformations, offer the potential to reduce the dimensions of public keys. Yet, this approach is not immune to vulnerabilities, as demonstrated by Sidelnikov and Shestakov [SS92], who successfully revealed underlying code structures.

Similarly, Reed–Muller codes and Low-Density Parity Check (LDPC) codes have undergone investigation. Reed–Muller codes, introduced by V.M. Sidelnikov [Sid94] in 1994, have faced adaptations of attack methodologies that expose code structures. Regarding LDPC codes, challenges arise due to the need for substantial code sizes to prevent the recovery of low-weight vectors from the dual code. Medium Density Parity Check (MDPC) codes present a promising avenue for research.

The Niederreiter cryptosystem, introduced by Harald Niederreiter in 1986 [Nie86], introduces syndromes and disguised parity check matrices, providing security comparable to the original McEliece system while enhancing message transmission and encryption speed. Techniques such as error specification, low-weight transformations, and specified variety transformations are explored to maintain the security and adaptability of code-based cryptographic schemes. Code-based cryptography remains essential, especially in the post-quantum era, where its resistance to quantum attacks and potential for efficient implementations position it as a prominent choice for secure cryptographic solutions.

4.3 Hash-Based Cryptography

Leslie Lamport introduced the concept of the one-time signature (OTS) [Lam79], an algorithm for generating signatures that are designed for single use. Despite initial doubts about the practicality of this concept, Ralph Merkle astutely transformed these reservations into a signature algorithm [Mer79] with the capability for multiple uses. This innovation marked the inception of the initial multi-time hash-based algorithm. Broadly speaking, hash-based signature algorithms can be categorized into two distinct groups:

(1) Stateful
(2) Stateless.

Both types of algorithms efficiently manage multiple instances of one-time signatures. However, in stateful hash-based algorithms, special care is required to prevent the unintended signing of multiple messages using a single OTS signing key. To achieve this, the system's state is updated after each signature to keep track of the used OTS keys.

The accuracy of handling this state is critical, as a flawed implementation could lead to serious consequences. In scenarios where multiple hardware components collaborate, strategic architectural decisions, such as those related to disaster recovery, become crucial. It is advisable to collaborate with experienced entities proficient in stateful hash-based cryptography to ensure the competent management of these complexities.

In distinction, stateless hash-based signatures eliminate the need for state management, simplifying their implementation process. However, this simplicity comes at a cost: these signatures are larger and require more computational resources. Unfortunately, standardization for stateless signatures is projected to be realized in 2–4 years, whereas their stateful counterparts are on track for standardization within a year, benefiting from real-world expertise to ensure secure implementation.

The primary contenders in hash-based signature schemes are the stateful Multi-Tree eXtended Merkle Signature Scheme (XMSSMT) and Hierarchical Signature System (HSS), alongside the stateless SPHINCS+. Born from the visionary minds of Leslie Lamport and Ralph Merkle in the late 1970s, hash-based schemes have evolved significantly since their inception by Merkle.

These schemes are gaining momentum as promising alternatives to existing signature methods, positioning themselves as suitable candidates for the impending advent of large-scale quantum computers. Their appeal lies in their robust security, even in the face of quantum-assisted attacks, setting them apart from other post-quantum signature methods. They skip resource-intensive mathematical operations, relying solely on secure cryptographic hash functions.

Hash-based cryptography involves the construction of signature algorithms rooted in the security of chosen cryptographic hash functions. Trusted hash functions like the Secure Hash Algorithm 2 (SHA-2) serve as foundational elements for hash-based signatures, offering resilience against formidable adversaries and potential quantum threats. In the unlikely event of a compromise in SHA-2, transitioning to an uncompromised hash function promptly restores the security of hash-based signatures, demonstrating a versatile crypto-agile system with minimal transition costs.

The application of hash-based cryptography involves the swift implementation of hash-based solutions, particularly in scenarios necessitating robust security measures, such as long-lived connected devices operating in challenging environments. The adoption of endorsed hash-based digital signature algorithms preemptively addresses future logistical and financial challenges. Hash-based cryptography embodies cryptographic constructs anchored in hash function security, representing a pivotal category within the field of post-quantum cryptography. Its utility spans various domains, including digital signature schemes like the Merkle signature scheme, zero-knowledge and computational integrity proofs such as zk-STARK, and range proofs through protocols like HashWires.

Hash-based signature schemes combine one-time signature schemes, such as the Lamport signature, with hierarchical Merkle tree structures. This amalgamation overcomes the limitations of one-time signatures by aggregating multiple keys within an organized framework. The Merkle tree structure efficiently calculates tree nodes using iterative hash functions and concatenation, facilitating this aggregation. Nonetheless, it is essential to note that hash-based signature schemes are constrained in their ability to securely sign a finite number of messages due to their reliance on one-time signature schemes.

Adoption is required of post-quantum cryptographic algorithms that can support at least 2^{64} worth of signatures, according to the NIST. As one of the three algorithms designated for standardized digital signatures, SPHINCS+ was chosen by NIST in 2022, following its discernment. The eXtended Merkle Signature Scheme (XMSS) and Leighton–Micali Signatures (LMS) serve as the foundation for NIST's standardization of stateful hash-based cryptography, which recognizes the difficulties associated with state management during implementation.

Hash-based signatures offer a well-established and robust method that remains resilient against quantum computer attacks, assuming the security of the underlying hash function. This solution is clear and standardized, particularly in its stateful variants, despite the problems involved in deployment. Multi-tree versions of these algorithms provide flexibility in striking a balance between size and speed. SPHINCS+, a stateless hash-based signature scheme, has gained NIST's endorsement and is progressing towards standardization, albeit with notably larger signatures.

4.4 *Isogeny-Based Cryptography*

Elliptic curve cryptography (ECC) is a widely recognized cryptographic technique known for its utilization of compact key sizes. Unlike RSA, which requires significantly larger keys, ECC achieves comparable security to 128-bit AES using a 256-bit key. However, ECC's vulnerability to quantum computing compromises its quantum resistance. Quantum computers can efficiently solve the elliptic curve discrete logarithm problem, rendering ECC non-quantum-resistant.

An alternative to ECC, known as isogeny-based encryption, shares its foundation in elliptic curves but follows a distinct approach. While ECC involves computations on elliptic curves, isogeny methods operate on networks of functions connecting these curves. Among these approaches, only one, namely Supersingular Isogeny Key Encapsulation (SIKE), maintains a connection to elliptic curves. SIKE builds upon isogenies between supersingular elliptic curves.

This cryptographic avenue, referred to as isogeny-based cryptography, emerged relatively recently, primarily in the 2000s. It draws upon the foundational concepts of ECC from the 1980s and has gained heightened attention in anticipation of the quantum computing era. Isogeny-based cryptography stands out as a promising post-quantum cryptographic approach due to its presumed resilience against quantum-based cryptanalysis, a trait that proves more challenging for other cryptographic methodologies.

The Supersingular Isogeny Diffie–Hellman (SIDH or SIKE) protocol is a cryptographic mechanism designed for secure key establishment in the post-quantum era. Operating along principles similar to Diffie–Hellman, SIDH utilizes walks within a graph of supersingular isogenies. Initially, SIDH displayed promise, boasting small key sizes and supporting perfect forward secrecy to protect communication sessions even if long-term keys were compromised. Unfortunately, a significant key-recovery attack on SIDH was discovered in July 2022 by Wouter Castryck and Thomas Decru [CD22], rendering it insecure even without relying on quantum computing.

Despite its potential, isogeny-based cryptography remains a specialized domain with limited expertise. The broader mathematical literature on isogenies extends beyond this specific community, including a range of results that are not exclusively tied to elliptic curve isogenies.

4.5 Unbalanced Oil and Vinegar Schemes

The cryptographic evolution of the Unbalanced Oil and Vinegar (UOV) scheme is an advanced iteration of J. Patarin's original oil and vinegar cryptographic concept [Pat97]. In multivariate cryptography, particularly in digital signatures, the UOV scheme is rooted in cryptography. Its security foundation lies in the complexities of an NP-hard mathematical challenge, and its experimental implementation involves solving a minimalist quadratic equation system, which is key for generating and validating signatures.

The NP-hard complexity of this system becomes evident when it comprises m equations and n variables. Notably, this challenge remains formidable even in the presence of quantum computation, particularly in scenarios where m and n converge in value. Nevertheless, the problem becomes more manageable when m significantly differs from n. Various signature schemes have emerged, harnessing the complexities of multivariate equations. Among these, the UOV scheme stands out, offering significant security benefits. Yet, it presents a notable drawback: the potential for unwieldy key sizes. In practice, m often takes a value twice that of n, resulting in substantial key dimensions due to the encoding demands for coefficients originating from numerous equations—this encoding requires nearly 200 kilobytes.

The 'unbalanced oil and vinegar' approach, intriguingly named, is a cryptographic methodology that shares foundational similarities with the RSA encryption scheme. Both procedures leverage challenges that are easy to enact yet difficult to unravel—essential characteristics for robust encryption. In UOV, the core revolves around deciphering the composition of a linear and nonlinear operator, setting it apart from RSA's reliance on prime number multiplication. A significant distinction lies in UOV's resilience against quantum-based threats, effectively countering potential forgeries by powerful quantum computers. This contrasts with the vulnerability of RSA to quantum-driven attacks.

The importance of the UOV algorithm lies in solving complex networks of multivariate polynomial equations within finite fields—an inherently NP-hard problem unless specific structural attributes offer simplification. One such variant, the LUOV scheme, extends the core principles of UOV and expands the spectrum of post-quantum digital signature capabilities. UOV's architecture introduces 'oil' and 'vinegar' variables, whose fusion with an invertible linear operator creates a distinct arrangement similar to a concealed blend. The term 'unbalanced' conveys the scheme's heightened security when the quantities of 'oil' and 'vinegar' variables diverge.

By inverting the linear operator and polynomials, a signatory familiar with the hidden structure generates a vector linked to a predefined value as a component of the UOV signature procedure. In contrast, a verifier equipped with only the aggregate arrangement evaluates a complex ensemble of multivariable polynomial equations during verification. Despite this, the verifier is incapable of forging signatures due to the intractable nature of system inversion. The scope of UOV's polynomial systems is moderate, with approximately one hundred equations and variables, but with more variables than equations. Despite their quadratic nature, these systems require several kilobytes for representation, which contributes to the comparatively larger size of public keys in comparison to conventional digital signature algorithms such as ECDSA.

The primary asset of UOV is its resistance to quantum technology–induced vulnerabilities. In contrast to established methods such as RSA and ElGamal, which rely on discrete logarithm challenges susceptible to quantum-based factorization via Shor's algorithm, UOV is resistant to quantum computing developments. Also, UOV's operational processes are comparatively simple, making them compatible with hardware with limited resources, such as smart cards.

The vast majority of signature generation and validation procedures involve the addition and multiplication of tiny values. However, the limited adoption of UOV is due to observable weaknesses. Even though there are a few known attack strategies, new attack vectors may emerge if the scheme obtains widespread adoption. Due to the ongoing requirement for exhaustive security evaluations, UOV is not yet available for commercial deployment.

5 Quantum Key Distribution (QKD)

The focus of our discussion is on the notion of Quantum Key Distribution and its importance within the framework of modern digital society, whereby safeguarding data is of priority across diverse sectors like e-banking, e-business, e-health, and e-government. The increasing reliance on digital communication and data storage has led to a pressing need to prioritize the preservation of information confidentiality and integrity. In this section, we will look at the importance of QKD in easing these security worries, especially now that quantum computers could pose a threat to current cryptographic techniques.

The section further examines the possible uses of QKD in industries such as healthcare and critical infrastructure, highlighting its prospective advantages in terms of long-term security. In the present digital setting, characterized by the electronic transmission and storage of sensitive information, ensuring data security has emerged as a principal consideration. Industries such as electronic banking, electronic business, electronic health, and electronic government primarily depend on secure communication channels to safeguard the confidentiality of transactions, personal information, and sensitive government processes. Nevertheless, the emergence of quantum computers presents a substantial challenge to traditional cryptographic methods that serve as the foundation for ensuring data security.

Definition 12 (Quantum Key Distribution) Quantum Key Distribution (QKD) is a cryptographic methodology using quantum mechanics to create a secure key between two parties, often called Alice and Bob, while identifying eavesdropping, with its security grounded in quantum properties like the uncertainty principle and no-cloning theorem.

The possibility of quantum computers compromising conventional encryption techniques, leading to the exposure of sensitive information and undermining the fundamental basis of digital trust, is a significant concern. The implementation of QKD in the healthcare sector presents a very persuasive argument. The safe and private maintenance of patient records throughout an individual's lifetime is important since these records include sensitive information that must be protected from unauthorized access and tampering. The need for a technique capable of guaranteeing long-term security becomes apparent in light of the potential of quantum computers to undermine current encryption technologies.

QKD has emerged as a remarkably promising solution in this particular situation, presenting a distinctive method of encryption that capitalizes on the laws of quantum mechanics to ensure the prolonged secrecy of data. Likewise, critical infrastructure, such as energy networks, encounters the task of ensuring the protection of communication channels that are essential for the effective operation of society. Quantum technologies can provide resilient solutions in this domain as well. Although QKD systems have been present in the commercial sector for several years, current research and development initiatives are concentrated on enhancing the compactness, cost-effectiveness, and long-distance capabilities of these systems.

These technological breakthroughs help in enabling the integration of quantum technologies by governmental bodies and enterprises, thereby ensuring the robustness of essential infrastructure against possible quantum-related risks. However, the integration of QKD systems into current network infrastructures poses a complex and diverse problem.

The successful execution of this undertaking necessitates the cooperation of a wide range of stakeholders, including telecommunication equipment makers, suppliers of essential infrastructure, network operators, providers of QKD equipment, specialists in digital security, and scientists. The participants are working together to integrate QKD technology into existing communication systems, aiming to achieve a smooth integration that is both compatible and enhances security measures.

QKD fundamentally centres on the safe dissemination and exchange of cryptographic keys that are required for the encryption and decryption of communications. The essential goal is to preserve the confidentiality of these cryptographic keys among the parties involved in communication. Quantum mechanics presents a paradoxical phenomenon in which the act of viewing or perturbing quantum systems results in the disturbance of their inherent quantum attributes.

By using this principle, QKD utilizes a secure communication protocol that is based on the principles of quantum mechanics, sometimes mistaken for quantum cryptography. This is a notable implementation of quantum cryptography. The process involves the use of a cryptographic protocol that facilitates the generation of a mutually agreed-upon secret key between two entities. The key is only accessible to the individuals involved in the communication and may be used to encrypt and decrypt messages.

One notable benefit of QKD is its capability to effectively identify and avert any unauthorized third party seeking to acquire the cryptographic key. The basis for this detection method is rooted in the basic principle of quantum mechanics, which states that the act of measuring a quantum system always disrupts it. As a result, every effort to intercept communications generates noticeable anomalies, serving as a protective measure against unauthorized entry. The security of encryption by QKD is based on the fundamental principles of quantum mechanics, which represents a divergence from conventional public key cryptography methods.

Public-key cryptography is based on the computational difficulty of certain mathematical functions. However, the security of this cryptographic system does not have mathematical proofs that establish the complexity of reversing these functions. On the other hand, QKD is characterized by its ability to provide demonstrable security grounded in information theory and the concept of forward secrecy. The difference highlights the resilience of QKD in the face of possible threats enabled by quantum computers. One significant constraint of quantum-key distribution is its need for an authorized classical communication channel.

This stipulation necessitates that either symmetric keys of enough length or public keys with satisfactory security levels have been previously exchanged. Therefore, it is possible to achieve authenticated and secure communication in real-world circumstances without the need for QKD by using well-established cryptographic techniques.

This prompts an inquiry into the cost-effectiveness of QKD, given its redundancy in replicating the capabilities of a stream cypher at a somewhat greater cost. Significantly, QKD is a protocol that generates and distributes cryptographic keys without the transmission of message data. The key that is created may be used afterwards with different encryption algorithms to ensure the safe transmission of communications through traditional communication routes. The one-time pad is often linked to QKD owing to its proven security when using a secret random key.

However, practical situations typically require the integration of QKD with symmetric key techniques such as the Advanced Encryption Standard (AES). The technical basis of QKD leverages the unique characteristics of quantum mechanical systems. The method described here is the foundation for quantum cryptography, a field that utilizes similar concepts of physics and technology to provide secure communication over dedicated connections.

One notable characteristic of QKD and quantum cryptography, in general, is their inherent capacity to identify the existence of eavesdroppers, a skill that is not inherent in traditional encryption. Quantum-resistant algorithms augment cryptographic security through their reliance on mathematical complexity. The integration of these algorithms into cryptographic protocols provides a mechanism to guarantee the secrecy, integrity, and authenticity of communications, even in anticipation of eventual advancements in quantum computing technology.

NIST is now involved in a rigorous selection procedure to discover algorithms that are resistant to quantum computing for standardization. The earlier stationed project highlights the importance of strengthening cryptographic infrastructure to effectively counteract the potential threats posed by advancements in quantum computing, which can compromise traditional encryption techniques.

QKD is considered to be the fundamental basis of quantum communication technology, providing an unsurpassed degree of data security. The security of quantum encryption is derived from the fundamental laws of quantum mechanics, distinguishing it from traditional encryption systems that depend on the complexity of mathematical algorithms. The importance of quantum cryptography arises from its inherent resistance to any breakthroughs in computer power that may pose a threat to traditional cryptosystems. The essential significance of QKD lies in its focus on maintaining the security of digital communication and data storage.

In a dynamic and swiftly advancing digital environment where the transmission of sensitive data occurs across many industries, QKD appears as a solution that has promise in addressing the vulnerabilities presented by quantum computers. The effects of QKD extend to other fields, including healthcare and critical infrastructure, hence emphasizing the long-term security advantages that QKD offers.

Despite the many difficulties and the need for verified traditional channels, QKD appears to be a popular technique for ensuring secure communication. It relies on the fundamental principles of quantum mechanics to generate cryptographic keys that are resistant to the possible threats brought about by quantum computing. The ongoing progress in research and development has the potential to significantly transform data security for both governments and enterprises via the incorporation of QKD into current network infrastructures.

6 Quantum Applications

In addition to its potential applications in cryptography, AI, ML, and communication as a whole, including satellite communication, quantum computing has a vast amount of potential for addressing a variety of problems. The utilization of quantum applications has the potential to significantly impact our understanding of the present computing setting and our perception of the digital world. Yet, these advancements will not only disrupt the existing socioeconomic status quo but also create a void in the technological spectrum, favouring massive technological enterprises and providing them with a competitive advantage over the entire application sector. This section will explore the practical applications of quantum technology across several domains.

In recent years, there has been a notable increase in the prominence of quantum computing, with AI emerging as a prominent field of application within this domain. Gradually but consistently, quantum computing is preparing itself for its day in the spotlight. However, despite the eventual predicted dominance of quantum computing, its imaginable consequences continue to be mostly speculative. Nevertheless, this observation mostly stems from the nascent stage of development in quantum computing rather than a failure to deliver on its potential. In this part, we will not provide complete coverage of all conceivable applications. Rather, we will aim to concisely provide the introductory aspects that we, as authors, deem relevant for students who are being introduced to the subject of quantum computing and its associated factors that directly impact digital resilience.

6.1 Quantum Machine Learning (QML)

AI or ML seems to be involved in the vast majority of recent technological developments. At present, ML includes many aspects of human cognitive capacities but not all of the human intellect. ML facilitates the enhancement of human productivity by using intelligent technology to promote collaborative efforts. This phenomenon may be analogized to granting technology with a heightened act of human characteristics. Recommendation systems and forecasting systems are prevalent across numerous areas, and they provide valuable advantages via the use of AI and ML technologies.

These systems enable the efficient execution of arduous tasks, the acquisition of information across several fields, and the optimization of labour-intensive activities. The use of quantum computing has become more prevalent due to the emergence of QML systems. The possible implications for our security and creativity remain speculative and require more study. Hence, it is a rational expectation that the use of QML systems would significantly propel advancements in medicine, physical sciences, healthcare, and other pressing societal problems.

We will attempt to provide a brief overview of the concept of QML and assess its prospect of outperforming traditional ML methods. QML includes the integration of quantum algorithms with approaches to ML, thereby presenting opportunities for improved computational capabilities. In this discussion, we will explore the distinctive characteristics of quantum systems that may give rise to computing benefits in comparison to conventional classical approaches.

QML has the potential to make a big change in the field by using quantum principles. It could make it possible to meet requirements that are hard for standard ML algorithms to meet. QML includes the integration of quantum algorithms within ML techniques. The term 'quantum-enhanced ML algorithms' often refers to the use and application of these algorithms on classical data, followed by their execution on a quantum computer. QML leverages the inherent properties of qubits and quantum phenomena to enhance computational performance and expand data storage capabilities. This concerns the amalgamation of conventional and quantum computing methodologies, particularly in sophisticated and delicate computational jobs.

In addition to processing normal data, quantum algorithms possess the capability to investigate quantum states. The use of conventional ML techniques for the examination of data obtained from quantum experiments has been extensively investigated in the academic domain. The present study investigates the similarities between physical systems and learning systems, specifically neural networks, and evaluates the potential for using both techniques interchangeably. Researchers are involved in the examination of a conceptual framework referred to as 'quantum learning theory', which centres on the acquisition and manipulation of information within quantum information. The intersection of quantum computing and ML has sparked inquiries into the possible enhancement of training and testing procedures via the use of quantum computers.

Roughly speaking, research is being done right now to find out how quantum computers could be used in different areas, such as improving quantum error-correcting codes, estimating system properties, and making algorithms work better. The integration of quantum computers, which use the principles of quantum mechanics for information processing, with ML greatly expands the range of available hardware options in this domain. Hence, QML contains investigations into both theoretical model quantum computers and practical near-term quantum devices, which show resemblances to specialized hardware such as application-specific integrated circuits (ASICs) or field programmable gate arrays (FPGAs).

Quantum computer systems have a resemblance to neural networks in their capacity to undergo retraining via the manipulation of physical parameters. This characteristic allows them the capability to address computationally challenging problems that are beyond the reach of conventional computing systems. For instance, a proficient quantum circuit has the potential to effectively categorize photographs by maintaining the visual information in the physical state of the quantum device and then performing measurements taken after the data retrieval process.

6.2 Computational Chemistry

One of the most formidable challenges in materials science is the discovery and optimization of catalysts and processes for the production of new materials or the improvement of existing ones. The complexity of these molecular interactions often surpasses the computational capacity of traditional computers, rendering them a prominent domain for the utilization of quantum computing. Quantum computers use the principles of quantum mechanics to model and comprehend complex quantum phenomena, thereby presenting a groundbreaking methodology for addressing computational chemistry challenges that were previously deemed unsolvable.

One of the most critical objectives in materials research is the achievement of a superconductor that operates at ambient temperature. This material would demonstrate a complete absence of electrical resistance at typical temperatures, leading to significant advancements in power transmission, transportation, and several other technical fields. At present, the operational efficacy of superconductors is contingent upon the attainment of very low temperatures, imposing noteworthy practical constraints.

The use of quantum computers in computational chemistry enables the modelling and prediction of material behaviour at the quantum scale. Through the use of simulations that model the interactions occurring between electrons and atoms, scientists can distinguish compounds that possess the capacity to manifest superconducting characteristics at elevated temperatures. This discovery has the potential to facilitate the development and production of ground-breaking materials, thereby advancing the realization of room-temperature superconductivity.

Carbon sequestration represents a crucial approach for addressing the challenges posed by climate change since it involves the capture and subsequent storage of carbon dioxide to effectively prevent its emission into the Earth's atmosphere. The development of highly effective catalysts that aid this process is necessary. Quantum computers can effectively simulate the complex dynamics of catalytic processes, which often include a multitude of intermediate stages and quantum phenomena.

Scientists may use these simulations to find catalysts that boost the effectiveness of carbon sequestration processes, making a valuable contribution to the mitigation of greenhouse gas emissions. Using computational chemistry and quantum computers together has the potential to speed up the search for and development of catalysts that can turn carbon dioxide into valuable goods, easing environmental worries in the process.

The integration of computational chemistry with quantum computing has the potential to enhance energy storage greatly. Lithium-ion batteries, although becoming widely used, encounter constraints in terms of energy density, charge-discharge rates, and safety. Quantum computers provide a unique opportunity to investigate innovative battery chemistry with great precision.

Through the process of modelling the behaviour of different materials and their interactions within a battery system, researchers can enhance the performance and longevity of batteries. This can introduce a new phase of energy storage solutions, facilitating the development of batteries that include extended durability, accelerated charging capabilities, and enhanced safety features. These batteries may be used across a wide range of applications, including portable electronic devices, electric cars, and large-scale energy storage systems for power grids.

6.3 Financial Portfolio and Asset Optimization

Financial portfolio and asset optimization is a fundamental practice within the finance sector that involves carefully managing assets to attain an optimal balance between predicted returns and related risks. The complexity of this work necessitates the careful consideration of several parameters and variables, making it a highly suitable candidate for the use of quantum computing. The use of quantum technology has the potential to enable the banking sector to achieve enhanced solutions and accelerate the optimization process.

The usual methodology for portfolio optimization involves the use of mathematical models and optimization techniques. Regardless, the extensive range of potential investment combinations and the complex interaction of factors sometimes make conventional computers insufficient for generating optimum solutions within a practical timescale. Quantum computing presents a novel approach to tackling this difficulty by using its ability to efficiently handle large data sets and simultaneously examine various possibilities.

The distinguishing characteristic of quantum computing is its capacity to do parallel calculations via the use of quantum bits, also known as qubits. The inherent parallelism of quantum computers allows for the simultaneous exploration of several alternative portfolio allocations, resulting in a considerable acceleration of the optimization process. Also, quantum computers have exceptional proficiency in resolving optimization problems, a quality that has significant relevance in financial portfolio optimization. This work involves the determination of the most advantageous combination of assets while simultaneously accounting for several restrictions, thus representing a crucial undertaking.

Risk assessment is a crucial component of financial portfolio optimization. Conventional methodologies often resort to simplifications and approximations as a result of computing constraints, potentially resulting in inferior risk-return characteristics. The computational capabilities of quantum computing may enhance the integration of more complete models for risk assessment. Quantum computers can enhance the precision of risk assessment of diverse investment combinations by simulating many market scenarios and evaluating their possible consequences on portfolio performance.

Also, the ability of quantum computing to effectively process extensive volumes of historical financial data can facilitate the making of more well-informed judgements. The analysis of financial markets poses challenges for classical computers due to the complex patterns and interdependencies that may take significant time to completely comprehend. Quantum computers can effectively handle and analyse extensive information, enabling the discovery of concealed insights that have the potential to inform portfolio optimization techniques via a more profound comprehension of market dynamics.

It is essential to recognize that quantum computing is now in its nascent phase and that the mainstream adoption of practical applications for financial portfolio optimization remains limited. The construction and upkeep of quantum computers present notable difficulties, such as the presence of error rates in quantum processes and the need for specialized algorithms that effectively harness quantum benefits. Nevertheless, with the maturation of quantum technology, whatever possible influence it may have on financial optimization becomes more perceptible.

6.4 Industrial-Scale Logistics and Scheduling

Modern industrial operations rely heavily on logistics and scheduling, which present a collection of optimization challenges. The emergence of quantum computing presents a paradigm shift with the potential to significantly improve the efficiency and efficacy of these essential activities. Two principal instances demonstrate the potential of quantum technology in transforming logistics and scheduling: the optimization of aircraft deployments for airlines and the maximization of industrial output for factories.

Airlines function in a dynamic operational context where the effective allocation of aircraft resources is of central importance to provide maximum service quality while simultaneously minimizing expenses. Traditional methods for aircraft scheduling include complex algorithms that aim to identify solutions that are close to optimal. Nevertheless, these algorithms often encounter challenges when confronted with the extensive array of variables and restrictions implicated. The distinctive parallelism of quantum computing offers a transformative answer to this particular difficulty.

Quantum computers have exceptional proficiency in concurrently investigating several possibilities, a phenomenon often referred to as quantum parallelism. When used in aircraft staging, quantum computers can simultaneously evaluate many deployment possibilities. The use of parallelism significantly decreases the amount of time needed to identify optimal solutions. Airlines have the potential to use quantum computing technology to analyse several variables, including flight pathways, demand from passengers, crew availability, and maintenance schedules, to identify the optimal approach to positioning their aircraft that is both cost-effective and service-oriented. This not only improves operational efficiency but also has a favourable effect on customer happiness and overall profitability.

Within manufacturing, managers of factories are constantly faced with the ongoing responsibility of optimizing production schedules to minimize expenditures and time constraints while simultaneously maximizing productivity. This task involves managing several factors, such as the availability of machines, staff schedules, material availability, and manufacturing targets. Conventional scheduling approaches often depend on heuristic algorithms, which may not consistently provide optimal answers.

The situation under consideration highlights the remarkable capacity of quantum computing to tackle optimization challenges. Factories have the potential to harness quantum technology to effectively assess several production schedule alternatives, taking into account various limitations and goals simultaneously. The quantum computer has the aptitude to comprehensively analyse the complex interrelationships among many elements and restrictions, allowing plant managers to make educated choices that result in heightened levels of production. Through the attainment of improved optimization, factories can diminish periods of inactivity, optimize the allocation of resources, and enhance the efficiency with which production objectives are met.

Understanding the unique capabilities and theoretical benefits of quantum computing is crucial for realizing the promise of quantum computing in logistics and scheduling. Nevertheless, the discipline is still in its early stages, and the practical applications have not yet achieved broad adoption. The focus of quantum computing research continues to revolve around problems associated with error rates, algorithm design, and hardware limits.

7 Challenges and Future Prospects

In this section, we are going to look at the difficulties linked to the advent of the quantum age along with the prospective developments that may lie ahead. We know that the beginning of the quantum era raises certain concerns, particularly with the vulnerability of quantum computers to external effects, which may lead to mistakes in their computational outputs. Scientists are making diligent attempts to improve the accuracy of their work. An illustrative example may be found in Google's aspiration to improve the calibre of quantum computers by the year 2030. One notable challenge arises from the temperature constraints associated with quantum computers, since the majority of these systems need an environment that is far colder than room temperature, approaching but not quite reaching absolute zero. Any departure from this specified temperature range has the potential to result in computational inaccuracies.

With the escalating use of quantum computers, there is a growing concern over their ecological effects. One of the key challenges arises from the limited availability of personnel with extensive knowledge in quantum mechanics, engineering, and statistics. There is a notable absence of knowledge in these disciplines. There is a notable absence of a collaborative network that effectively enables synergistic collaboration across various research groups. The strategic trajectory that India may adopt concerning quantum technology remains undetermined since there is a deliberation between promptly leveraging its immediate uses or contemplating its long-term advantages. The conversion of research initiatives into actual real-world applications in quantum technology is of importance for India.

7.1 Error Correction and Scaling Quantum Systems

This subsection will address the difficulties associated with the expansion of quantum computers. The initiatives to construct more robust and potent quantum systems from theoretical principles to real-world applications have been accompanied by a range of challenges, specifically in the areas of scaling up quantum computers and augmenting their stability and computational capabilities. This part examines the problems associated with quantum systems and investigates the current attempts to develop more resilient and powerful quantum systems.

One of the primary obstacles in quantum computing revolves around the problems of qubit stability and coherence. Qubits, which are the quantum equivalents of classical bits, exhibit a high level of fragility and are vulnerable to the influence of environmental perturbations and decoherence. As the number of qubits rises, preserving their coherence gets more complex, resulting in computational mistakes. The presence of quantum noise, often known as a phenomenon, is a substantial obstacle to the development of more extensive and fault-tolerant quantum computing systems. Tackling this difficulty necessitates advancements in qubit design, materials, and cooling techniques to provide a conducive setting for the precise control and manipulation of qubits while minimizing external disturbances.

Another significant obstacle that has to be addressed is the matter of quantum error correction. In contrast to conventional computers, quantum computers are prone to mistakes as a result of their inherent quantum properties. To get dependable quantum computing, the use of error-correcting codes and algorithms is essential. Nevertheless, the implementation of error correction is a hard task due to the need to include more qubits and use advanced control techniques. Quantum error correction codes and fault-tolerant architectures are now being investigated by researchers to address mistakes and enhance the overall dependability of quantum computing.

The advancement of quantum computers is closely intertwined with the progress in quantum interconnects since the scaling up of quantum computing systems relies heavily on the development of efficient interconnection technologies. Interconnecting quantum systems is often required in many applications, such as the distribution of entanglement or the execution of distributed quantum algorithms. The development of efficient quantum interconnects that effectively maintain fragile quantum states and provide dependable information flow is a complex undertaking. Researchers are now exploring several physical platforms, such as photons and superconducting circuits, to build dependable quantum communication channels.

Advanced quantum control and measurement methods are required to effectively operate powerful quantum systems. To achieve precise control and manipulation of qubits at the quantum level, it is essential to use new control methods. This includes the advancement of high-fidelity gates, which are fundamental operations executed on qubits, along with strategies to mitigate cross-talk among qubits. Also, because it is hard to measure quantum states accurately and effectively while keeping their fragile coherence, it is necessary to use advanced sensor technology and signal processing methods.

The collective involvement of academia, business, and government agencies is crucial in the attempt to develop quantum computers that are both scalable and reliable. Various research centres and businesses are now engaged in initiatives aimed at expanding the capabilities of quantum hardware. These efforts include the exploration of several physical manifestations of qubits, including trapped ions, superconducting circuits, and topological qubits. These efforts include the refinement of manufacturing procedures, the optimization of qubit designs, and the enhancement of quantum error correction capabilities.

In addition, the development of quantum software and algorithmic techniques is indistinguishably linked with improvements in hardware. As the size of quantum computers increases, it becomes necessary to develop novel algorithms that can effectively harness their capabilities. Scientists are now developing algorithms that make use of the distinct characteristics of quantum computers, resulting in faster processing times compared to conventional computers for certain types of problems. The establishment of a symbiotic relationship between advancements in hardware technology and advances in algorithms is important to fully harness the potential of quantum computing.

7.2 *Ethical and Societal Implications*

In this section, we will examine the ethical and societal implications of quantum technologies, including both their potential benefits and limitations. The discussion around quantum technology has emerged as a significant domain capable of bringing about notable transformations in several aspects of human existence. These worlds span many areas, including but not limited to computers, communication, sensing, and cryptography. As the further development of these technologies continues, it is essential to thoroughly analyse their ethical and social consequences, including both their potential advantages and disadvantages.

Quantum technologies have significant potential for effectively tackling critical computational and information processing concerns. For example, quantum computers can tackle problems that are now insurmountable for conventional computers. These jobs include activities such as the simulation of quantum systems, the optimization of complex processes, and the decryption of cryptographic codes.

Loosely speaking, the potential can facilitate notable progress in several domains, including drug development, materials research, and optimization of supply networks. Also, quantum communication provides a greater degree of security through QKD, guaranteeing the preservation of confidentiality for critical data even when subjected to formidable eavesdropping attempts. The phenomenon has the potential to have significant consequences in secure communication and data security within an ever-expanding digital landscape.

While quantum technologies offer various benefits, it is important to note that they also raise ethical dilemmas. One of the main concerns is the risk of quantum computers disrupting existing cryptography methods. This can lead to a breach of conventional encryption techniques, thus creating a significant ethical and cybersecurity challenge. The safeguarding of confidential data and communication is contingent upon the use of encryption protocols. The advent of quantum computers, with the ability to decrypt such encryption, has the potential to result in infringements against privacy, instances of corporate espionage, and unauthorized intrusions into individuals' personal information. The development of cryptographic approaches that are resistant to quantum computing is important to maintain the integrity of data security in the age of quantum technology.

Also, the development and implementation of quantum technologies give rise to concerns about accessibility and fairness. Throughout history, it has been observed that the introduction of new technology often amplifies preexisting disparities within societies. To mitigate the occurrence of a 'quantum divide', it is essential to undertake initiatives aimed at ensuring the equitable distribution of the advantages offered by quantum technologies, thereby avoiding their concentration within a limited subset of individuals and societies. This necessitates the resolution of concerns about education, training, and infrastructure to facilitate the active participation and advantageous outcomes of a heterogeneous workforce in the quantum revolution.

The assessment of possible environmental consequences associated with quantum technologies is an additional factor to be taken into account. The successful operation of quantum computing systems necessitates the establishment and maintenance of a meticulously regulated and isolated environment, which is crucial for preserving the fragile quantum states essential for computational processes. The process of cooling and maintaining these systems may be characterized by a significant demand for energy, perhaps leading to an escalation in overall energy consumption. Ensuring an equilibrium between the advantages derived from quantum breakthroughs and their corresponding environmental impact is of utmost importance, particularly within the framework of worldwide initiatives aimed at mitigating climate change and facilitating sustainable technology.

The development of quantum technologies has been predicted to give rise to significant ethical dilemmas in the areas of data privacy and monitoring. The application of quantum sensors, characterized by their great sensitivity, has the potential to extend the range of applications beyond what is now achievable with conventional sensors. This prompts inquiries about the ethical use of quantum sensing technologies, possible invasions of privacy, and the requirement for regulatory measures to ensure that the implementation of these kinds of sensors upholds the rights of individuals.

7.3 Challenges in Contemporary Cryptography

The fundamental premise that underlies multiple public-key cryptography systems is reliant on the complexity of two distinct problems. This debate primarily focuses on the problem of integer factorization, with a particular emphasis on the significant contributions made by RSA [Riv+78] and Rabin [Rab79]. Our next concern is the discrete logarithms inside prime fields. Diffie-Hellman [DH76], DSA, XTR [LV00], and CEILIDH [RS03] are significant cryptographic systems that use discrete logarithms inside prime fields. The definition of this concept may be broadened to include elliptic curves, which incorporate the use of cryptographic methods such as ECC, Suite B, and 25519 [Ber06].

Nevertheless, the cryptography environment is now overshadowed by the introduction of Shor's algorithm, which has been meticulously designed to use the potential of quantum computers. The programme demonstrates its efficacy in tackling these categories of challenges and identifying possible flaws in systems that were previously considered secure. Adding to the complexity of the problem is

Grover's method, a quantum advancement that improves the efficiency of generic search tasks.

In this ever-evolving cryptographic domain, several important events emerge in the foreground. The assessment of alternative models of quantum computing against the current cryptographic basis necessitates the implementation of a detailed analysis. The evaluation of cryptographic systems that are included in established standards or smoothly integrated into widely used products has considerably more significance. Concurrently, the emergence of these quantum algorithms necessitates the development of novel cryptographic concepts. It is essential to articulate these statements in a way that effectively mitigates the risks posed by the progress in quantum technology, ensuring the protection of confidential data in an era characterized by the emergence of quantum computing capabilities.

The pursuit of enhanced cryptographic systems requires a total understanding of both traditional and quantum-based methods of breaching security. We shall demonstrate a complete understanding of the complex relationship between cryptographic techniques and their corresponding parameters. This chapter around cryptography is seeing a significant shift as the traditional and quantum domains become interconnected in a nuanced dynamic. As cryptographers and researchers guide through their work, it is crucial to prioritize the enhancement of current systems, explore novel approaches, and understand the complex nature of attacks to protect information security.

8 Dangers to Conventional Cryptography

In the following sections, we will examine some mathematical problems that now serve as the foundation for modern public-key cryptosystems. These difficulties include the factorization problem and the discrete logarithm problem. These challenges are well recognized for their ability to generate trapdoors in public-key cryptosystems due to their inherent computational difficulty. These methods are anticipated to become obsolete in light of the implementation of Shor's factoring algorithm. In this discussion, we will provide a concise overview of Shor's algorithm to facilitate a rudimentary understanding of the implications it presents, not only to cybersecurity but also to the broader field of web technology, which heavily relies on RSA and Diffie–Hellman Key Exchange algorithms.

With RSA, scientists and mathematicians commonly believe that the mathematical operation of factorization corresponds to the decoding of encrypted data. The procedure involves identifying prime numbers p and q to calculate their product, generating the public value n. This technique is critical for ensuring RSA encryption's security and serves as the basis for implementing asymmetric cryptography. Current expert opinion asserts that the computational complexity linked to factoring large numbers increases exponentially. The theory states that conventional computing systems encounter inherent challenges in the factorization process, with complexity significantly rising as the magnitude of the factored numbers (referred to as n) increases. The notion is supported by the belief that the computational cost of factoring a specific number, n, is closely connected to the growth rate of its logarithmic function.

Nevertheless, do note that cryptography and computational complexity are continuously evolving, particularly due to the emergence of quantum computing. Quantum computers apply fundamental principles of quantum mechanics to perform certain types of calculations much more efficiently than conventional computers. These quantum computers can significantly impact cryptographic systems like RSA by quickly conducting integer factorization, especially using techniques like Shor's algorithm. Quantum technology introduces unknown challenges and complexities to data security.

Despite the absence of actual realization of Shor's algorithm so far, there is a growing understanding of the rapid advancements in quantum computing research and their potential implications. The NIST has also recognized this development and has therefore identified new quantum-proof algorithms for consideration. This section aims to initiate a preliminary examination of the prevailing mathematical challenges encountered in traditional public-key systems while also clarifying the significance of Shor's algorithm as a ground-breaking development in computer science and mathematics.

8.1 Factorization Problem

As previously mentioned, it is anticipated that quantum computers will have a significant impact on certain existing public key infrastructures (PKIs), such as the RSA scheme. The RSA system relies on the belief that the computational problem of integer factorization is impractical to solve, but quantum computers are likely to challenge this assumption. Consider a hypothetical scenario where a quantum computer is endowed with 4098 qubits. In this scenario, the quantum computer can break down an RSA encryption system that utilizes a 2048-bit key by using Shor's method. It is worth mentioning that the largest numerical value that has been successfully factorized using Shor's algorithm on a quantum computer is 56153. This number is obtained as the product of 233 and 241.

Formally, given a positive integer n, the task is to determine a prime factorization representation of n expressed as $n = \prod_{i=1}^{k} p_i^{e_i}$ where $p_1, p_2, \ldots, p_k$ are distinct prime numbers, and $e_1, e_2, \ldots, e_k$ are positive integers, ensuring that n is the product of these prime factors. In other words, the objective is to find p_i and e_i such that $n = \prod_{i=1}^{k} p_i^{e_i}$, where each p_i is a prime number, each e_i is a positive integer, and the primes p_i are unique in the factorization.

By considering the basic theorem of arithmetic, it can be deduced that the arrangement of prime factors in ascending order allows for the exclusive representation of every positive integer as the product of these primes. The task of integer factorization is of significant difficulty and serves as the fundamental basis for several modern security methods. The process of factorizing a positive integer, represented as n, involves identifying positive integers p and q that fulfil the requirement that their product is equal to n, with both p and q being greater than 1.

The variables p and q are the integral factors in what is often referred to as a factorization of the variable n. The basic premise of the problem of integer factorization may be expressed as follows; given a composite number n, one's objective is to find two integers x and y whose multiplication results in n. The significance and the importance of factorization lie in its capacity to compromise the security of RSA encryption when performed efficiently.

When discussing the problem of integer factorization, it is pertinent to remember that it is classified as an NP problem, although it is not considered NP-complete because of the absence of a known reduction proof. It is essential to take into consideration that there is no inherent mathematical argument for the complexity of factoring. In contrast, a convincing argument may be presented to demonstrate the distinction between P and NP; however, it is extremely improbable that factoring falls into the category of NP-hard problems.

Consider the scenario in which we encounter the challenging task of factoring an integer n consisting of d digits. The conventional approach, sometimes referred to as brute force, involves examining all potential prime candidates p that is less than or equal to the square root of n and afterwards determining if p is a divisor of n. Unfortunately, in the worst-case scenario, this method may require a time commitment that is approximately proportional to the square root of n, so adding an exponential component in terms of the number of digits, d, Fortunately, a more proficient approach, known as the quadratic sieve technique, which is generally considered a good way of factoring numbers which are not necessarily extremely large. Although this method is more efficient, it is not the fastest factorization algorithm that is currently available.

On the contrary, the General Number Field Sieve (GNFS), an algorithm with reduced runtime complexity that is generally regarded as the quickest for factorization problems in conventional computing, has improved runtime complexity. The GNFS has a particular exponential relationship with the cube root of d. Unfortunately, the exponential increase in latency restricts the practical applicability of classical factoring algorithms to numbers with a relatively small number of digits, typically a few hundred or fewer. The present standard of this field is the factorization of a 32-digit number, which required more than two thousand years of central processing unit (CPU) use.

The number field sieve technique, including both its special and universal variations, may be seen as an improvement over the simpler approaches of the rational sieve or quadratic sieve. This basic principle underlies the methodology. In using these techniques for the factorization of a large integer n, it is crucial to identify smooth numbers, which are numbers with small prime factors of the magnitude $n^{(\frac{1}{2})}$. The magnitudes of these smooth numbers demonstrate an exponential correlation with the scale of n, as shown in the following manner.

In sharp contrast, the generic number field sieve effectively identifies smooth numbers that exhibit a subexponential relationship with the magnitude of n. The numbers under consideration exhibit a higher probability of being smooth due to their comparatively lower magnitudes, especially when compared to the numbers analysed by previous algorithms. The effectiveness of the number field sieve is mostly determined by this inherent property. To achieve enhanced computational speed, the number field sieve algorithm requires the performance of calculations and factorizations inside number fields. As a result, the algorithm presents a range of complexities, which are in opposition to the relatively straightforward design of the rational sieve.

The input dimension for this method may be expressed as the logarithm base 2 of n or the number of bits in the binary representation of n. This dimension exhibits an exponential relationship with the logarithm of n. An element having an order of n^c, where c is a constant, demonstrates an exponential correlation with the logarithm of n. The computing time required by the number field sieve exhibits a super-polynomial growth pattern while remaining sub-exponential concerning the amount of input.

8.2 Discrete Logarithm Problem

Consider a finite cyclic set denoted as G, with a cardinality of elements represented by n, and let b function as a generator of G. The generator b produces each element g within G in the form $g = b^k$, where k is an integer. It is important to note that any pair of integers defining g exhibits congruence modulo n. Consequently, a function $\log_b : G \longrightarrow \mathbb{Z}_n$ (where $\mathbb{Z}_n$ denotes the ring of integers modulo n) is introduced. This function assigns to g the congruence class of k modulo n. Acknowledged as the discrete algorithm to the base b, this function stands as a group isomorphism.

In a broader perspective, a standard algorithm denoted as $\log_a(b)$, provides a solution to the equation $a^x = b$ within real or complex numbers. Analogously, within a finite cyclic group G, considering elements g and h, a solution x to the equation $g^x = h$ is denoted as the discrete logarithm to the base g of h within the group G. The historical significance of discrete logarithms in number theory and their initial usage in computations within finite domains is noteworthy.

The foundation for establishing public-key cryptosystems resides in the Discrete Log Problem (DLP). Various present-day cryptographic algorithms rely on the complexity of this problem for their security. The complexity arises notably from the utilization of the Pohlig–Hellman algorithm, compromising the security of the commonly chosen $\mathbb{Z}_p^*$ (where p is prime) if $p - 1$ comprises small primes. To mitigate this vulnerability, the recommendation is the application of a safe prime, i.e., a prime number equal to $2q + 1$. This precaution ensures the complexity of the discrete logarithm problem. However, even with a safe prime, the index calculus algorithm introduces a sub-exponential threat, necessitating the adoption of considerably large prime numbers, typically exceeding 1024 bits, for secure cryptographic systems.

Expressed formally, the discrete logarithm problem in a finite group G involves determining the smallest positive integer x such that $\alpha^x = \beta$, given $\alpha \in G$ and $\beta \in \langle\alpha\rangle$. Alternatively, in additive notation, $x\alpha = \beta$, where x is referred to as the discrete logarithm of β concerning the base α, denoted as $\log_\alpha \beta$. This formulation includes the computation of the order of α as a special case: $|\alpha| = \log_\alpha 1_G$.

A nuanced version of the problem involves determining $\log_\alpha \beta$ and indicating $\beta \notin \langle\alpha\rangle$ if applicable. This extended problem, while challenging, offers clear conclusions through deterministic algorithms such as the baby-steps giant-steps method. Similarly, the extended discrete logarithm problem and the vector form are crucial in computing the structure of finite abelian groups. The factorization of the group order into small primes facilitates the computation of discrete logarithms. Interestingly, the ease of computing discrete logarithms in additive groups of finite fields stems from the availability of multiplicative inverses via the extended Euclidean algorithm.

The complexity of solving discrete logarithm problems is contingent not only on the group's order or structure but also on the specific representation of the group. Even groups of prime order, isomorphic to $\mathbb{Z}/p\mathbb{Z}$, face susceptibility to efficient discrete logarithm computations due to their ring structure. While the DLP is commonly perceived as a formidable task, its complexity intertwines with the context of the approach, highlighting the interplay between group theory, algebraic structures, and cryptographic applications.

Intuitively, the symbol G signifies a finite cyclic set, characterized by elements arranged in a specific order. The symbol n denotes the cardinality of elements within the finite cyclic set G, while b acts as a generator, capable of producing all constituent elements through exponentiation. Within this notation, g represents an individual element of the cyclic set G.

The symbol k serves as an integer exponent in expressing an element g as a power of b, namely $g = b^k$. The function $\log_b$ is a mathematical mapping from an element g in the cyclic set G to the congruence class of its corresponding exponent k, where the modulus is n. This function establishes a correspondence between the elements of the set G and the integers within the modulo n. The notation $\mathbb{Z}_n$ signifies the ring of integers modulo n, categorizing integers into distinct groups based on remainders from division by n. The expression $\log_a(b)$ represents the conventional logarithmic function, solving equations of the form $a^x = b$ to ascertain the unknown exponent x in real or complex numbers.

The notation $\mathbb{Z}_p^*$ represents the multiplicative group of integers modulo p, where p is a prime number. The symbol F conventionally denotes a finite field, a mathematical structure resembling a field and notable for containing a finite number of elements. Additionally, the notation $\langle\alpha\rangle$ indicates the subgroup generated by the specific element α within a given group. Finally, $\mathbb{Z}/p\mathbb{Z}$ signifies the ring of integers modulo p, with p representing a prime number and encapsulating residue classes of integers modulo p.

8.3 *Shor's Algorithm*

For anyone with an insatiable curiosity about exploring quantum computing, Shor's factoring algorithm is an unavoidable subject of discussion. The algorithm in question has a prominent position among a limited number of essential quantum algorithms, hence serving as a noteworthy example of the computing benefits offered by quantum systems. Fundamentally, this algorithm carries out calculations using quantum principles, surpassing traditional approaches in terms of complexity and speed. Shor's technique is particularly remarkable since it accomplishes a quantum computation that seems to be essentially unachievable using conventional methods on a significant scale. Shor's factoring algorithm stands out among the principles of fundamental quantum algorithms. This difference is based on two strong reasons.

The method developed by Shor is a highly innovative quantum approach that has the potential to revolutionize cryptography and computational number theory. The quantum-based capacity to effectively factorize big integers represents a notable progression compared to classical equivalents, hence influencing both theoretical and practical domains. There is significant variation in the estimated number of qubits needed for Shor's factoring algorithm. Distinguishing between logical and physical qubit estimations is of importance. The number of qubits may be adjusted depending on the desired goals, with the trade-off being either increased execution time and circuit depth for a lower qubit count or enhanced runtime and reduced circuit depth for an increased qubit count. Various qubit numbers will provide distinct availability times and runtime benefits.

First and foremost, it exhibits an exceptional capacity to factor numbers at an accelerated rate as compared to all presently recognized classical algorithms. Different quantum algorithms described provide computing benefits, but Shor's algorithm stands out as a distinguished subset due to its exponential acceleration and practical usefulness. Significantly, it presents a legitimate risk to the prevailing cryptographic technologies that are depended upon to protect worldwide financial operations.

The fundamental basis of RSA encryption, which serves as the foundation for global financial security, is the multiplication of two very large prime integers. The products of prime numbers, which are only divisible by one and themselves, include inherent characteristics that make them very large and difficult to factorize using traditional techniques. The process of factoring may be seen as a kind of reverse engineering applied to multiplication, revealing the prime numbers used and perhaps facilitating the unauthorized decryption of Internet data.

The conventional method for computing the greatest common divisor (GCD) between a randomly generated number and a given target number is used to ascertain if the target number has been inadvertently factored. The feasibility of the proposed approach is higher for smaller numerical values, perhaps necessitating the use of a supercomputer for bigger numerical values. However, the use of a quantum computer would be essential for the processing of cryptographically secure numerical values.

The primary function of a quantum computer is to determine the period of a given number to factor using Shor's algorithm. The outcomes determine whether more experimentation is required for new random numbers or whether the factors have been identified. Upon performing the process of factoring, Shor's algorithm arrives at a result. As a reminder, the classical algorithm that exhibits the best asymptotic behaviour is the General Number Field Sieve (GNFS). This algorithm has a superpolynomial scaling complexity represented as,

$$O\left(\exp\left(h_c \frac{\ln n}{3(\ln \ln n)^{2/3}}\right)\right)$$

where h_c represents a constant factor. The algorithm's complexity is characterized by its dependency on the natural logarithm of n, the input number being factored. The GNFS is known for its efficiency in tackling large integer factorization problems. The remarkable effectiveness of Shor's technique arises from the use of a quantum computer to do several calculations simultaneously. Broadly speaking, the capacity allows the algorithm to efficiently explore several possibilities, leading to a significant increase in speed compared to traditional methods. It is worth mentioning that the runtime of Shor's method exceeds that of the most advanced classical factoring techniques, which experience exponential growth concerning the size of the input.

Although the overall procedure may seem straightforward, it includes many stages that may be dissected into individual instructional sessions. Executing the method might be considered a reasonably simple task after an assignment, although attaining a comprehensive understanding requires a substantial amount of in-depth research. Shor's factoring algorithm has three primary components:

(1) Classical Computing,
(2) Quantum Computation,
(3) Classical Computation.

On its whole, the process consists of six notable stages, whereby the quantum aspect is further divided into four subcomponents that may be comprehensively explained. Two subcomponents may be considered very straightforward; however, the other two subcomponents are deemed essential quantum subroutines.

Quantum phase estimation (QPE) is a fundamental quantum subcomponent that is critical in facilitating the modular arithmetic required for period determination. The inverse Quantum Fourier Transform (iQFT) is another important part. It helps turn the results of quantum modular arithmetic into classical information through the measurement process. The algorithm proposed by Shor begins with a series of classical steps, which are then followed by quantum computing to determine the period. The obtained quantum information is then converted back into classical form, and the algorithm concludes with further classical steps. If a solution is not identified, the algorithm is modified and then executed again.

The primary focus of attention is on an algorithmic approach for factorization that exhibits polynomial computational complexity. The method operates inside a partly quantum computing framework and demonstrates a computational complexity of $O((\log n)^2(\log\log n)(\log\log\log n))$. The complexity involved in this context involves both the early and end stages of classical computational preparation, with intermediate stages that selectively incorporate quantum computing. The principal aim of this method is to address the problem of determining an integer n and afterwards finding an integer p within the range of 1–n that is capable of dividing n. Shor's algorithm consists of two crucial components:

(1) The factorization problem may be strategically translated into a component of the order-finding problem, which can be handled successfully using classical computer approaches.
(2) Quantum-based algorithm designed to solve the complexities of the order-finding problem.

The Shor's algorithm is systematically presented as follows:

Determine if n is even, prime, or a prime power. If so, exit.

(1) Select a random integer $x < n$ and calculate $\gcd(x, n)$. If this is not 1, then we have obtained a factor of n.
(2) Quantum algorithm:
    (1) Select q as the smallest power of 2 such that $n^2 \leq q < 2n^2$.
    (2) Determine the period r of $x^a \pmod{n}$.
    (3) Measurement gives us a variable c which has the property $c^q \approx d^r$, where $d \in \mathbb{N}$.
(3) Determine d and r via the continued fraction expansion algorithm. d and r are only determined if $\gcd(d, r) = 1$ (reduced fraction).
(4) If r is odd, go back to step 2. If $x^r \equiv -1 \pmod{n}$, go back to step 2. Otherwise, the factors p and q are given by $\gcd(x^r \pm 1, n)$.

Although the complex nuances of the complexity and mathematical principles of the algorithm are not included in this discussion, a full depiction of the algorithm as a whole is shown to encourage future investigation. Likewise, individuals with a strong command of mathematical principles may engage in an in-depth exploration of quantum computing. Although the breadth of this material does not allow for a detailed explanation, a brief mention is made to pique interest. The actual use of Shor's algorithm encounters difficulties in preserving coherence across quantum states, despite its considerable promise. However, the potential progress in the development of increasingly powerful quantum computers may enable the practical use of Shor's technique as a means to factorize big numbers.

The runtime of Shor's algorithm is inversely correlated with the number of logical qubits and is contingent upon the specific qubit technology used, given the inherent variability in the operating speeds of quantum computers. The increase in the number of qubits leads to a longer runtime for algorithms while simultaneously accelerating the obsolescence of current cryptosystems. The acceleration of cryptosystem breaking by the use of more qubits remains a prospect that has yet to be realized in the foreseeable future. Despite its extensive duration and attempts to mitigate dangers, Shor's factoring algorithm continues to serve as a powerful instrument for increasing awareness about the possible vulnerabilities of sensitive data and communications to quantum computing. During this process, executives from diverse industries may identify many use cases of quantum computing that have the potential to provide substantial benefits for their organizations in the future.

It is necessary to distinguish between physical and logical qubits. Current qubits possess a physical nature and are susceptible to mistakes. The indistinguishability of quantum computing outcomes between right and wrong replies poses a significant challenge to achieving dependable solutions. Logical qubits refer to a group of interconnected physical qubits that work together to rectify mistakes, resulting in the creation of 'logical qubits' or 'perfect qubits'. The abstractions serve as representations of physically imperfect qubits inside quantum circuits, characterized by inherent noise.

Shor's factoring algorithm has played an essential role in raising awareness and recognition of quantum computing among the general public. The very fact of this emerging technical frontier has prompted national governments, companies, and the general public to recognize its existence. Throughout the subsequent decades, this algorithm has maintained its prominent position among quantum algorithms. Significant efforts have been continuously undertaken over a prolonged duration to enhance the resilience of worldwide financial institutions, safeguard national security, and harden all cryptographic programmes against their prospective ramifications. The current geopolitical situation highlights the significant risk that arises when any organization, particularly state actors, acquires sufficient quantum computer power to use Shor's technique before universally applicable quantum-safe encryption methods are developed.

Shor's algorithm has played an integral part in attracting substantial expenditures, amounting to billions of dollars, towards the development of quantum technologies, with particular emphasis on quantum computing. The ongoing effort has been looking into additional practical applications that exponential speed improvements might enhance for thirty years since its initial discovery. The fundamental assumption is that if one algorithm is capable of achieving such acceleration, it is probable that other algorithms might also exist with similar capabilities. Throughout an extended period, most prospective candidates get inspiration from Shor's algorithm. The process of factoring integers using Shor's method begins by randomly picking an integer that is less than the desired value.

9 Concluding Remarks

The present day provides an exceptionally thrilling opportunity to participate in the rapid expansion of quantum computing. The technology is grounded on the basic principles of quantum mechanics and is anticipated to provide substantial benefits in many industries like the military, finance, aircraft design, utilities, polymer design, ML, AI, computational chemistry, finance, and digital manufacturing. Quantum computation is predicated on the basic principles of quantum mechanics and is now undergoing a remarkable rate of growth. In recent times, there has been a notable surge in the attention of several big technological enterprises, including IBM, Microsoft, Google, Alibaba, Nokia, Intel, HP, and Volkswagen, towards delving further into quantum computing.

The increased level of interest may be attributed to the growing need for this specific technology. Quantum computing has the possibility of becoming a transformational technology given its potential for significant advancements, despite its current state of fast growth. The introduction of quantum computers has enabled the improvement of simulation accuracy in computational fluid dynamics. The advancements indicated above have enabled the introduction of production design processes that are cost-effective as well as the improvement of portfolio investment strategies via the use of sophisticated modelling tools that can effectively analyse complex financial markets.

On the other hand, quantum computers provide a significant obstacle to encryption, which is a crucial component of our digital presence. This section gives a detailed review of the possible consequences of quantum computing and quantum-resistant encryption, including the conclusions derived from the research conducted. A thorough examination has been undertaken to explore the difficulties related to pre-quantum cryptography methods, with particular emphasis on the RSA and ECC algorithms. Also, a thorough analysis has been done to see how well Shor's factoring algorithm works in decomposing number-theory functions, especially those that have already been explained.

Extensive research has been undertaken on lattice-based cryptography, which is now acknowledged as a very promising option owing to its capacity to provide resilient encryption in a quantum environment. Also, we have presented our recommended methodology, which we deem to be the most auspicious in terms of guaranteeing quantum security in the next era. The development of post-quantum cryptography also referred to as quantum-safe encryption has been promoted as a response to the susceptibility of cryptographic techniques to quantum computers. The primary objective of this initiative is to effectively mitigate and minimize the possible threats posed by quantum computing technology.

The attainment of improved computing efficiency in quantum computers is assisted by the utilization of a wide array of mathematical components, including operations that are beyond the capability of traditional computers. Nevertheless, it is crucial to recognize that post-quantum cryptography is anticipated to encounter challenges in performance. The encryption and decryption procedures will need more computational resources, while the transmission of extended keys and certificates will demand increased networking resources. When used with traditional computer systems and cloud-based architectures, quantum computing can tackle problems that were previously considered unsolvable, opening up limitless possibilities for the economy and society.

The significance of determining the potential of quantum computing in expediting responses to forthcoming pandemics, persistent health crises, and the spread of debilitating illnesses that impact a substantial global populace should not be underestimated. The use of sophisticated chemical simulations in medical research and development may facilitate the attainment of this objective. The transformative potential of quantum computing surpasses the constraints imposed by traditional computing methods, thereby possessing the capability to restructure social frameworks fundamentally. It offers a mechanism for resolving persistent obstacles and addressing the most urgent concerns that have afflicted human civilization. The continuous standardization process of post-quantum cryptography necessitates the availability of substantial and persuasive evidence to assess the security of a given solution in both quantum and binary computing, namely within cryptanalysis.

The areas of quantum computing and post-quantum cryptography are interconnected, and their anticipated influence on the next age of computing and security is significant. Quantum computing has the potential to make exponential progress in tackling certain problem categories, including some that are now infeasible to address with traditional computer approaches. This opportunity has great potential. Despite the progress made in this area, traditional encryption methods are at risk of being compromised by quantum computers, given the enormous advancements made in this sector.

The development and progression of post-quantum cryptography have resulted in the formulation and improvement of cryptographic algorithms to offer resilience against attacks originating from both conventional and quantum computing systems. The research on post-quantum cryptography is centred on the advancement of cryptographic methods that exhibit resistance to possible attacks from quantum computers. It is quite probable that quantum computing and post-quantum cryptography will maintain a significant level of interconnectedness and exert mutual effects in the foreseeable future. The escalating progress in quantum computing necessitates a heightened recognition of the importance of post-quantum cryptography.

To conclude, the emergence of quantum computation in the modern digital society presents both prospects and obstacles for ensuring digital resilience. Quantum-resistant cryptographic algorithms must be developed and implemented in a concerted effort to prevent the compromise of existing cryptographic systems. Concurrently, the optimization functionalities innate in quantum computation present opportunities to strengthen digital resilience across diverse domains. But to achieve a digital infrastructure that is quantum-resilient, technological, financial, and regulatory aspects must be meticulously evaluated. Constant investigation and cooperative efforts among the policy and scientific communities are indispensable for navigating the complex trajectory towards a digital future that is both secure and resilient in the era of quantum computation. The examination of the impact of digital resiliency on quantum computing takes into account several elements, including encryption, algorithms, and communication. In the era of quantum computing, continuing research and development activities in the area of quantum-resistant cryptography and algorithmic resilience are of the highest significance in terms of their ability to influence the strategies that are intended to strengthen digital resilience.

Chapter 8
Advancing Security and Resilience

1 Introduction

We discussed a multitude of subjects in the chapters that preceded. We began by defining digital security and then discussed its relationship to cybersecurity and the effect that laws and regulations have on it. Following that, we discussed several noteworthy topics, such as quantum computation and the resilience of minors. In this chapter, we will discuss the ongoing technological advancements and innovations. The serious after-effects of the COVID-19 pandemic on digital ecosystem utilization have generated an urgent need for research in the domains of cybersecurity and digital resilience, as is common knowledge. The expansion of online platforms and schemes ensued as a consequence of the widespread adoption of social media and digital services, including those associated with healthcare, shopping, and daily necessities. Due to this expansion, the danger of security-related problems affecting networked devices and systems has increased. As a result, the significance of cybersecurity research in detecting and mitigating these threats has grown.

It is also generally acknowledged that the growth of devices and systems has experienced a substantial upsurge, resulting in a substantial enlargement of the possible attack surface. Attack surface refers to the quantity of potential entry points an adversary could exploit to gain access to your system. Consequently, the rapid growth of seemingly intelligent devices and computing devices designed to simplify our lives has facilitated the ability of malicious actors to sow discord. It is important to bear in mind that modern threats are considerably more sophisticated in their complexity, possessing advanced features and attributes that demand substantial attention.

S. K. Shandilya et al., *Digital Resilience: Navigating Disruption and Safeguarding Data Privacy*, EAI/Springer Innovations in Communication and Computing,
https://doi.org/10.1007/978-3-031-53290-0_8

Consider a simple example of malicious software. Historically, it consisted solely of adware—software designed to display advertisements on a system—or, at most, data-stealing viral infections. In the past, the majority of antivirus providers utilized signatures to identify specific forms of infection. Malware now also uses polymorphism, a method by which it modifies and transforms its code to evade detection by signature detection. Undoubtedly, this has increased the difficulty of detection and identification. It is noteworthy that current advancements in payload delivery have added to the complexity of the matter, thereby emphasizing the criticality of cybersecurity research that concentrates on developing novel defensive mechanisms capable of effectively countering threats.

The prevalence of malicious actors who continually develop novel and increasingly sophisticated methods to gain unauthorized access to a variety of systems is a persistent source of concern. As aggressive strategies advance, a rapid growth of security vulnerabilities compromises systems and devices. Research is a two-way street, which includes both offence and defence. Therefore, research initiatives that seek to develop novel protective systems have become a necessity in this dynamic environment. There are substantial motivations for both academia and industry to invest in cybersecurity research; this must continue to be the case. For a broad understanding of the present condition of cybersecurity, it is critical to conduct an exhaustive analysis of existing laws and regulations.

Given the ever-evolving nature of cyber threats, regulatory frameworks must demonstrate the capacity to swiftly adapt to and resolve the challenges and demands that emerge with it. Besides, it has become obligatory to attempt to redefine legal and regulatory frameworks to address the increasing perils presented by cyber threats. Without international laws and severe regulations, research can only advance to a certain extent. By ensuring that legal frameworks remain current with the ever-changing collection of threats, it is feasible to improve the efficacy of preventing malicious activities.

More broadly, cybersecurity research is a necessity now more than ever, it provides a fundamental approach to identifying and reducing digital dangers by offering insights into potential attack vectors, current defence tactics, and undiscovered opportunities for advancement. Sophisticated software and the multiplication of connected devices have posed challenges to the detection and prevention of malware threats. Due to the complexity of countering established security measures caused by these methods of transmitting malicious payloads, research into effective defence mechanisms is fundamental. The development of innovative methods by malicious actors to breach systems illustrates the criticality of cybersecurity research.

The increasing vulnerability of devices to security threats, which has been shown by the growing dependence on digital platforms since the COVID-19 pandemic, underscores the criticality of novel protective measures. Post-quantum cryptography and adaptive detection mechanisms, the two most prevailing and advanced disciplines, have been the focus of research for some time, addressing emergent vulnerabilities and strengthening security. The cybersecurity research community is dedicated to the development of inventive defensive strategies that are resistant to adversaries. These strategies may incorporate adaptive defence mechanisms, post-quantum cryptography, and nature-inspired approaches. Strict investigation is critical for the preservation of effective cybersecurity protocols.

2 Paradigms of Computing

Consider the case when someone looks for anything on Google and how quickly it provides the information and how almost all of its information is available at such a fast response time. The solution is distributed computing in the form of the Colossus, the file system that is the backbone of Google's services. Another instance is the current hype around blockchains and decentralized systems. Based on architecture, there are several sorts of computer paradigms. Many computer architecture paradigms are one sector that cannot be overlooked when developing any form of digital resilience framework. There are several sorts of computing paradigms depending on categorization; however, this section will concentrate on centralized, decentralized, and distributed computing systems. Refer to Table 8.1 for a summary of the paradigms.

2.1 Centralized Computing

For many years, centralized computing has served as the fundamental basis of conventional computing paradigms. The system is dependent on a solitary, centralized server or processing unit serving as the central hub for data processing and task execution. Within the framework of this hierarchical paradigm, client devices or end users initiate the transmission of their requests to a central server. The central server then undertakes the processing of these requests and later facilitates the delivery of the corresponding results.

The usefulness of centralized computing, although historically significant, has been progressively apparent in technical environments and the escalating requirements of applications. Centralized computing, as opposed to distributed or decentralized computing, is characterized by the reliance on a solitary central processing unit or server to manage all computational operations and data storage. The model has exhibited prominence during the initial stages of computing and continues to be used in many situations. In a centralized computing system, the consolidation of data and processing capacity in a single location offers the potential for streamlined management and enhanced control. On the other hand, there are certain constraints in terms of scalability and fault tolerance associated with it. The occurrence of any disturbance to the central server has the potential to lead to widespread failures within the system, hence reducing its suitability for applications that are mission-critical or require great scalability.

Table 8.1 Comparison of centralized, decentralized, and distributed computing

	Centralized computing	Decentralized computing	Distributed computing
1. Architecture	Centralized systems have a single point of control, with all processing and decision-making concentrated in one location	Decentralized systems distribute control across multiple points, allowing more autonomy for individual components or nodes	Distributed systems have multiple nodes that work together but operate independently, enabling parallel processing
2. Control	In centralized computing, control is maintained centrally, which may lead to bottlenecks and a single point of failure	Decentralized systems distribute control, reducing the risk of a single point of failure but may still face coordination challenges	Distributed systems allow for shared control and are more resilient to failures as the workload is distributed
3. Scalability	Centralized systems may face scalability challenges, particularly as the load increases	Decentralized systems offer better scalability than centralized systems but may still encounter limitations	Distributed systems provide excellent scalability by adding more nodes to the network to handle increased loads
4. Communication overhead	Centralized systems typically have low communication overhead as communication is concentrated within a single location	Decentralized systems may have moderate communication overhead as components need to coordinate with each other	Distributed systems can have higher communication overhead due to the need for communication between nodes
5. Fault tolerance	Centralized systems are less fault-tolerant, as a failure in the central point can lead to a system-wide failure	Decentralized systems offer improved fault tolerance as failures are localized and may not impact the entire system	Distributed systems exhibit high fault tolerance, as failures in one node are isolated, and the system can continue functioning
6. Resource utilization	Centralized systems may face uneven resource utilization, with some resources potentially underutilized	Decentralized systems can have better resource utilization but may still face challenges in optimizing resource allocation	Distributed systems typically exhibit efficient resource utilization by distributing the workload across nodes
7. Data storage	Centralized systems may centralize data storage, leading to potential data bottlenecks and limited redundancy	Decentralized systems distribute data storage, improving redundancy and access times but may face consistency challenges	Distributed systems distribute data across nodes, providing redundancy, parallel access, and improved fault tolerance
8. Coordination complexity	Centralized systems have lower coordination complexity, as all components are controlled from a single point	Decentralized systems have moderate coordination complexity, requiring communication and coordination among components	Distributed systems may face higher coordination complexity due to the need for effective communication and synchronization between nodes

(continued)

Table 8.1 (continued)

	Centralized computing	Decentralized computing	Distributed computing
9. Implementation cost	Centralized systems may have lower initial implementation costs but may face higher operational costs	Decentralized systems may have moderate implementation costs, balancing initial and operational expenses	Distributed systems may have higher initial implementation costs but can lead to cost savings in terms of scalability and fault tolerance
10. Security	Centralized systems may be more vulnerable to security breaches due to the concentration of control and potential single points of failure	Decentralized systems offer improved security by distributing control, reducing the impact of a single security breach	Distributed systems have enhanced security as they distribute control and data, making it more challenging for unauthorized access and providing fault tolerance against attacks

2.2 *Decentralized Computing*

In contrast to the concept of centralized computing, decentralized computing adopts a dispersed methodology for data processing and computation. Decentralized systems are characterized by the distribution of data and computing duties among numerous nodes or devices, hence eliminating the presence of a singular point of control. The decentralized method is in line with developing technologies like distributed ledger technologies (DLTs), edge computing, and the IoT, which include the widespread distribution of decision-making authority and data processing.

Decentralized computing can be considered as an intermediary approach that bridges the gap between centralized and distributed computing paradigms. In the present model, the distribution of computing activities and data occurs among numerous nodes or servers, accompanied by the distribution of control and coordination. Decentralization offers an equilibrium between scalability and fault tolerance, as it possesses the potential to sustain operations even in the event of node failures. This methodology is frequently used in peer-to-peer (P2P) networks, blockchain technology, and specific distributed systems. The utilization of decentralized computing is becoming increasingly prominent across multiple areas as it possesses the capability to leverage the combined capabilities of a network while also ensuring resilience against failures and threats. The advancements have resulted in novel solutions within domains such as cryptocurrencies, content distribution, and distributed databases.

2.3 *Distributed Computing*

The architectural paradigm of distributed computing has become a crucial factor in redefining the digital resilience and cybersecurity sector. The establishment of security measures in distributed systems necessitates a thorough understanding of

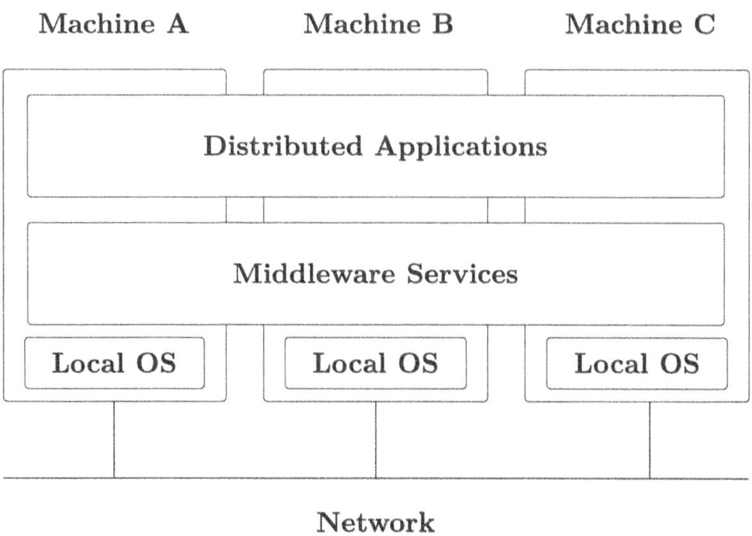

Fig. 8.1 Distributive computing

the fundamental principles underlying security in such systems. Figure 8.1 is a diagram showcasing the basic workings of distributive computing.

The successful safeguarding of a distributed system requires the integration of security measures at its initial development, rather than as a secondary consideration. The adoption of a security plan that incorporates several components, including hardware, software, and networking elements, is of importance. Conducting evaluations of open-source components, which are frequently used in the construction of distributed systems, is essential to detect potential security vulnerabilities. Due to the inherent decentralization of systems, it is crucial to establish a cohesive security policy that includes several organizational boundaries. Conducting regular and thorough reviews of the security of your distributed system is crucial for identifying potential vulnerabilities and implementing necessary improvements. Distributed systems have become a significant area of interest in the digital sector due to their extensive deployment and indispensable significance in various operational scenarios. The complex interdependence of computing nodes highlights their importance.

The importance of safeguarding distributed systems is a major concern since their growing dependence on communication, data exchange, resource allocation, and computing necessitates rigid security measures. The increasing reliance on these systems highlights the need for a thorough comprehension of the difficulties associated with safeguarding them from potential threats and vulnerabilities. The following section of this chapter is devoted to an extensive examination of security in distributed systems. The objective of this section is to outline the fundamental principles that influence the security environment, recognize significant issues associated with securing distributed systems, and elaborate on strategic frameworks and methodologies that are crucial for enhancing the security of these interconnected computational networks.

As the scale of distributed systems increases, the complexity of their behaviour gets more convoluted and difficult to predict or understand. The disparity is worsened by the inherent non-deterministic characteristics that are commonly found in large-scale distributed systems. These characteristics give rise to unforeseeable occurrences such as delays in message delivery, sudden failures of system components, or the execution of potentially detrimental actions by flawed or malicious machines.

The complex characteristics of extensive distributed systems present significant challenges in the domains of testing and simulation, principally attributable to their diverse nature and inherent unpredictability. As a result, there is a pressing requirement for theoretical frameworks that can effectively validate the authenticity of their characteristics, thus enhancing confidence in their practical implementation. The initial stage in developing a theory about distributed systems involves the creation of a model. This model is a conceptual framework based on mathematical principles that include all relevant characteristics of a large-scale distributed system.

A model should effectively include the complex interactions among the nodes comprising the system, the temporal order and arrangement of messages, and any other crucial aspects that affect the behaviour of the system. Once a model has been established, it functions as a powerful tool for inferring the behaviour of the system, identifying potential challenges or vulnerabilities, and developing methods to address these issues. This involves the development of algorithms designed to ensure the consistency and accuracy of data accessed by multiple users, as well as the design of robust systems that can continue to operate smoothly even when individual components are compromised.

The emergence of blockchain technology has been seen as a promising solution to address the challenges inherent in large-scale distributed systems. In essence, a blockchain operates as a decentralized database that is spread among a network of nodes. Every individual node inside the network maintains a replicated copy of the database, and any modifications made to the database require synchronization among all the nodes. One of the primary advantages of blockchain technology is its ability to promote trust and enhance security in an age characterized by decentralization. Transactions conducted on the blockchain undergo a validation process utilizing complex cryptographic methods, guaranteeing their resistance to tampering or alteration after their recording.

The applicability of blockchain for applications such as digital currencies is attributed to its capacity to prevent double-spending and maintain the integrity of the system, which is of crucial significance. It is important to note that blockchain possesses the capability to support smart contracts, which are autonomous agreements integrated within the structure of the blockchain. Smart contracts provide the ability to automate complex business procedures, such as supply chain management or insurance claim resolution, hence reducing dependence on intermediaries or external validators.

Distributed systems demonstrate their unique characteristics through a configuration characterized by multiple devices, each autonomously running its own set of programme. The increasing size and complexity of these systems pose a significant difficulty in terms of predicting or understanding their behaviour. Therefore, it is of importance to prioritize the development of theoretical tools that are specifically designed to model and rationalize the behaviour of these systems, to ensure their reliability and safety. In the given context, the emergence of blockchain technology presents a viable solution to the numerous issues faced by extensive distributed systems. It offers a way to establish trust and security in an age marked by decentralization.

2.4 Blockchain

Blockchain technology, which has its roots in the Bitcoin protocol, is a decentralized and extensively distributed system that functions over both public and private networks. Transparency is upheld through the dissemination of each transaction to a consortium of users, guaranteeing the accuracy of transactions and deterring illegal transactions. The consequences of this technology extend beyond banking, including various other domains like financial transactions, supply chain management, and digital identity verification.

The use of blockchain technology is important in maintaining the safeguarding of data, mitigating security concerns, and upholding the integrity of data. Removing intermediaries and optimizing operations enable the effective transmission of crucial company data and personally sensitive information. The implementation of blockchain technology facilitates the establishment of a digital ledger of transactions, enabling audits and enhancing comprehension of network activities. Refer to Fig. 8.2 for the figure of a basic blockchain, to get a better understanding of the workings of blockchain technology.

Blockchain technology offers organizations the opportunity to make accurate and dependable assessments by creating a credible digital repository that contains financial transactions, processes, activities, contracts, and other pertinent data. Once included in the blockchain, data assume a persistent state that exhibits a high degree of resistance against tampering or hacking attempts. The utilization of blockchain technology has promise in improving operational efficiency and optimizing organizational frameworks, while concurrently heightening reliance on data security and integrity. The adoption of blockchain technology by organizations in diverse industries and business sectors has been driven by its ability to address a range of difficulties. These challenges include quality assurance, accounting process management, contract processing, supply chain optimization, and data protection.

The distinguishing characteristic of blockchain technology is its capacity to facilitate safe, transparent, and decentralized methods of record-keeping. The system functions as a type of distributed ledger technology, which serves to authenticate and record transactions within an unalterable and decentralized database. Each block

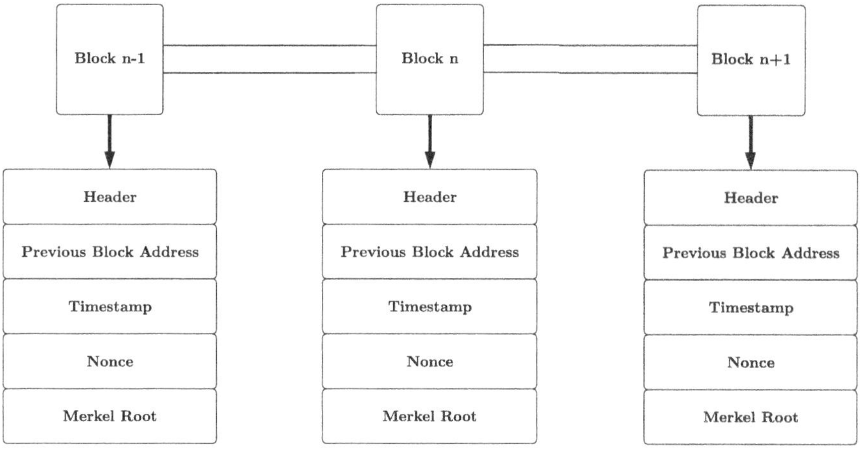

Fig. 8.2 Blockchain

inside the blockchain incorporates a cryptographic hash of the preceding block, leading to the establishment of an unalterable and uninterrupted ledger that comprehensively records all transactions taking place within the network. The fundamental basis of blockchain technology resides in its complexly crafted distributed database mechanism, which manages data and transactions within a network structure that lacks centralization. Distributed systems refer to interconnected computer networks that collaborate in a coordinated fashion to accomplish shared objectives. The primary distinguishing feature of blockchain technology is its capacity to distribute the ledger throughout the network, endowing every machine within the network with a duplicate of the ledger.

The significance of blockchain technology in enhancing distributed systems is highlighted by its decentralized and immutable ledger, the consensus mechanism it promotes, and the procedures of validation and verification. Blockchain technology, by the elimination of a central authority or intermediary, effectively provides a formalized degree of security and confidence inside dispersed networks. This characteristic renders it highly suitable for a wide range of applications, including but not limited to financial transactions, supply chain management, and digital identity verification.

3 Cloud Computing

Cloud computing, in its essence, refers to the provisioning of various computational services via the medium of the Internet, often referred to as 'the cloud'. This approach yields expedited innovation, pliable resource allocation, and the realization of economies of scale. Cloud computing can be defined as a wireless computing

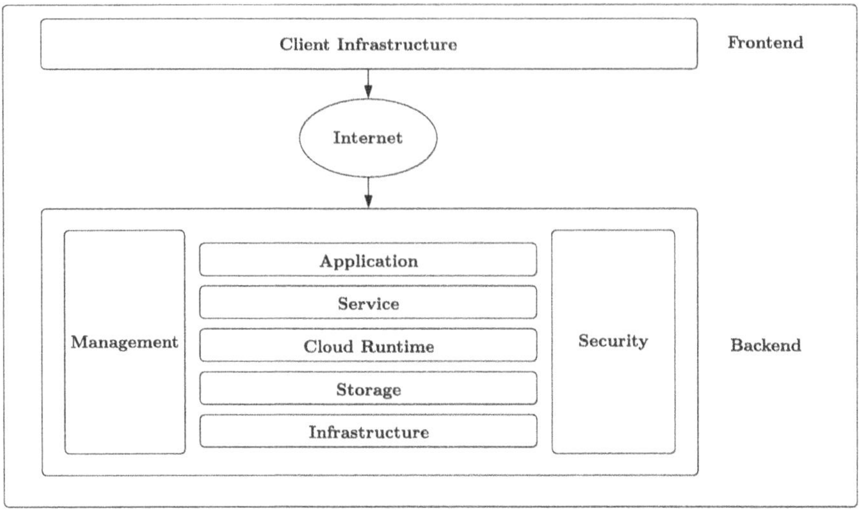

Fig. 8.3 Cloud computing paradigm

model that facilitates on-demand access to a shared pool of configurable computing resources, including servers, storage, databases, and software applications. Cloud computing functions by allowing companies to lease access to various resources from cloud service providers, avoiding the need to oversee their proprietary computational infrastructure or data centres. This approach has the advantage of sidestepping the initial expenses and complexities linked with the ownership and maintenance of IT infrastructure. Instead, enterprises remunerate for the resources they use on a consumption-based model. Figure 8.3 showcases the cloud computing paradigm to be referred to get a better look into the cloud technology.

The importance of cloud computing in this technology-driven age is transformative, redefining the paradigms of information access, storage, and utilization. The potential it embodies presents convenience and scalability, empowering individuals and businesses to harness computational resources with adaptability. However, this digital revolution necessitates a key focus on safeguarding and upholding the protection of the virtual systems upon which we rely. At the core of cloud computing lies a cardinal principle: the geographical location of the service and complex technical attributes, such as hardware specifications and operating systems, bear minimal significance to the end user. The metaphorical construct of the 'cloud' is adopted from earlier telecommunications network schematics, symbolizing the concept that geographical location is of little consequence; it is merely a conceptual aggregation of resources. While this simplification holds merit, for a multitude of users, the location of services and data remains a pertinent consideration.

Cloud computing, as a domain, includes an amalgamation of discrete constituent elements, each addressing distinct strata of the technological stack and catering to diverse usage scenarios. Let us explore some of the components to gain familiarity.

(1) **Infrastructure as a Service (IaaS):** IaaS extends the foundational building blocks of computing for rental, including physical or virtual servers, storage solutions, and networking infrastructure. This option garners interest from entities striving to craft applications from the ground up while retaining considerable control.
(2) **Platform as a Service (PaaS):** PaaS operates at a higher level than IaaS. It incorporates not only the underlying infrastructure like storage, networking, and virtual servers but also includes various tools and software requisite for developers to fabricate applications. This set might span middleware, database management systems, operating environments, and development utilities.
(3) **Software as a Service (SaaS):** SaaS involves the delivery of applications in the form of services, representing a manifestation of cloud computing that frequently interfaces with the common populace. The foundational hardware and operating system aspects are of little consequence to the end user, who accesses the service through web browsers or applications. This model often operates on a per-seat or per-user basis.

Cloud computing subsidizes a dynamic line for accessing computational resources, enabling heightened adaptability and scalability across a wide spectrum of industries.

3.1 Cloud Security

Cloud security is founded on three distinct operational environments within cloud computing: public cloud services, private clouds, and hybrid clouds. Cloud service providers (CSPs) administer public cloud services, which consist of infrastructure as a service (IaaS), software as a service (SaaS), and platform as a service (PaaS). Conversely, private clouds are supervised by a solitary entity, whereas hybrid clouds integrate elements from both public and private clouds. Two categories of cloud security mechanisms can be distinguished within this context: those supplied by cloud service providers (CSPs) and those implemented by consumers.

It is noteworthy to mention that security responsibility is seldom confined to the consumer or the CSP alone. In general, it involves a collective effort characterized by a model of shared responsibility. While not standardized, this model elaborates how the customer and the CSP are responsible for executing security duties. To ensure coverage, organizations utilizing cloud services must identify the security responsibilities delegated to their providers and those retained internally. The security responsibilities that are explicitly described between the CSP and the customer are outlined in this shared responsibility model. The level of customer responsibility typically increases from SaaS to PaaS to IaaS, with the service model SaaS, PaaS, or IaaS influencing the security responsibilities of CSPs. Inevitably, the accountability for servers and storage falls upon CSPs.

In addition to configuring the physical data centres, networks, and hardware components that support the infrastructure, they oversee and upgrade the underlying infrastructure. These responsibilities are predominantly associated with IaaS environments. CSPs take on additional obligations in PaaS environments, including the protection of runtime, networking, operating systems, data, and virtualization. In conjunction with these obligations, CSPs' SaaS packages contain application and middleware security. The precise distribution of security obligations may differ depending on the provider and the client. CSPs specializing in SaaS, for instance, might or might not provide customers with visibility into the security tools they employ. On the other hand, IaaS providers frequently incorporate security mechanisms into their offerings that grant customers access to and inspection of CSP security tools, frequently in the form of customer-alerting functionalities.

In IaaS cloud configurations, customers are typically accountable for the security of the application, middleware, virtualization, data, operating system, network, and runtime, in addition to the CSP's security controls. Customers can implement their set of tools into architectures such as Amazon Virtual Private Cloud (VPC)[1] or Microsoft Azure Virtual Network (VNet),[2] thereby introducing, replacing, or superimposing pre-existing cybersecurity mechanisms.

PaaS environments place the primary responsibility for application and middleware security on the clients, whereas IaaS environments involve comparatively lesser security obligations. Likewise, SaaS environments accommodate a reduced number of customer obligations. It is worth noting that regardless of the cloud delivery model, data security and identity and access management (IAM) consistently reside within the customer's sphere of influence. Also, encryption and compliance are customer obligations. However, further risk mitigation can be a complex process since CSPs have authority over the infrastructure in which consumer applications and data function. It is prudent to include IT security personnel in the early stages of evaluating CSPs and cloud services. Evaluation of the CSP's default security tools by security teams is required to determine whether additional in-house measures are required. To incorporate an organization's security tools into cloud environments, it is customary to implement virtual security instruments that operate on a network. This methodology grants security administrators the ability to customize particular security configurations and policy settings. Numerous organizations determine that it is economical to deploy comparable tools in both their local area networks (LANs) and public clouds, thereby facilitating the enforcement of security policies in both domains.

[1] docs.aws.amazon.com/vpc/latest/userguide/what-is-amazon-vpc.html.

[2] learn.microsoft.com/en-us/azure/virtual-network/virtual-networks -overview.

3.2 Advantages

Security is a requirement in cloud computing for safeguarding applications and data against malicious intent. Maintaining a robust cloud security posture not only ensures protection but also enables organizations to reap the well-established benefits of cloud computing. Likewise, cloud security itself offers advantages, including reduced upfront expenses, diminished ongoing operational costs, seamless scalability, heightened reliability and availability, and strengthened defence against Distributed Denial of Service (DDoS) attacks. Here are the prominent security benefits of cloud computing:

- **Lower Upfront Costs**: An advantage of cloud computing is the elimination of the need for dedicated hardware investment. This translates to substantial initial cost savings, while also allowing for enhanced security. Cloud Service Providers (CSPs) undertake security responsibilities, mitigating the necessity to use an internal security team to protect dedicated hardware.
- **Reduced Ongoing Operational and Administrative Expenses**: Cloud security contributes to decreased ongoing administrative and operational overhead. CSPs shoulder security needs, obviating the requirement to allocate resources for manual security updates and configurations. Expert CSP staff manage security matters, enhancing both security and cost-effectiveness.
- **Increased Reliability and Availability**: Ensuring swift and secure data access is crucial. Cloud security guarantees that authorized users can access data and applications promptly. This consistent availability empowers quick responses to potential security issues.
- **Centralized Security**: Cloud computing centralizes data, applications, and security endpoints. This centralized approach simplifies the management of security across diverse applications, devices, and data, strengthening efficiency in tasks such as disaster recovery planning, network event monitoring, and web filtering.
- **Greater Ease of Scaling**: Cloud computing accommodates agile scaling according to demands, accommodating expanded applications and data storage. Cloud security seamlessly scales alongside cloud services, allowing the integration of new features without compromising data security.
- **Improved DDoS Protection**: Distributed Denial of Service (DDoS) attacks pose significant threats to cloud computing. Cloud security effectively defends against these attacks by monitoring and mitigating them, thus strengthening cloud server resilience.

A lively café serves drinks without needing special home gear. Likewise, cloud computing helps users by giving them access to computer stuff online, without dealing with complicated server stuff. Cloud security means keeping cloud systems, apps, and data safe, so things stay private and available. It's a shared job between providers and users, showing a commitment to keeping things safe and working well in the world of tech.

In an analogy of the Cloud Cafe concept, cloud security operates as a critical guard, safeguarding customers' digital orders and personal preferences. Roughly, cloud security contains a set of policies, technologies, and practices that have been carefully constructed to form a protective barrier around cloud computing environments, applications, and data. The primary objective is to prevent unauthorized access, data theft, and potential destruction. With the escalating adoption of cloud computing, the significance of security concerns is magnified.

The shared responsibility model serves as a foundational principle, highlighting that while cloud providers are tasked with securing the underlying infrastructure, customers share the responsibility of actively safeguarding their applications, data, and user access. To adeptly traverse this complex environment, the security industry is dedicated to pioneering research and innovation, striving to create solutions beyond conventional security measures. Analogous to cafes diligently overseeing ingredient quality and order accuracy, cloud security utilizes a parallel set of measures aimed at guaranteeing the preservation of digital assets:

- **Continuous Monitoring and Threat Detection**: Comparable to a cafe manager overseeing the preparation of dishes, continuous monitoring and threat detection involve an assessment of the cloud environment. This involves the utilization of intrusion detection systems, security information and event management (SIEM) systems, and vulnerability scanners. These tools diligently observe and respond to potential security breaches, ensuring the integrity of the cloud environment.
- **Robust Authentication and Access Controls**: Analogous to cafes entrusting authorized personnel to handle orders, robust authentication and access controls serve a pivotal role in cloud environments. Effective access management is central to mitigating the risks associated with unauthorized access and potential data breaches. It ensures that only authorized users can access sensitive resources and data.
- **Data Encryption and Protection**: Much like cafes prioritize the confidentiality of customer preferences, cloud security centres focus on data protection. Encryption stands as a formidable defence mechanism against unauthorized entities. Whether during the transmission of data or while at rest within storage, encryption ensures that sensitive information remains inaccessible to unauthorized parties.
- **Disaster Recovery and Business Continuity Planning**: Similar to cafes planning for contingencies, cloud environments adopt strategies for disaster recovery and business continuity. These strategies guarantee uninterrupted operations even in the face of unforeseen challenges, thereby ensuring seamless service availability for users.
- **Compliance with Regulatory Requirements**: Just like cafes complying to health and safety regulations, cloud environments must conform to industry-specific regulations. Demonstrating compliance with regulations such as GDPR or HIPAA highlights a sense of accountability and responsibility in managing user data and promoting trust among customers.

- **Regular Security Assessments**: Analogous to cafes periodically evaluating their practices to maintain consistent quality, cloud security mandates regular security assessments and penetration testing. These assessments serve as a means to identify vulnerabilities and gauge the effectiveness of the implemented security controls, facilitating continuous improvement.

3.3 Security Challenges

At the customer's end, obligations include user management, securing cloud accounts, encrypting and preserving cloud-based data, and overseeing security compliance. Advanced challenges unique to cloud-native security and the various risk layers faced by present-day cloud-centric organizations include:

- **Increased Attack Surface:** The public cloud is an extensive and attractive target for hackers who exploit inadequately secured ingress points to disrupt cloud-based workloads and data.
- **Lack of Visibility and Tracking:** Cloud providers often retain control over the infrastructure layer in IaaS, and this lack of visibility extends to PaaS and SaaS models. Cloud customers struggle to distinguish and quantify assets within their cloud environments.
- **Ever-Changing Workloads:** Cloud assets are dynamically provisioned and decommissioned, rendering traditional security tools insufficient in safeguarding workloads that constantly shift.
- **DevOps, DevSecOps, and Automation:** In automated DevOps cultures, embedding security controls into code and templates during early development stages is critical. Post-deployment security modifications can compromise security posture and market readiness.
- **Granular Privilege and Key Management:** Loose configurations of user roles can grant unwarranted privileges. Misconfigured keys and privileges at the application level pose security risks.
- **Complex Environments:** Managing security uniformly across hybrid and multi-cloud settings necessitates tools that function seamlessly across various cloud providers and on-premise deployments.
- **Cloud Compliance and Governance:** Leading cloud providers comply to renowned accreditation programmes, but customers are responsible for ensuring their workloads align with regulatory standards. Continuous compliance checks and real-time alerts are essential due to the dynamic nature of the cloud.

Cloud security necessitates a complex relation of preemptive measures and responsive tactics, working in concert to safeguard the credibility and durability of digital assets within an environment defined by incessant advancement. The obstacles encountered in cloud security bear similarities to conventional cybersecurity concerns, including internal threats, data compromise, breaches, access management, deceptive tactics, malicious software, and additional challenges. The efficacy of data security in cloud contexts pivots on a confluence of elements: the classification and sensitivity of data, the structural blueprint of the cloud deployment, the suite of accessible security mechanisms, the profiles of sanctioned users, and the tiers of data accessibility. In the sphere of securing corporate data stored in the cloud, several best practices come to the fore:

- **Data Encryption**: Encryption should be applied to data in transit, data at rest, and data in use. This ensures that sensitive information remains confidential and inaccessible to unauthorized entities.
- **Identity Verification**: Utilizing two-factor authentication (2FA) or multifactor authentication (MFA) strengthens user identity verification, reducing the risk of unauthorized access.
- **Cloud-Edge Security Measures**: Implementation of cloud-edge security measures, such as firewalls, intrusion prevention systems (IPS), and antimalware solutions, provides an additional layer of defence against potential threats.
- **Ransomware Mitigation**: Isolating cloud data backups helps mitigate the risks posed by ransomware attacks, safeguarding critical data from potential encryption by malicious actors.
- **Data Visibility and Control**: Ensuring visibility and control over the location of data and imposing restrictions on data copying are essential practices in maintaining data integrity and preventing unauthorized dissemination.
- **Monitoring and Logging**: Logging and monitoring data access, additions, and changes facilitate the prompt identification of suspicious activities or unauthorized access attempts.
- **Leveraging Emerging Tools**: Consideration of emerging cybersecurity tools, such as network detection and response (NDR) and AI for IT operations (AIOps), can significantly enhance the overall security posture of the cloud environment.

By following the established best practices, organizations can proficiently handle the inherent challenges of cloud security, ultimately nurturing a secure and reliable digital infrastructure. Numerous tools commonly used in on-premises setups find equivalent utility in cloud computing, albeit with potential cloud-specific modifications. This toolkit contains fundamental components such as encryption, Identity and Access Management (IAM), single sign-on (SSO), data loss prevention (DLP), intrusion prevention and detection systems (IPS/IDS), and public key infrastructure (PKI). There exist specialized tools meticulously designed for the cloud environment, including:

- **Cloud Workload Protection Platforms (CWPPs):** These platforms ensure consistent safeguarding of workloads, including applications, virtual machines (VMs), and data.
- **Cloud Access Security Brokers (CASBs):** Positioned between cloud clients and services, CASBs enforce security policies, thus introducing an extra layer of security.

- **Cloud Security Posture Management (CSPM):** Including a set of security products and services, CSPM oversees cloud security by detecting misconfigurations and addressing compliance-related concerns.

Similarly, emerging cloud security paradigms such as Secure Access Service Edge (SASE) and zero-trust network access (ZTNA) are gaining prominence. Within this context, Security as a Service (SECaaS) emerges as a subset of software as a service, spanning categories such as IAM, DLP, web security, email security, security assessments, intrusion management, Security Information and Event Management (SIEM), encryption, business continuity/disaster recovery (BCDR), and network security. Services like firewalls as a service, cloud-based Virtual Private Networks (VPNs), and key management as a service (KMaaS) are included within these defined categories.

3.4 Research Avenues

The requirement to ensure the reliability and continuity of services in the face of potential disruptions has led to the exploration of research avenues aimed at augmenting the resilience of cloud-based systems. This pursuit is driven by the recognition that the reliance on cloud infrastructure has grown substantially across various domains, necessitating robust mechanisms to mitigate the impact of adverse events on the availability and functionality of cloud services. The research concerning the enhancement of resilience in cloud-based environments includes several essential directions that merit thorough analysis. These directions include various challenges, ranging from architectural considerations to algorithmic innovations. Understanding the mechanisms and potential solutions that can enhance cloud resilience requires an examination of these research trajectories.

Firstly, architectural design constitutes a fundamental domain for exploration. Research inquiries for the cloud include the crafting of resilient cloud architectures that can effectively accommodate fault tolerance and recovery mechanisms. Investigating the optimal distribution of resources, redundancy strategies, and failover protocols stand as integral aspects of this direction. An understanding of architectural choices that facilitate seamless adaptability to disruptions is a requisite goal.

Secondly, the analysis of fault detection and prediction mechanisms is critical. Advancements in this area necessitate the formulation of algorithms capable of swiftly identifying anomalies and deviations from the norm within the cloud ecosystem. By leveraging data-driven techniques and ML paradigms, it becomes feasible to preemptively recognize potential sources of failure, thereby allowing for proactive intervention and betterment of the situation.

Thirdly, the orchestration of resource allocation and load balancing represents a required route for research. The dynamic nature of cloud workloads calls for algorithms that can efficiently distribute tasks across available resources while adhering to constraints imposed by resource availability and QoS expectations.

An exploration of load-balancing strategies that can optimize resource utilization and alleviate performance bottlenecks is crucial. Likewise, data replication and synchronization merit dedicated analysis. Ensuring data integrity and consistency in distributed cloud environments demands innovative approaches for replicating and synchronizing data across geographically dispersed nodes. In-depth analyses of replication protocols, consistency models, and conflict resolution mechanisms are indispensable in this context.

4 Hardware Security

The urgency to solve security vulnerabilities has been heightened due to the rising interconnection and complexity of digital gadgets, which include personal computers and critical infrastructure components. All of these issues can be classified into two broad categories: obstacles connected to software and challenges related to hardware. Hardware security is distinguished by its interdisciplinary nature, as it includes the protection of both the physical and electronic components of computing and communication systems. This comprises both the tangible components and the software applications of these systems. Figure 8.4 showcases the components of hardware security.

The primary objective of implementing these security measures is to safeguard against unauthorized entry, manipulation, forgery, and other malicious actions that have the potential to compromise the accessibility, reliability, and confidentiality of hardware devices. An understanding of hardware security is crucial for the establishment of digital resilience, as it serves as the fundamental basis for the construction of dependable digital systems.

4.1 Significance

The significance of guaranteeing the dependability and safeguarding of hardware components has increased concomitantly with the extensive incorporation of digital systems into our everyday activities. The prominence of hardware as a target for attackers aiming to exploit software vulnerabilities is due to its essential involvement in the execution of software programs. The protection of hardware extends beyond the mere prevention of external threats. It also involves the establishment of resilient systems that can withstand both internal and external pressures. To maintain

Fig. 8.4 Hardware security components

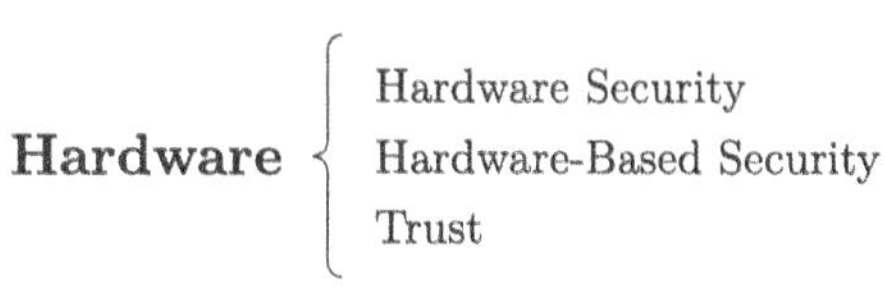

their operational effectiveness, digital systems must demonstrate resilience, even in the face of hardware failures, cyberattacks, or design deficiencies.

The comprehension of hardware security holds significance in digital technology. The escalating dependence on technology and the interconnectivity of gadgets have rendered hardware security a central problem. It is essential for both individuals and businesses to fully grasp the importance of preserving hardware components to safeguard sensitive information and effectively manage potential threats. Hardware security plays a vital role in ensuring digital resilience by offering numerous benefits in the protection of digital systems and data. There are several advantages associated with this.

- **Physical Protection:** Hardware security methods include inherent resistance against remote attacks due to their reliance on physical access to the device or component. The inclusion of an additional level of protection serves to enhance resilience against cyber threats.
- **Tamper Resistance:** The concept of tamper resistance is integral to numerous hardware security systems, as they are specifically engineered to impede attackers from engaging in physical manipulation or compromising the hardware without leaving behind any discernible signs of tampering.
- **Secure Key Storage:** The utilization of hardware security modules (HSMs) and trusted platform modules (TPMs) provides a robust solution for secure key storage, effectively safeguarding valuable cryptographic keys from potential vulnerabilities originating from software-based attacks.
- **Secure Boot and Firmware Verification:** The implementation of secure boot and firmware verification mechanisms in hardware can effectively establish a secure boot process, thereby restricting the execution of solely trusted firmware and software on a given device. Consequently, this serves as a preventive measure against the infiltration of malware and the unauthorized execution of code.
- **Cryptographic Acceleration:** Hardware-based encryption and decryption procedures are frequently characterized by superior speed and efficiency in comparison to software-based approaches, hence enabling enhanced security in communication and safeguarding of data.
- **Trust Anchors:** Trust anchors, specifically those that are hardware-based, such as hardware security tokens, serve as a reliable basis for authentication and access control. By utilizing these trust anchors, the potential for identity theft is mitigated, hence enhancing security measures.

Gaining a deep awareness of the complexities related to hardware security necessitates a thorough grasp of the diverse approaches and methodologies utilized in this field. Hardware security heavily depends on many technologies that have been specifically developed to provide physical protection. The methods discussed above include tamper-evident seals, secure memory compartments, and secure enclosures, which all serve to hinder unauthorized physical access to hardware. Authentication and authorization systems are critical components in several contexts. These strategies ensure that only authorized users or components can interact with the hardware, hence prohibiting the execution of unauthorized software or firmware. The achievement of this objective is facilitated through the deployment of access controls, which serve to guarantee that only individuals or components possessing appropriate authorization can interact with the hardware.

4.2 Security Challenges

In security, the utilization of hardware gives rise to a unique set of specific considerations. In comparison to software, which can easily undergo patching or updating procedures, hardware is susceptible to direct attacks due to its fundamental physical attributes. The utilization of specialized mitigation measures becomes crucial due to the presence of risks, including reverse engineering, supply chain attacks, and physical tampering. To effectively counter the continuous evolution of attacks over a protracted duration, security systems must exhibit robustness in the face of hardware component durability. This requirement necessitates that all activities about hardware security must consistently preserve a state of alertness and flexibility.

- **Obsolete Firmware:** The firmware in question is considered to be outdated and no longer in use. The firmware developed by domestic manufacturers of IoT and Industrial Internet of Things (IIoT) components may display a wide range of security vulnerabilities. The failure to prioritize patch management could potentially worsen the situation, hence increasing the likelihood of more vulnerabilities arising. The timely updating of firmware through the provision of security updates is crucial for protecting hardware environments that are susceptible to vulnerabilities.
- **Fault Induction:** The concept of fault induction refers to the deliberate introduction of faults or errors in a system or process. Adversaries intentionally induce or provoke hardware anomalies to impede the normal functioning of a system. Fault attacks possess the capability to compromise the overall security of a system by creating faults that facilitate unauthorized privileges or lead to the disclosure of confidential information. These kinds of attacks possess the capacity to exert an influence on interconnected devices that rely on compromised hardware.
- **Hidden Backdoors:** Hardware backdoors are covert vulnerabilities intentionally inserted into a device during its production phase. Threat actors can bypass authentication protocols to get unauthorized access. The challenges associated with mitigating hardware backdoors make them a significant channel for the installation of malicious software or harmful code, in contrast to software backdoors.
- **Eavesdropping Intrusions:** Eavesdropping attacks occur when an individual lacking adequate authority gains unauthorized access to equipment and covertly intercepts its information. Perpetrators can intermittently acquire unauthorized entry to diverse equipment, including card skimmers that are put on automated teller machines (ATMs) or point-of-sale (PoS) terminals, to extract sensitive

information. These attacks possess the capacity to compromise the integrity of data security and establish avenues for periodic transmission of data to the attacker.

- **Counterfeit Components:** The ongoing presence of counterfeit hardware poses a persistent and substantial threat, as it facilitates the convenient targeting of companies by malicious actors. Unauthorized devices that lack authorization from the original equipment manufacturer (OEM) have the potential to have backdoors and vulnerabilities. The vulnerabilities that have been identified possess the capacity to be exploited by malicious individuals, leading to unauthorized actions and unlawful penetration of organizational networks.
- **Inadequate Encryption:** The integration of enterprise devices into IP networks has become more prevalent, and it has been observed that a considerable proportion of these devices lack sufficient encryption requirements. The relevance of preserving operational technology devices interconnected with networks necessitates the use of robust encryption measures for both data at rest and data in transit. Insufficient implementation of encryption protocols can leave information vulnerable to unauthorized access by malicious entities within the network. Devices that lack encryption are subject to theft, potentially resulting in compromised data.
- **Unprotected Local Access:** The hardware used in the IoT and Industrial Internet of Things (IIoT) applications is often accessible via on-premises interfaces or local networks. Smaller organizations, in particular, may neglect the proper configuration and physical protection of these local access points. Insufficient control measures may render enterprise hardware susceptible to unauthorized individuals who can exploit and manipulate company systems with ease.
- **Tampering Attacks:** Tampering attacks refer to malicious activities that involve unauthorized alteration or modification of data or systems. These attacks involve the intentional alteration of a device's normal functions, hence enabling unauthorized users to bypass hardware restrictions. This goes beyond the act of covertly eavesdropping on conversations, as malicious individuals can influence the transmission between devices. Multiple methodologies can be utilized, including the introduction of malicious software or the exploitation of preexisting weaknesses, hence enabling man-in-the-middle attacks.
- **Persistent Default Passwords:** While it is true that companies that prioritize software security may occasionally overlook crucial hardware security practices, it is recommended to alter default passwords that come pre-installed on devices. Still, it has been observed that employees frequently persist in utilizing these passwords, particularly in the context of inexpensive IoT devices. In certain instances, default passwords can be physically inscribed on the device, hence granting unrestricted access to individuals near the device.
- **At-Risk Custom Hardware:** Many organizations rely on customized hardware solutions to meet their specific operational needs, such as data centres or speciality engineering projects. Manufacturers may not comply to rigorous criteria while evaluating the security of custom circuits and devices, in contrast

to mainstream hardware. This lack of compliance can potentially lead to the emergence of vulnerabilities.

Understanding the many methods of attack that can be manipulated by malicious actors is a critical part of hardware security. To mitigate the risk of physical component compromise during production and distribution, it is crucial to prioritize the maintenance of supply chain integrity through the implementation of verification procedures. Reverse engineering is a methodical process involving the disassembly and analysis of physical components to identify security flaws. This highlights the need to utilize effective obfuscation methods. The act of exploiting vulnerabilities through direct means requires the use of invasive probing techniques, which involve physically interacting with hardware components. This requires the integration of physical protective features together with tamper-evident designs.

4.3 Future

In recent times, there have been several inquiries conducted on the safety aspects of hardware components. The complex nature of hardware designs poses challenges in effectively identifying and addressing vulnerabilities. Ongoing research attempts in hardware security are focused on the identification of potential future threats and the development of innovative remedies.

- **Mitigating Side-Channel Attacks:** In cybersecurity, safeguarding against malicious attacks originating via alternative communication channels is important. Side-channel attacks, a matter of ongoing concern, refer to actions that exploit the inadvertent leakage of information through unanticipated side channels, such as power consumption or electromagnetic radiation. Academic researchers are currently engaged in the pursuit of novel cryptography methodologies and countermeasures at the hardware level. The primary objective of these attempts is to safeguard against potential attacks and enhance the overall security of hardware systems.
- **Detecting Hardware Trojans:** The identification of hardware Trojans, which refer to unauthorized modifications secretly introduced during the manufacturing process without the manufacturer's awareness, constitutes a significant subject of inquiry. Researchers are now engaged in the development of methodologies for the detection of hardware Trojans. Several approaches are being applied, such as scan-chain design, Bayesian modelling, and ML, to address the problem at hand.
- **Hardware for Post-Quantum Cryptography:** The potential widespread adoption of quantum computing could render existing cryptographic systems susceptible to security breaches. Research is currently underway to investigate the hardware implementation of post-quantum cryptographic algorithms, to guarantee long-term security in the quantum era.
- **Hardware-Assisted Virtualization Security:** The continuing study focuses on leveraging the capabilities of the underlying hardware to enhance the secu-

rity of virtualized systems. To safeguard virtualized systems from faults and vulnerabilities, the implementation of this approach involves the utilization of hardware-assisted isolation and secure memory management.

- **Supply Chain Integrity Verification:** Ensuring the monitoring and maintenance of the integrity and authenticity of hardware components is of importance within the supply chain. The research being undertaken suggests that to ensure the integrity of the supply chain, it is necessary to establish reliable verification methods. These methods may involve the use of blockchain-based solutions and hardware roots of trust.

Investing in research and implementing efficient hardware security solutions are crucial to mitigate emerging risks and challenges in modern computer systems. Enhancing the security measures of hardware components enables digital systems to maintain their reliability in the face of constantly evolving threats, hence enabling an augmentation of digital resilience.

5 Internet of Things (IoT)

The IoT has emerged as a crucial driver of innovation, in a period characterized by rapid technological progress, fundamentally transforming our interactions with the surrounding environment. The paradigm shift under consideration places significant emphasis on the concept of interconnectedness at an unparalleled magnitude.

The inception of the IoT may be traced back to the 1980s when a vending machine was connected to the Advanced Research Projects Agency Network (ARPANET). The word was initially created by Kevin Ashton in the year 1999. The concept of the IoT involves the integration of physical objects that are equipped with sensors, software, and network access. This integration facilitates the acquisition and transmission of data among various objects over the Internet. The IoT holds considerable potential to have a substantial impact since it can redefine several industries and bring about a revolutionary transition in our daily lives. However, the advancement encountered in this regard is not devoid of challenges, namely, about concerns of privacy and security. The increase in IoT devices has led to a substantial escalation in the potential for cyber vulnerabilities, necessitating the adoption of additional security measures.

The IoT refers to a dynamic network that facilitates the connection of physical objects, often referred to as 'things', that have been augmented with sensors, software, and the remarkable ability to communicate with other similar entities and external systems via the vast expanse of the Internet. However, the inherent significance of the IoT transcends mere connectivity. It functions as a catalytic force that is radically reshaping our interactions with technology, information, and the worldwide ecosystem. The IoT represents a paradigm shift that goes beyond conventional concepts of connectivity, ushering in disruptive times where the boundary between the physical and digital worlds becomes blurred. In the context

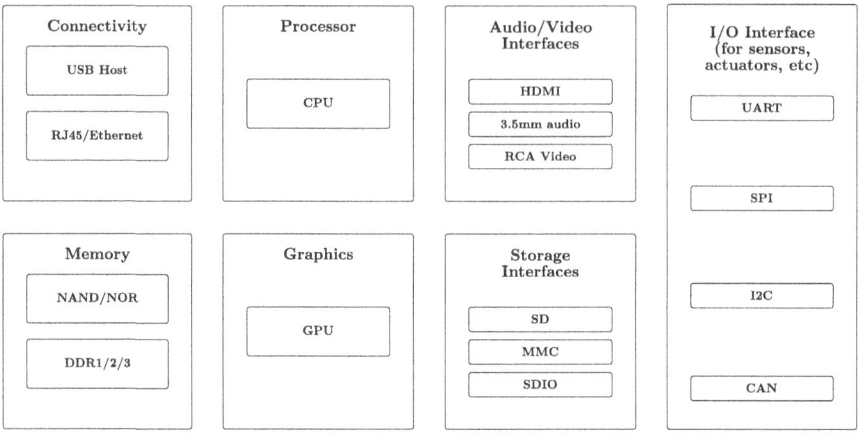

Fig. 8.5 IoT system

of this symbiotic relationship, commonplace objects transform, assuming the role of intermediaries for the transmission of information and awareness. Through their seamless integration inside digital surroundings, they promote a transformation in the very essence of human life. Refer to Fig. 8.5 to get a better look at the schematics of an IOT system.

5.1 Advantages

The IoT has made major strides, leading to an impact on our interactions with the surrounding environment. The advent of this interconnected network of gadgets and sensors has brought about a paradigm shift in technological advancement. In this section, we explore the numerous benefits that the IoT offers to individuals and society. These advantages include improved connection, data-driven decision-making, predictive maintenance, and heightened operational efficiency. Through these benefits, we can develop a better comprehension of the transformative impact that the IoT is having on various sectors and enhancing our day-to-day encounters.

- **Enhanced Connectivity:** The concept of enhanced connectivity refers to the improvement and expansion of communication and networking capabilities. The IoT facilitates a degree of interconnectedness that was previously inconceivable. The technology facilitates seamless communication across devices, resulting in improved efficiency and convenience across multiple domains of human existence. One instance of the application of smart homes involves the capability to automatically regulate heating and lighting systems in response to occupancy levels, hence resulting in energy conservation and enhanced comfort.

- **Data Collection and Analysis:** The process of data collection and analysis involves the accumulation of extensive quantities of data by IoT devices, derived from their immediate environment and operational activities. The data presented in this context holds significant value for businesses and organizations, as it facilitates decision-making based on data, optimization of processes, and perhaps the creation of novel services. One example of the application of IoT sensors in agriculture involves the provision of real-time data about soil moisture levels. These data enable farmers to make informed and accurate decisions on irrigation practices.
- **Predictive Maintenance:** Predictive maintenance is considered a significant benefit of the IoT due to its inherent capability. IoT-enabled machinery possesses the capability to autonomously monitor its performance and identify potential difficulties in advance, hence mitigating the risk of breakdowns. Not only does this practice decrease periods of inactivity, but it also improves overall efficiency and prolongs the operational lifespan of the equipment.
- **Operational Efficiency:** The operational efficiency of the IoT is primarily driven by the implementation of real-time monitoring and control systems. Industries such as manufacturing can derive advantages from the IoT through the monitoring of equipment performance, the optimization of supply chains, and the facilitation of timely maintenance. These benefits ultimately lead to heightened levels of operational efficiency.
- **Security:** IoT devices can include powerful security mechanisms, such as encryption and authentication, to safeguard data and devices against potential cyber threats. The significance of this matter lies in the fact that IoT devices frequently manage confidential data. In the healthcare sector, it is crucial to ensure robust security measures for IoT-enabled medical devices to protect sensitive patient data.

5.2 *Drawbacks*

With the increasing integration of the IoT into various aspects of our daily lives, it is crucial to carefully examine the challenges and drawbacks associated with this phenomenon. This section explores the negative aspects of the IoT, focusing on concerns related to data privacy, security vulnerabilities, difficulty in achieving interoperability, the issue of data overload, and the financial implications involved with its deployment. Through a careful examination of these drawbacks, we can acquire significant insights into the complex nature and inevitable compromises associated with this revolutionary technology.

- **Privacy Concerns:** The widespread presence of the IoT gives rise to substantial concerns over privacy. IoT devices can gather and disseminate substantial quantities of personal data, frequently without users possessing complete awareness or granting explicit agreement. This situation prompts inquiries regarding the

individuals or entities that possess the ability to retrieve this data and how it is utilized.
- **Security Risks:** The expansion of IoT devices poses a significant security threat as it expands the potential targets for malicious activities conducted by hackers. The exploitation of vulnerabilities in these devices can serve multiple objectives, such as initiating cyberattacks or illicitly acquiring confidential data. The continuous difficulty is in the assurance of security for IoT devices.
- **Interoperability Challenges:** One of the primary challenges in IoT ecosystems is the issue of interoperability, which arises due to the inclusion of devices from multiple manufacturers that utilize different communication protocols. The complexity of achieving flawless interoperability and compatibility across different technologies can give rise to integration challenges.
- **Data Overload:** As mentioned previously, the rapid increase in IoT devices has resulted in a substantial influx of data, which has the potential to inundate organizations. The effort of managing, storing, and analysing these data in a meaningful manner can require a significant allocation of resources.
- **Costs:** The implementation of IoT infrastructure, including the creation of devices and the establishment of supporting networks, can incur significant costs. The financial implications associated with this expense may serve as a barrier to the widespread implementation of IoT technology within specific businesses or geographical areas.

5.3 Security Challenges

The exponential growth of IoT devices has introduced a novel phase characterized by enhanced connectivity and ease. But, it has also posed several severe issues that necessitate our focused attention and innovative solutions. In this part, we analyse the varied issues encountered by IoT ecosystems. The significance of resilient solutions in the constantly changing IoT is highlighted by various problems, including cybersecurity vulnerabilities, data privacy issues, and the complex task of guaranteeing seamless device interoperability.

- **Cyber Vulnerabilities:** The exponential growth of IoT devices has resulted in the emergence of multiple vulnerabilities that can be exploited by malicious actors for cyberattacks. These devices possess potential weaknesses that might be exploited by hackers to attain unauthorized access or initiate attacks, hence necessitating the implementation of security measures.
- **Data Privacy:** The preservation of data confidentiality gathered by IoT devices is a challenge in data privacy. The act of gaining access to sensitive information without proper authorization can result in privacy breaches, which can have consequences for both individuals and organizations.
- **Malware and DDoS Attacks:** IoT devices are susceptible to malware infestations and distributed denial-of-service (DDoS) attacks. Devices that have been

compromised possess the capability to initiate extensive attacks, hence causing disruptions to Internet services and networks.

- **Authentication:** The problem lies in the authentication process, which aims to restrict access to IoT devices exclusively to authorized entities. Robust authentication measures are needed to prevent unauthorized access and safeguard against potential data breaches.
- **Integration Risks:** The integration of IoT devices with pre-existing systems may potentially give rise to security problems, provided that the process is not executed meticulously. Inadequately integrated devices possess the potential to generate vulnerabilities that malicious actors can exploit to compromise the entirety of the network.

5.4 Research Avenues

Within the dynamic domain of the IoT, it is crucial to emphasize the importance of prioritizing digital resilience and security. This subsection provides an introduction to the current research directions that are essential for enhancing the robustness of IoT ecosystems. In the process of adapting to this ever-changing context, we delve into the fundamental domains of continuous research and attentiveness, utilizing the potential of AI and ML for proactive defence, strengthening security measures, ensuring the privacy of data, and effectively managing the complexities of integration. In aggregate, these channels play a substantial role in the progression of IoT technologies and the safeguarding of digital resilience. In an ever-changing environment filled with possible threats, it is crucial to actively pursue ongoing research and be consistently alert. Maintaining a state of alertness regarding emerging threats and vulnerabilities is crucial to proactively anticipate and address new threats within the ecosystem of the IoT.

The application of AI and ML holds significance in safeguarding the security of the IoT. These technologies can analyse large amounts of data, enabling the prompt identification and recognition of possible threats before they reach significant levels. The implementation of this proactive approach assists in enhancing and strengthening digital resilience. Enhancing the security of the IoT necessitates a continuous attempt to improve and adjust measures of the dynamic and evolving set of risks. To sustain a competitive advantage over potential adversaries, organizations must consistently improve their security processes, implement requisite updates, and expand their defensive methods.

The preservation of user data and compliance with regulatory standards are of importance within the IoT ecosystem. The implementation of encryption, access controls, and anonymization techniques can effectively safeguard the integrity of data, hence reducing the potential for unauthorized access. The integration of IoT technologies with pre-existing frameworks poses a variety of complex issues. It is crucial to engage in proactive management and mitigation of potential interruptions and vulnerabilities that may arise during this procedure to ensure a seamless and secure integration.

6 Cyber-Physical Systems (CPS)

The field of CPS has experienced a significant advancement in terms of interconnection in recent years. The integration of digital and physical domains has exhibited its significance in diverse sectors including transportation, essential infrastructure, healthcare, and manufacturing. For reference, one would be advised to focus their attention on Figs. 8.6 and 8.7, which depict the schematic and interactions of CPS, respectively.

The operational dependability, security, and functioning of modern civilization are heavily reliant on the resilience of CPS in mitigating cyberattacks and disturbances. This statement carries significance in light of ongoing occurrences. CPS represent the integration of traditional physical processes with modern computing, communication, and control technology.

The convergence of these factors gives rise to the development of novel applications, including medical gadgets that enable continuous monitoring of patients, smart grids that enhance the distribution of energy resources, and autonomous cars

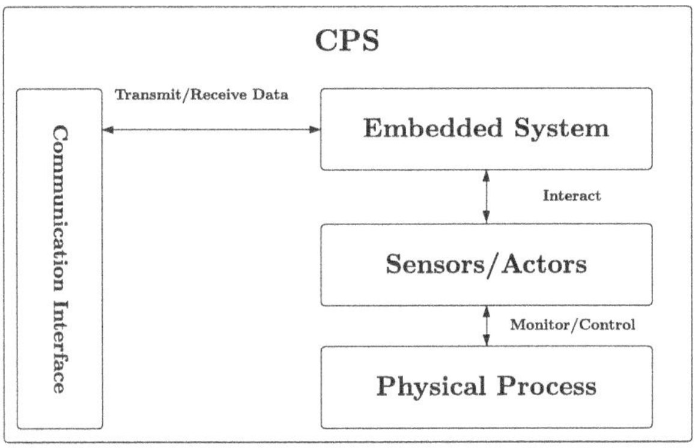

Fig. 8.6 Cyber-Physical Systems (CPS)

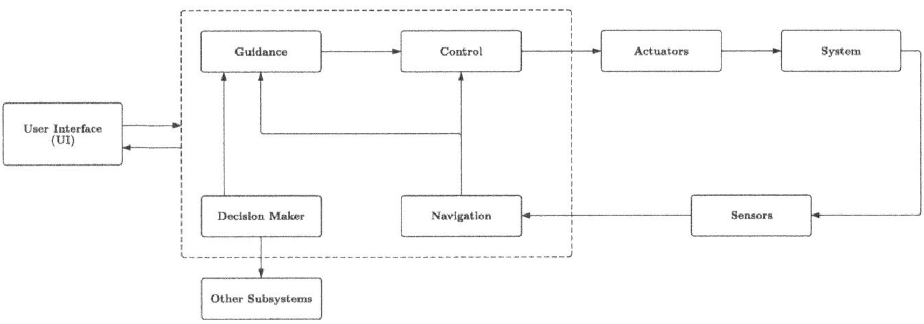

Fig. 8.7 Interaction of CPS

that effectively traverse complex real-world settings. The results produced by these systems are not exclusively determined by either the digital or physical components in isolation but instead emerge from the complex interaction between these two diverse categories of components. Still, as these systems advance and grow, the difficulties related to their cybersecurity and overall resilience also increase. When dealing with CPS, a common question is how it differs from IoT. To provide clarity, we have included a comparison in Table 8.2.

6.1 Significance

The recognition of the extensive interconnectedness between the digital and physical domains within CPS highlights the clear importance of digital resilience. In contrast to failures observed in typical computing systems, faults in CPS can have significant repercussions in the physical domain, as opposed to primarily impacting data and software. Ensuring the preservation of both functioning and safety is of significance for these systems. The concept of 'digital resilience' includes the strategies, mechanisms, and procedures that empower CPS to endure, adjust to, and recuperate from a diverse range of disruptions. The disruptions include cyberattacks executed by proficient opponents, complex technology faults, and other unanticipated catastrophes. There are notable significances of CPS for the enhancement of digital resilience:

- **Risk Assessment and Management:** Risk assessment and management are a crucial aspect of various fields and industries. It involves the identification, analysis, and evaluation of potential risks that may arise in a particular context. By systematically assessing these risks, organizations can develop strategies and implement measures to mitigate. The core principle of digital resilience lies in the adoption of an approach to assess and address potential threats. The procedure involves the identification of prospective vulnerabilities, a thorough understanding of their potential consequences, and the development of strategies to efficiently reduce these risks. The risk management methodology includes the identification and assessment of both physical vulnerabilities in the system's components and online vulnerabilities inside the digital domain.
- **Early Threat Recognition:** Early threat recognition refers to the ability to identify potential threats or dangers at an early stage. The recognition and alleviation of possible threats are an essential element of digital resilience. It is crucial to deploy reliable real-time monitoring systems inside the CPS environment, which make use of sensor networks, intrusion detection systems, and sophisticated algorithms capable of detecting anomalies and possible risks. The prompt recognition of a given circumstance allows for the prompt execution of corrective measures, hence reducing the potential magnitude of negative consequences.

Table 8.2 Comparison of CPS and IoT

	Cyber-Physical Systems (CPS)	Internet of Things (IoT)
1. Definition	CPS refers to systems that integrate computation, communication, and control to monitor and manage physical processes	IoT refers to a network of interconnected devices that communicate and share data to perform various tasks without direct human intervention
2. Integration of physical and cyber elements	CPS tightly integrates physical processes with computing and communication elements to create a unified system	IoT connects physical devices to the Internet, enabling them to exchange data and information
3. Control and automation	CPS emphasizes control and automation of physical processes through real-time feedback loops	IoT devices may involve automation, but the primary focus is on data collection, sharing, and remote monitoring
4. Real-time processing	CPS often requires real-time processing capabilities to respond quickly to changes in the physical environment	IoT applications may involve real-time processing, but it is not a strict requirement for all devices
5. Complexity and scale	CPS systems are often complex and may involve a smaller number of interconnected components	IoT involves a vast number of diverse devices, ranging from simple sensors to complex systems, contributing to scalability challenges
6. Industry focus	CPS is commonly associated with industrial and critical infrastructure applications	IoT has a broader application focus, including home automation, healthcare, agriculture, and various consumer-oriented domains
7. Dependence on connectivity	CPS systems may operate in offline mode, but connectivity enhances their capabilities	IoT devices heavily rely on Internet connectivity for data exchange and functionality
8. Safety and security concerns	CPS places a strong emphasis on safety due to its application in critical infrastructure	IoT faces security challenges, with a focus on securing a large number of diverse devices from potential cyber threats
9. Communication protocols	CPS often uses specialized communication protocols tailored to the specific application domain	IoT devices may use standard communication protocols like MQTT or HTTP for interoperability
10. Latency requirements	CPS may have strict latency requirements for real-time control and decision-making	IoT applications may tolerate higher latency, especially in scenarios where real-time responses are not critical

- **Adaptive Architectures:** Adaptive architectures refer to the design and implementation of systems that are capable of adjusting and responding to changing conditions and requirements. The design principles of resilient computer operating systems place a high emphasis on the qualities of adaptability and redundancy. These systems have been designed with particular engineering techniques to adapt their operations in real time to successfully handle disruptions. This ensures the uninterrupted functioning of critical operations, even in situations involving component failures or other disturbances. The accomplishment was achieved by the utilization of system architectures that hold the ability to adjust their operations in reaction to disruptions. The architectural ideas shown here enhance the capacity of CPS to navigate complex and uncertain settings while maintaining the integrity of critical operational elements. This is accomplished by the efficient redistribution of resources or the activation of alternative components.
- **Secure Communication Networks:** The topic of secure communication networks is of importance in information technology. The successful functioning of a Cyber-Physical System (CPS) relies on the construction of reliable and secure networks that enable communication between its digital and physical components. To address the potential threat of attacks that abuse communication channels, it is feasible to safeguard data transfers by utilizing protective measures such as redundancy, encryption, and authentication. Resilient communication networks frequently utilize these strategies. To guarantee the security and dependability of interactions among the interconnected components of CPS, it is crucial to establish safeguards that effectively protect the integrity of communication pathways.
- **Regulations and Standards:** The topic of regulations and standards is of significance in various fields and industries. The incorporation of regulations and industry norms greatly strengthens the resilience of CPS. These standards establish a collective framework for protocols and practice, guaranteeing uniform compliance with security and resilience measures across the whole sector. CPS engineers demonstrate the proficiency to design systems that follow established standards and successfully mitigate emerging risks, as they conscientiously follow industry best practices and assure compliance to established norms.

The capability of CPS to sustain continuous operations in the face of severe attacks or unanticipated technological barriers is a notable benefit. Resilient CPSs are purposefully designed to provide uninterrupted functionality despite disruptions, thereby mitigating possible disturbances that could otherwise result in prolonged periods of interoperability. The accomplishment of this objective is facilitated through the utilization of engineering principles and methodologies. The continuous and uninterrupted nature of this continuity results in the provision of uninterrupted services and sustained performance, hence boosting the overall dependability and reliability of the system. The advantages of developing digital resilience within CPS are manifold.

- **Reduced Downtime:** The recovery stages of traditional systems may include prolonged durations of system inactivity. However, it is important to note that resilient CPSs possess certain attributes, such as adaptability and redundancy, which effectively reduce the length of downtime. These systems exhibit the capacity to make real-time adjustments to their operations, including altering process paths, reallocating resources, or activating alternate components, in response to the current circumstances. The acceleration of the recovery rate due to this causal relationship serves to mitigate any disruptions to corporate operations. Due to the implementation of this efficient recovery mechanism, there has been a significant decrease in both operational inefficiency and negative consequences.
- **Improved Security:** The incorporation of enhanced security measures is an inherent benefit of digital resilience, hence further contributing to this advantage. Resilient CPSs possess the capacity to proactively identify and mitigate emerging threats through the integration of real-time monitoring and detection technologies into their architectural framework. As a result of monitoring, there is a capacity to mitigate any potential risks, strengthening the overall security posture of the system. The significance of this advantage is particularly critical given the increasing sophistication of cyberattacks, as the capacity to promptly detect and eliminate threats can significantly influence the outcome.
- **Sustaining Long-Term Survival:** The lasting operational sustainability of CPS can be linked to their inherent adaptability, which empowers them to effectively endure digital adversities and maintain robustness. These systems demonstrate a notable level of efficiency, while still being relevant within an ever-changing technical environment. Human beings have the inherent ability to adapt easily to novel situations and rapidly embrace creative technology breakthroughs. The usefulness of CPS lies in its ability to effectively navigate through dynamic environments filled with both risks and opportunities. This highlights the robustness of CPS in the digital domain.
- **Risk Management:** Risk management is a crucial aspect of every organization's operations. It involves identifying, assessing, and mitigating any risks that may impact the achievement of organizational objectives. By doing so, CPS engineers can proactively develop solutions aimed at minimizing risks and supporting the resilience of systems. This is achieved through conducting thorough evaluations of vulnerabilities and making estimates of future disruptions. The ability to identify potential vulnerabilities and anticipate future risks enables the development of countermeasures that reduce the likelihood and consequences of disruptions. The proactive technique discussed is in line with the broader concept of digital resilience, which asserts that problems should be detected and resolved early on to prevent their progression into more serious concerns.

6.2 Challenges

The concept of digital resilience is greatly influenced by CPS, which emphasizes the interconnectedness of various functionalities and the complex dynamics that emerge from the interaction between technology, finance, resources, and the achievement of optimal performance. The analysis of the obstacles associated with enhancing the resilience of CPS against the ever-changing threats carries great importance.

- **Design Complexities:** The difficulties linked to the planning and execution of digital resilience in CPS should not be underestimated. CPS is formed by the amalgamation of many technological elements, including digital control systems, communication networks, physical processes, and software applications. To attain equilibrium of diverse components while concurrently ensuring their durability, a thorough comprehension of many disciplines is required. Within the framework of the design process, a lack of thorough understanding regarding the complex interrelationships among various components has the potential to create vulnerabilities and points of failure.
- **The Financial Implications:** In the attempt to achieve digital resilience, significant financial investments are often necessary. The incorporation of redundant processes, real-time monitoring systems, improved security measures, and adaptive components all necessitate substantial initial investments. It is crucial to actively participate in continuous maintenance and upgrades to proactively remain ahead of constantly evolving threats and developments in technological capabilities. The justification for incurring costs related to better security and uninterrupted operation is in the benefits they offer. However, it is necessary to carefully assess their influence on the budget.
- **Resource Allocations:** Achieving a balance between the allocation of resources to support resilience and the allocation of resources for other critical domains may be a difficult task. While it is important to allocate resources towards improving redundancy and adaptability, it is essential to recognize the presence of a trade-off that needs to be considered. The prioritization of resources for resilience measures may result in a redistribution of resources from other essential domains, thereby affecting the overall effectiveness of the CPS system. To maintain the integrity of CPS operations, it is crucial to apply careful decision-making when seeking to enhance resilience, to minimize any potential negative effects on other essential components.
- **False Positives:** Real-time monitoring systems are necessary for the detection of potential threats. However, they provide difficulty due to the possibility of false positives and false negatives. The phenomenon of false positives arises when benign occurrences are mistakenly classified as possible threats, resulting in costly disruptions and inefficient allocation of resources during the implementation of response measures. On the other hand, a false negative occurs when the system is unable to correctly detect authentic threats, exposing it to potential harm. Monitoring systems encounter a continual issue in identifying

the ideal balance between sensitivity and specificity, hence achieving an optimal equilibrium between the two.

The strategies used by threat actors, including both individuals with evil intent and organized groups, are continuously evolving and improving. The actors use the continuously evolving state of technology to augment the intricacy of their attack vectors. The task of individuals tasked with identifying potential risks, formulating effective strategies to minimize these threats, and implementing defensive measures is significantly hindered by the inherent adaptability of these threats. The growing necessity for cybersecurity solutions that exhibit comparable adaptability to the ever-changing threats is becoming increasingly evident over time.

The existence of vulnerabilities inside CPS gives rise to an increased level of complexity as a result of the complex interaction between their digital and physical elements. Despite the improvements in functionality and efficiency that arise from this integration, it also gives rise to a complex network that is characterized by a multitude of possible vulnerabilities. The existence of a single flaw in a seemingly isolated element of the system has the potential to trigger a series of failures that spread throughout the entire system. With the complex network of interconnected links within CPS systems, it is essential to establish thorough security rules and guidelines that appropriately recognize the interdependence of these systems.

It is also essential to recognize that internal threats within a company are distinct and unparalleled, but it is indeed accurate that exterior threats are a noteworthy cause for concern. Individuals who have authorized access to CPS systems and possess insider knowledge can exploit their positions to breach security measures. This phenomenon has the potential to manifest either through intentional actions or unintentional occurrences. These individuals may include personnel, subcontractors, or anyone who possesses elevated system privileges. They may have an increased level of system privileges. To address internal threats emerging from within an organization, it is crucial to implement strong access control mechanisms and develop a security-focused culture among employees who possess varying levels of system access. The objective of this method is to improve the capacity to detect and fend against these threats. The task of achieving an ideal equilibrium between security and trust is a necessary undertaking.

The integration of real-time monitoring and data-gathering techniques has significantly contributed to the enhancement of threat detection capabilities in CPS. The implementation of these technological developments has concurrently generated concerns over the protection of personal privacy. The utilization of these strategies may infringe upon the privacy of individuals who engage with Child Protective Services (CPS) systems to differing extents. The difficulty of effectively balancing proactive threat detection with the protection of an individual's privacy rights presents a significant challenge. The integration of ethical problems holds significance in the advancement of technologies that offer efficient risk detection while simultaneously safeguarding the fundamental rights of humans.

6.3 Research Avenues

CPS is a relatively new topic, and its impact on digital resilience highlights the significance of the additional analysis. One potentially fruitful area of analysis is the examination of behavioural data within the context of CPS. The utilization of advanced approaches, namely, those that incorporate ML techniques, presents substantial opportunities for augmenting the security of CPS. These methodologies can identify and analyse behavioural patterns present in CPS systems, hence facilitating the identification of even minor deviations from established norms. The capacity to effectively detect abnormalities is of importance in the dynamic environment of CPS. Advancements in this field have the potential to substantially decrease the occurrence of both false positives and false negatives, hence enhancing the effectiveness of response attempts and facilitating a more precise distribution of resources towards authentic threats.

The continued analysis of secure communication protocols and encryption technology remains of importance. Adversaries consistently refine their methodologies for infiltrating communication channels, demonstrating their ability to adapt to the dynamic and evolving nature of cyber dangers. The preservation of data integrity and secrecy during transfers inside CPS necessitates the development of cryptographic underpinnings and communication protocols. The establishment of a safe CPS environment necessitates the utilization of robust encryption technologies and secure data transfer protocols.

An additional field of research that shows promise involves the augmentation of collaboration dynamics between humans and robots within complex CPS. The inclusion of human expertise in the assessment of complex situations can yield a substantial competitive edge. The achievement of this objective can be facilitated by the advancement of interfaces that are easy for users to navigate, the creation of powerful tools for visualizing data, and the implementation of decision-making procedures that incorporate human operators. The integration of many components can have a substantial impact on the reduction of time required for incident response and recovery, particularly in critical scenarios that necessitate swift decision-making. The integration of human intuition with computing efficiency is demonstrated to be a mutually beneficial and essential amalgamation.

The attainment of digital resilience necessitates the construction of regulatory and legislative frameworks. The task of achieving a harmonic equilibrium between promoting innovation and maintaining security is a complex one that necessitates thorough planning. The implementation of guidelines that support industry standards, ease the interchange of information, and promote the adoption of best practices helps develop a standardized approach to resilience in CPS across diverse industries. Regulatory frameworks must adapt in parallel with the advancement of emerging technologies and the emergence of new dangers, to maintain their efficacy in protecting CPS.

7 Autonomous Systems

Autonomous systems have emerged as significant advancements with the potential to revolutionize both businesses and societies in the current times. These systems include several applications, such as self-driving cars, unmanned aerial aircraft, computerized production lines, and advanced infrastructure. They incorporate AI and possess the ability to make independent decisions. Without a doubt, the integration of autonomy presents new challenges, particularly in ensuring their ability to withstand digital threats, despite the clear potential for enhanced skills, reduced human mistakes, and higher efficiency. Autonomous systems are dependent on advanced algorithms and sensors for seeing and engaging with their surroundings and are distinguished by their capacity to operate without continuous human oversight. The establishment of a robust digital infrastructure that can effectively survive a diverse set of cyber threats, technological challenges, and unanticipated events is important to achieve this degree of autonomy. The importance of the resilience of these systems in the face of setbacks and their capacity to maintain functionality becomes more pronounced as they become more widespread in critical sectors such as transportation, healthcare, and defence.

Autonomous systems exhibit the capacity to see and interact with their surroundings through the use of complex algorithms and sensors, hence functioning independently without the requirement of constant human supervision. The capability represents a defining characteristic that sets autonomous systems apart from non-autonomous systems. The establishment of a resilient digital infrastructure is of importance for the advancement of these systems, as it necessitates the ability to endure diverse cyber-attacks, technological glitches, and unanticipated occurrences. The increasing presence of autonomous systems, especially in key sectors such as transportation, healthcare, and defence, highlights the growing need for their resilience and ability to maintain uninterrupted functionality in the face of setbacks. The implementation of these technologies has the potential to bring about substantial transformations in several industries through the reduction of human dependency and the mitigation of potential errors. The capacity of autonomous systems to do real-time evaluations is attributed to their effective processing of significant volumes of data, facilitating the attainment of elevated degrees of precision and speed.

As a result, this phenomenon leads to increased levels of efficiency and productivity. For instance, the use of advanced sensors in self-driving vehicles allows them to proficiently manoeuvre complex traffic situations, surpassing the ability of human drivers. As a result, this reduces the probability of accidents that arise due to human mistakes. In a similar vein, unmanned aerial vehicles (UAVs) demonstrate the capacity to perform jobs with heightened precision and facilitate the swift delivery of medical supplies to regions significantly impacted by natural disasters. These attempts involve the analysis of remote areas.

The progress of autonomous systems requires a fundamental shift in the manner in which robots engage with and navigate through their environment. These systems achieve their objectives by the integration of AI, sensing technologies, and complex computational methods in an autonomous manner. They possess the ability to comprehend data, adjust to changing situations, and perform tasks without constant human oversight, hence demonstrating autonomy. The concept of autonomy can be analysed by breaking it down into its fundamental components, with the central component being the capacity to see, comprehend, and react to the surrounding environment without substantial human involvement.

The salient features of autonomous systems reside in their capacity to efficiently process and analyse substantial quantities of data in real time. These systems rely on sophisticated algorithms that allow them to make judgements by analysing patterns, trends, and sensor data. The feature confers autonomous systems with the capacity to navigate complex surroundings, detect barriers, and perform appropriate actions to achieve their goals. However, the increased capacity for making judgements also presents issues in comprehending the underlying reasoning behind these conclusions, which is usually referred to as explainable AI. The establishment of transparency and interpretability in the decision-making processes of autonomous systems is crucial to developing trust and responsibility in their development.

7.1 Significance

Within autonomous systems, the concept of digital resilience involves a more approach in contrast to traditional cybersecurity measures. The concept of digital resilience refers to the ability of a system to adapt, recover, and maintain its functionality in the face of difficult situations. The main focus of cybersecurity revolves around the mitigation of attacks and unauthorized access, while digital resilience relates to the assessment of a system's ability to effectively respond to these threats. To achieve digital resilience, it is crucial to implement methods that go beyond simply preventing data breaches. The strategies should be formulated in a manner that promotes efficient recovery and continuous functioning, particularly in the face of unforeseen obstacles. The increasing global dependence on autonomous systems has led to a heightened recognition of the importance of digital resilience as a crucial factor in effectively utilizing their capabilities.

To uphold the digital resilience of autonomous systems, it is crucial to embrace a proactive stance in the identification and mitigation of vulnerabilities, as well as in the preparation for a diverse set of potential outcomes. This involves not only the mitigation of cyberattacks but also the development of solutions for the prompt restoration of services following outages. To prioritize passenger safety, autonomous vehicles must possess the necessary capacity to successfully adapt to unexpected situations, such as sensor malfunction or abrupt weather changes. To accomplish this goal, it is crucial to establish systems that contain the capacity to identify deviations from the norm, formulate tactics to minimize the negative consequences of interruptions, and sustain consistent functioning.

The incorporation of autonomous systems in various sectors offers a multitude of benefits and has the potential to bring about substantial changes in both residential and professional settings. One of the primary benefits they offer is the capacity to mitigate the occurrence of accidents from human controllers, which is of significance, especially in transportation. Autonomous vehicles have the potential to greatly mitigate the frequency of accidents caused by variables such as driver distraction, fatigue, or decreased cognitive abilities. The rapid reaction in real-time situations is integral in improving street safety and reducing the death rate associated with vehicular crashes. The implementation of autonomous systems has the potential to significantly transform supply chains and logistics. Efficient and fast transportation of goods to remote or inaccessible regions can be accomplished through the utilization of drones equipped with autonomous navigation capabilities. The utilization of robots with sophisticated autonomous capabilities has promise for providing reliable and precise manufacturing processes in the industrial sector.

This has the potential to decrease the occurrence of errors and minimize the development of waste. Enhanced dependability and efficiency can offer substantial benefits to the medical sector, electronics company, and automobile production sector, making them extremely attractive fields. It is anticipated that these enhancements will lead to improved product quality and decreased production expenses. Even the exploration of outer space heavily depends on the effectiveness of autonomous systems. Autonomous robotic rovers and spacecraft demonstrate the ability to navigate difficult terrains and perform tasks under adverse circumstances, as evidenced by their utilization on the surface of Mars and remote asteroids. The ability of these systems to collect data, perform tests, and provide immediate evaluations has revealed hitherto unimaginable opportunities for investigating and obtaining knowledge.

7.2 *Challenges*

It is important to perform a thorough analysis of the difficulties and ethical problems involved with the adoption of autonomous systems because, even though it is impossible to dispute that autonomous systems give a multitude of benefits, it is necessary to do so. The moral dilemma that comes along with the act of deciding anything is, without a doubt, a significant obstacle to overcome. Because the functionality of autonomous systems is dependent on established rules and algorithms, this may give rise to difficulties in circumstances where definitive answers are lacking. An ethical dilemma that may arise for autonomous vehicles in the event of a collision is the decision-making process of prioritizing harm mitigation, either for the vehicle's occupants or pedestrians. To effectively address the ethical challenges currently at hand, it is crucial to use rigorous programming methodologies and establish a consensus over the hierarchy of values within these systems. It is only at that point that it will be feasible to effectively tackle the matter.

There will be substantial repercussions for society as a whole as a result of the transition towards an autonomously operating workforce. Even though autonomous systems have the potential to boost efficiency and productivity, there is also the possibility that workers may be replaced by robots. This argument has a particularly strong basis in reality for businesses that rely largely on individuals doing manual tasks. Manufacturing businesses and those that use truck drivers to staff their operations face the possibility of seeing their workforces shrink. It is a priority to put preventative measures into place to strengthen the skill sets of employees. This will enable employees to handle the ever-changing situation and reduce the negative effects of the issues they face.

The possibility of placing an excessive amount of reliance on modern technologies is one of the things that gives rise to worry. There is an increasing concern regarding the possibility of individuals acquiring an unhealthy dependence on these technologies as the integration of autonomous systems into individuals's daily lives becomes more universal. The reliance on this technology has the potential to lead to a decline in human cognitive capacities and skills in decision-making across a wide range of domains, which would be a negative consequence. The absence of human knowledge to intervene in the event of a malfunction or breakdown inside an autonomous system can result in adverse outcomes if the malfunction or breakdown occurs.

- **Threats posed by hackers:** are among the most significant obstacles to be overcome when it comes to ensuring the safety of autonomous systems. The software of the system could be hacked, communication channels could be intercepted, or sensor data could be manipulated. These are just a few of the many possible shapes that these attacks could take. An adversary can take control of an autonomous vehicle by exploiting a vulnerability, while adversaries in industrial settings can interrupt essential processes by exploiting a vulnerability.
- **Integrity of Data:** To make judgements, autonomous systems are dependent on receiving reliable data from sensors and other external sources. Any breach in data integrity, whether it is from erroneous sensor readings or corrupted data inputs, can have catastrophic effects. The difficulty of ensuring the validity and dependability of data is one that never goes away.
- **Concerns Regarding Privacy:** Autonomous systems frequently gather a massive quantity of data, which may include information on a user's location and behaviour. A major concern, particularly the development of rules such as GDPR and CCPA, is the safeguarding of this sensitive data against unauthorized access and the protection of user privacy.
- **Complexity:** Autonomous systems are vulnerable to both internal and external dangers due to the sophisticated architecture of these systems, which consist of various components and software layers that are interconnected with one another. The management of the security of these complex systems can be an intimidating task.

The autonomous system industry is experiencing major and complex issues in terms of information security. Not only is it essential to address these difficulties to preserve critical data and operations, but it is also essential to ensure the continuous development and use of autonomous technology. To ensure safe and reliable autonomous operations in a world that is becoming increasingly networked, digital resilience in autonomous systems calls for a proactive and interdisciplinary strategy that combines strict security measures, continuous monitoring, and user awareness.

When it comes to efficiently carrying out their functions, autonomous systems are extremely reliant on interconnected networks, sensors, and data processing capabilities. Even though these technologies come with many benefits, they also leave themselves open to several vulnerabilities that could be exploited by malicious parties. It is necessary, for the continued safe and dependable operation of autonomous systems, to have an understanding of these security concerns and the influence they have on digital resilience.

7.3 Research Avenues

Autonomous systems, including self-driving vehicles, unmanned aerial vehicles, and automated industrial processes, face a wide range of cyberattacks that pose a significant risk to their performance, data integrity, and user safety. These adversarial attacks can lead to even minor modifications in input data resulting in highly inaccurate decisions. For example, an adversarial attack could cause autonomous vehicles to misinterpret road signs or obstacles, potentially leading to dangerous situations. Given the increasing complexity of cyberattacks, it has become crucial to implement robust defence measures and intrusion detection systems.

The exponential growth of autonomous systems highlights their critical role in various domains, from transportation to industry automation. As they become increasingly integrated into everyday life and critical infrastructures, it is essential to address vulnerabilities and challenges that could jeopardize their functionality and reliability. To ensure the safety and robustness of autonomous systems, interdisciplinary collaboration and cross-domain research are crucial. Several key research areas are emerging to address these challenges:

- **Robust Sensor Fusion and Perception:** Effective perception of the environment is crucial for autonomous systems. Researchers are developing perception models equipped to detect misleading sensor inputs and correct them. This includes strategies to identify sensor abnormalities, integrate redundancy in sensor data fusion, and enhance overall data accuracy.
- **Architecture and Protocols:** Autonomous systems require cybersecurity architectures and protocols tailored to their unique attributes. Researchers are advancing secure communication frameworks, authentication systems, and encryption algorithms that can withstand complex cyberattacks. These efforts consider

factors such as the need for prompt communication, resource constraints, and the potential consequences of cyber intrusions.

- **Anomaly Detection and Intrusion Prevention:** Detecting anomalies and preventing intrusions are critical due to the dynamic and unpredictable nature of autonomous system environments. Ongoing research is focused on the development of advanced anomaly detection algorithms capable of identifying deviations from standard patterns, thus alerting to potential cyberattacks or system malfunctions. ML algorithms are necessary in adapting to dynamic threats.
- **Formal Verification and Validation:** Formal verification and validation involve rigorous examination of software and hardware components within autonomous systems to ensure accuracy and safety. Researchers are exploring formal methods that use mathematical reasoning and thorough testing to verify system functionality under various conditions, including exposure to cyberattacks or vulnerabilities.
- **Decision-Making:** Autonomous systems rely on complex ML techniques for decision-making. Ensuring transparency and interpretability in these decisions is essential for accountability and resilience. Research is focused on developing AI models capable of providing explanations for their decision-making processes, facilitating the identification and resolution of unexpected behaviours.

All-inclusive, the advent of autonomous systems presents numerous opportunities for the future but also poses significant challenges. These technologies have the potential to reshape industries, improve efficiency, and contribute to scientific research. However, addressing concerns related to ethical decision-making, workforce displacement, and over-reliance on technology is critical. Prioritizing digital resilience in autonomous systems is essential to ensure their continuous operation and safeguard them against evolving cyber threats. Interdisciplinary research and collaboration across various domains are key to developing effective strategies for securing and enhancing the resilience of emerging technologies.

8 Adaptive Defence

In the present day, which is characterized by rapid technological advancements, both organizations and individuals heavily depend on state-of-the-art applications such as fog and cloud computing, smart cities, the IoT, collaborative computing, and virtual and mixed reality environments. The need to ensure the security, reliability, and resilience of these systems in the presence of cyberattacks is unquestionable, therefore necessitating novel methodologies in cybersecurity and resilience.

The concept of adaptive defence within cybersecurity involves the deployment of a proactive and dynamic approach to safeguard computer systems and networks against many potential threats, including but not limited to malware, hackers, and other detrimental activities. The effectiveness of this approach relies on the continuous monitoring, analysis, and adjustment of security measures to successfully address the dynamic nature of cyber risks. Considerable advancements have been made in computational approaches, enabling the accurate replication of various natural phenomena, events, and organisms. The approaches discussed above include artificial neural networks, swarm intelligence, deep learning systems, and biomimicry. These systems possess noteworthy attributes that offer a range of innovative approaches to tackle emerging cybersecurity challenges.

Cybersecurity is presently experiencing notable transformations, comprising advancements in technology as well as methodological approaches. Adversaries exhibit an elevated degree of organization, whereas modern society and enterprises are progressively reliant on the Internet and digital platforms. The current defensive strategies and the individuals involved in strengthening security measures are ill-prepared to effectively address the sudden surge in cyberattacks. The conditions have led to the rapid adoption of AI, data science, and quantum computing techniques to effectively detect, mitigate, and reduce complex cyber threats. We describe adaptive defence as follows:

(1) **Threat Monitoring:** Continuously gather data and information from various sources to detect potential threats or anomalies in the system.
(2) **Analysis:** Use mathematical algorithms and statistical techniques to analyse the collected data and identify potential security breaches or unusual patterns.
(3) **Decision-Making:** Based on the analysis results, make informed decisions about whether a detected activity or behaviour represents a genuine threat or a false positive.
(4) **Adaptation:** If a genuine threat is detected, dynamically adjust security measures to respond to the specific threat. This may include updating firewall rules, applying patches, or isolating compromised systems.
(5) **Learning:** Continuously learn from past incidents and adapt security strategies to improve the overall defence posture. This involves feedback loops to refine detection and response capabilities.

The concept of adaptive defence is based on the concept of not only depending on pre-established security standards but also adjusting and responding to the dynamic nature of the threat environment. The primary objective is to reduce the duration between the detection of a cyberattack and the implementation of a response since this timeframe is integral in limiting the potential harm inflicted by such attacks. The concept of adaptive defence in cybersecurity refers to a proactive and responsive strategy that utilizes continuous monitoring, thorough analysis, informed decision-making, flexible adaptation, and ongoing learning processes. This method aims to safeguard against cyber threats in a manner that follows mathematical rigour.

8.1 Significance

The significance of adaptive defence and nature-inspired cybersecurity lies in their ability to effectively address the increasing complexity and constant evolution of cyber threats using a mathematical approach. The recognition of the continuous growth of the sector highlights the growing importance of adaptive defence strategies. In the given environment, it is apparent that cyber dangers are continuously evolving, growing increasingly complex and sophisticated over time. The incorporation of adaptive defence, along with its dynamic response mechanisms, allows the system to effectively adapt to the emergence of new threats, ensuring a proactive approach to cybersecurity.

The optimization of response time to cyberattacks is a crucial mathematical component within the framework of adaptive defence. This reduction can be expressed as a decrease in the amount of time that passes between the identification of an occurrence and the subsequent reaction. The decrease ultimately functions to reduce the potential damage inflicted by cyberattacks, emphasizing the practical need for implementing this specific approach. Similarly, adaptive protection strategies prioritize the mitigation of false-positive alerts. Through the implementation of thorough analysis and efficient decision-making protocols, this methodology ensures that security teams prioritize authentic risks, thus optimizing the effective allocation of crucial resources. The implementation of mathematical principles is required for enhancing the efficiency and functionality of cybersecurity procedures. Another salient attribute of adaptive defence is its continuous learning component.

Organizations can leverage knowledge acquired from past incidents to constantly enhance their cybersecurity posture. The implementation of a pedagogical approach aligns with the mathematical concepts of iterative refinement, hence augmenting the efficacy of the defensive strategy. Concurrently, Nature-Inspired Cyber Security (NICS) also holds significance. The approach draws inspiration from natural systems, which include the inherent attributes of resilience and flexibility. This strategy aims to replicate the adaptability observed in ecosystems when creatures gradually adapt to changes in their environment.

NICS utilizes mathematical optimization techniques, for example, genetic algorithms and ant colony optimization, to tackle complex optimization problems. The algorithms utilized in this particular context are designed to optimize various security settings and configurations, ensuring a robust and efficient defence strategy. Also, the utilization of nature-inspired ideas in cybersecurity aims to enhance the effectiveness of anomaly detection by closely examining and imitating the behaviours and interactions demonstrated by organisms in the natural environment.

The methodology enables the identification of anomalous activity patterns that may suggest the presence of a cyber-attack, in alignment with the mathematical rules that control the process of recognizing patterns. The mathematical characteristic of scalability holds significance in assessing the efficacy of nature-inspired methodologies. These systems inherently possess the capability to scale efficiently and demonstrate great compatibility with extensive and dynamic networks. The mathematical description of the efficacy of these systems in properly managing ever-increasing networks highlights their significance in safeguarding digital ecosystems from cyber threats.

Recent research in NICS has uncovered a promising pathway for the development of resilient security systems. The fundamental principle of NICS is centred around the development and deployment of security mechanisms that possess the ability to adapt and withstand malicious activities. These mechanisms are designed to effectively handle the inherent uncertainties, incompleteness, and ambiguity that are present in the information being processed. The objective of NICS is to develop a cyber defence framework through the utilization of nature-inspired tactics, including techniques such as camouflage, deception, and disguise.

The proactive strategy of adaptive defence in cybersecurity places significant focus on ongoing surveillance and flexible adjustment to newly emerging threats promptly. The major objective of this system is to promptly identify and respond to any security breaches, intending to minimize the impact of any resulting harm and prevent additional attacks. The utilization of this dynamic cybersecurity methodology is crucial. The rapidly changing threat is marked by adversaries of growing sophistication. The utilization of adaptive defence presents potential solutions for addressing the growing cyber threat. However, it is crucial to acknowledge and resolve the notable security challenges associated with this approach to ensure its efficacy and resilience.

The challenges arise from the complex structure of adaptive defensive mechanisms and cyber threats. One of the major obstacles encountered in adaptive defence pertains to the attainment of an optimal equilibrium between the significant threats and the mitigation of unnecessary disruptions arising from false positives. Due to the inherent volatility of cyber threats, routine activities may give rise to suspicions owing to rapid fluctuations in user conduct, network flow, or system settings. To mitigate the occurrence of both false positives, which may result in unnecessary defensive actions, and false negatives, which may overlook genuine threats, the utilization of complex algorithms and continuous refinement is necessary. The expansion of hostile ML poses a significant challenge within adaptive defence mechanisms.

Threat actors can exploit security systems powered by AI by providing them with carefully crafted data intended to deceive their algorithms. This phenomenon has the potential to facilitate the evasion of detection mechanisms, enabling malicious actors to exploit vulnerabilities without triggering defensive countermeasures. To address this issue, adaptive defence strategies must incorporate strong methods for identifying and mitigating adversarial attacks, thereby guaranteeing the continued efficacy and resilience of AI models. To effectively monitor, understand, and promptly respond to new threats in real time, adaptive defensive systems necessitate substantial computational resources.

The implementation and maintenance of such systems might pose challenges for locations or companies with limited resources. To mitigate the strain on infrastructure while ensuring fast reactions to potential threats, it is crucial to carefully maintain a delicate balance between efficient threat detection and system performance. Numerous firms have already implemented cybersecurity protocols. Integrating adaptive defence measures with outdated systems and ensuring seamless compatibility might pose significant challenges. The effectiveness of adaptive defence can be compromised by incongruent data formats, communication protocols, and operational processes. Consequently, strategizing and execution are needed to ensure the seamless integration of novel and pre-existing security protocols.

8.2 Security Challenges

The security issues associated with the implementation of adaptive defence and NICS necessitate a mathematical analysis to comprehend their inherent characteristics and their problems and issues. In adaptive defence, a complete examination of mathematical analysis uncovers numerous crucial concerns. The primary concern is the issue of false positives (FP). Despite the implementation of many strategies to mitigate its occurrence, the problem of FP continues to persist. The present mathematical dilemma revolves around achieving an optimal equilibrium between minimizing false alarms through the decrease of false positives and ensuring that legitimate threats are not overlooked. The maintenance of this equilibrium is essential since too aggressive false-positive reduction measures can result in the occurrence of false negatives. The false negative (FN) can be mathematically expressed as FN = 1 minus the true-positive rate. This equation serves to indicate the failure to accurately identify and detect legitimate threats.

Resource overhead (RO) is another noteworthy topic within adaptive defence. The optimal allocation of computing and operational resources is a complex mathematical optimization issue. Achieving an optimal equilibrium between security and efficiency necessitates the application of mathematical methodologies to guarantee the appropriate allocation of resources. Adversarial adaptation (AA) poses a significant difficulty within the adaptive defence. To effectively address the strategies adopted by cyber adversaries that possess the ability to adapt to defensive measures, the utilization of mathematical modelling and countermeasures becomes crucial.

The development of resilient mathematical tactics to maintain an advantage over adaptive opponents is of importance. Privacy concerns (PC) emerge when a significant volume of data is gathered for adaptive defence. The utilization of mathematical mechanisms is crucial to optimize the efficacy of gathered data, all the while upholding the limitations imposed by privacy considerations. The task of striking a balance between utility and privacy poses a mathematical difficulty in its own right. However, NICS encounters its distinct mathematical challenges. The performance of an algorithm, referred to as algorithm performance (AP), exhibits variability based on the particular problem being addressed and how it is implemented. Mathematical optimization is crucial to achieve optimal performance in a wide range of security jobs. The consideration of algorithm complexity (AC) becomes particularly pertinent in the context of complex algorithms such as genetic algorithms and neural networks. The assurance of the security features of these algorithms necessitates an examination and confirmation through mathematical analysis.

Algorithms may exhibit sensitivity to noisy or incomplete data, necessitating the implementation of mathematical modifications to effectively manage such data. Both adaptive defence and NICS are currently facing numerous mathematical and practical hurdles. The obstacles involve a range of issues, including the occurrence of false positives and false negatives, the allocation of resources, the adaptation to adversarial situations, concerns regarding privacy, the performance of algorithms, the complexity of the problem, the compliance to biological realism, the ability to scale, and the ability to withstand noise. To tackle these issues, it is crucial to engage in continuous research and development attempts. These attempts should be based on mathematical analysis and optimization. The optimization and fine-tuning of cybersecurity techniques to successfully address evolving threats and replicate natural processes necessitate a comprehension of their mathematical foundations. Therefore, the successful resolution of these difficulties relies on the ongoing examination and advancement of mathematical techniques within cybersecurity.

8.3 Research Avenues

Nature-Inspired Cyber Security (NICS) is an intriguing and rapidly growing field, capturing the attention of researchers in the domain of cybersecurity. This new field utilizes principles gained from biological and natural systems to design cybersecurity tactics that are more effective and robust. Throughout millions of years, natural processes have given rise to complex mechanisms that serve as defences against various forms of danger. NICS aims to use these mechanisms to create cybersecurity systems that are both resilient and flexible. The core principle of the NICS framework revolves around the use of knowledge derived from the natural world to develop security measures that exhibit qualities of robustness and longevity. The fundamental principles that underlie NICS involve the application of underlying concepts, namely, natural selection and adaptation. These ideas are utilized to develop systems that possess the capability to promptly detect and react to developing cyber threats. One of the notable benefits of NICS is its capacity to empower organizations to proactively safeguard against cyberattacks through the promotion of versatile and adaptive security measures.

To accomplish this objective, NICS utilizes sophisticated algorithms and ML techniques specifically designed to integrate environmental data and effectively respond to changing security risks. Cybersecurity includes a growing discipline known as adaptive defence and NICS. The core concept of this approach lies in the utilization of principles derived from the natural world to enhance the effectiveness of cybersecurity measures. The primary objective of NICS is to develop and promote flexible defence mechanisms that can effectively react to cyber threats. We aim to provide a comprehensive analysis of the various research opportunities and theme areas associated with NICS. The convergence of ML and NICS presents a promising domain for scientific analysis.

In anomaly detection, scientific attempts are focused on the analysis of innovative machine-learning methodologies. The strategies are derived from the immune system's extraordinary ability to distinguish alien entities. Therefore, the primary objective is to create autonomous models that are capable of detecting and classifying abnormal patterns present in network traffic or system behaviour. Likewise, the incorporation of evolutionary algorithms emerges as a prominent aspect. Genetic algorithms are utilized to optimize security settings and enable the adaptive reinforcement of defences across successive periods.

The objective is to investigate the potential application of nature's evolutionary mechanisms in the development of cybersecurity systems that possess self-enhancing capabilities. Biologically inspired defences serve a crucial role in NICS research. Within the framework of immunological methodologies, this analysis explores the replication of the human immune system. The human immune system can be seen as a fitting analogy for the evolution of cybersecurity systems that possess the qualities of self-repair and adaptability. Research efforts may involve the conceptualization and refinement of digital antibodies, which contain the autonomous capability to neutralize threats. Simultaneously, there is an ongoing examination of swarm intelligence.

The objective of this examination is to comprehend the group dynamics observed in organic swarms, such as bees and ants, and apply this knowledge to the design of distributed and self-organizing cybersecurity systems. The objective of the research conducted in this domain is to provide a comprehensive understanding of how swarm intelligence might enhance the ability to perceive and mitigate various threats. Adaptive network security includes various research avenues that investigate the dynamics of network topologies. Insights are derived from the examination of how organisms adapt their physical architecture as a means of protection. As a result, the research topic involves the examination of approaches to the dynamic reconfiguration of network topologies.

These models use simulations to represent ecological systems and include components like niche partitioning, mutualistic interactions, and predator-prey dynamics to enhance the robustness of ecological networks. The focus of research has shifted towards human-behaviour-centric techniques, particularly in the exploration of behavioural biometrics as a pioneering field. The examination of keystroke dynamics and mouse movements is conducted to assess their viability as authentication and intrusion detection systems inspired by nature. Analogies can be established between the immunological responses of humans and cybersecurity, which focuses on human-oriented approaches. This involves the examination of ideas such as security antibodies and the development of adaptive threat awareness.

The advent of quantum-inspired NICS poses a captivating horizon in scientific research. Quantum annealing has emerged as a prominent area of analysis, with scientific initiatives mostly focused on the exploration of quantum-inspired algorithms and computing. These efforts result in the creation of adaptive cryptographic systems designed to defeat quantum-based attacks, similar to the evolutionary adaptations observed in biological species responding to challenges. The phenomenon of quantum entanglement can be utilized to establish secure communication channels. Research initiatives are now exploring the utilization of quantum entanglement to establish communication channels that are secure and immune to breaches, thus emulating nature's utilization of entangled qualities for the conveyance of information.

The implementation of adaptive defensive strategies and the utilization of NICS present promising approaches for enhancing cybersecurity measures, particularly in response to the constantly changing nature of cyber threats. In this broad field, scientists have numerous opportunities to investigate various topics, including biologically inspired defence mechanisms, security paradigms affected by quantum principles, and techniques centred around human factors.

Together, thorough scrutiny is applied to the ethical and legal foundations, aiming to ensure a responsible and prudent advancement in the use of NICS approaches. The utilization of cybersecurity tactics inspired by nature exhibits considerable potential for the advancement of cybersecurity systems that possess enhanced adaptability, resilience, and efficacy. Organizations can enhance their ability to withstand the always-increasing threats in the digital ecosystem by utilizing knowledge derived from the natural world, thereby promoting innovative approaches to cybersecurity. With the increasing complexity of cybersecurity risks, it is crucial to examine novel approaches such as the NICS to uphold a proactive stance in safeguarding against prospective breaches.

The concepts of adaptive defence and digital resilience are strongly interconnected in this particular field. The concept of adaptive defence relates to the capacity of cybersecurity systems and practices to undergo evolutionary changes and adjustments to effectively counter emerging threats and weaknesses. The mentioned dynamic approach emphasizes the importance of maintaining constant monitoring, conducting thorough threat assessments, and implementing effective steps to prevent novel and emerging threats. The necessity of a proactive mentality and an understanding of the threat and potential threats are essential for organizations.

In contrast, digital resilience refers to the ability of an organization to withstand, adjust to, and recuperate from detrimental occurrences associated with cybersecurity. The assessment considers multiple dimensions of an organization's cybersecurity stance, including individuals, processes, and technology. Digital resilience includes a holistic and cohesive strategy in cybersecurity, intending to minimize the consequences arising from cyberattacks and vulnerabilities. The integration of adaptive defence and digital resilience offers a comprehensive framework for effectively managing cybersecurity risks and threats within a dynamic and rapidly evolving environment.

These methodologies enable enterprises to effectively recognize, address, and recuperate from cybersecurity incidents, a matter of importance in the digital age, characterized by the constant presence of cyber threats. The academic field of cybersecurity is currently focused on researching and developing the concepts of adaptive defence and digital resilience. These topics are considered crucial and continue to be actively explored on a global scale. These principles can enhance the safety and security of companies in the digital era. The concepts of adaptive defence and digital resilience are complexly interconnected and mutually supportive, presenting a remarkably efficient collaborative strategy for cybersecurity that is difficult to surpass.

9 Cyber-Security in an Open World: New Space

The modern space sector is currently in the midst of a tremendous revolution, which is characterized by a rapid shift towards digitalization and interconnectedness. The necessity for effective digital resilience and cybersecurity measures has never been more evident than it is now, as satellites, spacecraft, and ground-based systems become increasingly reliant on interconnected networks, sensors, and data processing capabilities.

The immensity of space, formerly distinguished by its remoteness, is now deeply made into the computerized structure of the twenty-first century, necessitating the implementation of thorough safety measures to protect against potential threats. Organizations in this sector can improve their digital resilience by implementing comprehensive security frameworks and proactive measures. This technique assures that space-based services and applications continue to operate in the face of ever-changing obstacles.

9.1 Significance

It is impossible to overstate the significance of digital resilience within the space industry. This matter transcends technological concerns and is linked to today's rapidly expanding demand for space-based applications and services. Numerous critical elements highlight the requirement and need to preserve digital resilience within the space sector. These elements comprise the following:

- **Vulnerability to Natural Catastrophes:** Space systems, which orbit high above the surface of the Earth, are vulnerable to the unpredictability of the forces that are produced by natural catastrophes. The integrity of space-based infrastructure is consistently at risk of a variety of threats, including solar flares, cosmic radiation, and impacts from micrometeoroids. The potential repercussions of these events, which may include a degradation of the system or even its total failure, underline the necessity of implementing effective digital resilience mechanisms.
- **Cybersecurity Dangers:** In this day and age, which is characterized by cyber warfare and advanced hacking tactics, the aerospace industry is not immune to the possibility of being targeted by hostile actors looking to exploit weaknesses.

Cyberattacks on space systems can have far-reaching implications, such as interfering with satellite communications and placing the safety of space missions at risk. As the nature of the online risks that space organizations face changes, so too must the digital resilience techniques they use.

- **Critical Infrastructure in Space:** Space-based systems have evolved into an essential component of the infrastructure upon which the operation of society is dependent. Space technologies provide the backbone of a diverse range of important services, ranging from global navigation and communication to the observation of Earth and scientific study. The degree to which we can preserve the dependability of these services despite the presence of obstacles is directly proportional to the degree to which these systems are resilient.
- **Ecosystem That Is Completely Connected:** The space industry is no longer a separate sector but rather an essential component of our whole digital ecosystem. Assets based in space are essential for a wide variety of fields, including disaster management, telecommunications, navigation, and weather forecasting. As a result of the potential for cascade impacts on terrestrial systems caused by disruptions in space activities, digital resilience is an issue of global concern.

Promoting digital resilience in the aerospace industry has emerged as an attempt that has assumed crucial importance. It involves not only protecting space systems from dangers but also assuring their capacity to anticipate, withstand, recover from, and adapt to a diverse range of issues. The path towards achieving digital resilience in space is characterized by a wide variety of obstacles and inventive solutions. These include cybersecurity policies, the development of advanced technology, specialized workforce training, and collaborative approaches.

9.2 Security Challenges

The space sector, characterized by its rapid growth, poses numerous complex challenges associated with its intersection with cybersecurity. Given the dynamic nature of society, which presents unique vulnerabilities and risks, it is crucial to implement robust cybersecurity measures. Businesses and other organizations may encounter substantial challenges while attempting to enhance their digital resilience in response to emerging threats.

9.2.1 Data Breaches

Data leaks are a prevalent issue within the expanding space industry. These breaches include the risks associated with illegal entry into confidential information, as well as the potential for the disclosure of personally identifiable data. To mitigate the effects of these threats, enterprises must adopt strict security protocols aimed at protecting their confidential data.

To achieve this objective, it is critical to establish rigorous protocols for safeguarding data. The incorporation of both encryption and access control mechanisms is necessary for the implementation of these protocols. It is crucial to conduct regular security audits and penetration testing to identify and rectify any potential vulnerabilities within the system.

9.2.2 Supply Chain Attacks

The nascent space industry is confronted with the prospective vulnerability of its supply chain as a result of its dependence on external vendors for crucial components. The susceptibility of these vendors to cyberattacks poses a possible threat of compromising the entirety of the supply chain, hence potentially resulting in the disruption of mission operations.

Given this context, companies must implement thorough due diligence protocols when evaluating prospective suppliers. This involves evaluating the degree of security and determining their conformity with the applicable security criteria. Ensuring vendors receive unequivocal security requirements is of significance, and continuous monitoring of compliance is critical.

9.2.3 Securing Ground Stations

Ground stations, which play a critical role in spacecraft communication and data collecting, are susceptible to potential cyberattacks. The vulnerabilities include potential consequences such as denial-of-service attacks and unauthorized access, exposing sensitive information to the danger of unlawful modification or theft. To mitigate the occurrence of these dangers, ground station operators are obligated to establish rigorous safety protocols. Access control, network segmentation, continual vulnerability assessments, and penetration testing are widely recognized as crucial measures for safeguarding information systems.

9.2.4 Satellite-Based Communication Systems

The susceptibility of essential infrastructure and services stems from their dependence on satellite-based communication technologies, resulting in vulnerability to potential attacks. Organizations tasked with overseeing the management of such systems should give precedence to the use of encryption and authentication methodologies to address potential vulnerabilities. It is crucial to conduct frequent security assessments and penetration testing to maintain the integrity of these systems and minimize the potential for any disruptions.

9.2.5 Natural Disasters

Natural disasters, which occur with regularity in the outer space environment, present a significant risk to the infrastructure and technical resources located within that domain. The potential problems and issues of the degradation or discontinuation of services and applications reliant on space infrastructure can be significant. Given the current circumstances, organizations must focus their resources towards the development and execution of comprehensive policies for disaster recovery and business continuity. The implementation of redundant and backup systems for essential infrastructure, coupled with thorough risk assessments and emergency response processes, is critical to minimize the detrimental impacts presented by these risks.

9.2.6 Workforce Challenges

The shortage of skilled and specialist personnel poses a significant hurdle to the development of digital resilience in the developing space sector. It is crucial to evaluate the obstacles posed by the technological aspects of the organization. The lack of a skilled workforce with the ability to successfully navigate the complexities of space technology and cybersecurity has become a significant barrier in current efforts to improve digital resilience in this field.

This presents a noteworthy challenge. The complicated integration of the cybersecurity industry with the space sector has created a substantial demand for professionals who possess a comprehensive understanding of both fields. Despite the growing public awareness and continuous research attempts, there is a significant scarcity of well-trained professionals in this particular subject. To rectify this discrepancy, it is crucial to enact focused educational interventions that equip students with the necessary skills to proficiently traverse the complex convergence of space technology and cybersecurity.

With the growing recognition and extensive attempts focused on this matter, the continued allocation of resources towards education and training is crucial, given the specialized nature of this specific area of inquiry. Given the ongoing growth and development of the space industry, it is crucial to develop a group of individuals who possess a thorough comprehension of its complexities. The enhancement of digital resilience and the protection of long-term security for space-centric systems and applications are of importance.

9.3 Research Avenues

The issue of ensuring the stability and security of space systems and their associated communication channels has emerged as a prominent priority, requiring concentrated efforts and specialized knowledge. The increasing demand for a holistic approach that includes continuous investments, specialist training, and collaborative tactics has arisen due to the distinct obstacles encountered by space systems.

- **Investment in Robust Technology and Infrastructure:** The development of strong infrastructure and the allocation of resources towards cutting-edge technologies are essential foundations for promoting the resilience of the space sector in the digital age.

  The establishment of digital resilience within the growing space sector is predicated upon two basic components. Space systems, as opposed to normal terrestrial systems, are specifically designed to function in an environment characterized by demanding circumstances such as extreme conditions, radiation exposure, and huge distances. To maintain continuous operations and effectively tackle these issues, organizations need to dedicate resources towards the research, development, and implementation of specialized hardware and software that are specifically tailored for space applications.

  This investment goes beyond the mere development of innovative technology; rather, it involves an ongoing process of improving existing technology and adapting to changing circumstances. The regular updating, patching, and upgrading of software are important to successfully react to the constantly changing nature of threats and properly address any potential security vulnerabilities that may arise over time. To ensure the enduring viability of their digital resilience, enterprises must exhibit a dedication to a proactive continuum of innovation and enhancement. This ensures that their efforts are in response to the continuously changing dangerous circumstances.
- **Training Cybersecurity Professionals for Space Systems Security:** Space systems are not only conventional systems located in space, rather, they are quite different and face distinct issues that necessitate specialized knowledge and experience. The establishment of robust security measures for space systems and their associated communication channels necessitates the expertise of cybersecurity professionals who possess an understanding of the complexities and subtleties inherent in space technology. Training programmes must be customized to adequately prepare experts in effectively managing the complexities associated with security in space. This involves equipping them with the necessary abilities to handle radiation-hardened equipment, encrypted communication protocols, and the isolation of critical functions.

  It is crucial for the training to have an understanding of the potential problems and issues that cyberattacks can have on space systems, as these repercussions can extend to a significant extent. The consequences of a successful attack can have significant implications for mission success and human safety, ranging from disruptions in satellite communication to compromised orbital motions. By providing cybersecurity professionals with the necessary training to understand the distinct characteristics of space technology and its vulnerabilities, organizations may enhance their defensive capabilities and effectively address possible threats.

- **Collaborative Approaches and Knowledge Sharing:** Space systems exhibit notable distinctions as compared to ordinary atmospheric systems, hence introducing distinct issues that require specific knowledge and practical proficiency to effectively address and minimize their impact. The inclusion of cybersecurity professionals possessing a thorough understanding of the complex elements and complexities of space technology is important to safeguard space systems and related communication networks. Professionals in this field are required to possess the requisite abilities to effectively navigate the complex aspects of security in space. Customizing the training programmes is necessary to provide professionals with the requisite skills and competencies.

  The training curriculum must have an understanding of the potential consequences arising from cyberattacks on space systems, given their capacity to yield substantial and extensive impacts. The potential problems and issues of a successful attack have the potential to undermine the accomplishment of a mission and pose a threat to human life through many mechanisms, including the disruption of satellite communications and the compromise of orbital motions. Organizations have the opportunity to strengthen their defensive capabilities and boost their ability to respond to potential attacks by offering specific training to their cybersecurity professionals. This training would enable these experts to have a comprehensive understanding of the distinctive characteristics of space technology and the accompanying threats it presents.

The preservation of digital resilience over an extended period is not limited to security concerns alone. Instead, it holds significance in influencing the course of space technology and the overall development of human civilization. The ability to withstand and recover from cyber-attacks will be essential for the success of missions and the general progress of the business, considering the continuous creation of new space applications. The applications stated above cover a broad range, including the use of satellite constellations for global connectivity and the progress of space exploration missions.

10 Computational Psychology and Cyberpsychology

A comprehensive strategy is necessary to uphold robust defensive measures. The significance of technology advancements notwithstanding, the comprehension of human behaviour has arisen as an essential element in the formulation of solutions for digital resilience. The intersection of computation and cyberpsychology serves as a means to address the complex nature of human psychology and the need to successfully navigate, mitigate, and respond to diverse digital threats. At its essence, computation and cyberpsychology labour to examine the fundamental correlation between human cognition, behaviour, and technology. Table 8.3 provides further elaboration on the subject matter.

Table 8.3 Computational psychology and cyberpsychology

	Computational Psychology	Cyberpsychology
1. Definition	Computational Psychology involves using computational models and simulations to study psychological processes and behaviours	Cyberpsychology explores the impact of digital technology on human behaviour, cognition, and mental well-being
2. Methodology	Computational Psychology employs mathematical and computational models to simulate and understand psychological phenomena	Cyberpsychology employs a multidisciplinary approach, integrating psychology, technology, and communication studies to investigate online behaviour
3. Focus	Computational Psychology focuses on developing and testing computational models to explain and predict human behaviour	Cyberpsychology focuses on the psychological aspects of online behaviour, including social interactions, cyberbullying, and Internet addiction
4. Applications	Computational Psychology applications include modelling memory processes, decision-making, and cognitive functions	Cyberpsychology applications include studying the impact of social media, online gaming, and virtual environments on mental health
5. Data sources	Computational Psychology often relies on experimental data and psychological theories to inform computational models	Cyberpsychology gathers data from online platforms, social media, and virtual environments to analyse digital behaviour patterns
6. Interdisciplinary nature	Computational Psychology is interdisciplinary, incorporating elements from psychology, computer science, and cognitive science	Cyberpsychology is highly interdisciplinary, integrating psychology, sociology, and technology studies to understand online behaviour
7. Simulation and modelling	Computational Psychology heavily relies on simulations and modelling to replicate and analyse psychological processes	Cyberpsychology emphasizes the analysis of real-world online behaviour, considering the impact of digital environments on individuals and societies
8. Cognitive processes	Computational Psychology often investigates cognitive processes such as perception, memory, and decision-making	Cyberpsychology explores the psychological aspects of online mental health, addressing issues like Internet addiction and cyber-related stress
9. Advancements	Computational Psychology advancements include the development of sophisticated computational models for understanding complex cognitive phenomena	Cyberpsychology advancements involve insights into the evolving nature of online behaviour and the psychological implications of emerging technologies
10. Ethical considerations	Computational Psychology considers ethical implications related to model accuracy, privacy, and the responsible use of simulations	Cyberpsychology addresses ethical concerns related to online privacy, digital well-being, and the psychological impact of online interventions

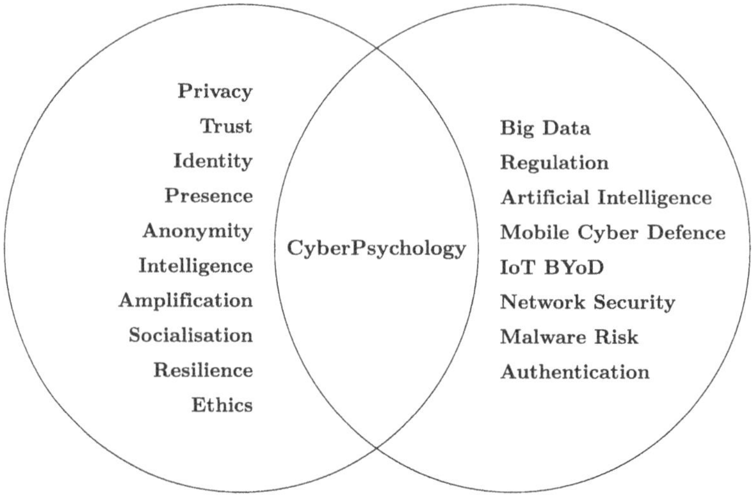

Fig. 8.8 Intersection of cyberpsychology with computational intelligence

Scientists in this field attempt to define the cognitive processes involved in individuals' perception, response, and interaction with digital environments, by uncovering the complex relation among several influencing elements. This observation, in turn, informs the creation of robust systems that can predict, adjust to, and bounce back from cyberattacks while ensuring a smooth and user-focused experience. Figure 8.8 displays a Venn diagram showcasing the intersection of Cyberpsychology with Computational Intelligence.

10.1 Significance

In today's interconnected sector, the role of cyberpsychology in enhancing digital resilience holds immense importance and should not be underestimated. The interdisciplinary nature of cyberpsychology is crucial in comprehending human behaviour, cognition, and decision-making within the domain of cybersecurity. The significance of cyberpsychology concerning digital resilience lies in its ability to address the psychological and behavioural aspects of individuals in the digital domain.

- **Human-Centric Approach:** The human-centric approach is a perspective that emphasizes the needs, experiences, and well-being of individuals as the central focus of cyberpsychology and emphasizes the central role of humans in the development and implementation of cybersecurity solutions. The statement acknowledges that digital resilience cannot be solely ensured by technology, as the success or failure of security measures is frequently contingent upon human actions and decisions. Cyberpsychology contributes to the enhancement of security measures by examining the cognitive processes and behavioural responses of individuals towards cyber threats. This understanding allows for the customization of security protocols to align with human behaviour, resulting in improved effectiveness and user satisfaction.

- **Behavioural Insights:** Behavioural insights include the study of human behaviour and decision-making processes, with a focus on understanding the underlying psychological factors. One of the fundamental contributions of cyberpsychology lies in its capacity to offer valuable insights into human behaviour inside digital contexts. Scholars in this particular domain investigate several elements like the assessment of risk perception, biases in decision-making, and the influence of emotions on choices about cybersecurity. This understanding enables organizations to develop enhanced security awareness programmes, training modules, and interventions that effectively target human vulnerabilities.
- **Social Engineering Defence:** The defence against social engineering involves countering the strategies used by cyber attackers, who frequently exploit human psychology. Cyberpsychology provides individuals and organizations with the knowledge of the functioning of these manipulative strategies. Individuals are necessary to enhance digital resilience by demonstrating awareness of and actively countering social engineering attempts.
- **User-Centric Design:** Security measures that prioritize user-friendliness are more likely to be adopted and followed by individuals. Cyberpsychology provides insights and guidance for the development of security interfaces and systems that are in line with human cognitive processes and patterns of interaction with technology. As a consequence, the implementation of these security measures does not hinder productivity or cause user frustration, so ultimately improving the overall digital resilience.
- **Training and Education:** Cyberpsychology provides valuable insights for the design and implementation of efficient cybersecurity training programmes. Organizations can enhance their training modules by gaining insights into the cognitive processes that impact decision-making, allowing them to effectively target and mitigate certain weaknesses and biases. This results in a workforce that is more knowledgeable and conscious of cyber-related issues, which is a crucial element in promoting digital resilience.
- **Preventive Measures:** Cyberpsychology contributes to the detection of initial indications and markers of prospective security breaches. Behavioural analysis and anomaly detection approaches utilize knowledge from cyberpsychology to identify anomalies in typical user behaviour. The provision of early warnings allows organizations to adopt proactive actions and effectively prevent or reduce cyber risks before their escalation.

10.2 Ethical Considerations

The preservation of digital resilience necessitates the careful management of security measures and the protection of individual privacy. Cyberpsychology plays a significant role in maintaining equilibrium by addressing ethical considerations about the collection of data, obtaining user consent, and safeguarding privacy rights. The compliance to ethical principles is important in ensuring the long-term digital resilience of security measures. The significance of cyberpsychology about digital resilience resides in its capacity to facilitate the connection between technology and human behaviour. By comprehending and mitigating the psychological dimensions of cybersecurity, organizations can develop security measures that are more efficient, user-centric, and morally sound. The use of this approach significantly improves the overall resilience of digital systems and empowers individuals to actively participate in the defence against cyber threats.

10.3 Psychological Factors

The establishment of digital resilience necessitates an examination of the psychological factors that underlie human behaviour inside the digital domain.

- **Human Decision-Making:** The examination of the decision-making process in the context of technology is a fundamental area of inquiry in computation and cyberpsychology. The analysis conducted by researchers focuses on the various factors that shape human decision-making in the face of potential cyber dangers. These factors include risk perception, cognitive biases, and the influence of emotions on the decision-making process.
- **Social Engineering and Manipulation:** Social engineering strategies are commonly used by cyber attackers to exploit cognitive flaws. These methodologies use psychological tendencies, deceiving individuals into revealing confidential information or engaging in actions that compromise security. An understanding of these strategies facilitates the formulation of countermeasures and the instruction of individuals in the identification and resistance of such manipulation.

10.4 Designing Resilient Systems

In addition to providing psychological insights, the idea of digital resilience includes the development and implementation of technological systems that possess the ability to handle malicious activities while maintaining user-friendliness and accessibility.

- **Interfaces for Human Cognitive Processes:** The topic of discussion revolves around interfaces designed to facilitate human cognitive processes. The development of resilient digital systems that maintain user-friendliness necessitates the design of interfaces that align with the cognitive processes of human beings.

  These tools are specifically engineered to provide information in a manner that is both lucid and comprehensible. These entities might be regarded as beneficial navigational indicators that provide clear guidance without confusion. In the context of expeditious security communications, such as alerts about prospective threats.

  The maintenance of this equilibrium is important to understand the seriousness of the situation without yielding excessive concerns. When these warnings are formulated to align with our innate cognitive processes and behavioural responses, there is an increased likelihood of making careful judgements during challenging circumstances.
- **Human-Centric Security Measures:** In the current environment of digital safety, the effectiveness of security measures cannot be exclusively dependent on their level of technical expertise or innovation. They must be carefully constructed, with due regard for the complexities of human behavioural tendencies. This particular viewpoint about security recognizes the inherent variety in individuals' levels of technology proficiency and the diverse range of strategies they use when doing tasks.

  This attempt aims to build a security plan that addresses the requirements of all parties involved. The usage of an individual's unique fingerprint as a method for unlocking a mobile device is being considered, wherein the individual utilizes a distinct trait that is personal to their own identity. The establishment of mobile device security has become a crucial aspect of individuals' daily routines.

  Another approach involves the enforcement of restrictions that can be adjusted based on individuals' usage of electronic devices. In circumstances of importance, regulations tend to grow more rigorous to protect one's interests, while avoiding unnecessary costs. These regulations can also be aware of an individual's geographical position and actions, allowing them to adjust properly while mitigating any possible interruptions.

10.5 Ethical Considerations

In consideration of the continuous advancements in technology, it is crucial to incorporate ethical and fair considerations while developing plans aimed at safeguarding the security of our digital environments. The ethical dimension gains significance within this particular setting as it involves the implementation of a set of standards that aid in balancing the need for security with the rights of individuals.

- **Balancing User Privacy and Security:** The delicate balance between safeguarding user privacy and ensuring security measures. The aim is to develop resilient protective measures to safeguard our data, while concurrently ensuring that these measures do not encroach upon our privacy rights. The attainment of this equilibrium necessitates the maintenance of a judicious level of surveillance, while concurrently strengthening our defences against conceivable cyber threats.

- **Ensuring Ethical Data Collection and Analysis:** Ensuring the Ethical Collection and Analysis of Data in the attempt to uphold safety and security, there are instances where it becomes crucial to collect information to acquire a deeper understanding. Just as healthcare practitioners conduct health evaluations to guarantee an individual's overall health and wellness. Just as medical professionals are obligated to acquire consent before accessing patients' health records, we must gain permission before collecting digital information.

  In the course of data collection, measures are used to safeguard against the instant identification of individuals based on the obtained information. The practice might be compared to the employment of pseudonyms instead of authentic names. In this approach, notwithstanding an individual's acquisition of access to the facts, their knowledge would remain devoid of the associated identity. We take measures to ensure the secure storage of personal information, thereby prohibiting unauthorized access and retrieval.

10.6 Research Approaches

The development of successful digital resilience measures requires the use of a research approach that can uncover the complexities of human behaviour within the digital domain.

- **Predictive Modelling and Behaviour Analysis:** Predictive modelling and behaviour analysis are two interconnected fields that have gained significant attention in academic research and practical applications. The behavioural analysis involves the examination of user behavioural patterns to detect any deviations that could indicate potential cyber threats. Predictive models can be developed by using these patterns to forecast potential dangers and proactively mitigate them.
- **Cognitive Biases Research:** Cognitive biases can be regarded as inherent tendencies of the human mind, characterized by recurring patterns of thinking that may sometimes result in suboptimal decision-making. Even within cybersecurity, these biases have the potential to impact our responses.

  By comprehending certain behavioural patterns, such as the inclination to prioritize recent occurrences (known as the availability heuristic) or the tendency to overestimate the probability of favourable outcomes (referred to as optimism bias), we can develop educational initiatives and security strategies that facilitate decision-making.

11 Consideration of Human Factors

Organizations should also prioritize the human aspects of cybersecurity by promoting a knowledgeable and proactive workforce. When considering network security within a commercial context, there is a tendency to primarily concentrate on the technical elements. This bias emerges from the tendency to see the ability to exert domination over controllable elements as complete and control. Proficiency in the implementation of firewalls, antivirus software, and encryption protocols is well-established. However, the concept of security goes beyond the domain of technology and also incorporates an interpersonal aspect that involves individuals. Individuals can function as fervent proponents of security, while also being vulnerable components within the security framework, particularly when occupying leadership positions.

Social engineering is a cunning tactic used by attackers to exploit this vulnerability. Social engineering refers to the deliberate manipulation of individuals to evoke the disclosure of confidential information that they would otherwise protect. Social engineering is characterized by its emphasis on psychological manipulation rather than exploiting technological weaknesses. It takes on different forms, with phishing, pretexting, and baiting being some of the most commonly observed manifestations.

Among these various kinds, phishing takes primacy. The act involves the dissemination of electronic communications, including emails, messages, or texts, which exhibit the appearance of originating from reliable entities, such as banking establishments or social networking platforms. Frequently, these forms of communication include hyperlinks that lead recipients to fraudulent websites that have been carefully crafted to closely resemble their legitimate counterparts. As a result, individuals who are unaware of the potential risks unintentionally grant attackers authorization to their accounts when they input their login credentials.

Pretexting, similar to phishing, exploits the act of creating a deceptive pretext to establish confidence with the intended victim. The act of assuming the role of a technical support agent or assuming false identities, such as police enforcement officers or government officials, is considered the modus operandi. It is noteworthy that specific occurrences have utilized voice-modifying technologies to imitate the voices of individuals with whom one has a close relationship or professional association. This deceptive tactic is used to manipulate victims into interacting with harmful links.

Consider a hypothetical case in which the Chief Technology Officer engages in telephonic communication and provides instructions for accessing a forthcoming hyperlink. This particular scenario is expected to evoke a high level of compliance. Baiting is a form of social engineering that involves enticing the victim with an attractive inducement, as opposed to manipulating them through a contrived pretence. These incentives may be presented in the form of electronic communications, specifically emails, which inform recipients of potential prizes or significant reductions in price for desired items. Perpetrators may use a deliberate approach by strategically positioning USB drives in public areas, often accompanied by enticing labels that offer significant incentives to individuals who connect them to their devices. Regrettably, these harmless behaviours lead to the victim's machine being compromised by malicious software.

The efficacy of these attacks is ascribed to the cognitive architecture inherent in the human brain. Existing research indicates that there is an innate tendency within the human brain to place trust in and actively seek out sources that are deemed trustworthy. Individuals in positions of authority or those who are viewed as trustworthy are especially adept at taking advantage of this reliance. Emotional stimuli, such as fear, greed, and curiosity, are significant factors that exert a crucial influence on social engineering exploits.

The experience of fear can lead individuals to engage in impulsive acts, as exemplified by the occurrence of phishing emails that use threats of account closure to elicit prompt responses. The concept of greed is effectively utilized by utilizing baiting strategies, which involve tempting individuals with enticing incentives to motivate them to take certain actions. Pretexting exploits the human tendency towards inquiry and the willingness to help others. The consequences of social engineering attacks are significant. In addition to monetary repercussions, those who have experienced victimization encounter adverse consequences such as reputational injury, privacy infringements, and, in certain cases, bodily injuries. Let us contemplate a hypothetical situation in which a healthcare establishment becomes a victim of a phishing attack, resulting in the compromise of patient records. Alternatively, let us envision a critical infrastructure provider succumbing to a pretexting attack, which then leads to a breach of their facility.

11.1 Human-Centric Approaches

The establishment of digital resilience necessitates the allocation of considerable focus towards the human element. Although technical defences are undoubtedly important, the implementation of a strategy necessitates a deep understanding of the pivotal role that individuals play in ensuring security. Human-centric approaches include a broad spectrum of strategies and methodologies that explore human behaviour, psychology, and relationships within the domain of cybersecurity. These techniques highlight the significance of security measures that possess both strong technical capabilities and are customized to account for the cognitive inclinations and vulnerabilities inherent in individuals.

The recognition of a human-centric perspective necessitates the comprehension that cybersecurity cannot be examined in isolation from the human element. It is the responsibility of human individuals to oversee, administer, and occasionally inadvertently bypass systems and protocols. Hence, the effectiveness of cybersecurity measures relies on their compatibility with the fundamental attributes of human conduct. Several methodologies, such as user-centred design, behavioural analysis, and psychological assessments, are crucial in the creation of security frameworks that emphasize user-friendliness, intuitive interfaces, and the capacity to anticipate and address vulnerabilities resulting from human factors.

11.2 Recognizing Human Error and Behaviour

Recognizing the inherent possibility of humans inadvertently exposing vulnerabilities or becoming susceptible to manipulation by threat actors is a fundamental first step in developing digital resilience. A cybersecurity strategy includes more than just a narrow focus on technological aspects. The statement recognizes the potential for errors in the process of decision-making, inadvertent disclosure of confidential information, or non-compliance with security protocols due to the complex nature of human behaviour. By acquiring a thorough comprehension of the many human errors, businesses can strategically devise security solutions that suitably target these vulnerabilities. The proposed course of action involves the incorporation of systems that incorporate redundancies and fail-safes to minimize the occurrence of potential faults. Also, developing a climate of responsibility and transparent communication might encourage staff members to disclose mistakes or potentially illicit behaviours.

11.3 Training and Education

The fundamental basis of digital resilience is in the provision of individuals with the requisite knowledge and competencies to participate in responsible digital engagements. Training and education initiatives are of importance in equipping personnel with the requisite skills and knowledge to effectively recognize possible threats, make educated decisions, and actively engage in the protection of hardware systems.

Regular training sessions include a wide range of issues, including basic cybersecurity maintenance practices and the development of advanced abilities in threat detection. It is crucial to sustain ongoing educational attempts to guarantee that consumers stay well-informed regarding rising threats and develop optimal methodologies. Organizations have the potential to improve the resilience of hardware systems through the implementation of training and education initiatives, aimed at promoting a knowledgeable and proactive user community.

11.4 Interdisciplinary Collaboration

The complex and diverse characteristics of cybersecurity require the involvement of multiple disciplines to effectively address its challenges. The recognition of the extensive scope of cybersecurity necessitates multidisciplinary collaboration, which enables the incorporation of insights from other fields such as psychology, sociology, communication, law, and technology. By incorporating insights from disciplines such as psychology, communication studies, law, and technology, organizations can establish robust defence mechanisms that are similar to employees' understanding, while also effectively managing the complex relationship between human behaviour and technological systems.

The facilitation of innovation is enhanced by collaboration across many academic fields since it enables the exchange and integration of varied ideas and methodologies. Psychologists bring useful insights into the behavioural cues that adversaries influence, while legal professionals possess the requisite experience to provide advice on the formulation of rules that properly address privacy concerns. The implementation of a collaborative approach ensures that security policies are not limited to technological concerns, but rather are strategically linked with the very nature of cybersecurity itself.

12 Metrics for Resilience

In the rapidly evolving field of cybersecurity, the concept of digital resilience has gained prominence as a key area of interest for enterprises aiming to effectively manage and mitigate the impacts of cyber disasters. To achieve this goal, it is crucial to develop robust measures that can effectively evaluate the level of resilience inside an organization. The indicators serve a pivotal role as tools for assessing the degree of preparedness, response capabilities, and recovery strategies for cyber threats. This section is dedicated to examining the essential elements of metrics about digital resilience.

12.1 Defining and Measuring Resilience

Digital resilience is a crucial element within modern cybersecurity strategies, including an organization's capacity to predict and endure adverse cyber occurrences, as well as effectively recover from them and adapt proactively. The concept of digital resilience holds significance in the operation of organizations, as it relates to their capacity to effectively manage and uphold data security in the face of cyber-attacks. This involves the reduction of operational disruptions and data breaches, ensuring the ongoing efficiency of the organization. The assessment of resilience is a complex undertaking that involves multiple dimensions of an organization's structure, activities, and reaction systems.

- **Technical Aspects:** The evaluation of digital resilience necessitates a thorough examination of an organization's technical safeguards. This involves evaluating the robustness of firewalls, intrusion detection systems, encryption techniques, and access controls. Technical metrics involve the process of quantifying the efficacy of several elements, including but not limited to threat detection, malware containment, and system recovery activities.

- **Organizational Aspects:** In addition to technical solutions, the organizational aspects of resilience are of equal importance. This involves evaluating the effectiveness of incident response plans, the comprehensibility of communication protocols during emergencies, and the congruence of cybersecurity tactics with broader organizational goals. The measurement of these organizational measures necessitates an understanding of an organization's culture, leadership dynamics, and decision-making mechanisms.
- **Evaluation:** To comprehensively evaluate digital resilience, it is crucial to establish a robust measurement framework that spans the key areas of prevention, detection, response, and recovery. Metrics should include an organization's ability to proactively mitigate cyber incidents, promptly identify breaches or anomalies, respond decisively to mitigate their impact, and expediently restore operations to minimize the periods of inactivity.

12.2 Quantitative Metrics

Quantitative measurements provide a fundamental basis for understanding the capabilities of an organization and evaluating its progress over a while. Several important quantitative metrics can be identified.

- **Recovery Time Objective (RTO):** This metric quantifies the maximum acceptable downtime an organization can endure after an incident before recovery must be achieved.
- **Recovery Point Objective (RPO):** RPO denotes the maximum tolerable data loss that an organization is willing to suffer after recovery. It governs data backup frequency and represents an organization's data-centric resilience strategy.
- **Mean Time Between Failures (MTBF):** The average time between system failures is calculated using MTBF. A shorter MTBF may suggest that system dependability and redundancy are needed.
- **Financial Losses:** The financial impact of cyber incidents provides a clear perspective on the tangible implications of breaches, which assists in resource allocation and risk assessment.

12.3 Qualitative Metrics

Qualitative metrics deals with the nuanced aspects of a company's resilience capabilities, often emphasizing human and cultural factors. Several important qualitative measures include the following

- **Effectiveness of Employee Training Programmes:** This measure assesses how well employees are prepared to recognize and respond to cyber threats. Regular training assessments demonstrate the organization's dedication to developing a cyber-aware staff.
- **Adaptability of Incident Response Plans:** Evaluating the adaptability and flexibility of incident response plans impacts an organization's ability to effectively address novel and emerging threats.
- **Quality of Communication During Crises:** Communication clarity, timeliness, and efficacy are crucial during and after cyber disasters. This indicator measures a company's ability to manage the public relations element of cyber resilience.

Qualitative metrics provide insights into an organization's culture, leadership, and the various aspects that impact its response to cyber incidents. Although challenging to measure in a purely objective manner, these factors offer valuable perspectives on the non-quantifiable yet equally significant aspects of resilience.

12.4 Challenges in Resilience Metrics

Organizations encounter a multitude of complex challenges when attempting to develop suitable metrics for assessing digital resilience.

- **Diverse Organizational Contexts:** Diverse organizational contexts refer to the various environments in which organizations operate. Effectively managing diversity across various organizational contexts poses a significant challenge. Organizations exhibit variability in terms of their size, industry classification, resource availability, and risk tolerance.

  Developing a set of universally applicable resilience indicators that effectively address the requirements of all the above-mentioned scenarios presents a formidable challenge. Metrics should possess the necessary flexibility to accurately reflect the unique characteristics and priorities of individual organizations, while still maintaining a certain level of comparability.
- **Rapidly Evolving threat:** The threat is undergoing rapid evolution. The dynamic and constantly evolving nature of cybersecurity is a significant obstacle when it comes to measuring resilience. New cyber dangers, attack vectors, and vulnerabilities are emerging at a rapid pace.

  The metrics that were created to mitigate risks in the past may become obsolete when confronted with future challenges. To maintain relevance, resilience metrics must possess the necessary adaptability to effectively capture the dynamic characteristics of cyber threats. This adaptability is crucial in ensuring that businesses are adequately prepared to mitigate and respond to future cyber-attacks.
- **Data Availability and Accuracy:** The topic of concern is the availability and accuracy of data. The effectiveness of metrics is contingent upon the quality

of the underlying data. The task of collecting accurate and suitable data poses challenges, particularly when it comes to the qualitative aspects of resilience.

The assessment of the effectiveness of staff training programmes, the evaluation of stakeholder communication, and the examination of event response plans' flexibility require the utilization of reliable and consistent data. To provide an impartial assessment of an organization's resilience capabilities, it is crucial to ensure the accuracy of the data collected.

- **Complex Interdependencies:** Complex interdependencies refer to complex relationships and connections between many elements or components within a system. These interdependencies often involve mutual influences, and in the interconnected digital environment, organizations heavily depend on complex ecosystems consisting of interrelated systems, processes, and third-party services.

  The challenge lies in formulating metrics that adequately include these complex interdependencies. A single incident within a specific component of an organization's digital infrastructure has the potential to trigger a series of subsequent events throughout interconnected systems. To provide a full assessment of an organization's resilience posture, it is crucial for metrics to incorporate and account for these interdependencies.

Addressing these issues necessitates a balance between standardization and flexibility. Resilience indicators should provide a framework for comparison while remaining flexible enough to suit each organization's specific circumstances. Overcoming these issues is critical to keeping resilience measurements current, accurate, and actionable in improving an organization's ability to withstand and recover from cyber disasters.

12.5 Role of Metrics in Resilience Strategies

As stated previously, metrics are necessary as tools that significantly influence the advancement and enhancement of efficient digital resilience measures. These indicators provide enterprises with valuable knowledge on how to effectively minimize cyber risks and enhance their overall security stance. There are several advantages associated with the use of metrics.

- **Informed Decision-Making:** The concept of being informed is of great importance. Metrics offer a quantitative basis for making well-informed judgements. Organizations can make strategic decisions by leveraging data-driven insights derived from their resilience measurements, as opposed to relying solely on intuition or assumptions.

  For instance, in cases where analytics indicate a prolonged average time for the identification of cyber incidents, organizational leaders should allocate additional resources towards enhancing their detection skills.

- **Benchmarking and Improvement:** One of the fundamental advantages of metrics is its capacity to provide a benchmark against industry standards and best practices. Organizations can assess and evaluate their resilience measures by comparing them to those implemented by their counterparts or by industry benchmarks.

  This methodology facilitates the identification of areas in which the organization demonstrates exceptional performance and areas that necessitate further improvement. Benchmarking is a valuable tool that facilitates the identification of precise development goals and the monitoring of progress over time.
- **Resource Allocation:** Metrics facilitate the allocation of resources following the priorities of an organization. If measures indicate a diminished rate of success in phishing awareness training, it may be prudent to spend resources towards enhancing training programmes and activities aimed at raising awareness. The allocation of resources guarantees their optimal utilization.
- **Continuous Improvement:** Continuous improvement is a concept that emphasizes the ongoing effort to enhance processes, systems, and outcomes. Resilience plans exhibit a dynamic nature, necessitating frequent refinement to align with the evolving nature of escalating threats and advancements in technology. Metrics facilitate ongoing enhancement by serving as a mechanism for feedback. By conducting regular assessments and analyses of metrics, organizations can identify trends, anomalies, and areas that require immediate attention. The utilization of this iterative approach ensures that resilience strategies remain adaptable and efficient.
- **Communication and Reporting:** Metrics serve as a means of facilitating effective communication and reporting between stakeholders with varying levels of technical expertise. The establishment of clear and comprehensible metrics is necessary in promoting effective communication regarding an organization's resilience posture, as it facilitates the development of a common language among stakeholders.

  By utilizing clearly defined metrics, organizational leaders, shareholders, regulatory bodies, and the wider public can gain valuable understanding regarding an entity's level of preparedness in the domain of cybersecurity. The implementation of such transparency promotes a sense of assurance and demonstrates a firm commitment to cybersecurity.

13 Concluding Remarks

From a scientific perspective, resilience can be understood as a conceptual framework that relates to the process of developing well-being and promoting positive development throughout an individual's lifelong learning trajectory. This process may initiate in typical settings that are favourable for human growth and development, or it may include utilizing tactics to overcome adversities and challenging situations. In recent decades, researchers and practitioners have implemented this concept in diverse contexts and disciplines, including but not limited to psychological and emotional development, community participation, environmental issues, physics, healthcare, and teaching. In the digital domain, it has only recently established its position. The concept of digital resilience includes a domain of study that has received limited attention in lifelong learning. In a progressively interconnected digital environment marked by the ongoing creation of complex and advanced cyber threats and attacks, the importance of digital resilience has become a subject of significance. In the present context, the term digital resilience refers to the capacity of an organization to actively foresee, withstand, react to, and recuperate from cyber calamities, all the while minimizing detrimental effects on its operations and the integrity of its data.

The increasing prominence of technology in everyday life has led to a growing recognition of the significance of understanding the problems and issues of digital resilience. The notion of digital resilience holds significant implications for a range of stakeholders, including corporations, governments, and individuals. A lack of adequate digital resilience has the potential to result in adverse financial consequences, harm to one's reputation, and the compromise of sensitive data. It is also integral to acknowledge the transnational character of cyber threats and recognize their capacity to impact not just national security but also international relations. It is essential to highlight that businesses are not solely concentrated on mitigating cyberattacks; instead, they are proactively making preparations for the potential occurrence of security breaches. The capacity to effectively sustain essential operations, protect sensitive information, and promptly reinstate functioning can have a substantial impact on an organization's viability and development. The notion of digital resilience is undergoing development in collaboration with the extensive dissemination of technologies. The process of evolution is complex and interconnected with the ever-changing cybersecurity. Understanding the prevailing patterns that are propelling this transformation is crucial, as it serves as the foundation for developing resilient cybersecurity solutions.

Ethical and societal factors are of importance when considering digital resilience. Ethical issues involve a range of factors, which may include but are not limited to user privacy, informed consent, and espionage practices. The need to find an equilibrium between the enforcement of robust security protocols and the safeguarding of individual rights has arisen as a prominent and complex concern. Besides, it is crucial to consider the wider societal problems and issues linked to cyber disasters. Incidents that specifically target critical infrastructure or result in large-scale data breaches have significant implications not only for the entities directly affected but also for society as a whole. These incidents highlight the interconnectedness of the digital domain with essential aspects of our existence. The continual development of digital resilience is characterized by a transition from reactive methods to proactive initiatives.

Numerous businesses are embracing the use of continuous monitoring, proactive threat hunting, and predictive analytics as strategies to detect vulnerabilities in advance of potential exploitation. Similarly, a significant paradigm shift is taking place in cybersecurity as it progressively assumes a pivotal role within the foundational framework of technological systems. This transition represents a change away from regarding cybersecurity as an incidental consideration and instead acknowledging its inherent and essential nature. In the present setting, collaborative approaches are increasingly acknowledged as essential means for promoting digital resilience. There has been an increasing prevalence of industry alliances, platforms for information exchange, and public–private partnerships.

Acknowledging the fact that cyber threats present difficulties that go beyond individual entities, these collaborative initiatives utilize collective intelligence to strengthen the overall resilience of the digital environment. By skillfully manoeuvring through these progressive advancements and diligently addressing the related factors, individuals and groups position themselves to effectively tackle the dynamic difficulties presented by a constantly evolving cyber environment. The concept of digital resilience extends beyond cyber resilience and embraces a wider range of disruptions that businesses may encounter. These disruptions include cyberattacks, natural disasters, and hardware failures. By adopting digital resilience strategies, businesses are better prepared to resist and recover from these many challenges. The essential elements of digital resilience consist of a strong cybersecurity strategy, prompt incident response, and thorough business continuity planning (BCP).

The integration of cybersecurity measures is a fundamental aspect of strengthening digital resilience. Cybersecurity refers to the set of methodologies, technologies, and protocols used to safeguard digital resources, computer systems, and information from unauthorized intrusion, breaches, and various forms of cyber threats. The attainment of digital resilience is a formidable task in the absence of a robust cybersecurity framework. Cybersecurity measures are implemented to detect, reduce, and address cyber threats, including but not limited to malware, phishing attacks, and data breaches. An organization that exhibits resilience not only proactively mitigates these threats but also implements ways to promptly recover from incidents. Digital resilience includes the capacity to sustain corporate operations and services amidst various disturbances, such as cyberattacks. The implementation of cybersecurity measures, including the establishment of backup and disaster recovery plans, is integral in safeguarding company continuity. The concept of digital resilience places significant emphasis on the qualities of adaptation and flexibility within the digital domain. To effectively mitigate cyber threats, cybersecurity measures must undergo continual evolution. To maintain resilience, organizations must modify their cybersecurity posture in response to evolving conditions.

The interconnectedness between cybersecurity and digital resilience is inherently linked to the practice of risk management. The primary objective of cybersecurity attempts is to reduce and overcome cyber threats. Conversely, digital resilience is a concept that incorporates various dangers, including cyber risks, intending to develop overall resistance against such risks. The establishment of effective communication and coordination among cybersecurity teams and other departments within an organization is essential to ensuring the attainment of digital resilience. The consequences of cybersecurity breaches can extend widely, and it is crucial to have a well-coordinated response to uphold resilience. The concept of digital resilience includes a cognitive orientation towards perpetual enhancement. Regular assessment and updating of cybersecurity measures are necessary to effectively address emerging threats and vulnerabilities. This practice aligns with the overarching objective of supporting an organization's overall resilience. Cybersecurity frequently crosses paths with the demands of regulatory compliance, particularly concerning compliance to data protection legislation. The fulfilment of these compliance criteria constitutes an integral component of attempts at cybersecurity and digital resilience.

Glossary

Access Controls: Security mechanisms that regulate resource access based on user roles, privileges, or permissions, ensuring proper authorization for data access.

Antivirus Software: Specialized software designed to detect, prevent, and eliminate malicious software (malware) from a computer system.

Artificial Intelligence: The replication of human cognitive processes, such as reasoning, self-correction, and learning, through the use of machines.

Big Data: Large and complex data sets that require advanced technologies to process, analyse, and extract insights from.

CCPA: California Consumer Privacy Act, a California law governing data privacy that grants individuals greater control over the personal information that organizations retain.

CERT-In: Computer Emergency Response Team – India, the Indian national nodal organization for cybersecurity, responsible for gathering, reviewing, and communicating cybersecurity incidents.

Cloud Computing: Internet-based delivery of computational services, encompassing storage, processing capabilities, and software, facilitates adaptable and expandable resource access.

Continuous Surveillance: An approach to cybersecurity that involves ongoing monitoring, analysis, and response to security threats in real time to detect and mitigate potential risks.

COVID-19 Pandemic: A global health crisis caused by the coronavirus.

Critical Infrastructure: Essential physical and virtual systems, networks, and assets, including power grids, transportation systems, and communication networks, which are crucial for societal functioning.

S. K. Shandilya et al., *Digital Resilience: Navigating Disruption and Safeguarding Data Privacy*, EAI/Springer Innovations in Communication and Computing,
https://doi.org/10.1007/978-3-031-53290-0

Cyber Incidents: Unplanned events that compromise the security or integrity of digital systems, often resulting from cyberattacks or technical failures.

Cyber Resilient Organization: An organization that possesses the capability to effectively manage and mitigate risks and threats arising from cyber disruptions.

Cyber Risks: Potential dangers and vulnerabilities associated with digital systems and online activities, including the threat of cyberattacks, data breaches, and other malicious activities.

Cyber Threats: Potential dangers and risks posed to digital systems and data by malicious activities, such as hacking, malware, phishing, and other forms of cyberattacks.

Cyberbullying: The act of using digital communication platforms, such as social media, to harass, threaten, or intimidate others.

Cybersecurity: The practice of protecting systems, networks, and digital information from cyber threats, including cyberattacks, data breaches, and unauthorized access.

Cybersecurity Awareness: The understanding and knowledge of potential cyber threats and the adoption of suitable steps to mitigate these risks within the digital domain.

Data Analysis: The systematic examination, cleansing, transformation, and modelling of data to reveal valuable insights, draw meaningful conclusions, and facilitate decision-making.

Data Breach: An incident in which sensitive, confidential, or protected data are accessed, stolen, or exposed without authorization.

Data Centre: A specialized facility designed to accommodate computer systems and their related components, including storage and telecommunications systems.

Data Privacy: Safeguarding personal and sensitive information from unauthorized access, utilization, or disclosure, in accordance with legislation and ethical norms.

Digital Divide: The disparity in access to and usage of digital technologies, particularly the internet and related resources, which can lead to social, economic, and political inequalities.

Digital Economy: An economy that is based on digital technologies, where economic activities are conducted through digital platforms, processes, and transactions.

Digital Resilience: An organization's or individual's ability to maintain the confidentiality, integrity, and availability of essential systems and information is achieved through a cohesive integration of technical, operational, and organizational actions while withstanding persistent threats, by implementing robust cybersecurity controls, effective incident response strategies, and a culture promoting cybersecurity awareness and education, all the while recognizing the inevitability of cyber incidents and necessitating rapid and efficient counteractions.

Digital Security: The practice of safeguarding digital systems, networks, and data against cyber threats, involving implementing various methods such as encryption, authentication, access control, and other security protocols.

Disaster Recovery: The process of restoring operations and data after a disruptive event, such as a cyberattack or natural disaster, to minimize downtime and data loss.

Encryption: The process of converting plaintext data into ciphertext using algorithms, enhancing data security by rendering it unreadable to unauthorized parties.

Firewall: Hardware or software implementation of a security system that monitors both inbound and outgoing network traffic in order to prevent unauthorized access to a system.

GDPR: General Data Protection Regulation, a thorough data privacy regulation that controls the handling and safeguarding of personal data belonging to EU residents.

Incident Analysis: The comprehensive investigation of a cyber incident's extent, impact, and root cause to gain an understanding of its dimensions and prioritize corrective actions.

Incident Remediation: The strategic orchestration of actions to contain a cyber incident, prevent further escalation, and restore affected systems and data. It includes measures like removing malicious software and addressing vulnerabilities.

Incident Reporting: The process of swiftly notifying relevant stakeholders, such as the IT department or incident response team, about a cyber incident to initiate the response and recovery procedures.

Incident Response: The organized approach and set of procedures taken to address and manage the aftermath of a cybersecurity incident or breach.

Information Technology Security: Sometimes used interchangeably with *cybersecurity*, it refers to measures designed to protect information technology systems and data from unauthorized access, attacks, and damage.

Intellectual Property: Original innovations of the mind, like inventions, artistic and literary works, symbols, names, designs, and images utilized in commerce, protected by patents, copyrights, and trademarks.

Internet of Things (IoT): An interconnected network of physical devices and objects that can communicate and exchange data over the internet.

IT Act: Information Technology Act, legislation in India that provides legal recognition to electronic transactions and cybersecurity.

KYC: Know Your Customer is a process mandated by the RBI to monitor and track customer data for enhanced protection against payment credential theft and fraud.

Machine Learning: A branch of AI concerned with making computers smarter and perform better through experience rather than static programming.

Malware: A type of software that is intentionally created to penetrate, harm, or obtain unauthorized access to computer systems or networks.

Nature-inspired Cyber Security: Nature-inspired cyber security (NICS) is a method of protecting computing systems by imitating nature's processes, behaviours, and phenomenons, for adaptivity and resilient behaviour.

NIST: A government agency for promoting technological innovation through the establishment of standards and guidelines across various industries.

Patch Management: The practice of applying timely updates (patches) to software and security features to maintain system security and resilience against new threats.

Perimeter Security: A traditional approach to cybersecurity focusing on protecting the boundaries of a network or system, typically through firewalls and intrusion detection systems.

Phishing Attacks: Malicious attempts to deceive people into exposing private information, often through fraudulent websites, emails, or messages, by impersonating trustworthy entities.

Physical Security: The practice of protecting physical assets, people, and resources through measures like locks, surveillance, access control, and security personnel.

PIPEDA: Personal Information Protection and Electronic Documents Act, a data privacy law that regulates how private sector organizations handle personal information in Canada.

Risk Assessment: An organized procedure of identifying, evaluating, and prioritizing potential threats and vulnerabilities that could impact an organization's activities and assets.

Social Engineering: Manipulative techniques used by cybercriminals to exploit human psychology, often tricking individuals into revealing confidential information or performing actions that compromise security.

Socio-Economic Sphere: The intersection of social and economic factors that influence and shape various aspects of society, including the distribution of resources, wealth, opportunities, and access to essential services.

Zero Trust Computing: Zero Trust is a security concept based on the doctrine of 'never trust, always verify', where no user or system is automatically trusted, and continuous verification is required for access to resources.

Zero-Knowledge Proof (ZKP): A zero-knowledge proof is a cryptographic primitive that enables one party to establish the truth of a statement without exposing any further information, in which the prover offers proof or a little bit of information, which may be checked by a verifier.

References

[AB18] Amina Adadi et al. 'Peeking inside the Black-Box: A Survey on Explainable Artificial Intelligence (XAI)'. In: *IEEE Access* 6 (2018), pp. 52138–52160. https://doi.org/10/gfvb5g.

[AC12] Kath Albury et al. 'Sexting, Consent and Young People's Ethics: Beyond Megan's Story'. In: *Continuum: Journal of Media & Cultural Studies* 26.3 (2012), pp. 463–473.

[AC15] Kim M. Anderson et al. 'Challenges and Opportunities of Using Digital Storytelling As a Trauma Narrative Intervention for Traumatized Children'. In: *Advances in Social Work* 16.1 (July 2015), pp. 78–89. https://doi.org/10/gsrr6n.

[Aïm+06] Esma Aïmeur et al. 'Machine Learning in a Quantum World'. In: *Advances in Artificial Intelligence*. Springer Berlin Heidelberg, 2006, pp. 431–442. ISBN: 978-3-540-34628-9. https://doi.org/10/d3mp2n.

[Ajt96] Miklós Ajtai. 'Generating Hard Instances of Lattice Problems'. In: *Proceedings of the twenty-eighth annual ACM symposium on Theory of computing*. 1996, pp. 99–108.

[Ali+21] Fatemeh Alizadeh et al. 'I Don't Know, Is AI Also Used in Airbags?' In: *I-Com* 20.1 (Apr. 2021), pp. 3–17. https://doi.org/10/gsm9mp.

[Alk+21] Rahaf Alkhadra et al. 'Solar Winds Hack: In-Depth Analysis and Countermeasures'. In: *2021 12th International Conference on Computing Communication and Networking Technologies (ICCCNT)*. IEEE, July 2021. https://doi.org/10/gsrskf.

[AM17] Kim M. Anderson et al. 'Digital Storytelling: A Narrative Method for Positive Identity Development in Minority Youth'. In: *Social Work with Groups* 42.1 (Dec. 2017), pp.43–55. https://doi.org/10/gsrr6p.

[Arr+20] Alejandro Barredo Arrieta et al. 'Explainable Artificial Intelligence (XAI): Concepts, Taxonomies, Opportunities and Challenges toward Responsible AI'. In: *Information Fusion* 58 (June 2020), pp. 82–115. https://doi.org/10/ggqs5w

[Asi88] Isaac Asimov. *I, Robot*. Doubleday, 1988, p. 192. ISBN: 9780385423045.

[Aus18] Office of the Australian Information Commissioner. *About the Notifiable Data Breaches Scheme*. https://www.oaic.gov.au/privacy/notifiable-data-breaches/about-the-notifiable-data-breaches-scheme. [Accessed 21-09-2023]. 2018.

[Bai+21] John (Jianqiu) Bai et al. *Digital Resilience: How Work-From-Home Feasibility Affects Firm Performance*. Tech. rep. National Bureau of Economic Research, Mar. 2021. https://doi.org/10/gsrr56.

[Bar+95] Adriano Barenco et al. 'Elementary Gates for Quantum Computation'. In: *Physical Review A* 52.5 (Nov. 1995), pp. 3457–3467. https://doi.org/10/chcw3n.

S. K. Shandilya et al., *Digital Resilience: Navigating Disruption and Safeguarding Data Privacy*, EAI/Springer Innovations in Communication and Computing, https://doi.org/10.1007/978-3-031-53290-0

[Bas+15] Len Bass et al. *Devops: A Software Architect's Perspective*. Addison-Wesley Professional, 2015.

[Bat17] Dhaya Sindhu Battina. 'Best Practices for Ensuring Security in Devops: A Case Study Approach'. In: *International Journal of Innovations in Engineering Research and Technology* 4.11 (2017), pp. 38–45.

[Bay08] Polly Baynes. 'Untold Stories: A Discussion of Life Story Work'. In: *Adoption & Fostering* 32.2 (July 2008), pp. 43–49. https://doi.org/10/gsrr6s.

[BB14] Charles H. Bennett et al. 'Quantum Cryptography: Public Key Distribution and Coin Tossing'. In: *Theoretical Computer Science* 560 (2014), pp. 7–11. https://doi.org/10/3d6.

[BC06] Virginia Braun et al. 'Using Thematic Analysis in Psychology'. In: *Qualitative Research in Psychology* 3.2 (Jan. 2006), pp. 77–101. https://doi.org/10/fswdcx.

[Bel97] John S. Bell. *Speakable and Unspeakable in Quantum Mechanics*. Repr. Collected papers on quantum philosophy. Cambridge [u.a.]: Cambridge University Press, 1997. 212 pp. ISBN:0521368693.

[Ben+17] Marcello Benedetti et al. 'Quantum-Assisted Learning of Hardware-Embedded Probabilistic Graphical Models'. In: *Physical Review X* 7.4 (Nov. 2017), p. 041052. ISSN: 2160–3308. https://doi.org/10/gcnkjz.

[Ben+18] Eli Ben-Sasson et al. 'Scalable, Transparent, and Post-Quantum Secure Computational Integrity'. In: *IACR Cryptol. Eprint Arch.* (2018), p. 46. http://eprint.iacr.org/2018/046.

[Ber+19] Daniel J. Bernstein et al. 'The SPHINCS+ Signature Framework'. In: *Proceedings of the 2019 ACM SIGSAC Conference on Computer and Communications Security*. CCS '19. London, United Kingdom: ACM, Nov. 2019, pp. 2129–2146. ISBN: 9781450367479. https://doi.org/10/gm5b69.

[Ber06] Daniel J. Bernstein. 'Curve25519: New Diffie-Hellman Speed Records'. In: *Public Key Cryptography - PKC 2006*. Springer Berlin Heidelberg, 2006, pp. 207–228. https://doi.org/10/cncrmg.

[Ber21] Luca Bertuzzi. *New EU Law Allows Screening of Online Messages to Detect Child Abuse*. https://www.euractiv.com/section/data-protection/news/new-eu-law-allows-screening-of-online-messages-to-detect-child-abuse/. [Accessed 26-09-2023]. 2021.

[Ber73] Elwyn Berlekamp. 'Goppa Codes'. In: *IEEE Transactions on Information Theory* 19.5 (Sept. 1973), pp. 590–592. https://doi.org/10/bm8vq2.

[Ber90] M. D. Berzonsky. 'Self-Construction Over the Life-Span: A Process Perspective on Identity Formation'. In: *Advances in Personal Construct Psychology: A Research Annual*. Ed. by G. J. Neimeyer et al. Vol. 1. Cambridge, MA: Elsevier Science/JAI Press, 1990, pp. 155–186.

[BG15a] Noam Ben-Asher et al. 'Effects of Cyber Security Knowledge on Attack Detection'. In: *Computers in Human Behavior* 48 (July 2015), pp. 51–61. https://doi.org/10/f7b49d.

[BG15b] Deborah Bunker et al. 'Digital Resilience and the Downside of Digitization'. In: *Mit Sloan Management Review* 56.4 (2015), pp. 61–68.

[Bha+20] Umang Bhatt et al. 'Explainable Machine Learning in Deployment'. In: *Proceedings of the 2020 Conference on Fairness, Accountability, and Transparency*. ACM, Jan. 2020. https://doi.org/10/ghppt9.

[Bha23] Deeksha Bhardwa. *AIIMS Ransomware Attack Led to New SOP on Cyber Breaches: Ex-Cybersecurity Chief Pant*. https://www.hindustantimes.com/india-news/aiims-ransomware-attack-led-to-new-sop-on-cyber-breaches-ex-cybersecurity-chief-pant-101688321198625.html. [Accessed 26-09-2023]. 2023.

[Bjö+15] Fredrik Björck et al. 'Cyber Resilience – Fundamentals for a Definition'. In: *New Contributions in Information Systems and Technologies*. Springer International Publishing, 2015, pp. 311–316. https://doi.org/10/f3nchp.

[BK22] Mary T. Breen et al. 'New Cybersecurity Regulations Are Coming. Here's How to Prepare'. In: *Harvard Business Review* (Aug. 2022). https://hbr.org/2022/08/new-cybersecurity-regulations-are-coming-heres-how-to-prepare.

[BL15] Alyssa N. Byers et al. 'Therapeutic Alliance with Youth in Residential Care: Challenges and Recommendations'. In: *Residential Treatment for Children & Youth* 32.1 (Jan. 2015), pp. 1–18. https://doi.org/10/gsrr6r.

[Bod+11] Deborah Bodeau et al. 'Cyber Resiliency Engineering Framework'. In: *Mtr110237, Mitrecorporation* (2011).

[Bod+15] Deborah Bodeau et al. 'Cyber Resiliency Engineering Aid–The Updated Cyber Resiliency Engineering Framework and Guidance on Applying Cyber Resiliency Techniques'. In: *Mtr140499r1, Pr* (2015), pp. 15–1334.

[Bos+18] Joppe Bos et al. 'CRYSTALS - Kyber: A CCA-Secure Module-Lattice-Based KEM'. In: *IEEE European Symposium on Security and Privacy (EuroS&P)*. IEEE. IEEE, Apr. 2018, pp. 353–367. https://doi.org/10/ggbscj.

[Boy15] Hugh Boyes. 'Cybersecurity and Cyber-Resilient Supply Chains'. In: *Technology Innovation Management Review* 5.4 (Apr. 2015), pp. 28–34. https://doi.org/10/gpxqrj.

[Boy20] Danah Boyd. *It's Complicated*. Yale University Press, Mar. 2020. https://doi.org/10/grw2rw.

[Boz+22] Elena Bozzola et al. 'The Use of Social Media in Children and Adolescents: Scoping Review on the Potential Risks'. In: *International Journal of Environmental Research and Public Health* 19.16 (Aug. 2022), p. 9960. https://doi.org/10/j398.

[BP21] Corneliu Bjola et al. 'Digital Propaganda, Counterpublics, and the Disruption of the Public Sphere: The Finnish Approach to Building Digital Resilience'. In: *The World Information War*. Routledge, 2021, pp. 186–213.

[Bre18] Thomas Brewster. *Revealed: Marriott's 500 Million Hack Came After A String Of Security Breaches*. forbes.com/sites/thomasbrewster/2018/12/03/revealed-marriotts-500-million-hack-came-after-a-string-of-security-breaches/. [Accessed 22-08-2023]. 2018.

[Bru17] Z. Bruckauf. *Adolescents' Mental Health: Out of the Shadows - Evidence on Psychological Wellbeing of 11-15-Year-Olds from 31 Industrialized Countries*. 2017. https://www.unicef-irc.org/publications/pdf/IRB_2017_12.pdf.

[BV97] Ethan Bernstein et al. 'Quantum Complexity Theory'. In: *SIAM Journal on Computing* 26.5 (Oct. 1997), pp. 1411–1473. https://doi.org/10/dkrq62.

[BW88] Michael Ben-Or et al. 'Completeness Theorems for Non-Cryptographic Fault-Tolerant Distributed Computation'. In: *Proceedings of the twentieth annual ACM symposium on Theory of computing - STOC '88*. ACM Press, 1988. https://doi.org/10/bx5d45.

[Cai+15] X.-D. Cai et al. 'Entanglement-Based Machine Learning on a Quantum Computer'. In: *Physical Review Letters* 114.11 (Mar. 2015), p. 110504. https://doi.org/10/gd7ff6.

[Can+96] Ran Canetti et al. 'Adaptively Secure Multi-Party Computation'. In: *Proceedings of the twenty-eighth annual ACM symposium on Theory of computing - STOC '96*. ACM Press, 1996, pp. 639–648. https://doi.org/10/dj366k.

[Car+20a] Juan Francisco Carías et al. 'Systematic Approach to Cyber Resilience Operationalization in SMES'. In: *IEEE Access* 8 (2020), pp. 174200–174221. https://doi.org/10/gsrr57.

[Car+20b] A. Carr et al. 'A Systematic Review of Reviews of the Outcome of Severe Neglect in Underresourced Childcare Institutions'. In: *Trauma, Violence & Abuse* 21.3 (2020), pp. 484–497.

[Car17] Kim Carter. 'Francois Raynaud on DevSecOps'. In: *IEEE Software* 34.5 (2017), pp. 93–96. https://doi.org/10/gsrr74.

[Cas+18] Luciana Cassarino-Perez et al. 'From Care to Education and Employment: A Meta-Analysis'. In: *Children and Youth Services Review* 95 (Dec. 2018), pp. 407–416. https://doi.org/10/gsrr6q.

[Cas16a] Defense Use Case. 'Analysis of the Cyber Attack on the Ukrainian Power Grid'. In: *Electricity Information Sharing and Analysis Center (E-ISAC)* 388.1-29 (2016), p. 3.

[Cas16b] Davide Castelvecchi. 'Can We Open the Black Box of AI?' In: *Nature* 538.7623 (Oct. 2016), pp. 20–23. https://doi.org/10/ggv7tg.

[Cav+16] Mary a Cavanagh et al. 'Social Media and Resilience in Youth: A Literature Review'. In: *Educational Psychology Review* 28.3 (2016), pp. 735–756.

[Cav22] Myriam Dunn Cavelty. 'Cyber-Security'. In: *Contemporary Security Studies*. Oxford University Press, Jan. 2022, pp. 422–436. https://doi.org/10/gsrr8m.

[CD22] Wouter Castryck et al. *An Efficient Key Recovery Attack on SIDH*. Cryptology ePrint Archive, Paper 2022/975. 2022. https://eprint.iacr.org/2022/975.

[Cho+16] Jin-Hee Cho et al. 'Metrics and Measurement of Trustworthy Systems'. In: *MILCOM 2016 - 2016 IEEE Military Communications Conference*. IEEE, Nov. 2016. https://doi.org/10/ggx8q6.

[CIS21] CISA. *Cyber-Attack Against Ukrainian Critical Infrastructure*. https://www.cisa.gov/news-events/ics-alerts/ir-alert-h-16-056-01. [Accessed 23-09-2023]. 2021.

[CIS22] CISA. *Cyber Storm: Securing Cyber Space*. https://www.cisa.gov/cyber-storm-securing-cyber-space. [Accessed 21-09-2023]. Mar. 2022.

[Cle+98] R. Cleve et al. 'Quantum Algorithms Revisited'. In: *Proceedings of the Royal Society of London. Series a: Mathematical, Physical and Engineering Sciences* 454.1969 (Jan. 1998), pp. 339–354. https://doi.org/10/dnzfpd.

[CN22] Hugh Collins et al. *IBM Unveils 400 Qubit-Plus Quantum Processor and Next-Generation IBM Quantum System Two*. https://newsroom.ibm.com/2022-11-09-IBM-Unveils-400-Qubit-Plus-Quantum-Processor-and-Next-Generation-IBM-Quantum-System-Two. [Accessed 24-08-2023]. 2022.

[Col17] Roger Collier. 'NHS Ransomware Attack Spreads Worldwide'. In: *Canadian Medical Association Journal* 189.22 (June 2017), E786–E787. https://doi.org/10/gq65j5.

[Com16] European Commission. *General Data Protection Regulation (GDPR) – Official Legal Text*. https://gdpr-info.eu/. [Accessed 21-09-2023]. 2016.

[Com21] European Commission. *Laying down Harmonised Rules on Artificial Intelligence (Artificial Intelligence Act) and Amending Certain Union Legislative Acts*. https://eur-lex.europa.eu/legal-content/EN/TXT/?uri=CELEX:52021PC0206. [Accessed 21-09-2023]. 2021.

[Com22a] European Commission. *Digital Markets Act: Rules for Digital Gatekeepers to Ensure Open Markets Enter into Force*. https://ec.europa.eu/commission/presscorner/detail/en/IP_22_6423. [Accessed 26-09-2023]. 2022.

[Com22b] European Commission. *The Digital Services Act Package, 2022*. https://digital-strategy.ec.europa.eu/en/policies/digital-services-act-package. [Accessed 26-09-2023]. 2022.

[Com22c] European Commission. *The Digital Services Act: Ensuring a Safe and Accountable Online Environment*. https://commission.europa.eu/strategy-and-policy/priorities-2019-2024/europe-fit-digital-age/digital-services-act-ensuring-safe-and-accountable-online-environment_en. [Accessed 26-09-2023]. 2022.

[Com23a] European Commission. *Data Protection: European Commission Adopts New Adequacy Decision for Safe and Trusted EU-US Data Flows*. https://ec.europa.eu/commission/presscorner/detail/en/ip_23_3721. [Accessed 26-09-2023]. 2023.

[Com23b] European Commission. *EU AI Act: First Regulation on Artificial Intelligence*. https://www.europarl.europa.eu/news/en/headlines/society/20230601STO93804/eu-ai-act-first-regulation-on-artificial-intelligence. [Accessed 21-09-2023]. 2023.

[Com23c] European Commission. *Proposal for an EPrivacy Regulation*. https://digital-strategy.ec.europa.eu/en/policies/eprivacy-regulation. [Accessed 26-09- 2023]. 2023.

[Com23d] European Commission. *Questions & Answers: EU-US Data Privacy Framework*. https://ec.europa.eu/commission/presscorner/detail/en/qanda_23_3752. [Accessed 26-09-2023]. 2023.

[Con96] United States Congress. *Health Insurance Portability and Accountability Act (HIPPA) of 1996*. 1996. https://www.congress.gov/bill/104th-congress/house-bill/3103.

[Cos+17] Craig Costello et al. 'Efficient Compression of SIDH Public Keys'. In: *Lecture Notes in Computer Science*. Springer International Publishing, 2017, pp. 679–706. https://doi.org/10/gsm65m. https://eprint.iacr.org/2016/963.

[Cou+01] Nicolas T. Courtois et al. 'How to Achieve a McEliece-Based Digital Signature Scheme'. In: *Advances in Cryptology —ASIACRYPT 2001*. Springer Berlin Heidelberg, 2001, pp. 157–174. https://doi.org/10/cnt9kh.

[Cra+14] Dan Craigen et al. 'Defining Cybersecurity'. In: *Technology Innovation Management Review* 4.10 (Oct. 2014), pp. 13–21. https://doi.org/10/gnx5x7.

[DB14] Nikesh S. Dattani et al. 'Quantum Factorization of 56153 with Only 4 Qubits'. In: (25th Nov. 2014). https://doi.org/10/gsm646. arXiv: 1411.6758 [quant-ph] (cit. on p. 480).

[DB18] Vedran Dunjko et al. 'Machine Learning & Artificial Intelligence in the Quantum Domain: A Review of Recent Progress'. In: *Reports on Progress in Physics* 81.7 (June 2018), p. 074001. https://doi.org/10/gd7gkw.

[DC21] József Dombi et al. *Explainable Neural Networks Based on Fuzzy Logic and MultiCriteria Decision Tools*. Springer International Publishing, 2021. https://doi.org/10/gsm9mt.

[Dep17] Department of Health and Department for Education. *Transforming Children and Young People's Mental Health Provision: A Green Paper*. 2017. https://assets.publishing.service.gov.uk/government/uploads/system/uploads/attachment_data/file/664855/Transforming_children_and_young_people_s_mental_health_provision.pdf.

[Dep18] Department of Health and Department for Education. *Mental Health and Behaviour in Schools*. 2018. https://assets.publishing.service.gov.uk/government/uploads/system/uploads/attachment_data/file/755135/Mental_health_and_behaviour_in_schools__.pdf.

[DH00] David Deutsch et al. 'Information Flow in Entangled Quantum Systems'. In: *Proceedings of the Royal Society of London. Series a: Mathematical, Physical and Engineering Sciences* 456.1999 (July 2000), pp. 1759–1774. https://doi.org/10/fm2nnz.

[DH76] W. Diffie et al. 'New Directions in Cryptography'. In: *IEEE Transactions on Information Theory* 22.6 (Nov. 1976), pp. 644–654. https://doi.org/10/cmtvwm.

[Die82] D. Dieks. 'Communication by EPR Devices'. In: *Physics Letters A* 92.6 (Nov. 1982), pp. 271–272. https://doi.org/10/dfvj8v.

[Dig19] Virginia Dignum. *Responsible Artificial Intelligence*. Vol. 2156. Springer International Publishing, 2019. https://doi.org/10/gsm9mw.

[DJ92] David Deutsch et al. 'Rapid Solution of Problems by Quantum Computation'. In: *Proceedings of the Royal Society of London. Series a: Mathematical and Physical Sciences* 439.1907 (Dec. 1992), pp. 553–558. https://doi.org/10/fjp4h9.

[Dor+17] Derek Doran et al. 'What Does Explainable AI Really Mean? A New Conceptualization of Perspectives'. In: *ArXiv Preprint ArXiv:1710.00791* (2017).

[DR99] Joan Daemen et al. 'AES Proposal: Rijndael'. In: (1999).

[Duc+17] Leo Ducas et al. *Crystals – Dilithium: Digital Signatures from Module Lattices*. Cryptology ePrint Archive, Paper 2017/633. 2017. https://eprint.iacr.org/2017/633.

[Dup19] Benoît Dupont. 'The Cyber-Resilience of Financial Institutions: Significance and Applicability'. In: *Journal of Cybersecurity* 5.1 (Jan. 2019), tyz013. https://doi.org/10/gpxqrg.

[Ebe+16] Christof Ebert et al. 'DevOps'. In: *IEEE Software* 33.3 (May 2016), pp. 94–100. https://doi.org/10/gd2xw5.

[Edu21] U.S. Department of Education. *Family Educational Rights and Privacy Act (FERPA), 1973*. https://www2.ed.gov/policy/gen/guid/fpco/ferpa/index.html. [Accessed 23-09-2023]. 2021.

[EF23] Jen Easterly et al. *The Attack on Colonial Pipeline: What We've Learned & What We've Done Over the Past Two Years*. https://www.cisa.gov/news-events/news/attack-colonial-pipeline-what-weve-learned-what-weve-done-over-past-two-years. [Accessed 22-08-2023]. 2023.

[Ein+35] A. Einstein et al. 'Can Quantum-Mechanical Description of Physical Reality Be Considered Complete?' In: *Physical Review* 47.10 (May 1935), pp. 777–780. https://doi.org/10/bgd7kp.

[EL14] Adel S. Elmaghraby et al. 'Cyber Security Challenges in Smart Cities: Safety, Security and Privacy'. In: *Journal of Advanced Research* 5.4 (July 2014), pp. 491–497. https://doi.org/10/gfrbvj.

[EM17] Thomas W Edgar et al. *Research Methods for Cyber Security*. Syngress, 2017.

[Fay18] Sharifah Yaqoub A. Fayi. 'What Petya/NotPetya Ransomware Is and What Its Remidiations Are'. In: *Advances in Intelligent Systems and Computing*. Springer International Publishing, 2018, pp. 93–100. https://doi.org/10/gsrsdj.

[Fei+88] Uriel Feige et al. 'Zero-Knowledge Proofs of Identity'. In: *Journal of Cryptology* 1.2 (June 1988), pp. 77–94. https://doi.org/10/dfb8d2.

[Fer15] Jason Ferdinand. 'Building Organisational Cyber Resilience: A Strategic Knowledge-Based View of Cyber Security Management'. In: *Journal of business continuity & emergency planning* 9.2 (2015), pp. 185–195.

[Fey82] Richard P. Feynman. 'Simulating Physics with Computers'. In: *International Journal of Theoretical Physics* 21.6-7 (June 1982), pp. 467–488. https://doi.org/10/cpkzhd.

[Fey86] Richard P. Feynman. 'Quantum Mechanical Computers'. In: *Foundations of Physics* 16.6 (June 1986), pp. 507–531. https://doi.org/10/b3kbv7.

[FS03] Niels Ferguson et al. *Practical Cryptography*. Vol. 141. Wiley New York, 2003.

[Gar97] Michael R. Garey. *Computers and Intractability. A Guide to the Theory of NP-Completeness*. W. H. Freeman, 1997, p. 338. ISBN: 0716710447.

[GB20] Kholekile Gwebu et al. 'Data Breaches in Hospitality: Is the Industry Different?' In: *Journal of Hospitality and Tourism Technology* 11.3 (2020), pp. 511–527.

[GE21] Craig Gidney et al. 'How to Factor 2048 Bit RSA Integers in 8 Hours Using 20 million Noisy Qubits'. In: *Quantum* 5 (Apr. 2021), p. 433. https://doi.org/10/gmbpbg.

[GG19] Beni Gómez-Zúñiga et al. 'Digital Resilience in Adolescence: A Systematic Review'. In: *Children and Youth Services Review* 104 (2019), p. 104390.

[GH21] Jelle Groenendaal et al. 'Cyber Resilience during the COVID-19 Pandemic Crisis: A Case Study'. In: *Journal of Contingencies and Crisis Management* 29.4 (May 2021), pp. 439–444. https://doi.org/10/gmbzgx.

[Gis+02] Nicolas Gisin et al. 'Quantum Cryptography'. In: *Reviews of Modern Physics* 74.1 (Mar. 2002), pp. 145–195. https://doi.org/10/bjxpf6.

[Gol+89] Shafi Goldwasser et al. 'The Knowledge Complexity of Interactive Proof Systems'. In: *SIAM Journal on Computing* 18.1 (Feb. 1989), pp. 186–208. https://doi.org/10/cqx3{km}.

[Gol+97] Oded Goldreich et al. 'Public-Key Cryptosystems from Lattice Reduction Problems'. In: *Advances in Cryptology —CRYPTO '97*. Springer Berlin Heidelberg, 1997, pp. 112–131. https://doi.org/10/dg9xb3.

[Gol97] Shafi Goldwasser. 'Multi Party Computations'. In: *Proceedings of the sixteenth annual ACM symposium on Principles of distributed computing - PODC '97*. ACM Press, 1997. https://doi.org/10/bz8c22.

[Gol98] Oded Goldreich. 'Secure Multi-Party Computation'. In: *Manuscript. Preliminary version* 78.110 (1998).

[Gov03] Australian Government. *Spam Act 2003*. https://www.legislation.gov.au/Details/C2016C00614. [Accessed 21-09-2023]. 2003.

[Gov06] Australian Government. *Do Not Call Register Act 2006*. https://www.legislation.gov.au/Series/C2006A00088. [Accessed 21-09-2023]. 2006.

[Gov10] Australian Government. *Healthcare Identifiers Act 2010*. https://www.legislation.gov.au/Series/C2010A00072. [Accessed 21-09-2023]. 2010.

[Gov12] Australian Government. *My Health Records Act 2012*. https://www.legislation.gov.au/Details/C2021C00475. [Accessed 21-09-2023]. 2012.

[Gov18a] Australian Government. *Telecommunications and Other Legislation Amendment (Assistance and Access) Act 2018*. https://www.legislation.gov.au/Details/C2018A00148. [Accessed 21-09-2023]. 2018.

[Gov18b] UK Government. *Data Protection Act 2018*. https://www.gov.uk/data-protection. [Accessed 21-09-2023]. 2018.

[Gov23a] U.S. Goverment. *Data Privacy Framework - Official Website*. https://www.dataprivacyframework.gov/. [Accessed 26-09-2023]. 2023.

[Gov23b] IT Governance. *The EU EPR (ePrivacy Regulation) | What You Need to Know*. https://www.itgovernance.co.uk/eprivacy-regulation-epr. [Accessed 26-09-2023]. 2023.

[Gov88] Australian Government. *Privacy Act 1988*. https://www.legislation.gov.au/Series/C2004A03712/Amendments. [Accessed 21-09-2023]. 1988.

[Gov97] Australian Government. *Telecommunications Act 1997*. https://www.legislation.gov.au/Series/C2004A05145. [Accessed 21-09-2023]. 1997.

[Gov99] Australian Government. *Electronic Transactions Act 1999*. https://www.legislation.gov.au/Series/C2004A00553. [Accessed 21-09-2023]. 1999.

[Gre+19] Alex Green et al. 'Keeping Children Safe Online'. In: *New Philanthropy Capital* 28 (2019).

[Gre17] Seena Gressin. 'The Equifax Data Breach: What to Do'. In: *Federal Trade Commission* 8 (2017).

[Gre18] Andy Greenberg. 'The Untold Story of Notpetya, the Most Devastating Cyberattack in History'. In: *Wired* 22 (2018).

[Gro96] Lov K. Grover. 'A Fast Quantum Mechanical Algorithm for Database Search'. In: *Proceedings of the twenty-eighth annual ACM symposium on Theory of computing - STOC '96*. ACM Press, 1996. ISBN: 978-0-89791-785-8. https://doi.org/10/fkzv3d.

[Gun+19] David Gunning et al. 'XAI-Explainable Artificial Intelligence'. In: *Science Robotics* 4.37 (Dec. 2019). https://doi.org/10/ggk3km.

[HA13] Lewis Herrington et al. 'The Future of Cyber-Resilience in an Age of Global Complexity'. In: *Politics* 33.4 (Nov. 2013), pp. 299–310. https://doi.org/10/f5f8n3.

[Ham+23] Simon Patrick Hammond et al. 'Using a Socio-Ecological Framework to Understand How 8–12-Year-Olds Build and Show Digital Resilience: A Multi-Perspective and Multimethod Qualitative Study'. In: *Education and Information Technologies* 28.4 (2023), pp. 3681–3709.

[Hau20] Kjell Hausken. 'Cyber Resilience in Firms, Organizations and Societies'. In: *Internet of Things* 11 (Sept. 2020), p. 100204. https://doi.org/10/gsrr55.

[Héd23] Mihály Héder. 'Explainable AI: A Brief History of the Concept'. In: *Ercim News* 134 (2023), pp. 9–10.

[Hee+23] Raoul Heese et al. *Explaining Quantum Circuits with Shapley Values: Towards Explainable Quantum Machine Learning*. 2023. https://doi.org/10/gs44w4. arXiv: 2301.09138 [quant-ph].

[Her17] Alex Hern. *WannaCry, Petya, NotPetya: How Ransomware Hit the Big Time in 2017*. https://www.theguardian.com/technology/2017/dec/30/wannacry-petya-notpetya-ransomware. [Accessed 21-09-2023]. 2017.

[HF17] Qi Huang et al. 'A Framework for Digital Resilience of Critical Infrastructure'. In: *Proceedings of the 2017 IEEE International Conference on Information Reuse and Integration (IRI)*. 2017, pp. 13–18. https://doi.org/10/grw2r2.

[HL16] Uwe Hasebrink et al. 'Comparing Children's Online Opportunities and Risks across Europe: Cross-National Comparisons for EU Kids Online'. In: *Children and Youth Services Review* 63 (2016), pp. 58–67.

[HM14] Wayne Harrop et al. 'Cyber Resilience: A Review of Critical National Infrastructure and Cyber-Security Protection Measures Applied in the UK and USA'. In: *Current and Emerging Trends in Cyber Operations*. Vol. 7. 2. Palgrave Macmillan UK, 2014, pp. 149–166. https://doi.org/10/gsrr6f.

[HN09] Lene Hansen et al. 'Digital Disaster, Cyber Security, and the Copenhagen School'. In: *International Studies Quarterly* 53.4 (Dec. 2009), pp. 1155–1175. https://doi.org/10/cwhqrz.

[Hof+09] Jeff Hoffstein et al. 'Practical Lattice-Based Cryptography: NTRUEncrypt and NTRUSign'. In: *The LLL Algorithm*. Springer Berlin Heidelberg, 2009, pp. 349–390. https://doi.org/10/d8bjhj (cit. on p. 486).

[Hsu18] Tony Hsiang-Chih Hsu. *Hands-On Security in DevOps: Ensure continuous security, deployment, and delivery with DevSecOps*. Packt Publishing Ltd, 2018.

[Hue+18] A. Huelsing et al. *XMSS: eXtended Merkle Signature Scheme*. Tech. rep. May 2018. https://doi.org/10/gsm65k.

[IBM19] IBM. *IBM Quantum System One*. https://research.ibm.com/interactive/system-one/. [Accessed 20-09-2023]. 2019.

[Ind00] Government of India. *India Code: Information Technology Act, 2000*. https://www.indiacode.nic.in/handle/123456789/1999. [Accessed 21-09-2023]. 2000.

[Ind02] Government of India. *Guidelines on "Know Your Customer" Norms and "Cash Transactions," 2002*. https://www.rbi.org.in/commonman/Upload/English/Notification/PDFs/30926.pdf. [Accessed 21-09-2023]. 2002.

[Ind08] Government of India. *Information Technology Act (Amendment), 2008*. https://www.meity.gov.in/writereaddata/files/itact2000/it_amendment_act2008.pdf. [Accessed 21-09-2023]. 2008.

[Ind11a] Government of India. *Information Technology (Intermediary Guidelines and Digital Media Ethics Code) Rules, 2011*. [Accessed 21-09-2023]. 2011. https://www.meity.gov.in/writereaddata/files/GSR313E_10511%281%29_0.pdf.

[Ind11b] Government of India. *Information Technology (Reasonable Security Practices and Procedures and Sensitive Personal Information) Rules, 2011*. [Accessed 21-09-2023]. 2011. https://www.meity.gov.in/writereaddata/files/GSR314E_10511%281%29_0.pdf.

[Ind13] Government of India. *National Cyber Security Policy, 2013*. [Accessed 21-09-2023]. 2013. https://www.meity.gov.in/writereaddata/files/downloads/National_cyber_security_policy-2013%281%29.pdf (cit. on p. 221).

[Ind20] Government of India. *National Cyber Security Strategy, 2020*. https://www.dsci.in/files/content/knowledge-centre/2023/National-Cyber-Security-Strategy-2020-DSCI-submission.pdf. [Accessed 21-09-2023]. 2020.

[Ind21] Government of India. *Information Technology (Intermediary Guidelines and Digital Media Ethics Code) Rules, 2021*. [Accessed 21-09-2023]. 2021. https://mib.gov.in/sites/default/files/IT%28Intermediary%20Guidelines%20and%20Digital%20Media%20Ethics%20Code%29%20Rules%2C%202021%20English.pdf (cit. on p. 224).

[Ind23] Government of India. *Digital Personal Data Protection Act, 2023*. [Accessed 21-09-2023]. 2023. https://www.meity.gov.in/writereaddata/files/Digital%20Personal%20Data%20Protection%20Act%202023.pdf.

[ITU20] International Telecommunications Union (ITU). *Guidelines for Parents and Educators*. https://www.itu-cop-guidelines.com/parentsandeducators. 2020.

[JM15] M. I. Jordan et al. 'Machine Learning: Trends, Perspectives, and Prospects'. In: *Science* 349.6245 (July 2015), pp. 255–260. https://doi.org/10/f7kcgq.

[Jus23] Department of Justice (DOJ). *Keeping Children Safe Online*. https://www.justice.gov/coronavirus/keeping-children-safe-online. 2023.

[Kal19] Laurence Kalman. 'New European Data Privacy and Cyber Security Laws'. In: *Communications of the ACM* 62.4 (Mar. 2019), pp. 38–38. https://doi.org/10/gsrr8n.

[Kia+21] Mazaher Kianpour et al. 'Systematically Understanding Cybersecurity Economics: A Survey'. In: *Sustainability* 13.24 (Dec. 2021), p. 13677. https://doi.org/10/gsm9mh.

[Kil92] Joe Kilian. 'A Note on Efficient Zero-Knowledge Proofs and Arguments (extended Abstract)'. In: *Proceedings of the twenty-fourth annual ACM symposium on Theory of computing - STOC '92*. ACM Press, 1992. https://doi.org/10/b7p96m.

[Kip+99] Aviad Kipnis et al. 'Unbalanced Oil and Vinegar Signature Schemes'. In: *Advances in Cryptology —EUROCRYPT '99*. Springer Berlin Heidelberg, 1999, pp. 206–222. https://doi.org/10/fpcvwq.

[Kit95] A. Yu. Kitaev. 'Quantum Measurements and the Abelian Stabilizer Problem'. In: (20th Nov. 1995). https://doi.org/10/gsm645. arXiv: quant-ph/9511026 [quant-ph].

[KL07] Jonathan Katz et al. *Introduction to Modern Cryptography: Principles and Protocols*. Chapman and Hall/CRC, 2007.

[KL19] Alexander Kott et al. *Cyber Resilience of Systems and Networks*. Springer, 2019.

[Kla+16] Günter Klambauer et al. 'DeepTox: Toxicity Prediction Using Deep Learning'. In: *Toxicology Letters* 280.80 (2016), S69. https://doi.org/10/gsngt5.

[Kos17] Jeff Kosseff. 'Defining Cybersecurity Law'. In: *Iowa L. Rev.* 103 (2017), p. 985.

[KS98] Aviad Kipnis et al. 'Cryptanalysis of the Oil and Vinegar Signature Scheme'. In: *Advances in Cryptology —CRYPTO '98*. Springer Berlin Heidelberg, 1998, pp. 257–266. https://doi.org/10/b3g8dx.

[KZ77] A. Korkine et al. 'Sur Les Formes Quadratiques Positives'. In: *Mathematische Annalen* 11.2 (June 1877), pp. 242–292. https://doi.org/10/dff9fj.

[Lam79] Leslie Lamport. *Constructing Digital Signatures from a One Way Function*. Tech. rep. CSL-98. Oct. 1979. https://www.microsoft.com/en-us/research/publication/constructing-digital-signatures-one-way-function/.

[Lee15] Newton Lee. *Counterterrorism and Cybersecurity Total Information Awareness. Total Information Awareness*. Springer, 2015, p. 489. ISBN: 9783319172439.

[Lei+19] Leonardo Leite et al. 'A Survey of DevOps Concepts and Challenges'. In: *ACM Computing Surveys* 52.6 (Nov. 2019), pp. 1–35. https://doi.org/10/ggmt4t.

[Len+82] A. K. Lenstra et al. 'Factoring Polynomials with Rational Coefficients'. In: *Mathematische Annalen* 261.4 (Dec. 1982), pp. 515–534. https://doi.org/10/c3j8hx.

[Li18] Jian-hua Li. 'Cyber Security Meets Artificial Intelligence: A Survey'. In: *Frontiers of Information Technology & Electronic Engineering* 19.12 (2018), pp. 1462–1474. https://doi.org/10/gk764c.

[Lid83] Rudolf Lidl. 'Niederreiter, Finite Fields'. In: *Encyclopedia of Mathematics and Its Applications* 20 (1983).

[Lip18] Zachary C. Lipton. 'The Mythos of Model Interpretability'. In: *Queue* 16.3 (June 2018), pp. 31–57. https://doi.org/10/gkbfrd.

[Liu16] Alan Liu. *Digital Resilience: Is Your Company Ready for the Next Cyber Threat?* Boca Raton, FL: Taylor & Francis, 2016.

[LK19] Igor Linkov et al. 'Fundamental Concepts of Cyber Resilience: Introduction and Over-view'. In: *Cyber Resilience of Systems and Networks* (2019), pp. 1–25.

[Llo+13] Seth Lloyd et al. *Quantum Algorithms for Supervised and Unsupervised Machine Learning*. 2013. arXiv: 1307.0411 [quant-ph].

[LM95] Frank T Leighton et al. *Large Provably Fast and Secure Digital Signature Schemes Based on Secure Hash Functions*. US Patent 5,432,852. July 1995.

[LN96] Rudolf Lidl et al. *Finite Fields*. Vol. 20. Cambridge University Press, Oct. 1996. https://doi.org/10/bv3d8g.

[Log17] LogRhythm. *A Technical Analysis of WannaCry Ransomware*. https://logrhythm.com/blog/a-technical-analysis-of-wannacry-ransomware/. [Accessed 21-09-2023]. 2017.

[Loy19] Octavio Loyola-Gonzalez. 'Black-Box Vs. White-Box: Understanding Their Advantages and Weaknesses from a Practical Point of View'. In: *IEEE Access* 7 (2019), pp. 154096–154113. https://doi.org/10/ggw69g.

[Lun+20] Scott M. Lundberg et al. 'From Local Explanations to Global Understanding with Explainable AI for Trees'. In: *Nature Machine Intelligence* 2.1 (Jan. 2020), pp. 56–67. https://doi.org/10/ggjtp4.

[LV00] Arjen K. Lenstra et al. 'The XTR Public Key System'. In: *Advances in Cryptology – CRYPTO 2000*. Springer Berlin Heidelberg, 2000, pp. 1–19. https://doi.org/10/cktjsq.

[Lyu+13] Vadim Lyubashevsky et al. 'A Toolkit for Ring-LWE Cryptography'. In: *Advances in Cryptology –EUROCRYPT 2013*. Springer Berlin Heidelberg, 2013, pp. 35–54. https://doi.org/10/gqvd55. https://eprint.iacr.org/2013/293.

[MA18] Puneet Mehra et al. 'Digital Resilience: A Precondition for Sustainable Development'. In: *Sustainable Development* 26.5 (2018), pp. 433–446. https://doi.org/10/gkjzwn.

[Mac+19] Mike Mackintosh et al. 'Preliminaries of Orthogonal Layered Defence Using Functional and Assurance Controls in Industrial Control Systems'. In: *Journal of Sensor and Actuator Networks* 8.1 (Feb. 2019), p. 14. https://doi.org/10/gj742h.

[Mah20] Batta Mahesh. 'Machine Learning Algorithms-A Review'. In: *International Journal of Science and Research (ijsr)* 9.1 (2020), pp. 381–386.

[Man80] Yuri Manin. 'Computable and Uncomputable'. In: *Sovetskoye Radio, Moscow* 128 (1980), p. 28 (cit. on p. 467).

[Mat+09] Tim Mather et al. *Cloud Security and Privacy. An Enterprise Perspective on Risks and Compliance*. O'Reilly, 2009. ISBN: 9780596802769.

[Mat12] John Matson. 'Quantum Teleportation Achieved Over Record Distances'. In: *Nature* (Aug. 2012). https://doi.org/10/gsrr63.

[MC17] Håvard Myrbakken et al. 'DevSecOps: A Multivocal Literature Review'. In: *Communications in Computer and Information Science*. Springer International Publishing, 2017, pp. 17–29. https://doi.org/10/ggbwkh.

[MC19] Jan Miller et al. *WannaCry, Software S0366*. https://attack.mitre.org/versions/v13/software/S0366/. [Accessed 21-09-2023]. 2019.

[McC04] Pamela McCorduck. *Machines Who Think. A Personal Inquiry into the History and Prospects of Artificial Intelligence*. A.K. Peters, 2004, p. 565. ISBN: 1568812051.

[McE78] Robert J McEliece. 'A Public-Key Cryptosystem Based on Algebraic'. In: *Coding Thv* 4244 (1978), pp. 114–116.

[Men+18] Alfred J. Menezes et al. *Handbook of Applied Cryptography*. CRC Press, Dec. 2018. https://doi.org/10/gqcwj8.

[Mer79] Ralph Charles Merkle. *Secrecy, Authentication, and Public Key Systems*. Stanford University, 1979.

[Mer82] Ralph C Merkle. *Method of Providing Digital Signatures*. US Patent 4,309,569. Jan.1982.

[MG02] Daniele Micciancio et al. *Complexity of Lattice Problems: A Cryptographic Perspective*. Vol. 671. Springer Science & Business Media, 2002.

[Mic01] Daniele Micciancio. 'Improving Lattice-Based Cryptosystems Using the Hermite Normal Form'. In: *Lecture Notes in Computer Science*. Springer Berlin Heidelberg, 2001, pp. 126–145. https://doi.org/10/cmp4z2.

[Mic05] Daniele Micciancio. 'Closest Vector Problem'. In: *Encyclopedia of Cryptography and Security*. Springer US, 2005, pp. 79–80. https://doi.org/10/bz48hm.

[Mit97] Tom M Mitchell. *Machine Learning*. 1997.

[MO16] Vaishnavi Mohan et al. 'SecDevOps: Is It a Marketing Buzzword? - Mapping Research on Security in DevOps'. In: *2016 11th International Conference on Availability, Reliability and Security (ARES)*. IEEE. IEEE, 2016, pp. 542–547. https://doi.org/10/gh7kdj.

[Mon+18] Grégoire Montavon et al. 'Methods for Interpreting and Understanding Deep Neural Networks'. In: *Digital Signal Processing* 73 (Feb. 2018), pp. 1–15. https://doi.org/10/gcvxrb.

[Mon+95] C. Monroe et al. 'Demonstration of a Fundamental Quantum Logic Gate'. In: *Physical Review Letters* 75.25 (Dec. 1995), pp. 4714–4717. https://doi.org/10/bm9qwt.

[Mul54] D. E. Muller. 'Application of Boolean Algebra to Switching Circuit Design and to Error Detection'. In: *Transactions of the I.r.e. Professional Group on Electronic Computers* EC-3.3 (Sept. 1954), pp. 6–12. https://doi.org/10/gsm65j.

[Mur+19] W. James Murdoch et al. 'Definitions, Methods, and Applications in Interpretable Machine Learning'. In: *Proceedings of the National Academy of Sciences* 116.44 (Oct. 2019), pp. 22071–22080. https://doi.org/10/ggbhmq.

[Mut+21] Rohit Mutalik et al. *Rockyou*. 2021. https://doi.org/10/gsscxt.

[Nad+22] D. P. Nadlinger et al. 'Experimental Quantum Key Distribution Certified by Bell's Theorem'. In: *Nature* 607.7920 (July 2022), pp. 682–686. https://doi.org/10/gsm65p.

[Nas16] Condé Nast. *Inside the Cunning, Unprecedented Hack of Ukraine's Power Grid*. https://www.wired.com/2016/03/inside-cunning-unprecedented-hack-ukraines-power-grid/. [Accessed 23-09-2023]. 2016.

[Nas21] Condé Nast. *A Year After the SolarWinds Hack, Supply Chain Threats Still Loom*. https://www.wired.com/story/solarwinds-hack-supply-chain-threats-improvements/. [Accessed 22-08-2023]. 2021.

[Nat89] United Nations. *The United Nations Convention on the Rights of the Child*. 1989. https://downloads.unicef.org.uk/wp-content/uploads/2016/08/unicef-convention-rights-child-uncrc.pdf?_ga=2.100996225.1478653946.1575647430-387733425.1575647430.

[NC12] Michael A. Nielsen et al. *Quantum Computation and Quantum Information*. Cambridge University Press, June 2012. https://doi.org/10/cgwx.

[Net+20] Nelson Novaes Neto et al. 'A Case Study of the Capital One Data Breach'. In: *SSRN Electronic Journal* (Jan. 2020). https://doi.org/10/gsms3m.

[New17] BBC News. *Massive Ransomware Infection Hits Computers in 99 Countries*. https://www.bbc.com/news/technology-39901382. [Accessed 21-09-2023]. 2017.

[Nie86] Harald Niederreiter. 'Knapsack-Type Cryptosystems and Algebraic Coding Theory'. In: *Prob. Contr. Inform. Theory* 15.2 (1986), pp. 157–166.

[OC11] Gwenn Schurgin O'Keeffe et al. 'The Impact of Social Media on Children, Adolescents, and Families'. In: *Pediatrics* 127.4 (Apr. 2011), pp. 800–804. https://doi.org/10/dvnkrm.

[OCo12] Mary Ellen O'Connell. 'Cyber Security without Cyber War'. In: *Journal of Conflict and Security Law* 17.2 (2012), pp. 187–209.

[OhM21] OhMyBahGosh. *RockYou2021.txt*. https://github.com/ohmybahgosh/RockYou2021.txt. [Accessed 23-09-2023]. 2021.

[Onl22] ET Online. *Cyber Attack on AIIMS Delhi's Servers Originated in China, Say Government Sources*. https://economictimes.indiatimes.com/news/india/cyber-attack-on-aiims-delhis-servers-originated-in-china-says-mohfw/articleshow/96222031.cms?from=mdr. [Accessed 26-09-2023]. 2022.

[Par03] UK Parliament. *The Privacy and Electronic Communications (EC Directive) Regulations, 2003*. https://www.legislation.gov.uk/uksi/2003/2426/contents/made. [Accessed 26-09-2023]. 2003.

[Par16] UK Parliament. *Investigatory Powers Act, 2016*. https://www.legislation.gov.uk/ukpga/2016/25/contents/enacted. [Accessed 26-09-2023]. 2016.

[Par17] UK Parliament. *The Payment Services Regulations, 2017*. https://www.legislation.gov.uk/uksi/2017/752/contents/made. [Accessed 26-09-2023]. 2017.

[Par18a] UK Parliament. *Data Protection Act, 2018*. https://www.legislation.gov.uk/ukpga/2018/12/contents/enacted. [Accessed 26-09-2023]. 2018.

[Par18b] UK Parliament. *The Network and Information Systems Regulations, 2018*. https://www.legislation.gov.uk/uksi/2018/506/contents/made. [Accessed 26-09-2023]. 2018.

[Par18c] UK Parliament. *UK GDPR Guidance and Resources, 2018*. https://ico.org.uk/for-organisations/uk-gdpr-guidance-and-resources/. [Accessed 26-09-2023]. 2018.

[Par23] European Parliament. *Proposal for a Regulation on Privacy and Electronic Communications |Legislative Train Schedule*. https://www.europarl.europa.eu/legislative-train/theme-connected-digital-single-market/file-jd-e-privacy-reform. [Accessed 26-09-2023]. 2023.

[Par70] James L. Park. 'The Concept of Transition in Quantum Mechanics'. In: *Foundations of Physics* 1.1 (1970), pp. 23–33. https://doi.org/10/b2xzjc.

[Par90] UK Parliament. *Computer Misuse Act, 1990*. https://www.legislation.gov.uk/ukpga/1990/18/contents. [Accessed 26-09-2023]. 1990.

[Pat97] Jacques Patarin. 'The Oil and Vinegar Signature Scheme'. In: *Presented at the Dagstuhl Workshop on Cryptography September 1997*. 1997.

[Pet22] Sergei Petrenko. *Cyber Resilience*. River Publishers, Sept. 2022. https://doi.org/10/gsrr54.

[Phi+21] P Jonathon Phillips et al. *Four Principles of Explainable Artificial Intelligence*. Tech. rep. Sept. 2021. https://doi.org/10/jmxg.

[Pol93] J. M. Pollard. 'Factoring with Cubic Integers'. In: *Lecture Notes in Mathematics*. Springer Berlin Heidelberg, 1993, pp. 4–10. https://doi.org/10/cb97f5.

[Poy+20] Omer Ilker Poyraz et al. 'Cyber Assets at Risk: Monetary Impact of Us Personally Identifiable Information Mega Data Breaches'. In: *The Geneva Papers on Risk and Insurance-Issues and Practice* 45 (2020), pp. 616–638.

[Pre+20] Thomas Prest et al. 'FALCON'. In: *Post-Quantum Cryptography Project of NIST* (2020).

[Pro17] Sandra Proske. *What You Need to Know About WannaCry Now*. https://blog.f-secure.com/what-you-need-to-know-about-wannacry-now/. [Accessed 16-11-2023]. June 2017.

[Pro23] National Commission for Protection of Child Rights (NCPCR). *Guidelines - National Commission for Protection of Child Rights, India*. https://ncpcr.gov.in/guidelines. 2023.

[PT91] Lyle Pursell et al. 'Gram-Schmidt Orthogonalization by Gauss Elimination'. In: *The American Mathematical Monthly* 98.6 (1991), p. 544. https://doi.org/10/d9h34d.

[Pub23] Office of Public Affairs. *Data Privacy Framework Program Launches New Website Enabling U.S. Companies to Participate in Cross-Border Data Transfers*. https://www.commerce.gov/news/press-releases/2023/07/data-privacy-framework-program-launches-new-website-enabling-us. [Accessed 26-09-2023]. 2023.

[R21] Mike R. *RockYou2021 Leak - 8 Billion Passwords Compromised*. https://www.wheelhouseit.com/rockyou2021-leak-over-8-billion-passwords-compromised/. [Accessed 23-09-2023]. 2021.

[Rab79] Michael O Rabin. 'Digitalized Signatures and Public-Key Functions As Intractable As Factorization'. In: (1979).

[Raj+20] Manav Raj et al. 'COVID-19 and Digital Resilience: Evidence from Uber Eats'. In: *SSRN Electronic Journal* (2020). https://doi.org/10/gpfmdv.

[Raj+22] Roshan N. Rajapakse et al. 'Challenges and Solutions When Adopting DevSecOps: A Systematic Review'. In: *Information and Software Technology* 141 (Jan. 2022), p. 106700. https://doi.org/10/gn6237.

[Ran+13] F Ann Ran et al. 'Genome Engineering Using the CRISPR-Cas9 system'. In: *Nature Protocols* 8.11 (Oct. 2013), pp. 2281–2308. https://doi.org/10/f5fqpc.

[Ree54] I. Reed. 'A Class of Multiple-Error-Correcting Codes and the Decoding Scheme'. In: *Transactions of the IRE Professional Group on Information Theory* 4.4 (Sept. 1954), pp. 38–49. https://doi.org/10/ctv9wg.

[Rei+12] Stephanie M. Reich et al. 'Friending, IMing, and Hanging Out Face-To-Face: Overlap in Adolescents' Online and Offline Social Networks.' In: *Developmental Psychology* 48.2 (2012), pp. 356–368. https://doi.org/10/gcmc83.

[Rep17] Data Protection Report. *WannaCry Ransomware Attack Summary*. [Accessed 21-09-2023]. 2017. https://www.dataprotectionreport.com/2017/05/wannacry-ransomware-attack-summary/.

[Riv+78] R. L. Rivest et al. 'A Method for Obtaining Digital Signatures and Public-Key Cryptosystems'. In: *Communications of the ACM* 21.2 (Feb. 1978), pp. 120–126. https://doi.org/10/fkhmsc.

[Riv90] Ronald L Rivest. 'Cryptography'. In: *Algorithms and Complexity*. Elsevier, 1990, pp. 717–755. https://doi.org/10/gqc4d4.

[Roc13] James Roche. 'Adopting DevOps Practices in Quality Assurance'. In: *Communications of the ACM* 56.11 (Nov. 2013), pp. 38–43. https://doi.org/10/gsrr73.

[Ros+20] Ribana Roscher et al. 'Explainable Machine Learning for Scientific Insights and Discoveries'. In: *IEEE Access* 8 (2020), pp. 42200–42216. https://doi.org/10/ghn6h8.

[Ros+21] Ron Ross et al. *Developing Cyber-Resilient Systems: A Systems Security Engineering Approach*. Dec. 2021. https://doi.org/10/gs43bb.

[Rot18] Ray Rothrock. *Digital Resilience: Is Your Company Ready for the Next Cyber Threat?* Amacom, 2018.

[Row+11] Dale C. Rowe et al. 'The Role of Cyber-Security in Information Technology Education'. In: *Proceedings of the 2011 conference on Information Technology Education*. ACM, Oct. 2011. https://doi.org/10/cmwfqm.

[RS03] Karl Rubin et al. 'Torus-Based Cryptography'. In: *Advances in Cryptology - CRYPTO 2003*. Vol. 12. Springer Berlin Heidelberg, Nov. 2003, pp. 349–365. https://doi.org/10/cw29sw.

[Rud19] Cynthia Rudin. 'Stop Explaining Black Box Machine Learning Models for High Stakes Decisions and Use Interpretable Models Instead'. In: *Nature Machine Intelligence* 1.5 (May 2019), pp. 206–215. https://doi.org/10/gf4tp9.

[Sal12] Nathan Alexander Sales. 'Regulating Cyber-Security'. In: *Nw. UL Rev.* 107 (2012), p. 1503.

[Sam+19] Wojciech Samek et al. *Explainable AI: Interpreting, Explaining and Visualizing Deep Learning*. Vol. 11700. Springer Nature, 2019.

[Sam+22] S.Saniya Samrin et al. 'Design of Logic Gates Using Reversible Gates with Reduced Quantum Cost'. In: *Global Transitions Proceedings* 3.1 (June 2022), pp. 136–141. https://doi.org/10/gsrr64.

[San+18] David E Sanger et al. 'Marriott Data Breach Traced to Chinese Hackers'. In: *The New York Times* (2018), A1–L.

[Sat23] Satori. *Communication of Data Breach: Repairing Your Reputation and Staying Compliant*. https://satoricyber.com/data-privacy/communication-of-data-breach. [Accessed 25-08-2023]. 2023.

[SC02] Masahide Sasaki et al. 'Quantum Learning and Universal Quantum Matching Machine'. In: *Physical Review A* 66.2 (Aug. 2002), p. 022303. https://doi.org/10/c76q2k.

[Sch+14] Maria Schuld et al. 'An Introduction to Quantum Machine Learning'. In: *Contemporary Physics* 56.2 (Oct. 2014), pp. 172–185. https://doi.org/10/gfr9rk.

[Sch+17] Daniel Schatz et al. 'Towards a More Representative Definition of Cyber Security'. In: *The Journal of Digital Forensics, Security and Law* 12.2 (2017), p. 8. https://doi.org/10/gfgrwt.

[Sch35] E. Schrödinger. 'Discussion of Probability Relations between Separated Systems'. In: *Mathematical Proceedings of the Cambridge Philosophical Society* 31.4 (Oct. 1935), pp. 555–563. https://doi.org/10/dp2k57.

[Sch36] E. Schrödinger. 'Probability Relations between Separated Systems'. In: *Mathematical Proceedings of the Cambridge Philosophical Society* 32.3 (Oct. 1936), pp. 446–452. https://doi.org/10/bfwh52.

[Sec23] IBM Security. *Cost of a Data Breach Action Guide*. 2023. https://www.ibm.com/reports/data-breach-action-guide.

[SF13] Neil Selwyn et al. 'The Politics of Digital Literacy: A Strategic Approach'. In: *The Routledge Handbook of Digital Literacy in Society*. Routledge, 2013, pp. 27–38.

[Sha+11] Marzieh Shariati et al. 'Enterprise Information Security, a Review of Architectures and Frameworks from Interoperability Perspective'. In: *Procedia Computer Science* 3 (2011), pp. 537–543. https://doi.org/10/cf7r3c.

[Sho94] P.W. Shor. 'Algorithms for Quantum Computation: Discrete Logarithms and Factoring'. In: *Proceedings 35th Annual Symposium on Foundations of Computer Science*. IEEE Comput. Soc. Press, Nov. 1994, pp. 124–134. https://doi.org/10/czq562.

[Sid94] V. M. Sidelnikov. 'A Public-Key Cryptosystem Based on Binary Reed-Muller Codes'. In: *Discrete Mathematics and Applications* 4.3 (1994). https://doi.org/10/dh4sxw.

[Sim97] Daniel R. Simon. 'On the Power of Quantum Computation'. In: *SIAM Journal on Computing* 26.5 (Oct. 1997), pp. 1474–1483. https://doi.org/10/b77gj3.

[Sin13] Peter W. Singer. *Cybersecurity What Everyone Needs to Know Paper*. Oxford University Press, USA, 2013, p. 306. ISBN: 9780199918119.

[SK22] Maria Schuld et al. 'Is Quantum Advantage the Right Goal for Quantum Machine Learning?' In: *PRX Quantum* 3.3 (July 2022), p. 030101. https://doi.org/10/gqqs64.

[Slo19] Joe Slowik. *Implications of IT Ransomware for ICS Environments*. https://dragos.com/blog/industry-news/implications-of-it-ransomware-for-ics-environments/. [Accessed 21-09-2023]. 2019.

[SN13] Rossouw von Solms et al. 'From Information Security to Cyber Security'. In: *Computers & Security* 38 (Oct. 2013), pp. 97–102. https://doi.org/10/f5gkvv.

[SP18] Maria Schuld et al. *Supervised Learning with Quantum Computers*. Springer International Publishing, 2018. https://doi.org/10/gsm8b7.

[Sri+12] Madhan Kumar Srinivasan et al. 'State-Of-The-Art Cloud Computing Security Taxonomies'. In: *Proceedings of the International Conference on Advances in Computing, Communications and Informatics*. ACM, Aug. 2012. https://doi.org/10/gsm9k9.

[Sri+19] Jangirala Srinivas et al. 'Government Regulations in Cyber Security: Framework, Standards and Recommendations'. In: *Future generation computer systems* 92 (2019), pp. 178–188. https://doi.org/10/gsg3wp.

[SS92] Vladimir Michilovich Sidelnikov et al. 'On Insecurity of Cryptosystems Based on Generalized Reed-Solomon Codes'. In: *Discrete Mathematics and Applications* 2.4 (1992). https://doi.org/10/dppn4s.

[Sta18] State of California. *California Consumer Privacy Act (CCPA)*. Cal. Civ. Code §§1798.100 et seq. https://leginfo.legislature.ca.gov/faces/billTextClient.xhtml?bill_id=201720180AB375. 2018.

[Sta22] The Wire Staff. *5 AIIMS Servers Hacked, 1.3 TB Data Encrypted in Recent Cyberattack, Govt Tells RS*. https://thewire.in/government/aiims-servers-cyberattack-ransomware-rajya-sabha. [Accessed 26-09-2023]. 2022 (cit. on pp. 342–344).

[Ste94] Jacques Stern. 'A New Identification Scheme Based on Syndrome Decoding'. In: *Advances in Cryptology —CRYPTO' 93*. Springer Berlin Heidelberg, 1994, pp. 13–21. https://doi.org/10/czb2sn.

[Sto+04] G Stoneburner et al. *Engineering Principles for Information Technology Security (a Baseline for Achieving Security), Revision A*. Tech. rep. 2004. https://doi.org/10/gsm9mj.

[Sto09] Brad Stone. *In Facebook, Investing in a Theory*. https://www.nytimes.com/2007/10/04/technology/04facebook.html. [Accessed 23-09-2023]. 2009.

[Tan+16] Jun Tang et al. 'Ensuring Security and Privacy Preservation for Cloud Data Services'. In: *ACM Computing Surveys* 49.1 (June 2016), pp. 1–39. https://doi.org/10/gmgjns.

[Tea17] Counter Threat Unit Research Team. *WCry (WannaCry) Ransomware Analysis*. https://www.secureworks.com/research/wcry-ransomware-analysis. [Accessed 21-09-2023]. 2017.

[Tha+15] Kutub Thakur et al. 'An Investigation on Cyber Security Threats and Security Models'. In: *2015 IEEE 2nd international conference on cyber security and cloud computing*. IEEE. 2015, pp. 307–311. https://doi.org/10/gsrr8p.

[Tha13] David Thaw. 'The Efficacy of Cybersecurity Regulation'. In: *SSRN Electronic Journal* 30 (2013), p. 287. https://doi.org/10/gjhqmk.

[Tik11] Eneken Tikk. 'Ten Rules for Cyber Security'. In: *Survival* 53.3 (2011), pp. 119–132. https://doi.org/10/fj8bwm.

[Tim+21] Yenni Tim et al. 'Digital Resilience: How Rural Communities Leapfrogged into Sustainable Development'. In: *Information Systems Journal* 31.2 (2021), pp. 323–345.

[TR09] Himanshu Thapliyal et al. 'Design of Efficient Reversible Binary Subtractors Based on a New Reversible Gate'. In: *2009 IEEE Computer Society Annual Symposium on VLSI*. IEEE, 2009. https://doi.org/10/b7zg7q.

[Tru01] C. A. Trugenberger. 'Probabilistic Quantum Memories'. In: *Physical Review Letters* 87.6 (July 2001), p. 067901. https://doi.org/10/bqdsb8.

[Tru02] C. A. Trugenberger. 'Phase Transitions in Quantum Pattern Recognition'. In: *Physical Review Letters* 89.27 (Dec. 2002), p. 277903. https://doi.org/10/bbvh66.

[Tur50] A. M. Turing. 'Computing Machinery and Intelligence'. In: *Mind* LIX.236 (Oct. 1950), pp. 433–460. https://doi.org/10/b262dj.

[UKC19] UKCIS. *Digital Resilience Framework: A Framework and Tool for Organisations, Communities, and Groups to Help People Build Resilience in Their Digital Life*. London: Untied Kingdom for Internet Safety, 2019. https://assets.publishing.service.gov.uk/government/uploads/system/uploads/attachment_data/file/831217/UKCIS_Digital_Resilience_Framework.pdf.

[Uni22] United States Congress. *American Data Privacy and Protection Act (ADPPA) of 2022*. Accessed on September 22, 2023. 2022. https://www.congress.gov/bill/117th-congress/house-bill/8152/actions.

[UNI23] UNICEF. *9 Tips for Parents and Guardians to Keep Children Safe Online*. https://www.unicef.org/zambia/stories/9-tips-parents-and-guardians-keep-children-safe-online. 2023.

[Uni70] United States Congress. *Fair Credit Reporting Act (FCRA) of 1970*. Public Law 91-508, 84 Stat. 1127, 15 U.S.C. §1681 et seq. https://www.ftc.gov/legal-library/browse/statutes/fair-credit-reporting-act. 1970.

[Uni86] United States Congress. *Electronic Communications Privacy Act (ECPA) of 1986*. Public Law 99-508, 100 Stat. 1848, 18 U.S.C. §§2510–2523. 1986. https://www.congress.gov/bill/99th-congress/house-bill/4952.

[Uni88] United States Congress. *Video Privacy Protection Act of 1988 (VPPA)*. Public Law 100-618, 102 Stat. 3195. 42 U.S.C. §§2001aa-2001aa-12. 1988.

[Uni98] United States Congress. *Children's Online Privacy Protection Act (COPPA)*. U.S. Code. 15 U.S.C. §§6501–6506. 1998. https://www.ftc.gov/enforcement/rules/rulemaking-regulatory-reform-proceedings/childrens-online-privacy-protection-rule.

[Uni99] United States Congress. *Financial Modernization Act (FMA) of 1999*. 106th Congress, 1999–2000. 1999 (cit. on p. 169).

[US-17] US-Cert. *Indicators Associated With WannaCry Ransomware*. https://www.us-cert.gov/ncas/alerts/TA17-132A. [Accessed 21-09-2023]. 2017.

[Van+19] Antonius J Van Rooij et al. 'Digital Resilience: Is It Different from Coping?' In: *Journal of Behavioral Addictions* 8.4 (2019), pp. 688–690.

[Van+20] M. H. Van Ijzendoorn et al. 'Annual Research Review: Umbrella Synthesis of Meta-Analyses on Child Maltreatment Antecedents and Interventions: Differential Susceptibility Perspective on Risk and Resilience'. In: *Journal of Child Psychology and Psychiatry* 61 (2020), pp. 272–290.

[VB21a] Michael Veale et al. 'Demystifying the Draft EU Artificial Intelligence Act'. In: (July 2021). https://doi.org/10/gns2s9.

[VB21b] Michael Veale et al. 'Demystifying the Draft EU Artificial Intelligence Act — Analysing the Good, the Bad, and the Unclear Elements of the Proposed Approach'. In: *Computer Law Review International* 22.4 (Aug. 2021), pp. 97–112. https://doi.org/10/gqfdmj.

[Ven00] D Ventura. 'Quantum Associative Memory'. In: *Information Sciences* 124.1-4 (May 2000), pp. 273–296. https://doi.org/10/dx67w8.

[Ver+19] M. Verhoeven et al. 'The Role of School in Adolescents' Identity Development: A Literature Review'. In: *Educational Psychology Review* 31.1 (2019), pp. 35–63.

[VL21] Giulia Vilone et al. ‘Classification of Explainable Artificial Intelligence Methods through Their Output Formats’. In: *Machine Learning and Knowledge Extraction* 3.3 (Aug. 2021), pp. 615–661. https://doi.org/10/grbf5v.

[WA13] Martin Weller et al. ‘Digital Resilience in Higher Education’. In: *European Journal of Open, Distance and E-Learning* 16.1 (2013), pp. 53–66.

[Wan+16] H Wang et al. ‘Machine Learning Basics’. In: *Deep Learning* (2016), pp. 98–164.

[Wan+19] Qian Wang et al. ‘Digital Resilience: A Protective Mechanism for Youth and Emerging Adults against Problematic Internet Use’. In: *Journal of Educational Computing Research* 57.6 (2019), pp. 1315–1337.

[Wan+20] Qian Wang et al. ‘Digital Resilience among Chinese University Students: A Latent Profile Analysis’. In: *Current Psychology* 39.3 (2020), pp. 794–805.

[WC20] Jing Wu et al. ‘A Systematic Review of Educational Digital Storytelling’. In: *Computers & Education* 147 (Apr. 2020), p. 103786. https://doi.org/10/gg93nt.

[Wei09] Matt Weir. *The RockYou 32 Million Password List Top 100*. https://reusablesec.blogspot.com/2009/12/rockyou-32-million-password-list-top.html. [Accessed 23-09-2023]. 2009.

[Wie+14] Nathan Wiebe et al. *Quantum Algorithms for Nearest-Neighbor Methods for Supervised and Unsupervised Learning*. 2014. arXiv: 1401.2142 [quant-ph].

[Win11] J. R. Winkler. *Securing the Cloud Cloud Computer Security Techniques and Tactics. Cloud Computer Security Techniques and Tactics*. Syngress/Elsevier, 2011, p. 290. ISBN: 9781597495929.

[Wit14] Peter Wittek. *Quantum Machine Learning. What Quantum Computing Means to Data Mining*. Elsevier Science & Technology Books, 2014, p. 176. ISBN: 9780128009536. https://doi.org/10/gctpfv.

[WL13] Wenye Wang et al. ‘Cyber Security in the Smart Grid: Survey and Challenges’. In: *Computer Networks* 57.5 (Apr. 2013), pp. 1344–1371. https://doi.org/10/f4vfph.

[Wri16] Dawn J Wright. ‘Toward a Digital Resilience’. In: *Elementa* 4 (2016), p. 000082.

[Wu06] Q. Wu. ‘Commercialization of Digital Storytelling: An Integrated Approach for Cultural Tourism, the Beijing Olympics and Wireless Vas’. In: *International Journal of Cultural Studies* 9.3 (2006), pp. 383–394.

[WZ82] W. K. Wootters et al. ‘A Single Quantum Cannot Be Cloned’. In: *Nature* 299.5886 (Oct. 1982), pp. 802–803. https://doi.org/10/fcf8cd.

[Yao86] Andrew Yao. ‘How to Generate and Exchange Secrets’. In: *SFCS ’86 Proceedings of the 27th Annual Symposium on Foundations of Computer Science*. 1986, pp. 162–167.

[Yin+13] Juan Yin et al. ‘Lower Bound on the Speed of Nonlocal Correlations without Locality and Measurement Choice Loopholes’. In: *Physical Review Letters* 110.26 (June 2013), p. 260407. https://doi.org/10/ggz346.

[YR15] Tarun Yadav et al. ‘Technical Aspects of Cyber Kill Chain’. In: *Communications in Computer and Information Science*. Springer International Publishing, 2015, pp. 438–452. https://doi.org/10/gktc7n.

[Zha+19] Nianbo Zhao et al. ‘Digital Resilience: A New Perspective on Coping with Stress in the Digital Age’. In: *Frontiers in Psychology* 10 (2019), p. 2706.

[Zho21] Zhi-Hua Zhou. *Machine Learning*. Springer Singapore, 2021. https://doi.org/10/gsrr66.

[Zou+18] Yixin Zou et al. ‘I’ve Got Nothing to Lose: Consumers’ Risk Perceptions and Protective Actions After the Equifax Data Breach’. In: *SOUPS @ USENIX Security Symposium*. 2018.

Index

A

AIIMS ransomware attack, 303–305
Artificial intelligence (AI), 44, 122, 123, 128, 137, 138, 182–185, 228–230, 235–239, 250, 263–266, 308, 313–399, 410, 416, 437, 438, 456, 474, 485, 494, 495, 499, 500, 502
Autonomous systems, 50, 235, 280, 324–326, 494–499

B

Business continuity, 15, 19, 75, 109, 205, 207, 208, 210, 472, 475, 510

C

Cloud computing, 5, 9, 40, 78, 79, 84, 132, 224, 315, 329, 370, 467–476, 499
Colonial pipeline attack, 301–303
Computational psychology, 512–518
Computing paradigms, 412, 461–468
Cyberattacks, 4, 43, 138, 279, 331, 477
Cybercrimes, 6, 44, 85, 87, 105, 131–133, 138, 139, 156, 157, 163, 187–192, 200, 201, 241, 243–248, 258, 263–269, 289, 304, 305
Cyber-physical systems (CPS), 44, 50, 486–493
Cybersecurity, 3, 43, 128, 259, 282, 313, 402, 459
Cybersecurity laws, 187

D

Data breaches, x, 6, 14, 52, 53, 64, 79, 90, 91, 99, 101, 110, 128, 131, 140, 151, 165, 174–176, 178, 194, 208, 209, 212, 214, 216, 217, 219–221, 224, 226, 228, 294–300, 308, 364, 472, 485, 495, 508–509, 522, 527, 528
Deep learning (DL), 315–320, 323, 336, 343, 368, 397, 500
DevSecOps, 114–122, 473
Digital resilience, 3, 43, 130, 241, 306, 313, 401, 459
Digital resilience frameworks, 212–215, 217, 257–262, 269–271, 277, 278, 461

E

Encryption, 5, 52, 130, 280, 364, 401, 470
Explainable AI (XAI), 373–384, 386, 387, 495

G

General Data Protection Regulation (GDPR), 77, 80, 83, 91, 98, 128, 150–152, 156, 161–163, 176–178, 185, 217–220, 225, 228, 239, 255, 298, 372, 472, 497
Government agencies, 6, 111, 133–135, 163, 195, 204, 279, 444

H

Hardware security, 99, 476–481

S. K. Shandilya et al., *Digital Resilience: Navigating Disruption and Safeguarding Data Privacy*, EAI/Springer Innovations in Communication and Computing,
https://doi.org/10.1007/978-3-031-53290-0

Hash-based cryptography, 430–432
Human factors, 27, 211, 239, 506, 519–522

I
Incident response, 10–11, 14–16, 20, 67, 70, 75, 76, 78, 86, 87, 89, 101–114, 118, 128, 196, 211, 212, 216, 302, 332, 397, 493, 523, 524, 528
Internet of Things (IoT), 5, 40, 44, 50, 85, 224, 280, 324, 349, 423, 463, 478, 479, 481–485, 487, 488, 499

L
Lattice-based cryptography (LBC), 403, 425–427, 456
Legal frameworks, 28, 30, 127–128, 131–134, 137–139, 144, 164, 168, 170, 172, 175, 176, 184, 185, 188, 189, 191, 194, 195, 200, 222, 235–237, 372, 460

M
Machine learning (ML), 64, 70–71, 93, 128, 235, 264, 265, 313–399, 402, 416, 437–439, 456, 475, 480, 485, 495, 499, 502, 505
Marriott breach, 294–298
Minors, 26, 102, 142, 148, 153, 154, 183, 217, 241–278, 395, 459, 493, 498
Multi-layered security, 74–75
Multinational corporations (MNCs), 174, 263–266, 290

N
New space, 507–512
Non-governmental organizations (NGOs), 258, 263, 266–269
NotPetya attack, 284, 289–294

O
Oil and vinegar cryptography, 433

P
Post-quantum cryptography, 398, 425, 428, 431, 456, 457, 460, 461, 480

Q
Quantum computing, xii, 123, 138, 139, 223, 398, 401, 402, 404, 407–417, 419–423, 427, 432, 434, 437–448, 452–458, 480, 500
Quantum key distribution (QKD), 409, 434–437, 445

R
Regulatory compliance, 28, 30, 31, 81, 85, 106, 115, 117, 202, 215–221, 232–234, 339, 364, 381, 384, 529
Regulatory landscape, 127–240
Resilience metrics, 524–525
Responsible AI (RAI), 383–393
Rockyou breaches, 298–301

T
Threat intelligence, 50, 63–71, 138, 139, 195, 225, 398
Trustworthy systems, 76–87

U
Ukraine powergrid attack, 282–285

W
WannaCry attack, 103, 286–290, 292, 308

SPRINGER NATURE

GPSR Compliance

The European Union's (EU) General Product Safety Regulation (GPSR) is a set of rules that requires consumer products to be safe and our obligations to ensure this.

If you have any concerns about our products, you can contact us on ProductSafety@springernature.com

In case Publisher is established outside the EU, the EU authorized representative is:

Springer Nature Customer Service Center GmbH
Europaplatz 3
69115 Heidelberg, Germany

The manufacturer's authorised representative in the EU is Springer Nature Customer Service Centre GmbH, Europaplatz 3, 69115 Heidelberg, Germany. If you have any concerns regarding our products, please contact ProductSafety@springernature.com

Printed and bound by CPI Group (UK) Ltd, Croydon, CR0 4YY
28/09/2026
02224061-0009